Testing of Metals for Structures

Testing of Metals for Structures

Proceedings of the International Workshop 'Needs in Testing Metals' held by RILEM (The International Union of Testing and Research Laboratories for Materials and Structures) and organized by the Istituto di Tecnica delle Costruzioni, Universita degli Studi di Napoli 'Federico II'.

Naples, Italy
May 29–31, 1990

EDITED BY
F.M. Mazzolani

CRC Press
Taylor & Francis Group
Boca Raton London New York

CRC Press is an imprint of the
Taylor & Francis Group, an **informa** business
A TAYLOR & FRANCIS BOOK

CRC Press
Taylor & Francis Group
6000 Broken Sound Parkway NW, Suite 300
Boca Raton, FL 33487-2742

First issued in paperback 2019

ISBN-13: 978-0-419-15810-3 (hbk)
ISBN-13: 978-0-367-86365-4 (pbk)
ISBN-13: 978-0-442-31503-0 (USA)

A catalogue record for this book is available from the British Library

Library of Congress Cataloging-in-Publication data available

Publisher's Note
This book has been produced from camera ready copy provided by the individual contributors.

The publisher has gone to great lengths to ensure the quality of this reprint but points out that some imperfections in the original may be apparent

Visit the Taylor & Francis Web site at
http://www.taylorandfrancis.com

and the CRC Press Web site at
http://www.crcpress.com

Contents

Preface

It is well known that RILEM (Réunion Internationale des Laboratoires d'Essais et Recherches sur les Materiaux et les Constructions/The International Union of Testing and Research Laboratories for Materials and Structures) has, since 1947, promoted the unification of testing methods for measuring and modelling the properties and performance of materials and structures. This activity is developed by means of numerous Technical Committees (around 60) mainly devoted to non-metallic materials, such as mortar, concrete, masonry, plastic, timber, geotextiles and so on. The field of metals was until recently covered by only one Committee, TC 83-CUS 'Characterization of ultimate limit state performance of metals', which dealt with pre-codification of tension testing for determining stress–strain relations under monotonic and uniaxial loading.

A recent proposal to create a new Technical Committee dealing with the characterization of fatigue performance of metals called the attention of the RILEM Technical Advisory Committee to the lack of existing information in the field of metal testing in general.

As a direct consequence, RILEM decided to revise its strategy in this field and asked me to organize an exploratory workshop in order to identify the 'Needs in Testing Metals'. This workshop was held at the Engineering Faculty of the University of Naples on 29–31 May 1990.

Participation was limited to invited outstanding experts coming from all over the world, who presented around 40 papers on the general subject 'Needs in Testing Metals', covering not only the experimental procedures for testing metallic materials (steel and aluminium), joints and structural components, but also the analytical models to interpret their behaviour and to be used in the calculations. These papers have been collected in this book, where the overall matter has been subdivided into the following main subjects:

1. Monotonic behaviour of materials and joints.
2. Low cycle fatigue behaviour of materials, joints and structural components.
3. High cycle fatigue behaviour of materials, joints and structural components.
4. Non-destructive methods of inspection.
5. Geometrical and mechanical imperfections.
6. Behaviour of composite structures.
7. Behaviour of thin-walled structures.

These subjects cover behavioural phenomena (Parts One, Two and Three), testing methodologies (Parts Four and Five) and structural typologies (Parts Six and Seven).

Referring to these different ranges, the presented papers were strongly directed to the purpose of the workshop giving a clear answer to the main problems dealing with:

- the assessment of experimental equipments and of unified testing procedures;
- the identification of physical properties and behavioural parameters which have to be recorded during the test;
- the definition of analytical models to interpret testing results.

From the analysis of the current methodologies and their comparison on an international level, some unquestionable needs were emphasized:

- to assess new procedures in order to cover evident deficiencies in testing and/or in modelling of materials, joints and structures;
- to improve existing procedures for newly identified purposes;
- to unify testing methods and to harmonize different international codifications.

The outcome of the workshop was discussed during the General Reporters' Session and the following proposals were formulated for the establishment of new Technical Committees:

- to establish a new Committee on the behaviour of joints, both monotonic and cyclic;
- to establish a new Committee on high cycle fatigue, dealing with the definition of new testing procedures and the standardization of reporting results;
- to explore the opportunities for creating a new Committee in the field of non-destructive testing procedures;
- to establish a new Committee on structural imperfections, which is a very new subject not yet covered by other organizations;
- to establish a new Committee on composite structures dealing with new testing procedures for existing as well as for new typologies;
- to cover the needs of identifying testing methods in the field of cold-formed thin-walled structures by means of a new *ad hoc* Committee.

The General Reporters were asked to formulate in an appropriate way the above proposals for new Technical Committees, after having contacted the International Organizations which already work in the same subject area as well as possible participants in the future Committee activities. It was particularly recommended that these people, as well as being expert in the field, should also be strongly motivated.

In conclusion, the Naples Workshop can be considered a big success because of the very wide active participation of outstanding experts and the excellent level of their contributions. In addition, the most encouraging result has been the complete acceptance of all the proposals previously mentioned. In fact, the RILEM General Council in Brighton, UK on 13 and 14 November 1990 decided to create the following new Committees, which are strictly in line with the workshop outcome.

1. Geometrical and mechanical imperfections of metal components and structures (Chairman: J. Rondal, Belgium).
2. Characterization of ultimate limit state performance of metal joints (Chairman: A. De Martino, Italy).
3. High cycle fatigue behaviour of metallic materials, joints and structural components (Chairman: B. Atzori, Italy).
4. Non-destructive inspection of materials and structures (Chairman: G. Di Chirico, Italy).
5. Composite structures (Chairman: M. Mele, Italy).
6. Test methods for thin-walled metal structures and members (Chairman: T. Pekoz, USA).
7. Testing of metallic structural materials for determining stress–strain relations under cyclic loading (Chairman: B. Kato, Japan).

I am sure that the papers collected in this book can be considered as the basic sources from which the work of these new Committees will start.

In closing, I would like to congratulate my colleagues, who have the responsibility of guiding these Committees together with those who have accepted invitations to participate in their membership. I look forward to a new workshop in some years time, not on *needs*, but on *achievements* in metal structure testing and models.

F.M. Mazzolani
Naples, December 1990

PART ONE

MONOTONIC BEHAVIOUR OF MATERIALS AND JOINTS

1 EQUIPMENT AND SPECIMENS FOR BIAXIAL TESTS

A. DE IORIO
Istituto di Costruzione di Macchine, Faculty of Engineering, University of Naples, Italy

Summary

Design criteria for specimens obtained from plates or planes laminates for biaxial tests and for a machine for carrying out such tests are described. Using those criteria, two research groups are performing a study, by means of both FEM and BEM methods, to completely define the optimal geometry of the specimens tests and of the machine for the biaxial tests.

1 Introduction

The need to verify, through multiaxial tests, some strength criteria for orthotropic materials subjected to a biaxial stress state, and the few data available on this subject induced us to start a study aiming to design and make a test machine able to stress, along different complanar directions, specimens purposely conceived and manufactured.

In the field of civil engineering, much work, both numerical and experimental, has been carried out on the characterization of concrete in biaxial and triaxial states of stress, from which excellent results have been obtained [e.g. 1 - 7]. However, these results cannot be applied quantitatively or even qualitatively, to our studies, because of the great difference between the materials involved (concrete and metal plates or laminates) and because of the different kinds of load used: almost exclusively compression for concrete in a biaxial regime, and usually traction for the plates or laminates.

In this paper, project ideas are presented, both for the machine and for the specimen for biaxial tests on these laminates. At the same time, we are working on the design of the machine (with G. Castellano and G. Godono) and on the optimization of the specimen geometry with FEM and BEM techniques (with R. Esposito and A. Soprano).

Testing of Metals for Structures. Edited by F.M. Mazzolani.
Published by Taylor & Francis, 2 Park Square, Milton Park, Abingdon, Oxon, OX14 4RN.
ISBN 0 419 15810 3

2 Specimen Geometry

The specimen geometry shown in Fig. 1 is essentially a result of the following considerations.

1) For a biaxial stress state, with principal stresses in the load directions, it is necessary to have cross-shaped specimens with arms in orthogonal positions.
2) To induce fracture in the central zone of the specimen, it is necessary to reduce the reacting section orthogonal to each load, making cuts in all four arms of the cross.
3) To avoid notch effects which can appreciably modify the stress state in the central zone of the specimen, we must introduce holes at the end of the cuts on the axes of the arms.

To be sure that the fracture is produced in the symmetry planes of the specimen, containing the cuts, and not in the roots of the arms, it is convenient to have a central hole and suitable fillets between the arms. The hole geometry, whether circular or not, and that of the fillets must not appreciably modify the uniformity of stress distribution in the fracture sections and the biaxiality of the stress state.

It is evident that we cannot completely exclude the notch effect due to the holes and that the fact increasing hole dimensions tend to transform the stress state in the fracture zone from biaxial to monoaxial.

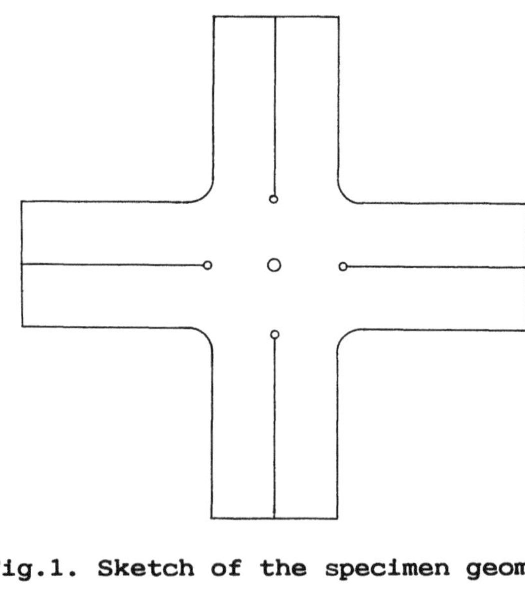

Fig.1. Sketch of the specimen geometry

The most convenient criterion is to find the largest dimension of hole compatible with biaxiality of the stress state, and then to modify the hole geometry to minimize the notch effect.

Once these results are achieved, one must measure the amplitude
ofthe shifting from both the uniformity and the perfect biaxiality of the stress in the minimum resistant sections, before carrying out experiments to check the numerical results and the strength criteria assumed for the materials under consideration.

3 Test Machine

To carry out biaxial tests on the usual cross flat specimens, of the kind already described, we need to design a test machine that allows us to apply the loads in two mutually orthogonal directions which does not induce loads in the specimen during the test.

A sketch of the machine we propose, which meets these requirements, is presented in Fig. 2. It consists of two frames welded together in a cross position and lying horizontally on four supports.

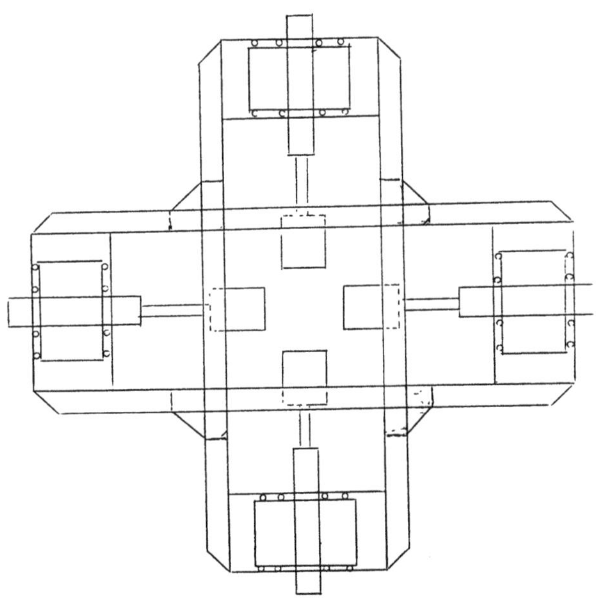

Fig.2. Sketch of the test machine

On the terminal beams of each frame are placed some sliding supports, to which are fixed the actuators, two for each frame, whose stroke, at least 250 mm, is used both for testing and to adjust the position of the grips according to the dimensions of different specimens.

Each pair of actuators are fed by their own hydraulic group, with a single circuit. The grips, of mechanical type, are composed of two halves that can move independantly so the two parts of each arm of the cross specimen can follow the deformation of the central zone of the specimen with no appreciable reaction. Furthermore, the grips are designed to fix specimens with an arm width not greater than 100 mm and a thickness falling in the range 0 - 6 mm.

The whole machine is designed to bear static loads up to 250 kN in each load direction.

The instrumentation for the load and position controls has been selected, for the moment, for static tests with no fatigue loads. Later, on the basis of the results obtained with regard to verifying the strength criteria, and once the test system has been shown to meet the design requirements, we shall study modifications to the machine for biaxial fatigue tests.

4 Conclusions

We have presented design ideas for specimens and a test machine for biaxial tests with plates and laminates. The design criteria here suggested form the basis on which two groups of researchers are developing the necessary investigations to resolve the problem completely. After the first experimental tests, we shall try to adapt or transform the test machine to carry out biaxial fatigue tests with the same kind of specimens.

5 References

[1] Kupfer, H.B. et al., "Behavior of Concrete under Biaxial Stresses", Journal Eng. Mech., ASCE, Vol. 99 No. EM4, 1973 pp. 853-866.

[2] Tasuji, M.E. et al., "Stress-Strain Response and Fracture of Concrete in Biaxial Loading", ACI J., Vol. 75, No. 7, 1978, pp. 306-312

[3] Hiramatsu, Y. et al., "Design and Production of Stiff Testing Machine and Test Results of Concrete", J. The Society of Material Science, Japan, Vol. 24, No. 260, 1975, pp. 447-454.

[4] Suzuki, K. et al., "Study on the Stiff Testing machine and Complete Stress-Strain Relation for Concrete", Proc. Annual Conv. of AIJ, Sept. 1978, pp. 1541-1542.
[5] Tanigawa, Y. et al., "A New Type of Stiff Testing Machine and Complete Stress-Strain Curve of Concrete", Trans. of AIJ, No. 260, 1977, pp. 9-19.
[6] Buyukotzurk, O. et al., "Stress-Strain Response and Fracture of a Concrete Model in Biaxial Loading", ACI J., Vol. 68, No. 8, 1971, pp. 590-599.
[7] Okajima, T., "Critical Tensile Strain of Plain Concrete under Biaxial Stresses", J. The Society of Material Science, Japan, Vol. 22, No. 232, 1973, pp. 33-37.

2 HIGH STRENGTH STEEL – RESEARCH ON THE BASIC PHENOMENA CONCERNING NEW DESIGN MODELS IN THE PROBABALISTIC APPROACH

E. HEMERICH and V. MILCIC
Civil Engineering Institute, Faculty of Civil Engineering,
University of Zagreb, Yugoslavia

SUMMARY

The paper deals with the use of high-strength steels in engineering structures from the probabilistic point of view, and shows a model how to solve this task and bring high-strength steels ($f_y > 450$ N/mm^2) in standard use.

1 INTRODUCTION

The development of the metal structures industry and the application of steel in this area depends on systematic introduction of high strength materials. The century-long experience with law-carbon milde structural steels shows that new advancements can be made and new weldable, fine-grain micro alloyed structural steels of high strengths ($f_y > 450$ N/mm^2) can be broadly introduced. However, the whole proces should be accelerated becouse the present development cannot rely on empirical experience only /1/.

The basic problem in introducing all new kinds of steel into practice is to determine changes of the parameters of individual variables determining the actual degre of structural safety in a given combination with respect to the parameters of the steel in nowadays use.

This task can now be solved thanks to the intensive scientific development of probabilistic methods where the degree of structure safety is numericaly quantitatively determined. This, in fact, precisely definies the objective of research of a structural engineering and defines its method of work.

The use of these method in structural engineering determines the safety of built structures (calibration), which serves as a basis to define the given safety degree of structures from new materials /2/. The practival development of probabilistic methods for safety verification has gone so far that this concept has been introduced into the new generation of codes (EUROCODE).

Still, due to a series of circumstances, the introduction of high strength steel into normal use is not only the problem of application of

Testing of Metals for Structures. Edited by F.M. Mazzolani. © 1992 RILEM.
Published by Taylor & Francis, 2 Park Square, Milton Park, Abingdon, Oxon, OX14 4RN.
ISBN 0 419 15810 3

new reliability concepts. It is a well-known fact that actual stresses cannot be calculated because of presence of residual stresses due to rolling, welding, straightening, caused by stress concentration near holes and welds, because of uncontrolled settlement of supports and inevitable geometric imperfections or deviation from mathematical model of stress analysis, so we often say that steel structures are safe due to materials plasticity /4/.

Just because plasticity of high strength steels is decreased in relation to the steels in use, the problem of their safety has not yet been solved. The new EUROCODE 3 limits the use of steel to the lowest strength which suits the quality with the nominal yield point f_y=355 N/mm^2 according to EN 10025, and accordingly excludes high strength steels from use.

The importance of this topic lies in the fact that most developed countries are preparing scientific projects to define the conditions for wide application of these steels /5/.

The objective of this paper is to point to characteristic dilemmas in assessing real safety of steel structures with yield point $f_y > 450$ N/mm^2 and enhance international co-operation with an aim to avoid obstacles and find safe solutions for introducing such steels into regular structural practice.

2 PROBLEM DEFINITION

Structural steels used so far have a ratio between strength and yield point of $f_u/f_y \sim 1.5$. Modern fine-grain structural steels have this ratio below 1.5. The decrease of this ratio at increased strength results in decreased plastic properties. On the other hand, it is a fact that the behaviour of these steels has been studied inadequately and there is no clear insight into the safety degree of structures to be build from these modern steels. The proposal EUROCODE 3, Final Draft (December 1988) has limited the use of steel qualities to those with specified minimum ulimate tensile strength to the specified minimum yield strength to the ratio of min 1.2., and has introduced additional requirements for plastic properties of the material:

- the elongation at failure on a gauge length of $5.65 \sqrt{A_o}$ (where A_o is the original cross section area) is not less than 15%
- the stress-strain diagram shows that the ultimate strain ε_u corresponding to the ultimate tensile strength f_u is at least 20 times the yield strain ε_y corresponding to the yield strength f_y.

It was also found that these criteria are met only by steels up to max f_y=355 N/mm^2 (Fe E 355 according to EN 10025).
Scientific arguments for such limitation are either non-existant or poorly studied. So we can justly ask how to formulate the requirements for plastic properties of these materials with an aim to introduce high-strength steels into wide practice without unargumented limitations, on condition that the structures made from such materials have a given safety degree in use. Naturally, the solution of this problem becomes more complicated

when we consider the consequences of possible technological degradation of the basic material due to thermic regime in the manufacture and use, especially due to welding. In this, we should certainly consider the requirements specific for different fields of application of high-strength steels in engineering structures

- building construction - buildings, towers, aerials
- bridges
- hydraulic construction - hydraulic equipment penstocks, off-shore structures
- equipment construction - pressure vessels, tanks, cranes

3 SOLUTION MODEL

When we generally consider characteristic phenomena of failure to which we can associate particular resistance models in the limit state equations governing the description of structural mechanism of failure of engineering structures, we can talk about

- tensile problem (plastic, brittle or mixed rupture)
- compression problem (stability)
- flexural problem (transformation into a mechanism)
- fatigue problem (repeated fluctuation of stresses).

It is clear that all members of a limit state equation which are connected with action effect description, as well as the members describing structural resistance are subjected to physical laws of random changes. The safety of a structure is therefore a random value, and the degree of its reliability can only be expressed as probability of failure p_f. The theoretical develop ment of probabilistic methods to calculate the probability of failure has shown that the ex act probabilistic approach (Level III) is very complex because it requires complicated mathematical operations and complete statistical data for all basic variables. There are many reasons why this procedure is not applicable in practice. That is why, after the realized safety degrees of built structures have been determined by simplified probabilistic methods (level II FOSM method) and were expressed with reliebility index β, it was possible to set the model for a new generation of codes (level I, semi-probabilistic methods of I order) /6/. The term "semiprobabilistic" means that the limit state equation is formally deterministic, but that its members (partial safety coefficients and characteristic values) were obtained with level II probabilistic method.

This method, as well as the way it was accepted for the calculation of safety elements in the new version of EUROCODE 3, are presented in /4/.

The essence of this application in EUROCODE is to distinguish two parts in the limit state equation: a part representing the function of the action effect (loading) and the part representing the resistance function. This is possible if the comparison level is determined, i.e. the desired safety (reliability index β and β_{min}). The result of this approach is the fact that, in determining the partial safety coefficients in limit state equations it will be possible to study action and resistance separately. This approach of separate study of resistance behaviour provides a systematic use of experimental laboratory investigations. A complex behaviour of the elements due to the above reasons in the Introduction requires experimental verification. The procedure described in /4/ provides a possibility for experimental verification of theoretical resistance models and shows a way, given a safety degree β, how to determine partial coefficient of a structure resistance, i.e. it gives information whether our theoretical resistance function (model) adequately corresponds to behaviour in reality.

The presented considerations are a solution model applied in preparing the new edition of EUROCODE 3, and presents a model to be applied when introducing high strength steels into new codes.

4 CHARACTERISTIC PROBLEMS

If we compare the characteristics of bearing capacity failure we can see similarities and differences in the behaviour of high strength steels im comparison to the classical steel construction materials.

In a tensile problem where the failure appears as a result of the lack of tensile strength, there appears an essential phenomenon. Ir order to provide equal safety of a net cross-section of a tensile member EUROCODE 3 requires a failure of net cross-section by using the ultimate tensile strength or gross cross-section failure, by reaching the yield strength.

According to researches from /7/, the behaviour at tensile capacity of net cross-section in high strength steels with corresponding toughness is similar to the behaviour of steel with lower strength. This provides a corresponding increase of bearing capacity. Due to the relatively high ratio of yield point to strength in these materials, although the failure appears in higher levels, it is not connected to plastic deformations of the gross cross-section. This reflects the basic difference in the behaviour of these materials with the following consequences:

- adequate safety can be achieved introducing additional criteria, i.e. raising the toughness criteria to a higher level so that properties of high strength steels can be utilized. This can be done by using fractomechanical methods, J integrals, by setting the criteria connected to the beginning of the stable crack growth: /7/

- Joints and conections should be studied and simple regulations should be made to control the bearing capacity of the joints where the material no longer shows great plastic deformations.

At the same time, probability density curves and characteristic statistical parameters should be precisely determined for each kind of steel and for each property of the material, primarily for the yield point and for the toughness.

It can generally be said for stability problems that the popular trend to build lighter structures has influenced a strong research and scientific process with an objective to study the phenomena (local and global stability and models of their interaction when defining their optimum bearing capacity) and the behaviour in the post-critical stage in order to formulate the numerical resistance models more realistically.

It should be noted that high-strength steel elements generally have more favourable treatment in the stability problem because the level of residual stresses is lower than in classical steels. On the other hand, there are natural limitations and the yield point level does not essentially determine their bearing capacity.

It can accordingly be considered that the resistance models in the new EUROCODE 3 are formulated so that they can be used for high strength steels in stability problems. A verification procedure with a controlled experiment is however still necessary.

With regard to flexure, the basic problem is to determine how much the use of high strength steels with decreased plastic properties can limit the rotational characteristics of a girder. According to document /5/ which presents the research programme of rotational characteristic of an I-beam in materials with different yield to strength ratios f_y/f_u (profiles are tested which have previously been cold formed to a cca 10% deformation and their ratio is therefore $f_y/f_u \sim 1.0$), the analysis includes interactions of different characteristic mechanisms of failure due to the bending moment (presented with the curve M - ϕ), transversal force and local buckling. Research will show whether there are objective limitations depending on the f_y/f_u ratio which could influence the size of the plastic rotation angle, or rather a redefinition of the rotation angle minimum requirement must be done for a cross-sections to be classified as "plastic-plastic", i.e. for those in a group with plastic cross--sectional resistance with plastic structural behaviour due to formation of plastic hinges. Although the research should be extended and the data should be sistematized, all resuts point to the fact that high strength steels will probably be used without special limiations.

In fatigue, it is a well-known fact that the advantages of statical strength in high strength steels can not be utilized in this respect because both working and permanent strength are practically independent on the kind of steel and are equally high for ordinary and high-strength steels.

5 CONCLUSION

The structural problem of using high-strength steels in engineering structures is a very topical research task.

Accordingly, in the situation when "RILEM" as an international organization of research laboratories is considering to extend its activity and study the application of metals, this theme should certainly be included. Considering the scope of the task and the character of "RILEM", special attention should be paid to quantitative assessment of fracture, i.e. the theme which would connect the classical laboratory research, with the new fracto-mechanical area taking into consideration the statistical characteristics of the phenomena with regard to their application in the probabilistic model of safety treatement.

REFERENCES

/1/ Milcic,V. Problematika dimenzioniranja konstrukcija iz celika visokih crstoca, Referat Z-27, VI.Kongres JDGK, Bled, Septembar 1979.

/2/ Milcic,V., Hemerich,E. Calibration of Steel Structures in view of Yugoslav Conditions of Actions and Resistances, International Conference on Steel Structures, Budva, September 28 - October 1, 1986, Part II, pages 599-609.

/3/ Vogel,U. Principles of ultimate strength design, Second International Colloquium on Stability, ECCS, 1976.

/4/ EC3-87-G10B1 - Background document for chapter 2 of Eurocode 3 draft, Aachen, September 8, 1987.

/5/ Sedlacek,G., Dahl,W., Spangemacher,R., Hubo,R. Untersuchung der Auswirkungen der unterschiedlichen Streckgrenzenverhälnisse auf das Rotations und Bruchverhalten von I-Trägern, PROJET 169, RWTH-Aachen, Juli 1988.

/6/ Genelral principles on reliability for structures, ISO/DIS 2394, 1984.

/7/ Dahl,W., Hesse,W. Auswirkung der Beurteillung von Stählen auf die Anwendung im Hoch-und Anlagenbau, "Stahl und Eisen" 106 (1986) Heft 12, pages 695-702.

3 FAILURE TESTS WITH STEEL FRAMEWORKS

M. IVÁNYI
Department of Steel Structures, Technical University of Budapest, Hungary

SUMMARY

Experimental and theoretical investigations are shown, making the analysis of plastic deformation capacity of frames with due regard to interaction of strength and stability phenomena possible, including the effect of local buckling of plate elements of sections on the response of the whole structure.

1 INTRODUCTION

The increasingly powerful experimental and computational tools of structural design require a well-defined design philosophy. As its basis generally the concept of limit states is accepted in many countries. Limit states are usually divided into the two following groups:

1st group - limit states of carrying capacity
(ultimate limit states)
2nd group - limit states of serviceability.

Checking serviceability at working load level the traditional and well tried out methods of structural analysis are used. The quite frequent tests on original structures show as a rule a relatively good accordance with calculated stress patterns or even better with predicted deflections (apart from the occasional effect of lack of fit at connections.

On the other hand, structural response in the vicinity of peak load can be extremely complex. The early and very simple methods of plastic limit analysis (based on the concept of rigid-plastic materials which are in principle completely insensitive regarding any forms of initial imperfections) are confined to a very limited class of structures, built up of bulky elements. As soon as global and even more as local instability plays role in failure,

- the effect of initial geometric imperfections is enhanced,
- the residual stresses (remaining regularly latent at lower loads, interacting with growing active stresses) result in premature plastic zones, and (last but not least)
- the usual and widely accepted tools of analysis - as beam theory (based on Bernoulli-Navier hypothesis), small deflection theory of plates and so on - can not describe exactly enough the structure's response at failure.

The simplified model is not elaborate enough to reflect real structural behaviour, so that a secondary, more detailed local model is introduced to depict the mostly critical part of the structures, by which more realistic quality parameters can be deduced from the already known primary parameters.

Testing of Metals for Structures. Edited by F.M. Mazzolani.
Published by Taylor & Francis, 2 Park Square, Milton Park, Abingdon, Oxon, OX14 4RN.
ISBN 0 419 15810 3

Because of the interaction of local and global behaviour this pattern can not be followed in the case of hyperstatic structures, as the additional information gained by the secondary, local model is to be fed back to the calculation of primary parameters as well. For this purpose, if - as very often - the secondary model can be analysed by numerical methods or only experimentally, the results have either to be re-interpreted to obtain mathematically treatable, simple enough rules, or the secondary model has to be simplified to get well-usable results. In both cases the validity of accuracy has to be proved by (usually very expensive) failure tests on the whole structures.

2 TEST PROGRAM

Between 1974-1989 extended experimental research project was carried out in the Laboratory of the Department of Steel Structures, Technical University, Budapest.

The **first part** of the program: Fig 1. gives a brief summary of the additional tests on stub columns, frame corners, plates elements, simple beams. (HALÁSZ, IVÁNYI 1979)

The **second part** of the program: Fig 2. gives a brief summary of the full-scale tests and dimensions of the specimens, indicating the loads and the characteristics of the loading process.

The **third part** of the program was a representative part of a multi-purpose, pinned, pitched roof industrial hall: a building section consisting of 3 frames, bracings with pinned elements, light gage purlins and wall beams with corrugated steel sheeting (Fig. 3.) (IVÁNYI, KÁLLÓ, TOMKA 1986.)

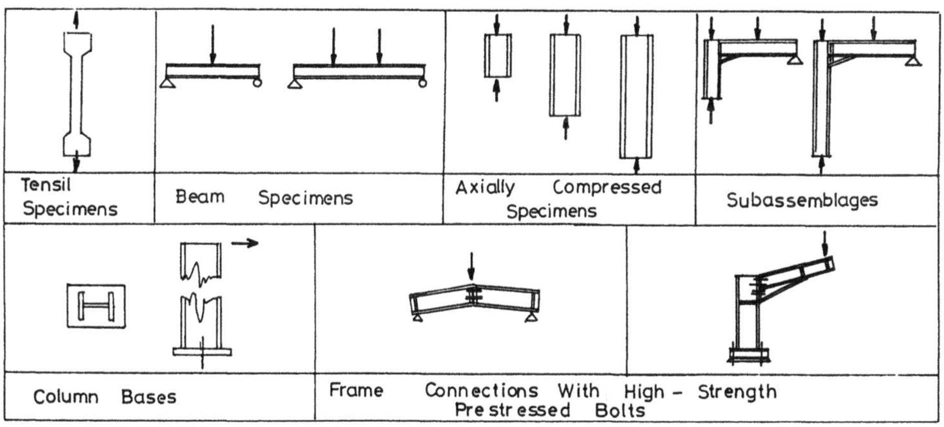

Fig. 1

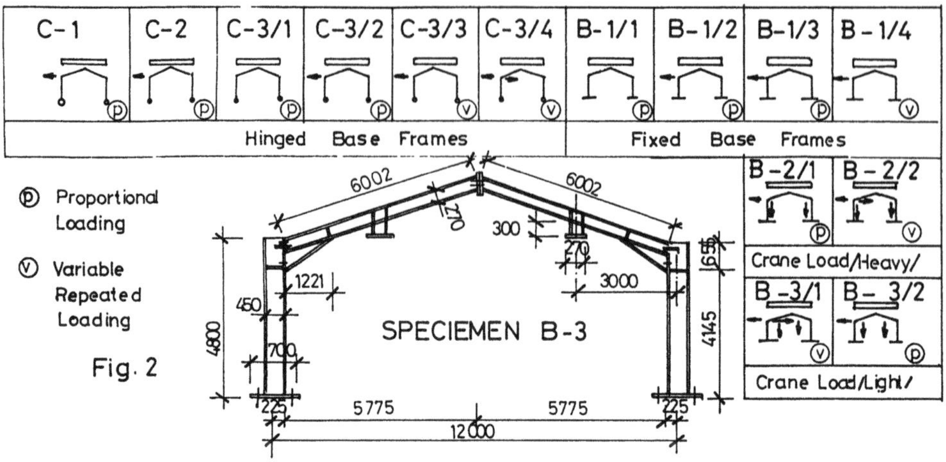

Fig. 2

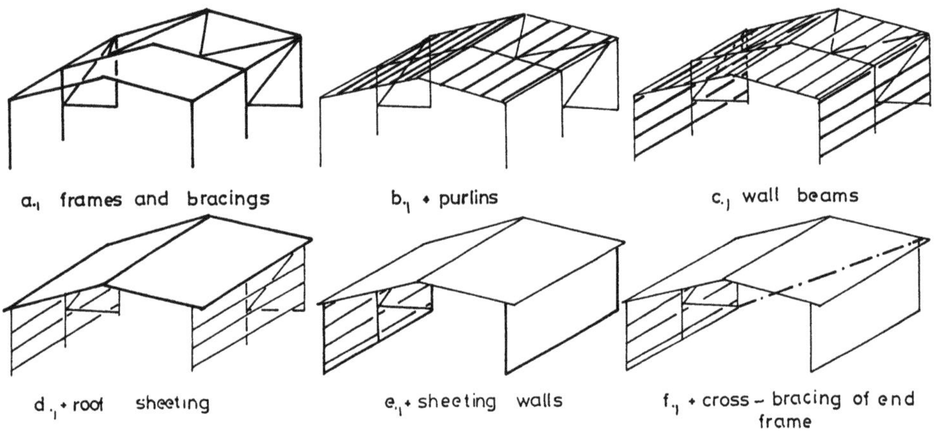

Fig. 3

3 COMPUTATIONAL MODEL

The traditional concept of plastic design of steel structures is based on the assumption, that under gradually increasing static loads plastic zones develop and grow in size and number, and eventually cause unrestricted, increasing deflections: thus loading to the onset of ultimate limit state of the structure. The concept was first introduced by KAZINCZY (1914) by establishing the concept of the "plastic hinge". Some basic questions are still discussed. Among them are the effects of the difference between ideal-plastic constitutive law and actual behaviour of steel material and the consequence of local instability (plate buckling, lateral buckling). Joining into the international research in this field we tried to introduce the concept of "interactive plastic hinge", which can substitute the classical concept of plastic hinge in the traditional methods of limit design, but can reflect the effect of phenomena like strain-hardening, residual stresses, plate buckling and lateral buckling (IVÁNYI 1983).

The element of the bar is considered to be built up of plate elements (following the pattern of steel structures) instead of a compact section. Then the behaviour of the "plastic hinge" can be characterized by tests with simply supported beams. Based on these tests a yield-mechanism for the bar-element can be introduced (Fig 4.), giving foundations for a mechanism curve: defining thus the descending branch of the moment-rotation diagram (Fig. 5.). Similar results can be obtained for the case of different loading conditions (IVÁNYI 1979/a, 1979/b).

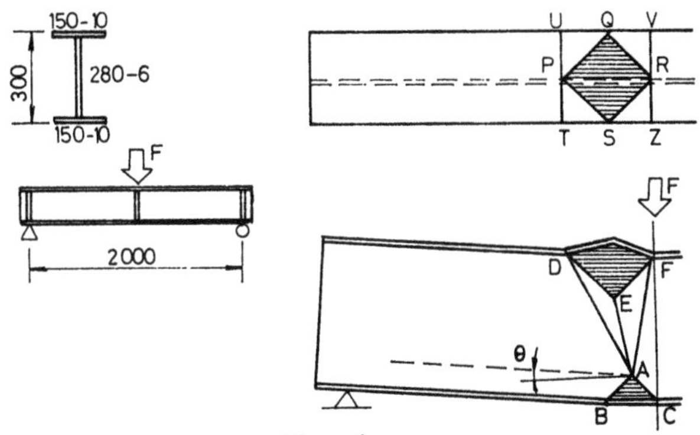

Fig. 4.

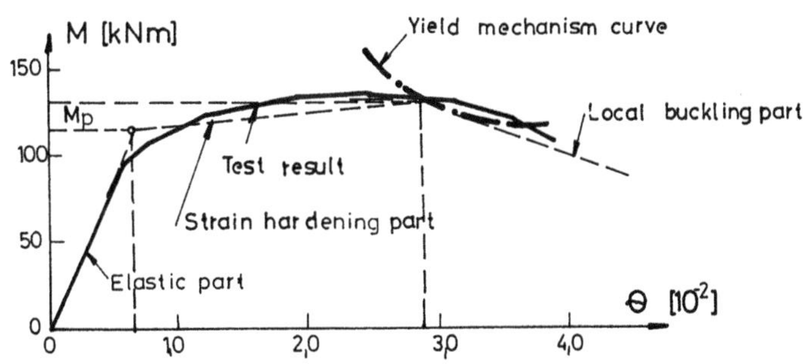

Fig. 5.

This information can be used to extend the model of "equivalent cantilever" suggested by HORNE (1960) including the effect of residual stresses and strain-hardening (Fig. 6), with the descending branch representing the effect of plate buckling, arriving thus to the concept of "interactive plastic hinge" (IVÁNYI 1983/a,b, 1985/a).

Supposing now the whole structure to be built up of elastic bars and "interactive plastic hinges", its response can be analyzed by using the original stiffness matrix and taking into account the effect of plastic hinges by inserting "kinematic loads" representing the concentrated rotations. Thus analysis consists of two steps: that of the

(a) elastic bars, and

(b) interactive plastic hinges.

The governing equations of the structure following the displacement method (SZABÓ, ROLLER 1978)

$$\underline{\underline{G}}^{x} \cdot \underline{s} + \underline{g} = 0$$

$$\underline{\underline{G}} \cdot \underline{u} + \underline{\underline{F}} \cdot \underline{s} + \underline{t} = 0$$

with following notations:

- $\underline{\underline{G}}$ - geometric matrix
- $\underline{s}$ - vector of load actions
- $\underline{g}$ - load-vector
- $\underline{u}$ - displacement vector
- $\underline{t}$ - vector of "kinematic loads"
- $\underline{\underline{F}}$ - flexibility matrix.

The "stiffness" of the interactive plastic hinge (Fig. 6) at different stages is given by the stiffness characteristic

$$c = \frac{\Delta M}{\Delta \theta}$$

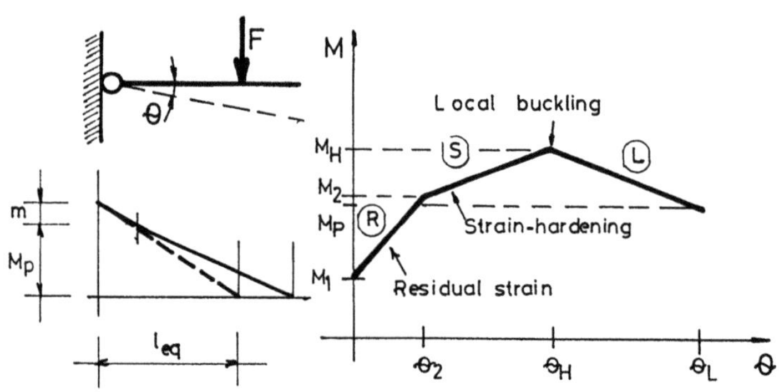

Fig. 6.

The load-deflection diagram can be obtained by a step-by-step approximation. Starting with an appropriate value of $\underline{q}$ load is increased by $\Delta\underline{q}$ increments - till in a potential plastic hinge moment reaches the value M_p: then the "kinematic load" is inserted, with due regard to the corresponding stiffness characteristic. Calculation was carried out by PDP 11/34 Typ Computer.

The interactive plastic hinge model is suitable for computer computations (IVÁNYI 1985/b) (BAKSAI, etc. 1985)

4 RESULTS OF THEORETICAL AND EXPERIMENTAL INVESTIGATIONS

4.1 Frame Structures under proportional loading

Concerning the experimental frame C-3/2, the relation of load-deflection curve developes according to Fig. 7. At the side of horizontal load, the first inelastic hinge develop due to the residual stresses and deformations in the cross section beneath the frame structure wedging up and this hinge develops at 52% of the maximal frame load. At 97% of the maximal load, the zone L describing the effect of plate buckling develops also in this cross section, i.e. in the frame cross section an "unstable" state - a descending characteristic curve - develops.

Fig. 7. introduces the characteristic load-displacement curve of the frame sturture in the case, too, when the basis of the computations is the traditional plastic hinge.

The results well shown that the presence of residual stresses influence in a major way the range of limited plastic deformation, however, mainly because of the cross section geometry of experimental beam the maximal load bearing values computed with the traditional (elastic-ideally plastic) hinge as well as those obtained by the interactive hinge coincide with the experimental results.

4.1.1 Effect of fabrication and erection

Effect of incorrect geometry was investigated by introducing different initial lateral displacements. Consequences are illustrated by Fig. 8.

The effect of the different values of residual stresses is shown in Fig. 9. The medium curve was in coincidence with test results.

The presented method for the complex analysis of frameworks takes several effects into consideration. (Fig. 10.)

4.1.2 Effect of structural details

It seems worthwhile to draw attention to the occasional decisive role a minor differences in structural details on failure as well. Some of the results are reproduced below.

(i) Column bases

Column bases were fixed or hinged. The hinges were not ideal: columns could have been supported by a bigger base-plates. The corresponding moment-rotation diagram of column base was checked experimentally, (Fig. 11.) its adaption to an interactive plastic hinge, indicating the load-deflection diagrams obtained by different end conditions. (Fig. 12.)

(ii) Frame corners

Elements are connected by tensile bolts. Fig. 13. and 14. show the test results regarding their behaviour: they were taken into account in the analysis.

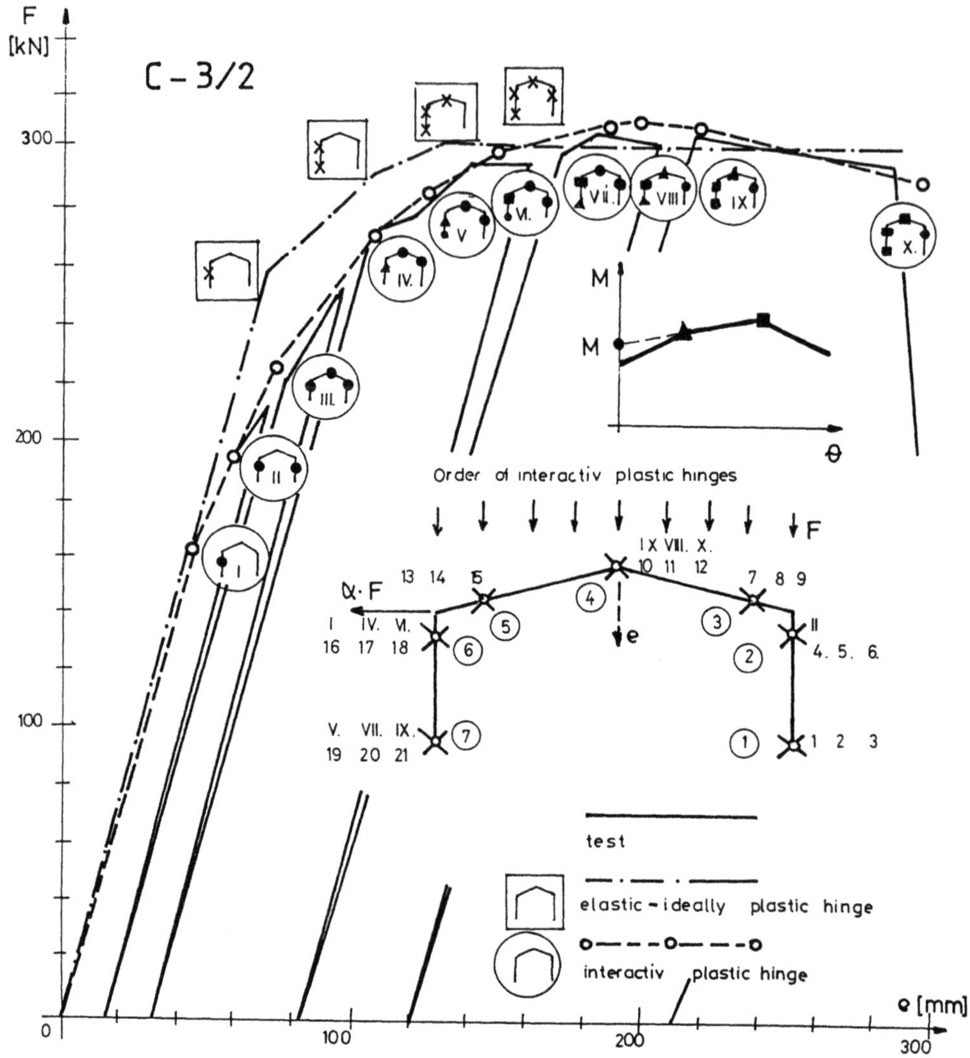

F [kN]
C-3/2
300
200
100
0
M
M
θ
Order of interactiv plastic hinges
F
α·F
e
test
elastic-ideally plastic hinge
interactiv plastic hinge
e [mm]
100
200
300

Fig. 7.

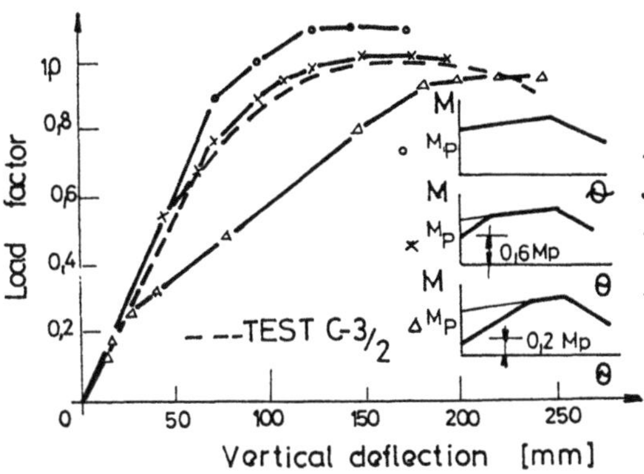

Fig. 8.

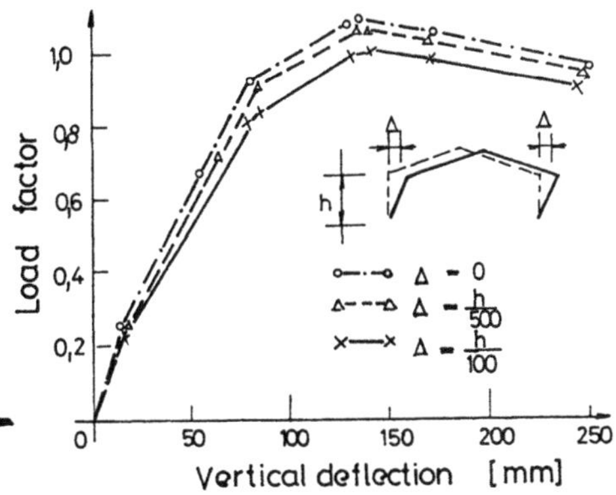

Fig. 9.

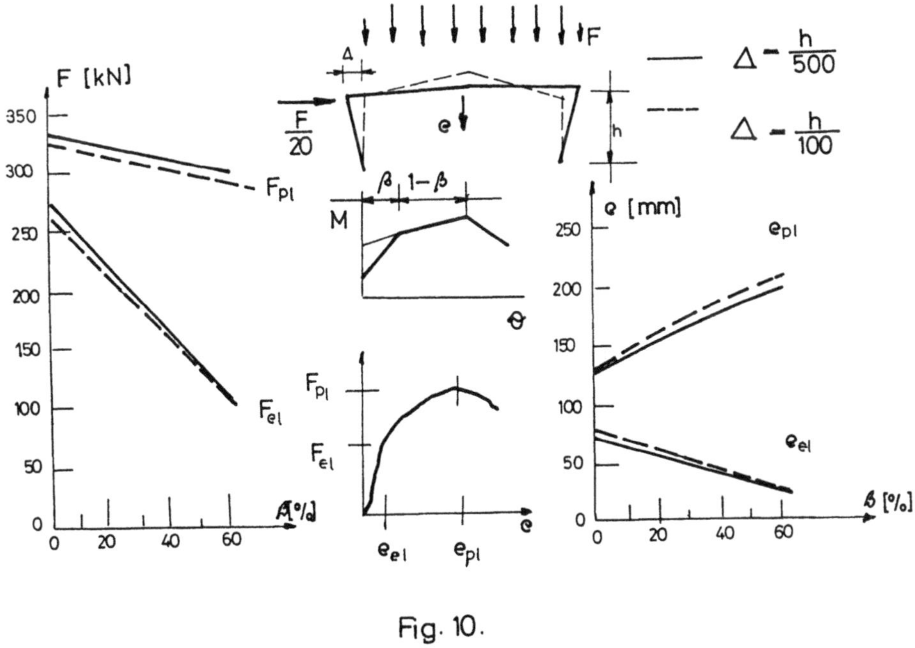

Fig. 10.

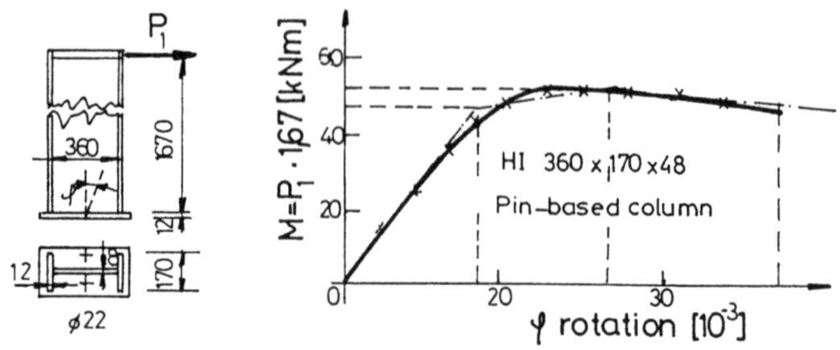

Fig. 11.

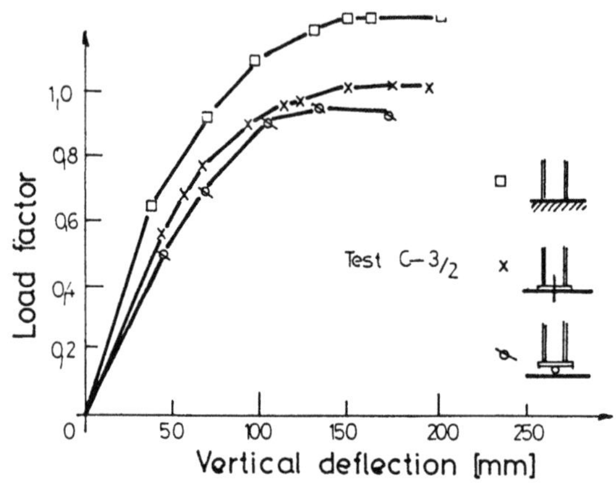

Fig. 12.

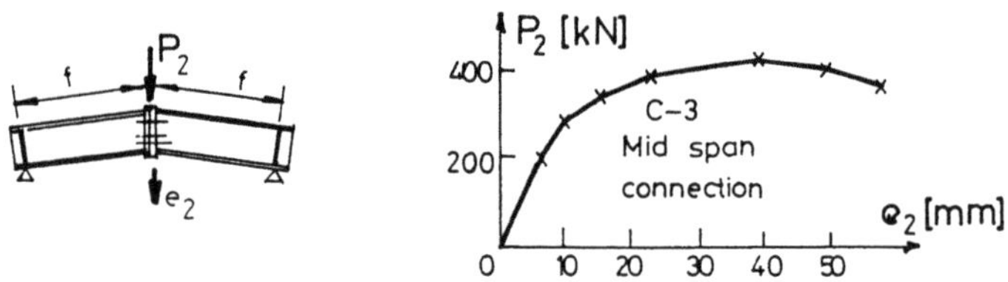

Fig. 13.

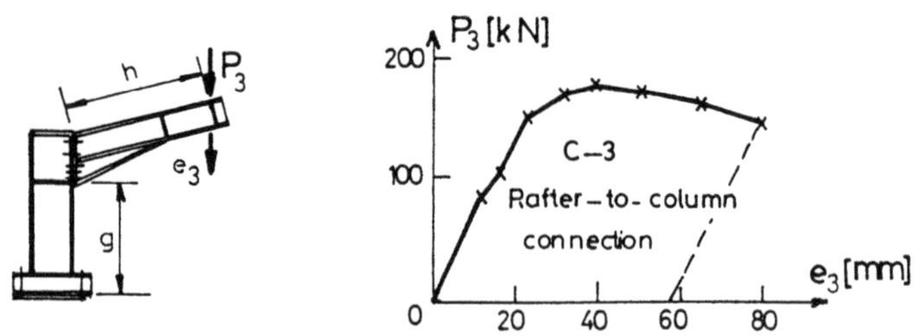

Fig. 14.

(iii) Lateral supports

Spacing and efficiency of lateral supports proved to be of basic importance. Different types of supports were applied (Fig. 15.) Their effect is illustrated by Fig. 16.

The importance of adequate spacing of lateral supports and their efficiency in preventing the rotation of cross-section around the bar-axis has to be emphasized as purlins and rails connected to tension flanges often can not regarded fully effective in case of thin webs. Not only the load carrying capacity can thus be substantially reduced (as by elastic lateral buckling in case of frame C-3/1 in Fig. 16.), but the yield plateau in the load-deflection diagram can be to short (as in the case of frame C-1 in Fig. 16.), rendering the structure sensitive against initial imperfections.

4.2 Building under proportional loading

4.2.1 Cross bracing of the end-frame

Measured eaves deflections from uniform horizontal loads are shown on Fig. 17. representing the effect of both semi-rigid cross bracing of the end frame.

4.2.2 Horizontal and vertical bracing system

Load-displacement diagrams of incremental collapse tests are shown in Fig. 18. Ultimate loads are influenced by local loss of stability, previous loadings and the layout of frame-horizontal and vertical bracing connections.

4.2.3 Some remarks

The actual behaviour of hall structures is more favourable than that of generally taken into consideration in design practice. At strength analysis, and at stability analysis in plane of the frame we generally neglect the effects of semi-rigid column bases, roofing sheeting, end walls.

Our measurement results showed that limit, with which these effects adventageously affect both internal forces and the rigidity (mostly that of the horizontal displacement) of the structure.

The actual rigidity of the hall structures, the semi-rigid connection of the structural elements adventageously affect the value of load intensity belonging to the loss of stability perpendicular to the plane of the frame.

It also should be mentioned that, beside adventageous effects, such effects occur that disadventageously effect the load-bearing capacity of the structure. These are the residual stresses due to welding, imperfections due to manufacturing and assembling, eccentricity.

There is no doubt that they must be taken into consideration according to their significance. Its method and the values of the factors (such as fictious eccentricity) applied in the calculations are mostly contained in our present specifications. In cases not regulated or not adequetly known, the decision of the designer governs. We would like to mention that economic design can not be imagined such as taking the disadventageous circumstances practically in full, into consideration, and totally neglecting the adventageous ones, or taking effect into consideration, considerally underestimated.

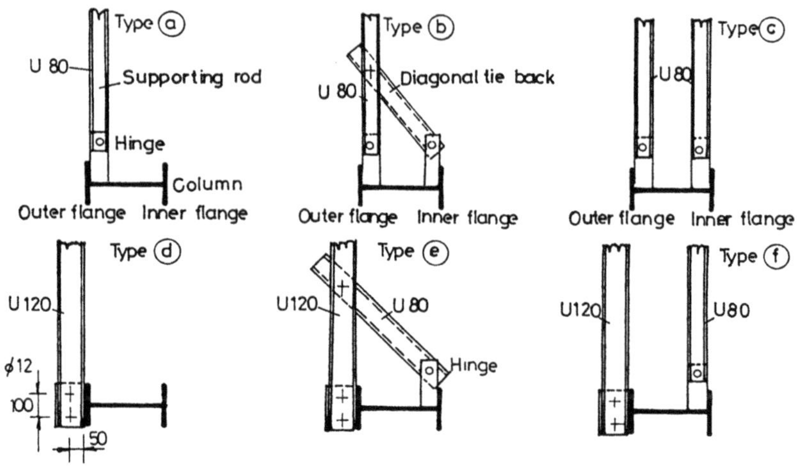

Type (a)
U 80
Supporting rod
Hinge
Column
Outer flange
Inner flange
Type (b)
U 80
Diagonal tie back
Outer flange
Inner flange
Type (c)
U 80
Outer flange
Inner flange
Type (d)
U 120
ϕ12
100
50
Type (e)
U120
U 80
Hinge
Type (f)
U120
U80

Fig. 15

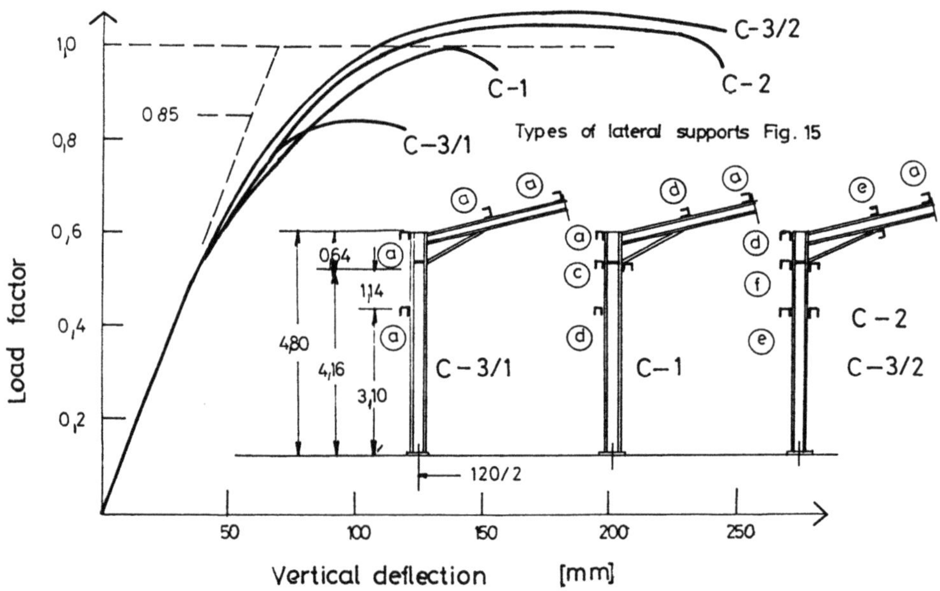

1,0
0,8
0,6
0,4
0,2
0.85
Load factor
C-3/2
C-2
C-1
C-3/1
Types of lateral supports Fig. 15
0,64
1,14
4,80
4,16
3,10
120/2
C-3/1
C-1
C-2
C-3/2
50
100
150
200
250
Vertical deflection
[mm]

Fig. 16

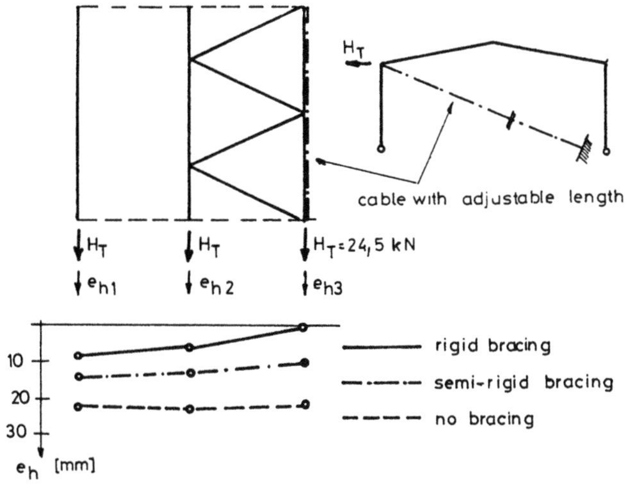

H_T
cable with adjustable length
H_T
H_T
H_T = 24,5 kN
e_{h1}
e_{h2}
e_{h3}
10
20
30
e_h [mm]
rigid bracing
semi-rigid bracing
no bracing

Fig. 17.

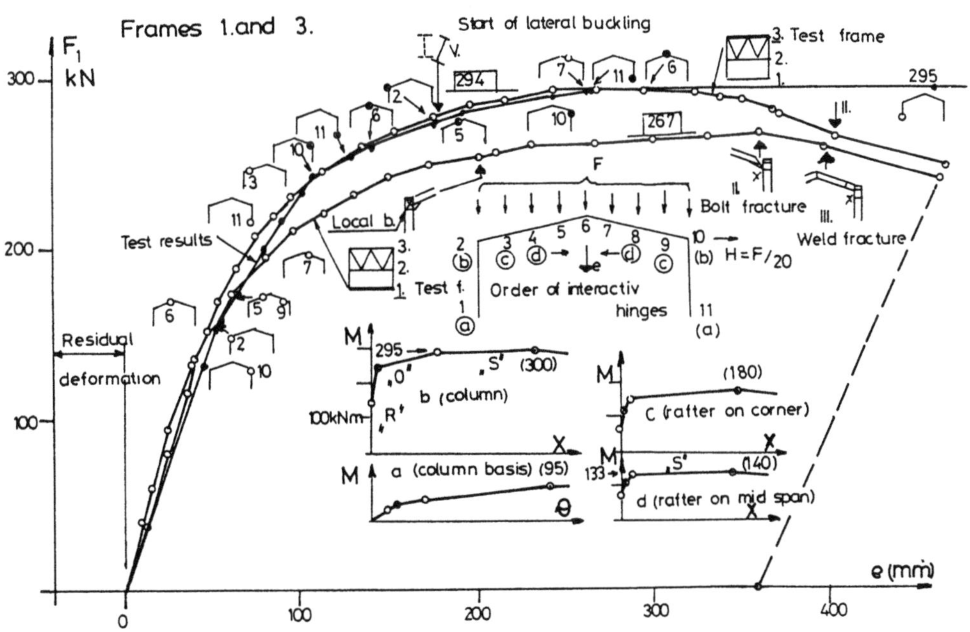

Frames 1.and 3.
F_1 kN
Start of lateral buckling
3. Test frame
2.
1.
295
294
267
Test results
Local b.
3.
2.
1. Test f.
Bolt fracture
Weld fracture
H = F/20
Order of interactiv hinges
Residual deformation
295
(300)
b (column)
100kNm
a (column basis) (95)
(180)
C (rafter on corner)
133
(140)
d (rafter on mid span)
e (mm)
0
100
200
300
400

Fig. 18.

5 CONCLUSION

Experimental and theoretical investigations have been carried out in connection with the elastic and the plastic load-bearing study of frame and hall structures, with the steel material strain hardening, the residual stresses and plate buckling taken into account.

A method has been presented for the investigation of frame structures applying the steps of known, traditional methods so that the structure behaviour can be analyzed during the entire process of loading. Certain effects determining the structure behaviour (e.g. residual deformation, steel material strain hardening and plate buckling) have been into consideration with the aid of the interactive plastic hinges. The interactive hinge was incorporated into an investigation method operating with the structure matrix-calculation method. The results of the elaborated method has been compared with the experimental investigation of full-scale structures.

6 REFERENCES

BAKSAI, R., IVÁNYI, M., PAPP, F.: 1985
Computer Program for Steel Frames Taking Initial Imperfections and Local Buckling into Consideration.
Periodica Polytechnica, Civil Eng., Vol.29., Nos 3-4, pp. 171-185

HALÁSZ, O., IVÁNYI, M.: 1979
Tests with Simple Elastic-Plastic Frames.
Periodica Polytechnica Vol. 23., Nos 3-4, pp. 151-182

HORNE, M.R.: 1960
Instability and the Plastic Theory of Structures.
Transactions of the EIC, Vol.4., No.2., pp. 31-43

IVÁNYI, M.: 1979/a
Yield Mechanism Curves for Local Buckling of Axially Compressed Members.
Periodica Polytechnica, Civil Eng., Vol.23., No.3-4., pp. 203-216

IVÁNYI, M.: 1979/b
Moment-Rotation Characteristics of Locally Buckling Beams.
Ibid. pp. 217-230.

IVÁNYI, M.: 1983/a
Interation Between Strength and Stability Phenomena. Role of Plate Buckling (in Hungarian) Thesis presented for the degree of Doctor of Sciences, Budapest

IVÁNYI, M.: 1983/b
A New Concept in Limit Design of Steel Structures.
Colloquium on "Stability of Metal Structures",
Paris, 16-17 Nov., Final Report, pp. 259-264.

IVÁNYI, M.: 1985/a
The Model of the "Interactive Plastic Hinge"
Periodica Polytechnica, Civil Eng., Vol.29., No.3-4., pp. 123-146

IVÁNYI, M.: 1985/b
Load-Deformation Relationships for Simple Steel Frames with Unstable Elements.
Ibid. pp. 147-170

IVÁNYI, M., KÁLLÓ M., TOMKA, P.: 1986
Experimental Investigation of Full-Scale Industrial Building Section.
Colloquium on "Stability of Steel Structures"
Hungary 25-26 Sept., Final Report.

KAZINCZY, G.: 1914
Experiments with Fixed-End Beam. (in Hungarian)
Betonszemle, 2.68.

SZABÓ, J., ROLLER, B.: 1978
Anwendung der Matrizenrechnung auf Stabwerke
Budapest, Akadémiai Kiadó

4 ON THE DETERMINATION OF MOMENT–ROTATION CHARACTERISTICS OF BEAM-TO-COLUMN JOINTS

P.A. KIRBY
Department of Civil and Structural Engineering, University of Sheffield, UK
R. ZANDONINI
Department of Structural Mechanics and Design Automation, University of Trento, Italy
J.B. DAVISON
Y.R.M. Anthony Hunt Associates, Sheffield, UK
(formerly University of Sheffield)

SUMMARY

This paper discusses the performance of beam-column joints in multistorey construction, distinguishing between behaviour assumed in conventional "simple" and "rigid" design models and true response. It begins by proposing definitions for the main terms, i.e. connection, joint, node, nodal zone, moment and rotation. It then sets out factors which must be considered before conducting tests to determine experimentally representative moment rotation characteristics, focussing attention on the specimens, instrumentation, loading, reporting and interpretation of the results.

1 BASIC PHILOSOPHY

Structural steel systems are constructed from members which are fastened together. Historically very simple theoretical models have been used in both the analysis and the design of frameworks where the interactions between the component members were usually assumed to be either pinned or perfectly rigid whilst it is recognised that most practical configurations lie between these extremes.
Recently there has been renewed interest in the concept of semi-rigid (or partial restraint) design which seeks to utilise the degree of continuity inherent in joints including those typical of simple design and normally considered as connections which resist shear only. Thus, in addition to joint strength, rotation capacity and joint stiffness under all stress-resultants are potentially important design parameters. This requires that the characteristics of joint behaviour be properly identified and defined in order to assure a reliable and consistent use of these parameters. In order to achieve this goal, standard testing and reporting procedures need to be established - as they have been for the other factors involved in structural analysis and design e.g. material properties, geometrical tolerances and member strengths.

This paper will discuss the basic characteristics of joint behaviour in the context of the present state of knowledge and highlights the problems related to the definition of such behaviour, including its incorporation into frame analysis. This in turn requires a proper definition of terms such as joint, connection, rotation etc which have previously had broad interpretations. An attempt is made to set out the general requirements to be met when both testing joints and reporting test data, in order to maximise their possible use.

Testing of Metals for Structures. Edited by F.M. Mazzolani.
Published by Taylor & Francis, 2 Park Square, Milton Park, Abingdon, Oxon, OX14 4RN.
ISBN 0 419 15810 3

2 NODAL ZONE BEHAVIOUR

At the outset it is pertinent to identify the type of construction considered. This paper is directed principally at multistorey beam and column construction. It is however recognised that similar considerations will apply to other configurations e.g. bracing systems and truss-type construction in addition to the rotational response of column foundations.
Figure 1 shows a typical beam and column arrangement; the dotted line bounds the region which contains all of the local interactions. This is known as the nodal zone. Under stress, the deformation in the region occurs as the accumulation of the distortions of the individual components: column, beam and connection. This means that for essentially the same connection (e.g. beam size, column size, angle size, bolt size and spacing) the overall behaviour may be quite different being dependent upon the orientation of the beam i.e. whether it frames into the major or the minor axis of the column.
To illustrate this consider a beam attached to a column by top and seat cleats firstly connected to the column web (Fig. 2) and secondly to the column flange (Fig. 3).

In the first case the in-plane joint rotation of the end of beam will be due to distortions which arise from the following sources:

(i) The distortion of the angles under shear and bending.
(ii) The slip of the bolts in the holes until intimate bearing occurs, followed by subsequent elongation of bolt holes.
(iii) Shear deformation of the bolts (particularly those in the beam).
(iv) Elongation of the bolts in tension (those through the column web in the top cleat).

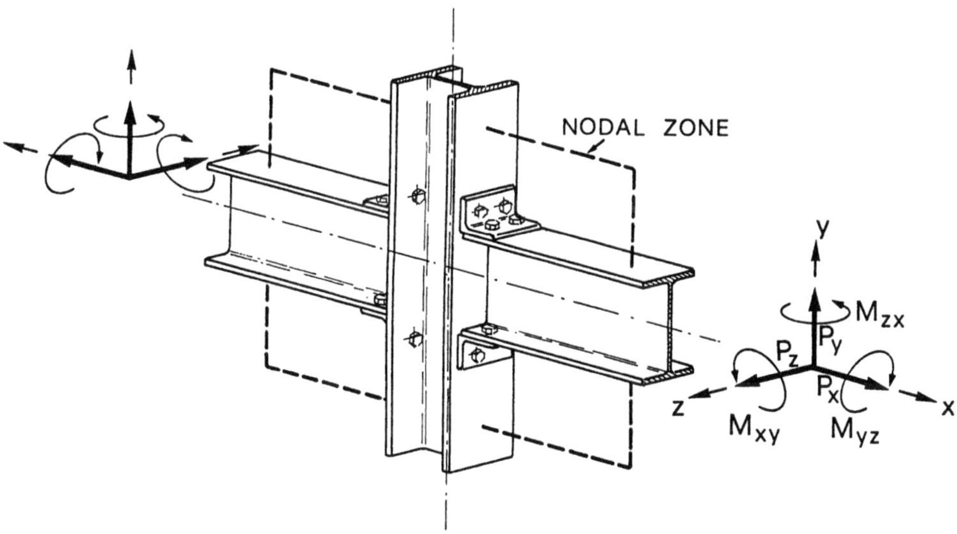

Fig. 1. *Typical beam column arrangement*

(v) Possible local distortion of the column web (likely to be greater when the flange cleats are narrow compared with the depth of the column web).

(vi) Initially, there is likely to be a clearance between the beam end and the column web. During the first stages of loading the beam rotation will occur approximately about its centroidal axis with the upper clearance increasing and the lower clearance decreasing. Eventually the lower beam flange will come into bearing with the column web resulting in a stiffening of the joint characteristic. The extent to which this may be relied upon in design is questionable but the existence of such effects must be recognised.

When considering the same basic components for a beam framing into the column flanges, effects (i) - (iv) and (vi) can be seen to be present again. However, the distortion of the column web, effect (v), which is essentially web bending for beams framing into the minor axis is replaced by shear deformation. This may occur together with either or both web buckling at the lower beam flange level and possible web tearing at the upper flange level. Additionally local bending of the column flange may occur.

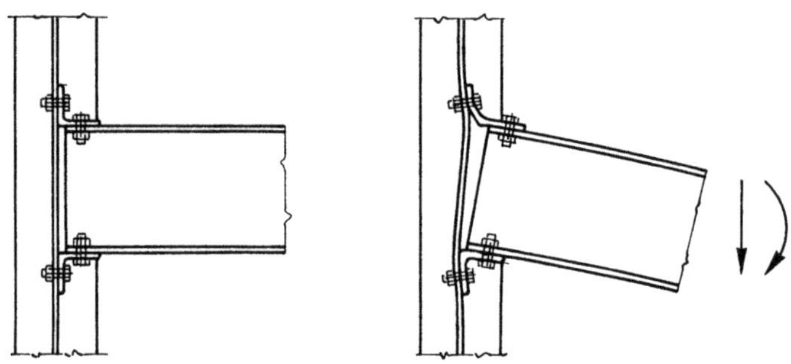

Fig. 2. *Beam framing into column web*

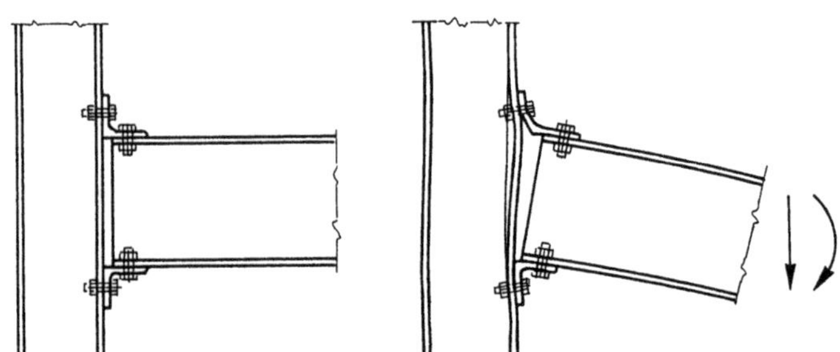

Fig. 3. *Beam framing into column flange*

It is instructive to note that the above examples correspond to a beam framing into one side of a column only. If members frame in on both sides (as indicated in Figs. 4a and 5) and these are subjected to equal moments (as in the commonly used cruciform test) then the contribution of the individual components may be affected. To illustrate this consider the contribution of the column web: due to symmetry, web bending is eliminated in weak plane joints, as well as the shear distortion in strong plane joints. In this latter case, the line of force, F, from the lower flange will pass more or less directly across the column web. Possible web buckling will therefore occur in a different manner to the case of a single sided connection, Fig. 4b, where the force F will tend to disperse to a greater extent. Besides column web behaviour, other differences in the response arise from sources such as the bolt elongation and the distortion of the other connection elements.

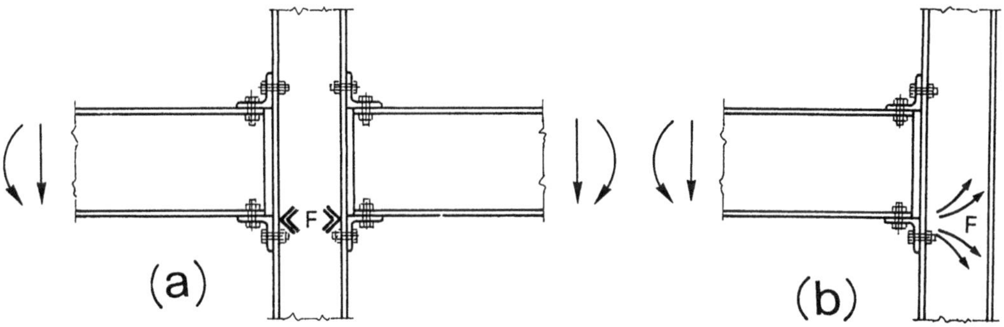

Fig. 4. *Beams framing into column flanges*

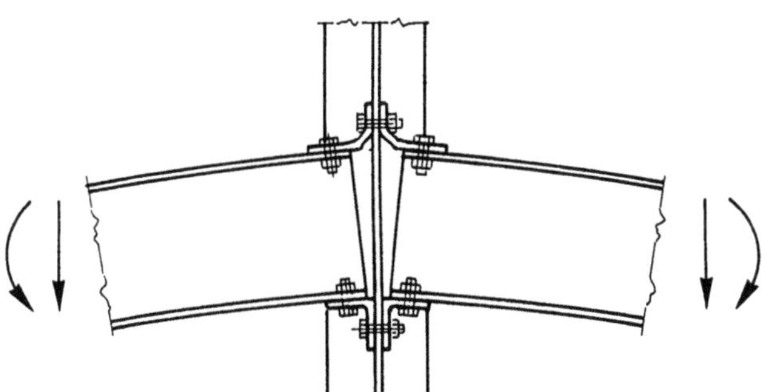

Fig. 5. *Beams framing into column web*

There is a further important consideration which concerns the number and orientation of the beams framing into the column which can occur in a variety of ways as indicated in Fig. 6, which shows in plan the possible combinations of beam members framing into a column section and which are variously described as one, two, three and four way framing. Clearly this feature will have an influence on the nodal zone responses. Firstly the presence of other connection elements, e.g. end plates used to attach the minor axis beams in the "four way" case, will stiffen up the web and thus affect the moment rotation behaviour of the major axis joints. This does not seem to have been considered in most of the current research, the exception being at Liegel, neither has the possibility of the beam members being attached to the column with their centrelines at slightly different heights.

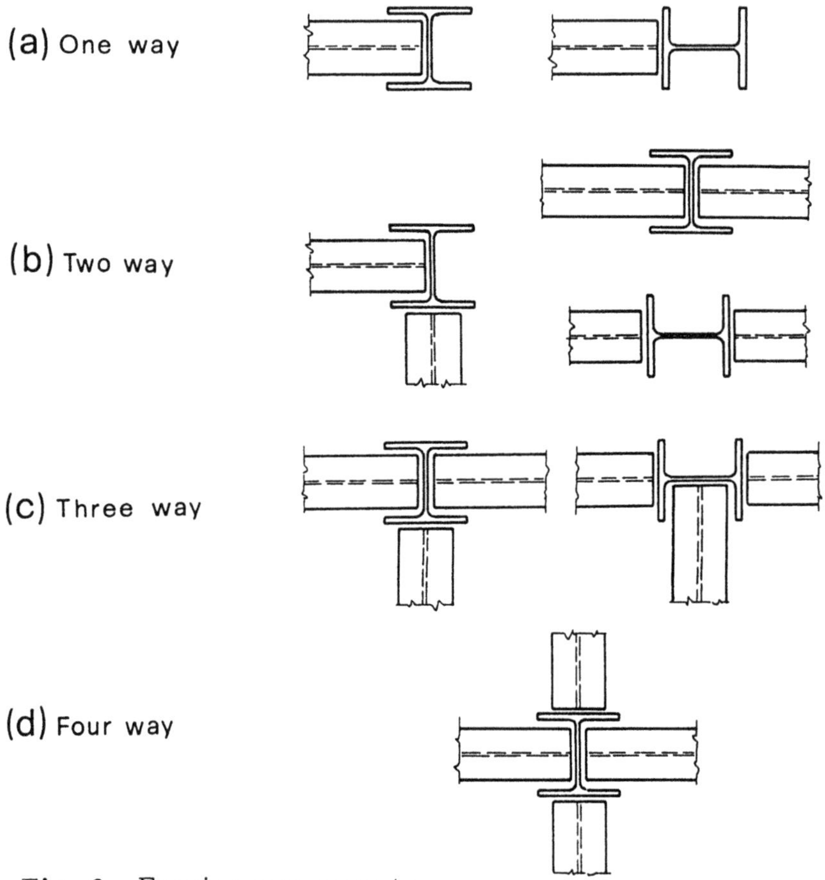

Fig. 6. *Framing arrangements*

3 DEFINITIONS

Before attempting an analytical representation of the form of behaviour described above it is vital to introduce a number of definitions. It is essential that, despite historical equivalence, a clear distinction is made between the terms node, connection and joint; the following definitions are here proposed:

- A **node** is "the location where the axes of two or more members meet".

- A **nodal zone** extends to cover the node and the adjacent lengths of the members which experience local deformations due to the interaction.

- A **connection** is "the physical components which mechanically fasten the beam(s) and column(s)", and is usually concentrated at the location where this fastening action occurs.

- A **joint** is here defined as "the connection and the corresponding zone of interaction between any two members".

Thus as a consequence of these definitions a connection is a component of a joint, whilst a node, which has particular significance for analysis, may comprise several joints. A schematic representation is shown as fig. 7.
In the context of normal multistorey construction a typical nodal zone will contain up to 4 joints with 4 connections. At any beam to column node there may be a number of joints each of which will, in general, resist actions with six components as indicated on Fig. 1. As discussed earlier the distortions which result are complex and too complicated for use in frame analysis. It is therefore convenient to refer to a simplified global representation and to adopt six generalised displacement components. Broadly speaking these displacements define the position of some cross-section in the beam relative to the centreline of the column. In the context of the action and deformation response for beam and column frames, only in-plane and out-of-plane moments (M_{xy}, M_{zx}) and their corresponding rotations are significant with the former being the most dominant. The response of the joint to the force components corresponding to the two shear forces P_y, P_z and the twisting action M_{yz} is such that the flexibility of the joint is so low as to be negligible when compared with that of the members themselves. Additionally axial deformation is normally insignificant in terms of structural response.

Thus, in practice, the problem becomes very much simplified in that the in-plane action is dominant although out-of-plane stiffness can become significant when full lateral restraint is not present. Furthermore Richard (2) has demonstrated that some limited interactive effects may occur but current investigations seem to indicate that these interactions may normally be neglected for practical applications and each action may be treated independently. Similar considerations apply to connection behaviour.

It should be noted that joint behaviour is not the sole component of the nodal response. In addition to the joint rotation, a "rotation" of the whole nodal zone is also to be considered. The former displacement component depends on the moment transmitted by the joint, whilst the latter arises due to the imbalance of joint moments. With reference to the case of beams framing into the column flanges, Fig. 8 identifies the first component, which is mainly associated with bending deformation, as ϕ_M and the second, which is related to shear distortion, as ϕ_V.

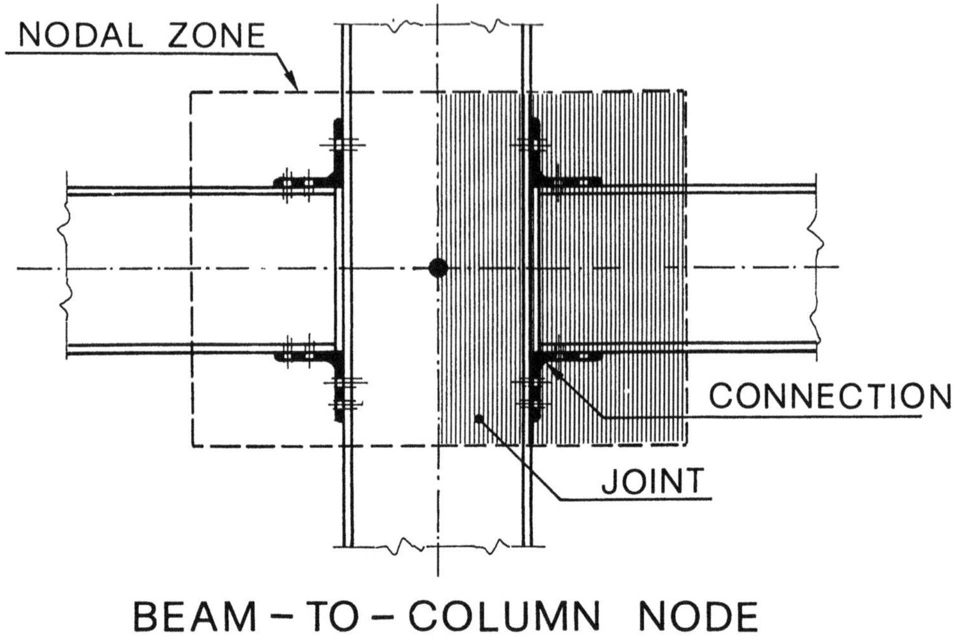

Fig. 7. *Definitions of: nodal zone, joint, connection*

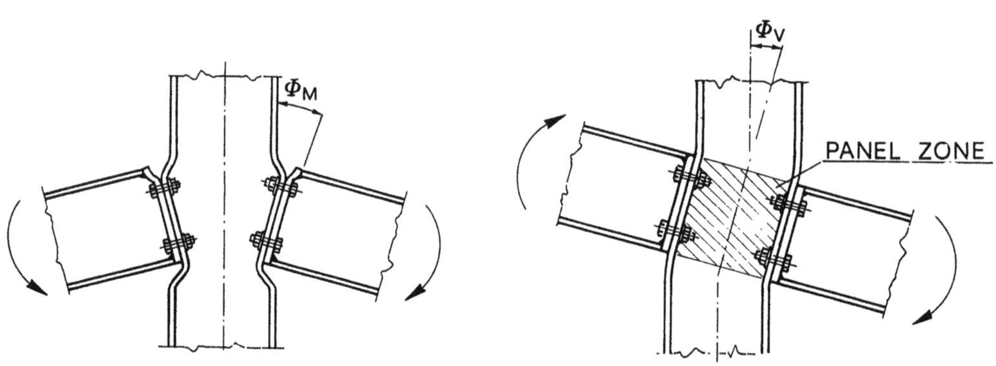

Fig. 8. *Joint rotation ϕ_M and shear rotation ϕ_V*

4 ANALYSIS AND GENERALISED FORCE AND DISPLACEMENT COMPONENTS

Joint deformation occurs as a result of complex interactions of local distortions in the components such as those outlined in the previous sections. The incorporation of this behaviour into frame analyses requires a knowledge of the overall joint response in terms of forces (usually moments) and displacements (usually rotations) in a form such that joint constitutive laws which are compatible with theoretical assumptions (generally classical beam theory) may be formulated.

A consistent definition for the relationship between the force component M and the corresponding displacement parameter ϕ is required. In the following, unprimed symbols are taken to correspond to in-plane action whilst the primed quantities M' and ϕ' correspond to out-of-plane action.

The joint rotation component ϕ_M(or ϕ'_M) should be taken as the relative rotation between the appropriate plane tangential (or normal) to the column web at point A and the tangential (or normal) plane to the beam axis at point B. Points A and B are identified on Fig. 9a, which shows a pair of identical beams framing into the flanges of a column, and subject to symmetrical loading.

Under any given loading the values of M and M' vary with distance along the beam centreline and it is therefore important to define the position at which their values are determined. As the deformation is frequently dominated by the distortion of the connecting elements and the fasteners (the exception arises in out-of-plane actions where substantial flexure of the column web may occur) the most appropriate location would seem to be the outer face of the column. Moreover, it is the value of the moment at this section which is utilised when checking the capacity of the connection detail and which dominates the rotational response, including the nodal response under moment imbalance.

The nodal rotation ϕ_V, due to moment imbalance, may be defined as the change in angle of the straight line joining the points of the column axis at the beam flange levels. In the case of beams framing into the column flanges, this represents the shear deformation of the web in the so-called panel zone (Fig.8). Rotation ϕ_V does not imply a rotation of the beam relative to the column. Thus, ϕ_M and ϕ_V are quite different in nature as is apparent from the figure.

In practice it is usually convenient for frame analysis to represent the joint behaviour by modelling each beam to column attachment as a single unit, and by summing the two rotation components ϕ_M and ϕ_V.

When the joint is treated as a unit the actual global distortion arises as the accumulation of component deformations which are usually distributed in a non-uniformmanner along a length of the joint. This may be represented in one of two ways; either by concentrating the effect at a single location, A or B, or by assuming a uniform flexibility over a length, usually A to B. The former is likely to be more appropriate when the contribution of the connection to the total rotation is dominant (3).

This representation may be refined by using models which isolate one or more of the individual components resulting in a global model which contains more than a single element. This usually implies that methods of analysis more sophisticated than simple beam theory are needed. Some such idealisations are shown as Figures 9(b)-(e). In all of these the beam elements, represented as the portions

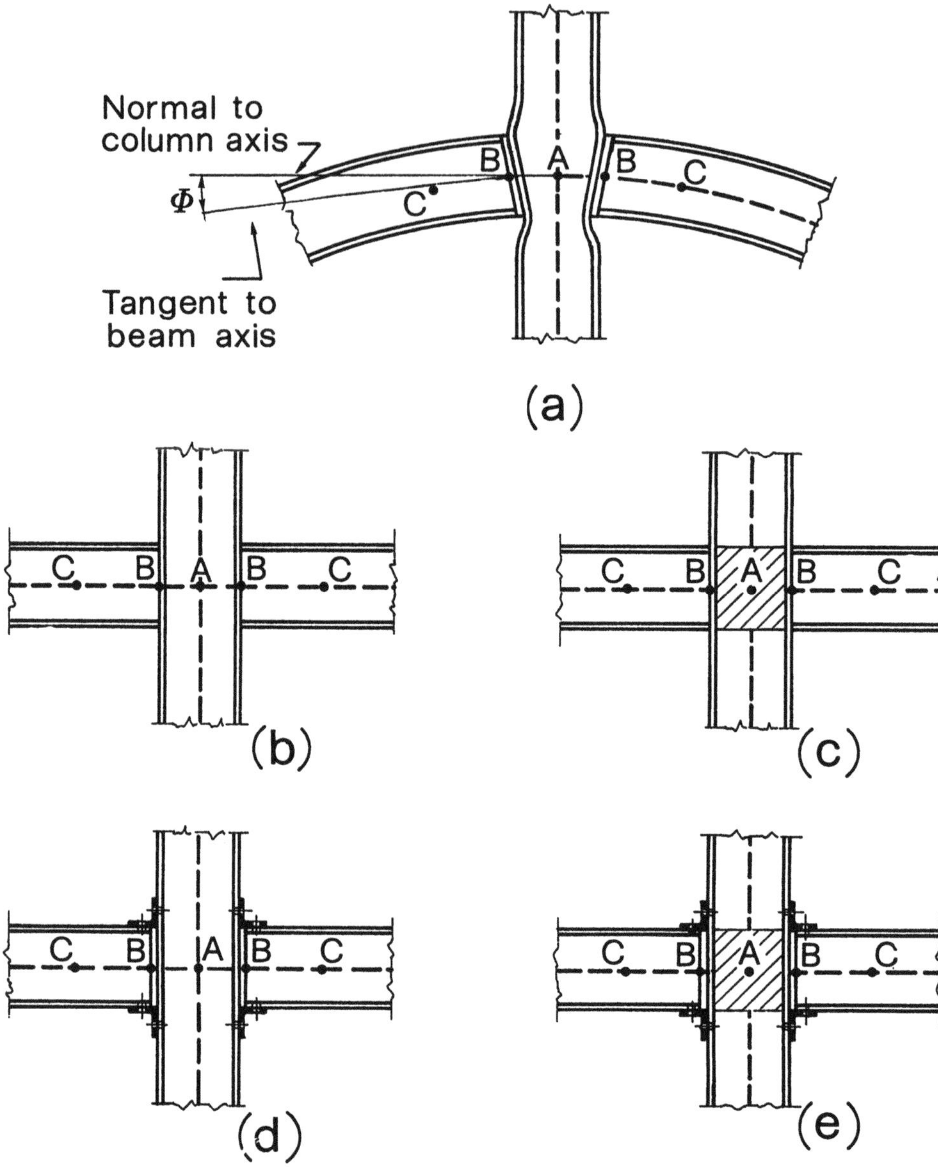

Fig. 9. *Modelling of joint response*

BC, are modelled as linear elements. In Figs. 9(b) and (d) the panel zone segments BAB are modelled as linear elements, whilst in Figs. 9(c) and (e) they are described as plate elements.

The use of these sophisticated models allows a more correct approximation of joint behaviour under non-symmetric moment but requires a definition of the component response which is consistent with the definition of the overall joint response. It also implies that in the derivation of joint moment rotation characteristics, usually experimental, the information is obtained in such a way that the required components may be derived. Care must always be taken to ensure that the representation of joint response is compatible with the procedure of the frame analysis. For example consider test rotations which include the influence of partial plasticity due to the presence of high moments for a moment resistant connection, then, if this is used in a computer program which also incorporates this plastic zone behaviour as part of beam response, there will clearly be an unwanted doubling of the effect.

5 EXPERIMENTAL DETERMINATION OF MOMENT-ROTATION BEHAVIOUR

5.1 Introduction

It is clear from the foregoing that care must be exercised in adopting a moment-rotation relationship which is appropriate to the analytical procedure which is to be employed. This requires that a clear assessment of the experimental method used is undertaken which, in turn, implies a full understanding of the differences resulting from alternatives in testing and of the importance of test records enabling identification of all test features.
It was during the accumulation of test records to be used in the establishment of a computerised database (17) that it became evident that the standard of much of the work conducted to date fell below ideal, often because it was carried out as a subsidiary operation in a more extensive study. This influenced the choice of test arrangement, loading process, instrumentation, coupons taken for material properties, bolt tightening procedures and/or welding details. Many tests were reported in an incomplete manner.

In section 2 reference was made to differing joint responses which would result from tests on joints with nominally identical connection elements but with the beams framing into the column flange or the column web. Similar reasoning was also applied to one way and two way in plane arrangements: both contain equal flexibilities arising from distortion of the connection elements but the latter may well contain significant flexibility arising from deformation of the column web. Similar arguments will apply to joints in which stiffeners are welded between the column flanges whilst even the presence of end plate connections of secondary beams framing into the column web will substantially modify the panel zone distortion. If a test is conducted in which more than one deformation component occurs, with modern testing, instrumentation and datalogging capabilities it seems little more than prudent to identify the individual components of distortion arising from the various sources so that the M-ϕ characteristics for variations of the tested joint may be assessed without recourse to a further test.

Such considerations, together with difficulties experienced during the interpretation of some earlier tests, highlight the need for the establishment of specific standards including the organisation of the experimental investigation together with more stringent requirements for reporting.

Some of the basic aspects of joint testing are here discussed more extensively. It should be recognised that in general all joints are potentially liable to undergo three distinct rotations which are in-plane, out-of-plane and twisting. However, tests on flexible joints such as web cleat connections have indicated that the torsional stiffness of the joint is so high relative to that of the connected member that its contribution to the torsional flexibility of the element in a frame may be disregarded.
Out of plane stiffness may have some influence when lateral-torsional buckling of beams occurs or when full three dimensional frame response is considered. Therefore the following sections will be restricted to in-plane and out-of-plane experiments.

5.2 Test Arrangements

Two basic joint test configurations have been widely used and to illustrate these a simple two way node comprising two joints subject to in-plane action will be considered. The first configuration is the cruciform arrangement of Fig. 10 which relies upon symmetry. This has the advantage that it is simple to employ and also gives two M-ϕ curves, one for each joint on either side of the column centreline. It does however suffer from the shortcoming that any variation from perfect symmetry (particularly if moment shedding is involved) can cause violent fluctuations in the relationship obtained (5). The second is the cantilever arrangement of Fig. 11 which does not rely on symmetry. This configuration requires more extensive apparatus and provides information from which only a single M-ϕ characteristic may be obtained for the joint but which is felt to be more complete as to the nodal response especially in the context of component distorsion as discussed in section 5.4.

When using the cruciform arrangement the load causing bending may be introduced as a single force 2F onto the column as indicated in Fig. 10 causing reactions of value F at the beam ends and a shear of value F in each connection. The distance of the end reaction may be adjusted to allow the ratio of shear to moment in the connection to be varied. A practical feature which has been noted concerns the undesirability of locating the beam force F close to the column face. It has been observed that this can allow a significant component of F to

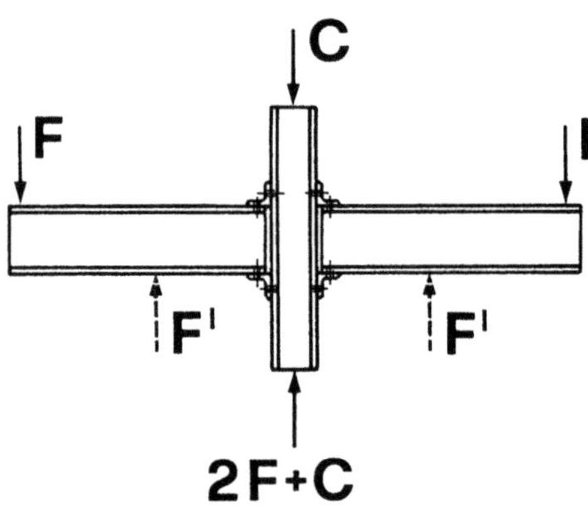

Fig. 10. *Cruciform test arrangement*

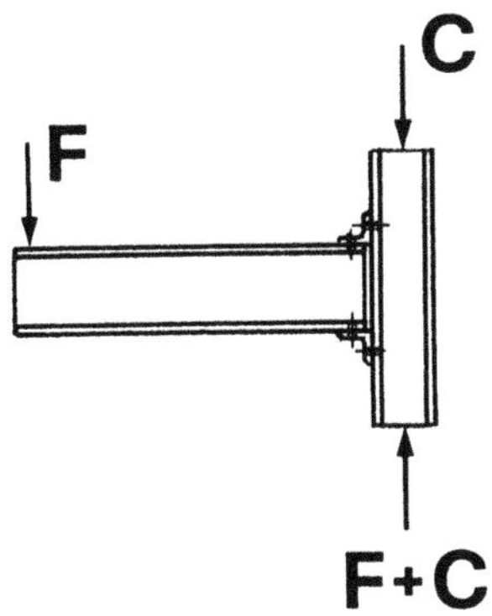

Fig. 11. *Cantilever test arrangement*

be transferred across the interface in diagonal compression rather than in bending through the connection components. This should be kept in mind when test arrangements are conceived ensuring that the connection is experiencing the intended value of moment and not some lower level due to this effect.

Similar considerations apply to the cantilever arrangement. In both cases an axial load, C, may be introduced into the column to determine the influence of its presence on joint response. Alternatively for the cruciform arrangement a four point loading system may be used with the column force 2F being replaced by two equal components (shown as F'in Fig. 10) applied to the beams outside the nodal region which then experiences pure flexure. Under this loading condition there is no shear in the connections.

Clearly, if the cruciform arrangement is employed, only a proportion of any distortion occurring in the panel zone of the column will appear in the response of either individual joint and the shear panel deformation will be absent. This would seem to be appropriate where the test is intended to simulate conditions which exist at a joint when a pair of symmetrical beams subject to identical loading framing into both sides of an interior column. It does not however closely represent a single beam framing into an exterior column nor to an interior column where either or both the geometry or the moments are unsymmetric. The cantilever arrangement would be more suitable for the exterior column situation but care needs to be taken. Similar considerations apply to out-of-plane action where it has been found that support conditions are of even greater importance than for in-plane action due to the increased flexibility of the column section.

It was during a recent series of tests using the cantilever arrangement that the importance of the support condition became evident. For example, restraining the column along its rear flange over its entire length or a length corresponding to the beam depth and with or without stiffeners was found to influence the moment rotation relationship obtained. A paper discussing the work (4) also demonstrates that the cantilever arrangement can be employed to simulate the symmetrical arrangement by the inclusion of appropriate stiffeners over half of the column web, a method which has the advantage over the cruciform system where perfect symmetry is required to avoid problems associated with unwanted unloading and moment shedding.

A further arrangement, which is an extension of the cantilever system, is shown in Fig. 12a in which a length of column is restrained, usually in a pin-ended manner, at its ends. After the application of a load at the free end of the beam the assemblage distorts as indicated in Fig. 12b. This arrangement thus has the advantage of more closely representing the actual beam edge column arrangement to be found in a real structure including the variation of response which will occur as a result of differing stiffnesses of upper and lower column lengths. It has the disadvantage however of requiring a more extensive test rig and difficulties may arise as the arrangement may be sensitive to the end conditions of the column.

5.3 Instrumentation

During the conduct of a test it is necessary to measure both the loads applied to the specimen, and inducing moments in the joint(s), and the distortional response from which the joint rotation may be computed.

The measurement of forces by load cells is straightforward and many standard commercial devices are readily available.
The determination of joint rotation however is more difficult. A variety of options are available and these may be classified into four main groups which are:

- (i) member slopes using dial gauges or LVDT's;
- (ii) rotation devices using gravity;
- (iii) optical devices using mirrors;
- (iv) rotation devices using lever arms and LVDT's.

Many researchers have used the most straightforward method of determining joint response by monitoring the changing geometry of the specimen by strategically positioned linear measuring devices, dial gauges or nowadays more usually linear variable displacement transducers from which the member profiles and hence joint

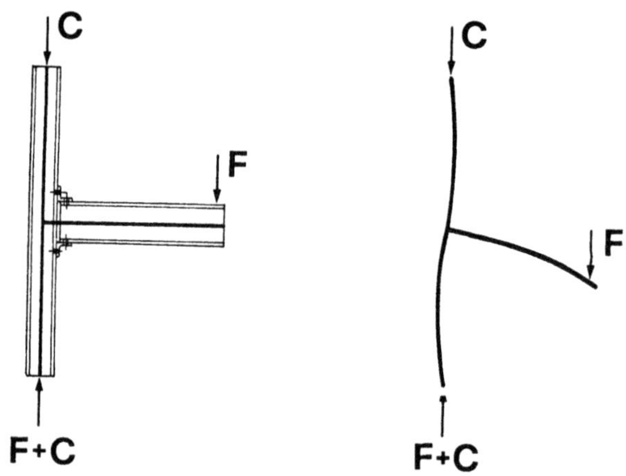

Fig. 12. *Extended column–cantilever test arrangement*

rotations may be computed. It has the advantage that the instrumentation is simple and normally readily available, but care must be taken in siting the devices so that unwanted local effects, such as local flange buckling are not included in measurements. The disadvantages are that the response must be in a single plane and that access in the joint area must be good which may not always be the case, as for example in a four way framing situation.

A number of rotation measuring devices which employ gravity as a reference have been used successfully. Bubble tubes (6), in which the electrical output varies according to the angle of inclination, were successfully used by Wood in frame tests at the Building Research Establishment (7). A second device, relying on gravity has also been used on frame tests at BRE (8). It consists of a dumbell weight hanging from a peg attached to the specimen at the location where the rotation is required and perpendicular to the plane of rotation, by a thin metallic strip. Strain gauges are bonded to the strip and connected to monitor the bending moment which is dependant upon the angle of rotation. The signal is normally calibrated directly in terms of rotation and any single measurement is best computed by averaging a number of readings taken at intervals of about one half of the natural frequency of the device to eliminate vibration effects.
A further instrument that falls within this category is the accelerometer which is commercially available. It is smaller than the dumbell but is considerably more expensive. All of the devices which rely on gravity are limited in their operation to measurements which have large components in the vertical plane. Being mounted on the specimen they are however insensitive to translational motion.

Another technique which avoids direct contact between specimen and instrument is the one which employs mirrors mounted on the specimens in the plane in which the rotation measurement is required. As the specimen rotates through an angle the reflection of an incident ray of light is displaced by twice the angle of rotation multiplied by the distance to the measurement plane. The system is relatively insensitive to translational motion and can accommodate rotations in all planes in which the direction of the incident ray lies. However a suitable line of sight must be established and maintained throughout the test and the method is not directly suitable for automatic data recording (9,10).

The final type of instrumentation which does not rely upon gravity is suitable for automatic data recording but does require physical attachement to the specimen is a device used by O'Heachteirn (11) for distortions associated with lateral torsional buckling. He employed a bar fixed to the specimen at the location where the rotation was required having its length perpendicular to the axis of rotation. Three inextensible strings were fastened to the ends of the bar and connected to individual LVDT's fixed in space. From measured displacements the in-plane movements of the attachment point, translational as well as rotational, may be computed. A further development of this system by Gibbons et al. (12) involving two bars and seven strings has been successfully employed to monitor the six degrees of motion of full three dimensional response in both subassemblage and frame tests.

All devices falling within categories (ii) to (iv) measure displacements and rotations at a point. Care must then be taken to check that significant localised distortions do not occur.

The positioning of the instrumentation as well as the accuracy of the measurements should always be thoroughly assessed in the preliminary design of the tests. Based on the experience to date it would seem that a general accuracy of about 1% is attainable. This requires particular care in selecting rotation measuring devices with sufficient resolution at very low rotations. This is a range important for the definition initial tangent stiffness.

5.4 M-ϕ Characteristics

It can be appreciated that joint response involves a number of complex interactions which contribute to the development of any characteristic even under monotonic loading. During a test, as moment is increased, the resulting relationship will be of the form shown in Fig. 13 from which a number of significant features may be identified. Initially, it is likely that the relationship will be sensibly linear, with a stiffness C_i, up to a moment M_e with a corresponding rotation ϕ_e. Non-linear effects will result in the characteristic losing stiffness to a value of C_{red}. Subsequently the joint will possibly achieve some notional plastic moment M_p, with a corresponding rotation ϕ_p, when plastic deformations will cause significantly increasing rotations. The joint will now have a new reduced "plastic stiffness" C_p. Eventually, at some ultimate moment M_u, stiffness vanishes although further rotation may occur up to an ultimate rotation ϕ_u at which the moment resisted begins to fall. If at any point along the curve unloading is permitted the relationship is usually linear with an unloading stiffness C_{unl}, close to the initial value C_i.

Analytical and experimental studies on both subassemblages and frames (6,7,8,13)have shown that even under monotonically increasing static loading column failure is frequently accompanied by moment reversal. Whilst some joints experience only increasing moment others unload and it is the latter, with their associated advantageous unloading stiffnesses, which are most beneficial in terms of frame strength and stability.

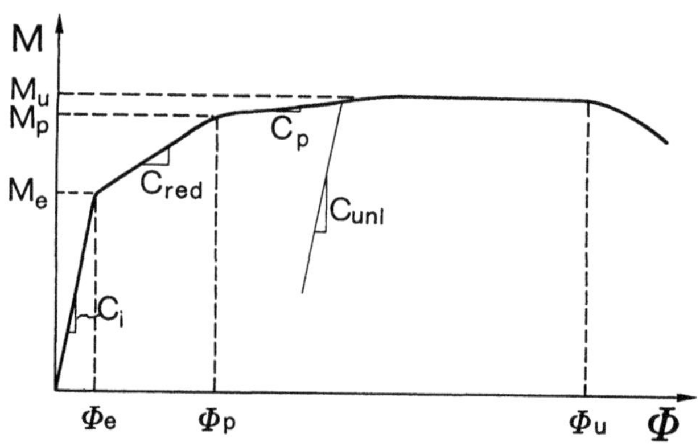

Fig. 13. *Idealised moment rotation characteristic*

These general features need to be borne in mind when conducting and evaluating joint tests. In addition some more specific items which require particular attention are :

(i) The initial slope of the characteristic, C_i, up to a rotation of the order of 5 milliradians is very important. Care should therefore be taken to include sufficient readings in the early stages of the test (5,16).

(ii) Sufficient data should be acquired to enable the values of M_e, ϕ_e, C_{red}, M_p, ϕ_p, C_p, M_u and ϕ_u, to be determined where appropriate for the particular joint.

(iii) The unloading stiffness C_{unl} is of particular importance and tests should include unloading, at least to almost zero moment, although ideally the full cyclic procedure (14) indicated in Fig. 14 is to be preferred.

(iv) In general it is not possible to monitor the beam rotation at location B of Fig. 9(a). Therefore this is usually determined by measurements at location C, which implies that a correction should be applied, and be consistent with an analytical model which uses simple beam theory. The appropriate value is that rotation which would occur from C to B due to beam flexure within the region. This has an additional advantage in that it avoids potential errors which may arise as a result of local distortions near the end of the beam due to the local stress concentrations which occur as a result of the presence of the connection. This procedure will generate the appropriate rotational stiffness of the joint if the analytical model concentrates this effect at B. If this concentration is to be made at A then a further correction for the region B to A is required.

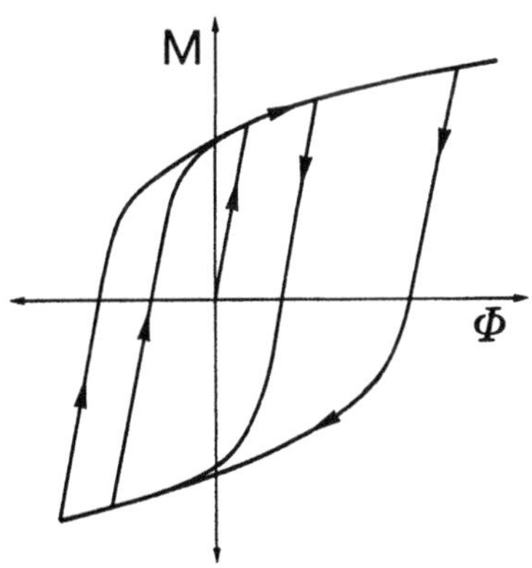

Fig. 14. *Slow cyclic loading procedure*

(v) During testing the loading rate should be sufficiently slow to permit all component actions to fully develop so that every point on the characteristic obtained corresponds to static conditions.

With modern measurement and datalogging techniques it is the organisation of the rig and specimen which is time consuming. Hence, it is strongly recommended that tests be conducted in such a way that the contribution of the various elements of the joint may be isolated. In this way tests conducted with one set of arrangements may be reassessed to permit their use in a different way e.g. including or excluding panel zone deformations. This approach permits various component deformations to be combined arithmetically which may then be used to produce moment rotation characteristics appropriate to different conditions, considerably enhancing the scope of each individual test. This concept of component distortions, which requires careful and consistent definition of moment and rotation, is here described in the context of experimental procedures to develop moment rotation relationships. It has a parallel equivalent in analytic techniques which are being developed for example by Tschemmernegg (15).

5.5 Practical Observations on Moment Rotation Responses

(i) In practice joints do not occur in isolation but as part of a frame. The most fundamental question which needs to be addressed has been partially answered by recent work in the United Kingdom (13), which has demonstrated good consistency between in-plane joint response measured in an isolated joint test and the behaviour determined when a nominally identical joint formed part of a complete frame.

(ii) Most previous work has concentrated on in-plane action but full three dimensional response is now receiving attention at Sheffield and this has involved out-of-plane joint testing (4,14).
It was during the conduct of these tests, with their greater flexibilities, that the importance of support conditions became evident. In cantilever tests, Fig. 11, it is possible to support the column flange remote from the beam over the depth of an equivalent beam so that the stress distributions throughout the section are similar to those occurring in the corresponding cruciform tests. Alternatively the column flange may be clamped over an extended length which will tend to spread the stresses over a larger region of the column. The extended column plus cantilever arrangement gives rise to a further possibility which more closely simulates an exterior column joint. Though in-plane tests will, in general, be less sensitive to this effect, in principle it is present and should at least be considered during the design of any test arrangement.

(iii) A further feature, observed during the same tests, concerns the way in which the rotation is measured. This is also discussed in reference 4. The evaluation of test results has confirmed the concept of component distortions where the total flexibility of a joint may be obtained as the accumulation of a number of contributory flexibilities. This philosophy has also been used in Milan (16), where, for a series of test on bolted end-plate connections, the overall connection rotation was considered as the addition of the rotation arising from bolt elongation and that due to end-plate deformation.

(iv) When a beam rotation is measured at some distance from the connection, it is important to be able to deduce the true joint and connection response. This requires the ability to identify and subtract from the measured rotation, the contribution arising from the bending of the beam length up to the column face. This acquires increasing importance as the joint stiffness and strength increases. The significance of this effect may be readily seen from Fig. 15, which relates to a test from the same Milan program, where rotations were measured at the connection location and a distance equal to the beam depth.

6 CONCLUDING REMARKS

The interest in determining full joint behaviour has increased rapidly in recent years. It is apparent from an appraisal of available publications that there is a general lack of consistency; yet simultaneously, an increasing number of codes are allowing semi-rigid approaches implying usage of joint response. There is therefore a vital need for the establishment of an agreed set of definitions of the terms involved (node, joint, connection, rotation etc) in addition to recommendations on the design and conduct of joint tests. Such definitions are essential to enhance the effectiveness of much needed database systems.

Reporting of test data is also important. It should be clear and comprehensive in order to give an adequate assurance of reliability and to permit any required reassessment to be made. In addition to geometrical material and loading parameters the report should include the assumed definitions of moments and rotations together with an assessment of the component contributions where appropriate. This should enable the user to ensure that the characteristic

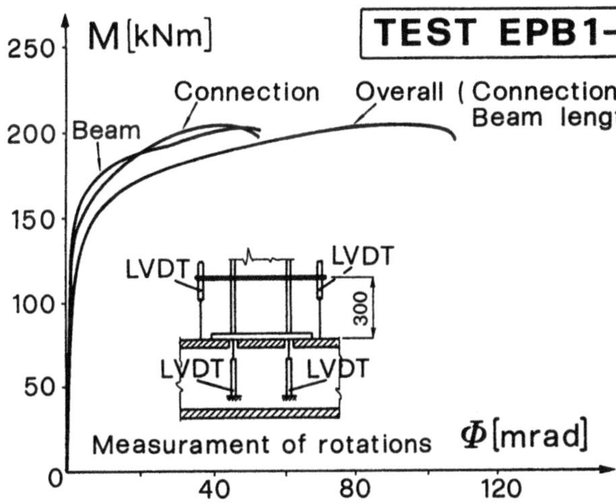

Fig. 15. *Component distorsions*

employed in any analytical process is consistent with the manner in which the test data has been assembled. To this aim the first two authors contributed to the production of standardised test data sheets for the reporting of results from three linked research projects funded by the European Coal and Steel Community (ECSC). These sheets may be used for tests on joints consisting of bare steel members or on joints incorporating composite action with concrete slabs.

This paper is regarded by its authors as a contribution towards a fuller discussion of the many facets contained in this problem which hopefully will lead to the establishment of an agreed standard for the conduct and reporting of moment rotation tests.

Acknowledgements

This paper is based on experience gained from experimental work carried out in the UK and in Italy with funding from SERC (the Science and Engineering Research Council) and CNR (the National Italian Research Council) as a Progetto Bilaterale, and from ECSC (European Coal and Steel Community).

References

1. Janss, J., Jaspart, J.P. and Maquoi, R., "Strength and Behaviour of In-Plane Weak Axis Joints and of 3-D Joints", State of the Art Workshop on Connections-Strength and Design of Steel Structures, Cachan, France, May 1987, pp. 60-68.

2. Richard, R.M., Gillett, P.E., Kriegh, J.D. and Lewis, B.A., "The Analysis and Design of Single Plate Framing Connections", AISC Engineering Journal, Second Quarter, 1980, pp. 38-52.

3. Anderson, D., Bijlaard, F., Nethercot, D.A. and Zandonini, R., "Analysis and Design of Steel Frames with Semi Rigid Connections", IABSE Survey, 1987.

4. Celikag, M. and Kirby, P.A., 'Out of Plane Moment Rotation Response for Common Joints', International Colloquium on Bolted and Special Structural Connections, Moscow, May 1989, Vol.2, pp.136-142.

5. Davison, J.B., Kirby, P.A. and Nethercot, D.A., "Rotational Stiffness Characteristics of Steel Beam to Column Connections", Journal of Constructional Steel Research, Vol. 8, 1987, pp. 17-54.

6. MacLachlan, D.F.A., Squire, A.M. and Planer, G.V., "Remote Indication of Angular Displacement", The Engineer, August 1962, p.235.

7. Wood, R.H., Needham, F.H. and Smith, R.F., "Tests on a Multi-Storey Rigid Steel Frame", The Structural Engineer, Vol. 46, No. 4, April 1968, pp. 107-119.

8. Kirby, P.A., Davison, J.B. and Nethercot, D.A., "Large Scale Tests on Column Subassemblages and Frames". Connections in Steel Structures: Behaviour Strength and Design, ed. Bjorhovde, R., Brozzetti, J. and Colson, A., Elsevier Applied Science Publishers, 1988, pp. 291-299.

9. Aggarwal, A.K. and Coates, R.C., "Moment-Rotation Characteristics of Bolted Beam-Column Connections", Journal Constructional Steel Research, Vol.6, No.4. 1986, pp.303-318.

10. Gent, A.R. and Milner, H.R., "The Ultimate Load Capacity of Elastically Restrained H-Columns under Biaxial Bending", Proc. Inst. of Civil Engineers, No.41, 1968, pp.685-704.

11. O'Heachteirn, P., "An Experimental Investigation into the Lateral Buckling Strength of Plate Girders", PhD Thesis, University of Sheffield, England, 1983.

12. Gibbons, C., Kirby, P.A. and Nethercot, D.A., "A Mechanical Electrical Device for Measuring the 3- Dimensional Deformation of Steel Columns", Journal of Strain Analysis (in press).

13. Davison, J.B., Kirby, P.A., and Nethercot, D.A., "Semi Rigid Connections in Isolation and in Frames", State of the Art Workshop on Connections - Strength and Design of Steel Structures, Cachan, France, May 1987, pp. 195-202.

14. Celikag, M. and Kirby, P.A., "Standardised Method for Measuring Three Dimensional Response of Semi Rigid Joints", State of the Art Workshop on Connections - Strength and Design of Steel Structures, Cachan, France, May 1987, pp. 203-210.

15. Tschemmernegg, F., "On the Non-linear Behaviour of Joints in Steel Beams", State of the Art Workshop on Connections - Strength and Design of Steel Structures, Cachan, France, May, 1987, pp. 158-165.

16. Zandonini, R. and Zanon, P., "Semi-rigid and Flexible Connections: An Experimental Investigation", Steel Structures: Recent Research Advances and their Application to Design, Part II, Ed. Hajdin and Sekulovic, September 1986.

17. Allen P.H., Kirby P.A. and Nethercot D.A., "Semi-Rigid Action in Steel Frame Structures", 4th Report to ECSC, September, 1989.

5 DESIGN-ORIENTED MATERIAL CONTROL IN ALUMINIUM STRUCTURES

D. KOSTEAS
Institute for Civil Engineering III, Steel Structures, Technical University of Munich, Germany

Abstract

Quality assurance procedures practiced in the research centers of the aluminium industry keep product quality at an unvarying level. An array of established test methods provide standardized values of physical and mechanical properties often quoted in brochures, handbooks and codes. Design recommendations furthermore identify "standard" alloys and provide characteristic or design values for several limit state assessments of structural components.

Design problems may though arise when non-standard alloys or tempers are used and/or when limit values for more sophisticated or optimized assessment procedures are needed. The latter case covers also the wide area of fatigue with only partially identifiable values for loading, component manufacturing and quality characteristics, correlation between test and real structure as well as in-service conditions.

Primary Product	End Product
Control Procedures in DIN or other codes	DIN 8561 Testing of welders
Sample size and properties controlled for: Chemical Composition, Metallurgic Anal., Physical and Mechanical Properties	DIN 8563 Quality Control in Welding, esp. Part 30: Weld Classification DIN 8570 Weld Tolerances
DIN 50049 defines: Art Certificate, Controlling Authority or Expert	Demands on the Manufacturer and depending on application: see outline on "Controlling Institutions"

Fig. 1: Codes for Material and Quality Control

Testing of Metals for Structures. Edited by F.M. Mazzolani. © 1992 RILEM.
Published by Taylor & Francis, 2 Park Square, Milton Park, Abingdon, Oxon, OX14 4RN.
ISBN 0 419 15810 3

State Construction Control Autherly with Acknolodgement Labs and IfB, Berlin	Application area of DIN 4113 *Load Carrying Structure* IfB Rules
TÜV Technical Control	Pressure Vessels AD Working Sheets inHP series *Arbeitsgemeinschaft* Druckbehälter, TÜV Essen
Ship Classification Agencies	Welding in Ships (Germanischer Lloyd Codes, vol LIII, ch. 6-7)
German Federal Railways	Rail Cars, DS 952 code or special Rules for "Light Structures in Transport Vehicles" (1970)
Federal Aviation MOD	Civil aircraft Military applications

Fig. 2: Controlling Institutions and Demands on Manufacturer/Welding

Alloy defined in	*Certificate demanded*
DIN 4113,p. 1	
"standard alloy"	DIN 50049-2.2
alloy mentioned only in DIN 1725,p.1-2 (chem. compos.)	material quality stated after DIN 50049-3.1B and acknowledged lab controls applicability for intended use and states design stresses based on properties after DIN 1745/6/8
all other alloys	check applicability with "general structural certificate" after IfB (tests on large components under service-like conditions)

State Construction Control Authorities themselves or thru control engineers check design calculations and construction plans.- Manufacturer is responsible to provide proof of his welding on the basis of certificate after IfB Rules or DIN.

Fig. 3: Material and Quality Control for Application Area of DIN 4113

CERTIFICATE	*DELIVERY*	*CONDITIONS*
DIN 50049-2.1 -2.2 -2.3 thru manufac-turer himself	as demanded by client or also after codes and techn. rules	continuous tests during manufac-turing and data from tests on deliv. material
DIN 50049-3.1A -3.1B -3.1C by internal or external autho-rity, not in-volved in manufacturing procedure	after codes and technical rules ans as demanded by client	data from tests on delivrd material
DIN 50049-3.2A -3.2C as above and by internal and external authority	as above	as above

Fig. 4: Material Control Certificates after DIN 50049

PRIMARY PRODUCTS

Chem. composition, Hardness, Tension or Compression Yield and Ultimate (true or relative) Strength, Shear Strength, Modulus of Elasticity or Shear, Temperature Effects, Buckling and other Ultimate or Serviceability State Carrying Capacity, Bending and Charpy, Internal Pressure, Fatigue, Corrosion, NDE (X-ray, ultrasonic, dye-penetration for cracks) or destructive testing methods (welds), Metallography

CONNECTIONS

as above and special tests for different weld types, filler metal, resistance welding, defects and imperfections as well as for adhesive joints

The above mentioned properties are the ones most often considered in structural engineering design. Usually a number of these is already fixed when using standard materials. Depending on the application and governing environmental conditions the one or other property becomes more significant and must be investigated by specific tests which may or not be already fixed in codes.

Fig. 5: Material Quality Control and Design Properties

DEVELOPING DESIGN STANDARDS FOR WELDED ASSEMBLY IN ALUMINIUM

Estimating / Calculating Strength and Performance in Service

Welding and the Material

Static Design

Fatigue Design

Information and know-how Recording and Dissemination

Fig. 6: Areas of Interest for Design Standards

STATIC DESIGN
Design rules rather well established

Further development areas:

Simplification and harmonization of weld design rules

Stability of structures

Multiaxial state of stress

Estimation of structural reliability and safety

Fig. 7: Static Design-areas of Development

FATIGUE DESIGN
Design rules still in development

Some problem areas:

Component/Full-size behavior
Residual stresses
Long-life region behavior
In-service behavior
Load spectra effects

Fracture mechanics parameters

Crack development and propagation

Reliability and safety concepts

Fig. 8: Fatigue Design - Areas of Development

Fatigue design data
Documentation, Validation
Evaluation, Presentation

Interrelation for information
on fabrication, design and
quality control
(fitness-for-purpose)

Interface to user

TUM tools development
ALDABA and ALFABET

Fig. 9: Recording and Dissemination of Information

6 INVESTIGATION BY TESTING OF THE STRUCTURAL RESPONSE OF SEMI-RIGID JOINTS

J.P. JASPART and R. MARQUOI
Département M.S.M., Civil Engineering Institute, University of Liege, Belgium

SUMMARY

During the last decade semi-rigid joints has focused the attention of several European researchers, regarding not only their actual behaviour but also on how and to which extent they alter the stability of frameworks. That is testified by the number of papers, dealing with this subject, presented at this Napoli Workshop. Extensive investigations have been and are still carried out in Liège on these topics either experimentally, or theoretically and numerically; bare steel- and composite joints are of concern. Accounting for the scope of the Workshop, present paper is restricted to several aspects of the testing.

1 INTRODUCTION

The knowledge of the actual behaviour of beam-to-column joints is of paramount importance; it is indeed likely to influence very much both idealization and discretization of the structure before performing the structural analysis, i.e. determining the stress resultants.

In the past, the current methods of structural analysis allowed only either for rigid joints or for pin-ended connections; the detailing of the joints was made in view to comply as much as possible with the corresponding assumptions.

A rigid beam-to-column joint should not allow for any relative rotation between the axis of the respective intersecting members; as a consequence, any external bending moment applied on this joint distributes amongst the connected members according to the member flexural stiffnesses. In contrast, a beam that is pin-ended to columns should freely rotate at the ends without transferring any bending to the column; this rotation at the ends is depending on the beam loading only.

Such extreme behaviours are respectively represented, in a bending - relative rotation curve, by the axis of ordinates for a rigid joint and by the axis of abscissae for a pin-ended beam-to-column connection.

Of course no joint is never either fully rigid or actually pinned. More especially, a presumed rigid joint always allows fo a relative rotation, even when the column web panel is transversely and diagonally stiffened and the connection of the beam made with stiff end-plate and preloaded bolts; however the rotation of such a joint remains usually so

Testing of Metals for Structures. Edited by F.M. Mazzolani. © 1992 RILEM.
Published by Taylor & Francis, 2 Park Square, Milton Park, Abingdon, Oxon, OX14 4RN.
ISBN 0 419 15810 3

small that the assumption of a rigid joint is quite acceptable. Presumed pin-ended connections, for their own, are always somewhat restrained because the actual connecting devices are never proper hinges; the corresponding constructive detailing is identified to hinges for sake of conservatism. Let us mention in addition that it is distinguished between full strength and partial strength joints; in the first case the joint is able to transmit the full resistance of the connected members, while in the second one, the connection is weaker than these members.

Rigourously speaking, any joint is ever neither fully rigid, nor pinned; therefore it should be termed as semi-rigid (fig. 1).

Modern steel construction aims at not only material savings but more especially at cost savings. In this respect, the detailing of the joints is simplified : the number of stiffeners in the joint is reduced, as well as the length of fillet welds; that results in an appreciable decrease in labour cost and consequently favours the global economy of the project. At the structural viewpoint, that leads to beam-to-column joints which are much more simple to execute but exhibit a fully non-linear behaviour and therefore a M-ϕ curve, which is sometimes very far from the characteristic curve of a rigid joint.

Such a semi-rigid and partial strength behaviour of the joints shall affect both the strength and the stability of steel frames. The corresponding structural response shall be non-linear; the non-linearity of the joint behaviour superimposes to possible material and geometric non-linearities. Several computer programmes allowing for those kinds of non-linearity are yet available. Most of them substitute each semi-rigid joint either by a notional linear rotational spring (constant stiffness whatever the moment amplitude) or better by a notional non-linear rotational spring; sometimes a piecewise linear characteristic curve is substituted for the actual non-linear one (fig. 1). The M-ϕ characteristic curve of such a spring represents the global response of the joint. Actually there are several components of the joint deformability, whose respective contributions are governed by many

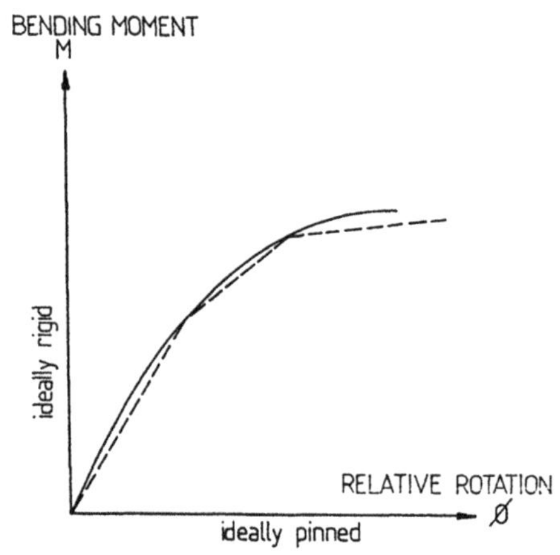

Figure 1

parameters. The relative amounts of these contributions are depending on the loading of the joint; therefore the response of a joint within a structure may be somewhat different from that of a joint belonging to a subassemblage tested in specified laboratory conditions. As a consequence, tests must be carried out so that all the sources of deformability be measured separately; the corresponding results are aimed at being used to calibrate physical models which, theoretically based, should allow the prediction of all the deformability contributions in a wide range of variation of the governing parameters.

Present paper is dealing with in-plane semi-rigid joints under gravity static loading only.

2 SOURCES OF JOINT DEFORMABILITY

The sources of deformability are being clearly identified first. For the reasoning, let us assume an in-plane strong axis beam-to-column joint, for which the end cross-section of the beam is connected, whatever the manner, to one flange of the column so that any loading experienced by the beam in the plane of the framework induces bending about the strong axis of the column cross-sectional area (fig. 2.a).

A *joint* is the whole region concerned by the assemblage of the beam(s) with the column. It is composed by the very end portion of the beam(s), the facing adjacent portion of the column, as well as by all the connecting accessories (end plate, cleats, bolts, welds,...) required by a specified type of connection. The deformability of a strong axis beam-to-column joint consists mainly in two parts, which are respectively fed by several contributions :

a) The deformability of the *connection area* associated to the following phenomena :
   - Deformation of the connection elements : end plate or cleats, bolts, rivets, welds,...
   - Slip and/or hole clearance ;
   - Deformation of the column web, across its depth, in the so-called tensile and compression zones, i.e. in the regions where the forces carried over by the beam(s) have to diffuse into the column web (this effect, termed "trapezoïdal effect", is the result of the respective lengthening and shortening of the column web depth).

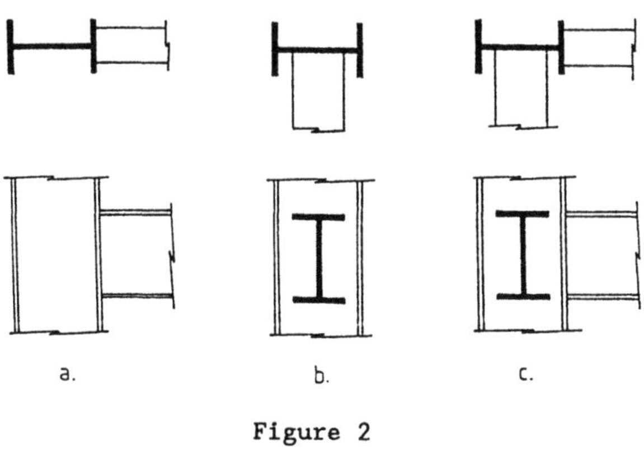

a. b. c.

Figure 2

b) The shear deformation of the *column web panel*, which is subject to a complex shear stress distribution in the region of the joint.

In a weak axis beam-to-column joint (fig. 2.b), the end cross-section of the beam(s) would directly be connected, whatever the manner, to the web of the column; any transverse load experienced by the beam in the plane of the framework should result in bending about the weak axis of the column. For such a joint the aforementioned "trapezoïdal effect" vanishes and the second source of deformability results mainly from the out-of-plane deflection of the column web when experiencing the bending moment transmitted by the beam(s).

For a 3 D-joint (fig. 2.c), beams are connected at about the same location to a single column, respectively onto the web and one (or two) flange(s) of this column, so that the deformability results roughly from the combination of the sources identified above, respectively to both axis, with account taken of the possible interactions.

Let us reason on a strong axis beam-to-column joints. The contribution θ of the deformability of the connection area to the joint relative rotation ϕ is defined by the difference between rotations θ_b and θ_c (fig. 3.b). That of the shear deformation of the column web is represented by the difference between θ_c and θ_f, where θ_f is the flexural rotation of the column (fig. 3.b).

Regarding the shear force in the column web panel, it shall be stressed that it results from : i) the combined action of equal but opposite forces F_b in the beam flanges, the resultant of which is statically equivalent to the bending moment in the beam, and ii) the shear force resulting from the moment distribution in the column.

Because both sources of deformability are dealing with the response of two different zones of the beam-to-column joint, their contributions shall be assessed with reference made to their respective appropriate loadings. As these latter are not identical, it must therefore be accounted for both deformability sources separately and laboratory tests have to be instrumented in such a way that measurements of the deformability components be possible.

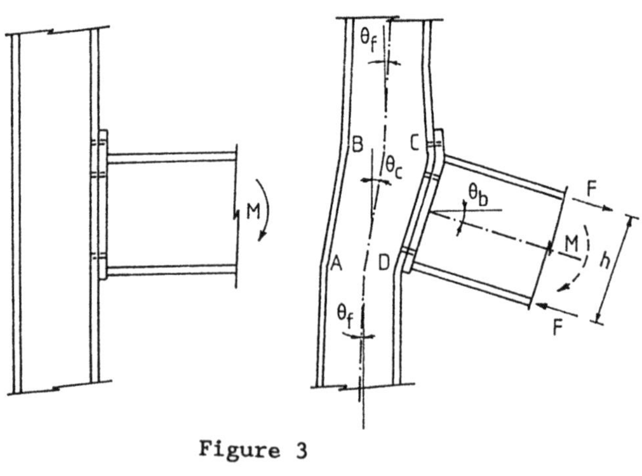

Figure 3

Though tests have been conducted in Liège on all above mentioned kinds of joints, only strong axis beam-to-column joints are of concern in the following, except where it is stated otherwise. The scope is restricted to hot-rolled sections only.

3 TEST SPECIMENS AND TESTING ARRANGEMENTS

Because only the response of in-plane semi-rigid joints is of concern, the experiments are of course conducted not on whole structures but on *structural subassemblages* composed of portion(s) of beam connected to a portion of column. These subassemblages must be designed so that to allow for realistic types of loading, while having sizes which prevent external load introduction effects from alterring the response of the joint itself.

Shear deformation of the column web panel is expected to be more pronounced in tee-joints, where a single beam is connected to the column (fig. 4.a). In contrast, the "trapezoïdal effect" exhibits as well in tee-joints as in cruciform joints, where two beams are connected to the column at the same level and are both similarly loaded (fig. 4.b). Therefore it shall be distinguished between tee-joints, which are representative of outer joints in a real framework, and cruciform joints which correspond to inner joints of this structure (fig. 5).

Because the attention is focused on both strength and deformability ability, there are many governing parameters of several natures; therefore only tests on full-scale specimens are conceivable, because the sole likely to give accurate and realistic information regarding the behaviour of joints in real structures.

As already mentioned above, the test specimens are subassemblages. The height of the column is chosen so that it represents roughly the depth of one storey. The beam is connected at mid-height of the column, so that the ends of the latter may be considered as points of contraflexure in the columns of a sway frame subject to horizontal loads.

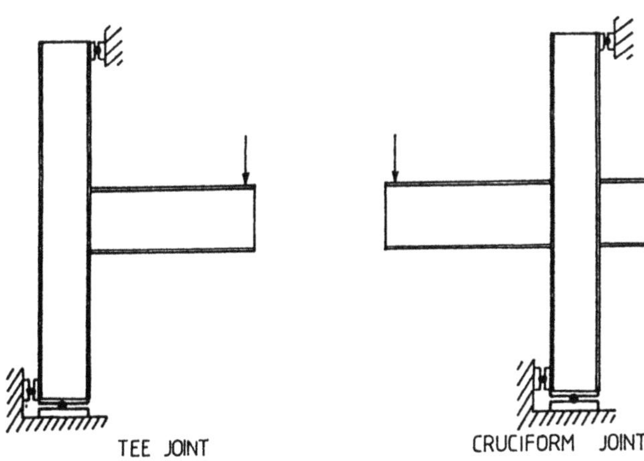

Figure 4

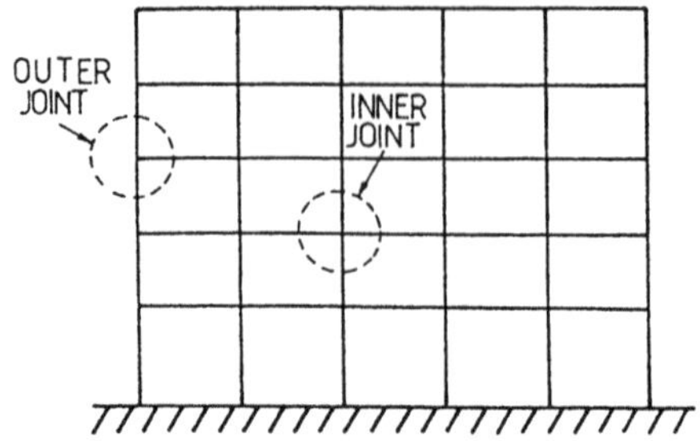

Figure 5

For sake of testing, it is thus sufficient to apply axial load at the ends of the column. Bending in the column as a result of the loading of the beam(s) will be produced by the horizontal support reactions at the ends of the column.

What about the length of the beam(s) in tee- or cruciform joint specimens ? It is determined in view to allow for bending-to-shear ratios in the connection similar to those encountered in practice. In this respect, it is while mentioning that some tests reported in the literature do not care at all for such condition, which should however be considered as a requisite. Bending in the beam is produced by point load(s) applied at the end of the cantilever(s).

When semi-rigid composite joints, the steel column is connected with the composite beam(s), composed of a hot-rolled section associated to a reinforced concrete slab, by means of shear connectors. Slab has a spatial structural behaviour. Because only in-plane tests are carried out the composite test specimens must be carefully designed in order to get an appropriate composite action. In this respect, the distribution of direct bending stresses across the slab width must be quasi uniform in the joint cross-section. Because the bending moment is produced by point load(s) at the end of the cantilever beam(s), the width of the slab in the test specimens cannot be smaller than the regular effective width, which is aimed to allow for the aforementioned required stress distribution. This condition must also be adopted as a prerequisite; it is checked by means of measurements, during the tests, of the direct stresses in the steel reinforcements in several cross-sections, and especially in the joint cross-section.

The test specimen is inserted in a steel testing rig, that is fixed on the 1.5 m thick concrete testing floor of the laboratory. This rig is aimed at providing the reactions required by the loads to be applied to the test specimen.

The ends of the column are supported by devices which are able to support vertical and horizontal reactions while providing no moment restraint; they can thus be considered as ideally hinged.

Guide-plates are located all along both the beam(s) and the column in view to prevent from any spatial behaviour and out-of-plane displacement of the subassemblage whose in-plane behaviour is investigated only.

Load is applied at the end of the cantilever(s) and possibly in the column by means of hydraulic jacks fitted with load cells. The load is increased up to the joint collapse or to the maximum deflection of the cantilever ends according to what is first reached. Unloading are carried out during the tests in view to compare the instantaneous stiffness with the initial one.

4 INSTRUMENTATION AND MEASUREMENTS

The tests are instrumented so that the measurements allow for determining the amplitude of all the components of the joint deformability at any level of the loading. It is useful to perform measurements of horizontal and vertical displacements as well as of rotations in appropriate sections; for this purposes, electronic transducers are used. The applied load as well as the support reactions are also measured by means of load cells.

As it is required to identify and measure separately all the components of the joint deformability, it is of paramount importance that the measurements be scheduled accordingly. In other words, it is necessary to perform "redundant" measurements which shall allow for computing a specified rotation by at least two different manners. Doing so warrants to get results even when one transducer is misfunctioning or when something wrong is likely to occur during the test. The "direct" measure is of course preferable and the most thrustable but the searched information must also be deducable from "undirect" measures, when necessary.

A peculiar attention must be paid to avoid second-order effects which could arise during the tests and could affect appreciably the measured values; possibly they must be properly accounted for.

Last the testing frame, which is anchored on the testing floor of the laboratory and is aimed at resisting the applied load and the support reactions, cannot be undeformable, whatever its rigidity. Therefore, because rather small quantities are to be measured, one must avoid to use this rig as support for the measurement devices. Accessories independent of the testing rig are therefore clamped on the testing slab whose rigidity can be considered as infinite. Following measurements are made for tee-joints (Table 1):

a) vertical and horizontal displacements at the end(s) of the beam and the column ;
b) rotations of the beam and of the column ;
c) rotation associated to the load introduction deformability of column web ;
d) rotation due to the slip at the junction cleat-beam flange ;
e) proper deformation of the upper flange cleat ;
f) proper deformation of the column flange in the tension zone of the joint.

The last three measurements are only performed for flange cleated connections.

In view to compute the rotation of the beam and of the column, six measurements were made (Table 1). The transducers (5) on the column flange are located symmetrically to the axis of the beam in the plane of the column web and on the outer flange of this column. Their spacing is as large as allowed by the beam depth. Transducers (4') are pointing to

Table 1

INSTRUMENTATION

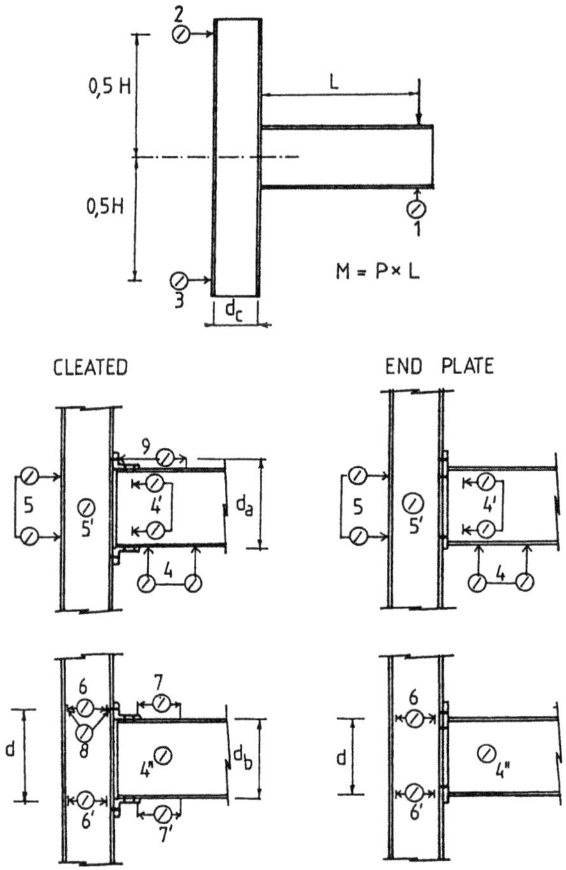

ROTATIONS

MAIN COMPONENTS

— CONNECTION

$\theta = \theta_b - \theta_c$ (fig 3b) θ_b by ④ ④ or ④'

θ_c by ⑤ or ⑤'

— SHEAR OF COLUMN WEB PANEL

$\gamma = \theta_c - \theta_f$ (fig 3b) θ_f theoretically evaluated

or

$\gamma = \phi - \theta$

— JOINT

$\phi = \theta + \gamma$ or $\phi = F[(1),(2) \text{ and } (3)]$

COMPONENTS OF CONNECTION DEFORMABILITY

— LOAD INTRODUCTION

$\theta_{li} = [⑥ - ⑥']/d$

— SLIP

$\theta_s = [⑦ - ⑦']/d_b$

— CLEAT/END PLATE & COLUMN FLANGE

$\theta_d = \theta - \theta_s - \theta_{li}$

— COLUMN FLANGE

$\theta_{cf} = [⑧ - ⑥]/d$

— CLEAT incl. BOLT ELONGATION

$\theta_a = \theta_d - \theta_{cf}$

$\theta_a = [⑨ - ⑦]/d_a$

their small stitches welded transversely into the web; they are also located symmetrically to the beam axis. Transducers (4) are located at the lower flange of the beam as near as possible to the connection cross-section.

The measurements of the rotation resulting from the slip at the interface between the flange cleat and the beam flange is made by measuring the relative displacement between two points located at a so small spacing as possible, respectively on the cleat and on the flange on both upper and lower flanges respectively (Transducers 7 and 7' in Table 1).

Direct measurements of the deformation of the column web in the compression and tension zone are made in view to get the deformation curve associated to the "trapezoïdal effect" in the column web panel (Transducers 6 and 6' in Table 1).

The computation of the rotation due deformation of the column flange results from the measurement of the relation displacement (8) between two points located respectively on one of the bolts connecting the upper cleat to the column and the column web (Table 1).

The rotation due to the deformation of the cleats (including the bolt elongation) is deduced from measurements of the relative displacements between the two points located at a so small spacing as possible respectively on one of the bolts connecting the upper cleat to the column and on the upper beam flange, (Transducers 9 in Table 1).

When cruciform joints, the deflections and the rotations are determined accordingly for both beam-to-column connections.

Composite connections require in addition the measurement of the slip between the concrete slab and the adjacent steel flange onto which the slab is connected by studs.

Strain measurements are aimed at investigating the stress distribution in the beam cross-section. For bare steel beam-to-column joints, only some tests have been fully instrumentd in this respect; it has soon appeared that strain measurement in both flanges of the beam in two cross-sections is quite sufficient. For composite joints, strain measurements not only in both flanges of the steel beam but also in the reinforcements have been made in several cross-sections in view to explore the stress distribution, more especially in the concrete slab because of the shear lag effect. As soon as it has been demonstrated that a uniform direct stress distribution in the slab in the cross-section corresponding to the connection was reached, the following tests have been conducted with a less important instrumentation.

The measurement of the beam rotation (fig. 3) has revealed to be rather dubious in most of the tests performed. That is the consequence of: i) the out-of-plane deformation of the beam web during the joint loading as a result of the dissymetry of some connection types or, ii) the distance between the real connection cross-section and the section where the rotation measurement is performed. However that does not at all prevent from determining the rotation of the beam because of the redundancy of the measurements performed. It has been anyway concluded that a direct measure by means of potentiometrical rotating transducers is far preferable and such devices have been purchased.

One must also stress the fact that the determination of the rotations may be questionable when beam and/or column behave inelastically. That is especially important when computing rotation θ_f.

5 DATA ACQUISITION

Measurements have been recorded during tests by means of a Hewlett-Packard data logger, which is able to measure up to 200 different parameters (transducers, strain gauges,...) at a time.
This data logger is automatically commended by a software running on a PC computer. This software - SCANPACK - was developed in the Liège laboratory ; it is especially useful for tests on structures and allows to follow, in real time on the screen, the evolution of any reference measurement versus the applied load. All the measurements are stored in

view to further analysis of the results. This package stores in addition all the information regarding the test conditions (connections of measuring devices, calibration factors,...); that is especially useful for repetitive tests.

6 EXPERIMENTAL CHARACTERISTIC CURVES

Because of the appropriate instrumentation, the following characteristic curves, associated to the different components of the joint deformability have been recorded :

a) the joint relative rotation curve ;
b) the connection relative rotation curve ;
c) the column web panel rotation curve ;
d) the moment-deflection curve.

For what concerns especially the deformability of the strong axis connection, it is composed by the addition of the following curves :

a) the load introduction rotation curve ;
b) the connection slip rotation curve (cleated connections) ;
c) the cleat deformability rotation curve (cleated connections) ;
d) the column flange rotation curve (cleated connections) ;
e) the column flange and end plate deformability rotation curves (end plate connections).

Figure 6 presents the different curves recorded for a specified test specimen.

Readers interested in knowing more about the test results are begged to refer to publications and reports listed in the following section.

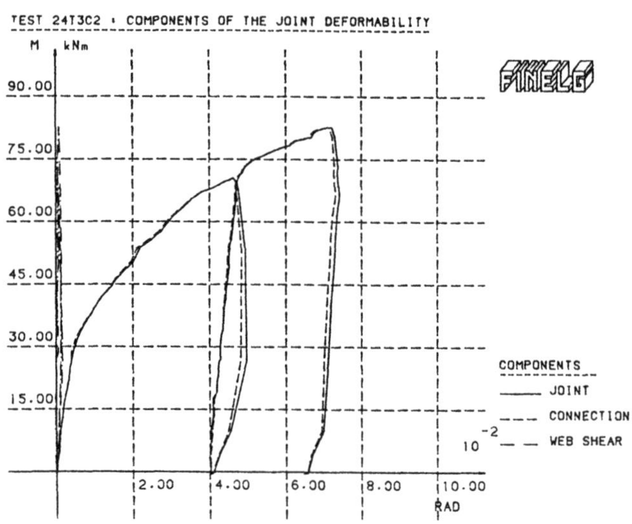

Figure 6.a.

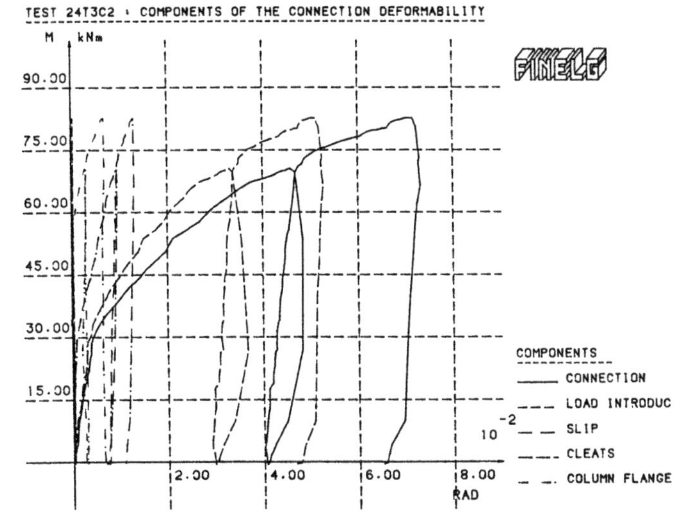

Figure 6.b.

7 REFERENCES

1. Semi-rigid action in steel frame structures
   CEC Agreement N° 7210-SA/507 - Committee F8
   Technical report n° 1
   Technical report n° 2
   Technical report n° 3
   Technical report n° 4
   Beneficiary : ARBED Luxembourg
   Technical support : University of Liège (MAQUOI - JASPART).

2. JANSS, J., JASPART, J.P. and MAQUOI, R. :
   Experimental study of the non-linear behaviour of beam-to-column bolted joints.
   Connections and the Behaviour, Strength and Design of Steel Structures, Elsevier Appl. Sc. Publ., 1988, pp. 26-32.

3. JANSS, J., JASPART, J.P. and MAQUOI, R. :
   Strength and behaviour of in plane week axis joints and of 3 D joints.
   Ibidem, pp. 60 - 68.

4. ALTMANN, R., MAQUOI, R. and JASPART, J.P. :
   Experimental study of the non-linear behaviour of beam-to-column composite joints. Intern. Colloquium, Stability of Steel Structures, Budapest, 1990, Vol. IV, pp. 3 - 10.

5. ATAMAZ SIBAI, W. and JASPART, J.P.:
   Etude du comportement jusqu à la ruine des noeuds complètement soudés.
   Rapport interne IREM 89/7, Lausanne et Rapport interne, MSM N° 194, Liège, Octobre 1989.

7 WHAT MATERIAL SCIENTISTS AND STRUCTURAL ENGINEERS NEED TO KNOW ABOUT FUNDAMENTAL MECHANICAL PROPERTIES OF METALS

M. MARINCEK
Emeritus Professor, Ljubljana, Yugoslavia

Summary

In the introduction general problems are discussed about the constititutive laws of metals in connection with computer aided analysis of joints, components or structures as regards their plastic, fatigue and fracture behaviour, particularly the problem of the ductility and constraint effects. Then the significance of the true fracture stress and strain obtained with the simple tension test is emphasized. These are particularly important in connection with the fracture mechanics parameters for crack initiation and propagation by monotonic and cyclic loading.

1 Introduction

For the computer simulation of the mechanical behaviour of plane, notched or cracked bodies relating elastic, plastic and fractural response, appropriate data about the constitutive laws of the materials are necessary. Fundamental are data about the uniaxial response curve (USRC), stiffness (Young's modulus, Poisson's ratio), strength (suitably defined yield strength), the shape of the USRC (in the simplest case expressed by one strain-hardening exponent) and the fracture ductility (based on the fracture strain, obtained from the reduction in area in the case of necking). Changes of stiffness, strength, shape of the USRC and fracture ductility with temperature and strain rate are also important experimental data.

In a nonuniaxial stress state the laws about the effects of the triaxiality (constraint factor: maximum principal stress divided by the von Mises stress) on plastic and fracture behaviour of the stress element of the material have to be known, and the degradation of the material during straining may also need to be considered. The simplest case of the constraint effect is proportional loading as a result of different

Testing of Metals for Structures. Edited by F.M. Mazzolani. © 1992 RILEM.
Published by Taylor & Francis, 2 Park Square, Milton Park, Abingdon, Oxon, OX14 4RN.
ISBN 0 419 15810 3

possible nonproportional loading paths. The initial anisotropy of the material and the anisotropy as a consequence of straining also have to be considered. A special problem is the role of the constraint factor in the region of the transition from ductile to brittle fracture.

In order to gain confidence in the possibilities offered by computational inelastic mechanics, systematic international experimental and numerical round-robins on the determination of the characteristic load-displacement relationship of different elementary strength cases are necessary, taking into account different representative shapes of the USRC and different fracture ductilities of the material. First, simple un-notched examples such as tension tests, bend tests (plane stress, plane strain), torsion tests (round, rectangular, full and hollow sections) and shear tests should be investigated. Additional experimental and numerical tests with notches of different severity are important.

Knowledge about the micro-mechanical effects enable not only an explanation of the course and length of the USRC, but also of the behaviour of the material under the influence of different constraint factors, and in this way also the explanation of different size effects. Chemical composition gives important information about the metal, but the thermal and/or mechanical treatment can essentially change the structure and hence the mechanical properties of the materials. Microvoids, microcracks and impurities are essential in relation to the fracture ductility. Decreasing the volume fraction of inclusions increases the fracture strain. This is especially important in the case of high strength metals, because they tend to be less ductile at higher strengths.

To obtain a fundamental understanding of plasticity and fracture it is necessary that material scientists, metallurgists and engineers work closely together. It appears impossible for either macroscopic mechanics or metallurgical micromechanics of fracture to achieve this aim separately.

Inelastic problems in structural engineering can be divided into small strain problems, medium strain problems and large strain problems.

Small strain problems in general include the determination of the damage limit state of components and systems (before the ultimate limit state is reached the structure loses its serviceability due to plastic deformation, e.g. a certain degree of plastic deflection) and the inelastic compression instability at the ultimate limit state. In the above cases it is not the elastic stress limitation but the plastic strain limitation which has to be considered if the material is not ductile enough, particularly in the presence of strain concentrations and/or repeated loading. Welded joints offer clear evidence of the role of ductility. Without defects they are strained during the welding process up to 2% strain and there is a large area under residual stresses in the level of yield strength.

A long tension member with constant cross-section and with no notch effect is the simplest example of the medium strain problem in the case of extreme overloading (e.g. seismic). The tensile strength determines the carrying capacity of the member and only the uniform strain defines its fracture elongation, when fracture with necking occurs (ductile instability). In the case of biaxial loading the constraint factor affects the ductility. The minimum fracture ductility occurs in the plane strain state.

Large strain inelastic problems occur in the forming of ductile metals, in the presence of notches and in the field of elastoplastic fracture mechanics, which deals with cracked bodies under general yielding.

In thin plates in the vicinity of the crack a plane stress state exists, and large localized straining of the material causes considerable blunting of the root of the crack. Strain and stress concentrations occur in the root of the crack and before the fracture ductility of the material is exhausted the crack does not propagate.

On the other hand, inside thick plates in the vicinity of the crack the plane strain state exists. The strain concentration is again in the root of the crack, here less blunted, because the constraint factor reduces the fracture ductility of the material. But the stress concentration is behind the crack front, because there is such a high constraint factor that the material is considerably harder than in the monoaxial stress state. Therefore the crack can initiate and propagate by a cleavage mechanism, if the maximum normal stress in the stress concentration reaches the cohesive strength of the material. The released elastic energy of the body is the driving force, which leads to explosive fracture according to Griffith's energy instability condition.

Fracture toughness (e.g. J integral) is the material parameter in fracture mechanics, in addition to the crack opening displacement (CTOD). Different methods and specimens are used for the determination of the fracture toughness of a material. However, these fracture toughnesses are not an elementary mechanical property. The fracture toughness of the material alone is therefore of more fundamental value. It is represented by the area under the USRC, where fracture strain has the decisive role besides the yield strength and the shape of the USRC. But the CTOD may be directly related to the fracture strain of the material.

2 Material parameters obtained by simple testing

The basis for the computer simulation of the inelastic behaviour of any joints, components or systems is thus the uniaxial stress response curve. The USRC expressed with the related nominal (engineering) stress - linear (engineering) strain up to the tensile strength, respectively uniform strain, is sufficient for smaller strain problems, while for large strain problems the relation between true stress and logarithmic strain up to fracture has been customary for a long time. In the small strain range both presentations of the USRC are nearly identical. Different USRCs in typical directions determine the anisotropy.

In the following section, improvements in the current international standard ISO 6892-1984 for the simple tension test and in the corresponding national standards are proposed:

1 For the round house USRC (without plastic plateau) instead of the yield strength defined by 1.2% plastic strain, the definition of yield strength with equal elastic and plastic strain is recommended. In this way all normalized USRCs (with yield strength and the belonging elastic strain as the norm) intersect in the ordinate 1 for the normalized stress and in the abscissa 2 for the normalized strain. In this manner the normalized Ramber-Osgood's USRC has only one parameter, the strain hardening exponent. Similarly, for the proportionality limit stress the criteria with the plastic strain of 5% of the belonging elastic strain is recommended. In order to determine the strain hardening exponent in the small strain range the additional characteristic stress with the plastic strain at a level of 50% of the belonging elastic strain is also proposed.

2 For the sharp knee type USRC the length of the plastic plateau (Luder's strain) has to be recorded. In the standard tension test the determination of the lower yield strength can give a value up to 20% higher than the static yield strength. A comment on this should be given in the standard.

3 It should be obligatory to register the uniform strain (at the maximum load) in the tensile test, because the uniform strain is the only measure in the calculation of the ductility of long tension members without the notch effect. It should be given in linear (not logarithmic) strain. Methods for measuring the uniform strain should be

recommended in the standard. It is clear that the tensile strength - yield strength ratio alone is not enough for the assessment of the strain hardening characteristics of the USRC.

4 The decreasing part of the load - extension curve in the tension test must be better exploited because it involves the most significant fracture properties of ductile metals: the true fracture ductility and the true fracture strength. ISO 6992-1984 only treats the determination of the average fracture elongation (which represents no scientific value for the ductility) and the determination of the reduction of fracture area. But the reduction of area should be transferred into the linear fracture strain, assuming incompressibility at the plastic deformation. Further, it should not be a problem to measure the fracture force in the tensile test and to divide this force by the fracture area in order to obtain the true fracture stress. True fracture stress and strain enable the determination of the strain hardening exponent of the USRC in the range of large strains, from the uniform strain up to fracture. It is known that part of the USRC determined in this way is not accurate, because due to the constraint effect in the neck the material becomes harder in the centre and distribution of stresses in the neck is not uniform. Additionally also, the strain rate in the neck is increased by the constant movement of the crosshead of the tensile machine. The above proposed manner for the determination of fracture stress and strain is thus a provisional approximation. It seems that the Bridgman correction formula is the best, but the necessary determination of the radius of the contour of the neck is cumbersome. After the necessary investigation the procedure for correction of the USRC due to the effect of the neck should be given in a separate standard. The same standard should also include instructions on how to determine the shape of the USRC, in respect of a suitable mathematical expression for it, and how to determine the fracture toughness of the material alone, integrating the area under the USRC.

The simple bend test offers the cheapest way to determine the approximate fracture ductility, according to ISO 7438-1985 for the parent material and ISO 5173 and 5177 for welded butt joints. Unfortunately the present standards for the bend test require only a yes/no answer as the result, because for a prescribed angle of bend and diameter of mandrel the absence of the crack, visible without the use of magnifying aids, is considered as evidence that the test piece withstood the test. Only very poor statistical analysis about the ductility can be made on the basis of such results. It is therefore proposed that the bend test should be continued by pressing further on both legs directly, until visible cracks appear. Then the bending fracture ductility can be determined by measuring the local fracture strain in a simple manner (e.g. using mm lines).

It is particularly useful to determine the local bending fracture strain by bend testing of welded butt joints because of heat affected zones. When the transverse bend test is performed the ductility of the softest zone can be determined. When the longitudinal bend test is performed the ductility of the hardest zone can be measured. In the case of a rectangular cross-section specimen having a width: thickness ratio of more than 5:1, a plane strain fracture strain can be obtained. The bend test is also very suitable for determination of the fracture ductility of thicker plates in the Z-direction (lamellar tearing).

Torsion tests of full round specimens show no complications with necking. There is only a certain disturbance at large strains, because the pure tangential stress state has additional longitudinal tensile stresses in the middle of the cross-section and compressive stresses near to the surface. Computer simulation of the torsion test for large strains and for different shapes of the USRC could give expressions for the necessary corrections for the experimental determination of the USRC and of the pure tangential fracture strength and ductility. Tangential fracture strength obtained in this way can be compared with the shear strength obtained by a simple shear test of the bolted connection. In any case, the shear strength should correlate better with the true fracture strength of the tensile test and not with the tensile streng th, which has been for a long time the material property referred to for the dimensioning of bolted shear connections.

Notched round tensile specimens are suitable for the determination of the limit strength of the metal under the influence of a high constraint factor, with brittle fracture as the consequence. The cohesive strength can be determined, when computer simulation of the specimen behaviour previously enables the relation between the load and the stress peak of the stress concentration behind the root of the notch to be expressed adequately. The supposition exists that cohesive strength is the most fundamental mechanical property, because practically it does not depend on temperature, strain rate or constraint factor, as the ductile fracture strength and strains do. The cohesive strength of ductile metals can be obtained with unnotched specimens only at extreme low temperatures or extreme high rate of strains. It is useful to find how the cohesive strength correlates with the true fracture strength obtained with standard tensile testing.

3 Fatigue and uniaxial stress response curve (USRC)

At the present, time fatigue strength in respect of fatigue life predictions are based mainly on experimental test data. In the future, fatigue strength calculations will be more and more based also on computational predictions. This has to be founded separately on the local strain approach for determination of the crack initiation time and on the fracture mechanics approach for determination of the crack propagation time. It is important to distinguish between homogeneous fatigue on one side and localized fatigue at the root of notches and sharp cracks on the other side. In the first case the fracture initiation time can reach 90% of the total life, in the second case the fracture initiation time can vanish. Unfortunately, present experimental fatigue data mostly give no information about crack initiation time, or about the final (critical) crack length in respect of area. It would therefore be very desirable to have such information from experiments for the sake of comparison with computational results.

Fatigue computations are based on solving equations which contain certain material constants. Thus the scatter of the experimental results due to the material properties can also be assessed with the help of parametric computational testing.

In any case, experimental fatigue testing will be necessary also in the future, not only because of the possibility of comparisons between experiment and theory. Experimental fatigue testing is particularly inevitable in the case of environmental conditions, which produce corrosion fatigue and stress corrosion cracking, because the crack tip chemistry (also in connection with creep) is still an enigma. Therefore corrosion technologists could also be interested to know how corrosion resistance is related to USRC, especially with the true fracture strain of the material.

It is known that at low cycle fatigue some metals show expressive cyclic softening or hardening degradation. The stabilized stress-strain cyclic curve (USRC) also has to be defined with the stiffness, yield strength and shape of the curve. In low cycle fatigue experiments without notches it is important not only to determine the crack initiation time (cycles) depending on plastic strain range. Information about the gradual decreasing of the fracture ductility would also be very important. The Manson-Coffin model for the plastic strain range - life (cycles) relationship clearly indicates that the true fracture strain is a decisive material parameter. Using Neuber's rule for the determination of the plastic strain range at plastic fatigue in the root of the notched bodies, it is possible, with the use of the Manson-Coffin relationship, to calculate the crack initiation time for notched samples.

At high cycle fatigue, crack propagation actually represents a gradual and very localized low cycle (plastic) fatigue. It is therefore logical that crack propagation rate should also depend on the fracture strain of the material. Consequently, the fracture strain of the material should never be omitted in reports about experimental fatigue testing. Anistropy in the fracture ductility can also have an influence on crack initiation, propagation and final fracture (e.g. flattened MnS in rolled steel).

Because the critical fatigue crack length depends also on the fracture toughness, it is important that the critical fatigue crack length can also depend on temperature and rate of loading in the way that the fracture toughness does (lower temperature, impact).

It is worth emphasizing that, due to fatigue, not only is knowledge about the residual carrying capacity (or residual safety) of joints and components important. The residual ductility of joints and components and therefore of the whole structure should also be treated seriously, especially in the assessment of seismically or impact overloaded structures.

The most frequent discontinuities which decrease fatigue strength and ductility are defects of different severities in welded joints like porosity, lack of penetration and cracks in the worst case. These defects also belong to the topics of quality assurance and non-destructive testing. Problems of defect tolerances are readily soluble using both local strain and fracture mechanics approaches. Emphasis must be given to the significance of the fundamental mechanical properties described with the USRC of individual zones due to welding, which express the inhomogeneity. The welding simulator (e.g. Gleeble) can help in the determination of the parameters of the USRC of individual narrow heat affected zones in the weldment. Fillet welding represents a complicated example of weldments. It has severe strain concentrations in the root. It needs a theoretical treatment of its inelastic and fracture behaviour.

There is a need for an internationally created atlas

of residual stresses. There is much knowledge of the influence of residual stresses also in fatigue and fracture. The atlas of residual stresses could help in the wider quantification of their influence. The use of computer software for the determination of the distribution and the magnitude of residual stresses from different origins is possible.

There is also a need to introduce a simple unified classification of metals in the form of a matrix of different strengths and different ductilities, with additional information about the shape of the USRC. Fracture strain should be expressed in linear and not in logarithmic strain, also to enable a more direct comparison with the elasto-plastic fracture parameter CTOD. The present systems of classifying metals according to the chemical composition, microstructure, thermomechanical treatment or application are certainly important. However, from the point of view of mechanics of materials and structures, the proposed classification according to yield strength and fracture ductility, with appropriate information about the type of the USRC, has many advantages.

4 Concluding remarks

As a matter of urgency, the international standard for the simple tension test, and the corresponding regional and national standards, should be improved in order to obtain also fracture parameters of the unidirectional stress response curve (true stress-strain curve). The standard for the simple bend test also has to be improved as regards the determination of fracture strain. This is important in connection with the present intense international activity in building computerized materials data bases and the desire to encourage their uniformity and therefore the ease of data comparison and data interchange.

The problem of the damage and ultimate limit state of metal structures in connection with the inelastic and fracture behaviour of the material is so complex that only wide international co-operation of all institutions dealing with material science, structural mechanics, and structural engineering in relation to metal structures in the whole engineering area, can give soon the worldwide acceptable knowledge for public use. That knowledge should be organized in the form of international standards, codes and technical information sheets as design aids for practice and for modern regular and continuing education in engineering. Organizations like IUTAM, IIW, RILEM, IABSE, ECCS, ESIS, ASTM etc. can contribute in this direction. Therefore, the need for the integration of metal structures in Civil and Mechanical Engineering has to be recognized.

8 TESTING FOR AND REPRESENTATION OF JOINT MOMENT–ROTATION BEHAVIOUR

D.A. NETHERCOT
Department of Civil Engineering, University of Nottingham, UK

SUMMARY

A review is made of the current position regarding testing of steel beam to column joints for the purpose of determining moment-rotation (M-ϕ) behaviour. The approach taken is not to attempt an exhaustive cataloguing of all relevant contributions. Rather, it is to appraise the present status of the subject, to discuss some of the more important recent findings concerning those aspects of M-ϕ behaviour that are most relevant to the subsequent study of semi-rigid joint action on frame performance and to highlight those factors requiring particular attention in future testing. The question of acceptable representations of the test data for application to frame studies is also addressed.

Introduction

It is becoming increasingly accepted that the joints in steel framed structures do not behave according to the idealised models of "pinned" and "rigid" but in a semi-rigid and partial strength fashion. Depending on the exact form of detail used, behaviour approaching full strength, rigid or negligible moment transfer, rotationally pinned may be achieved but a more realistic view of the situation is that all joints possess some degree of moment capacity and some degree of rotational stiffness. This is best treated through the fundamental property of the moment-rotation (M-ϕ) characteristic, required for use in any analysis and design procedure that recognizes semi-rigid joint action, see Fig. 1.

Despite numerous attempts to treat the problem analytically - even by means of complex finite element approaches - there seems no realistic alternative, for the present, to at least a limited degree of testing of actual joints. In order that the results of such tests can be properly used, the experiments need to be designed in such a way that the data produced is properly representative of the key indices of behaviour, may be readily and unequivocally utilised etc.

Merely working with an experimentally obtained M-ϕ curve in analysis or design is clearly impractical. Therefore numerous attempts have been made to represent individual results or collections of results in a form that is more suitable for subsequent use. Initially this consisted of providing simple curve-fits based on the use of an increasingly more sophisticated set of equations. More recently insights into the most important facets of the behaviour of particular joint types have been used to establish semi-empirical behavioural models containing certain free parameters, the values of which are based directly on test results.

Testing of Metals for Structures. Edited by F.M. Mazzolani. © 1992 RILEM.
Published by Taylor & Francis, 2 Park Square, Milton Park, Abingdon, Oxon, OX14 4RN.
ISBN 0 419 15810 3

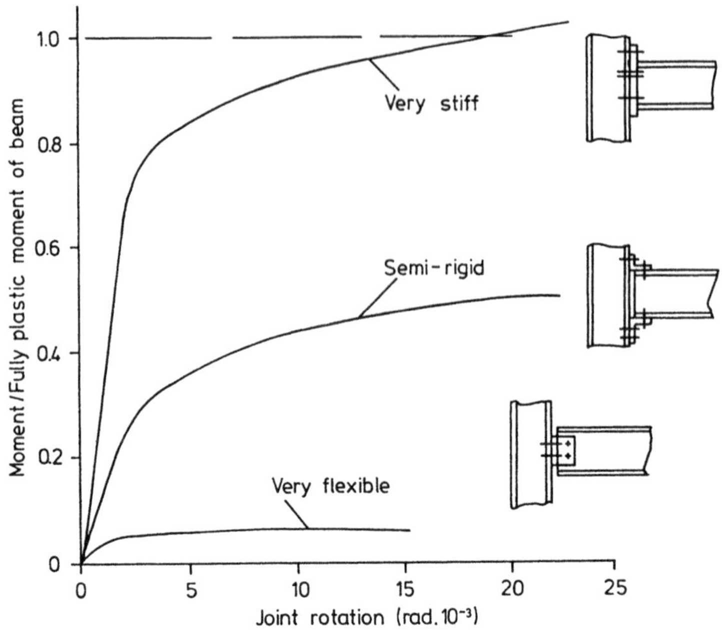

FIG. 1 TYPICAL MOMENT-ROTATION (M-ø) CURVES

It is the purpose of this paper to review and comment upon certain aspects of both the testing of connections to generate M-f characteristics and the subsequent representation of these results in a manner that renders them suitable for use in either analytical studies of the performance of steel frames with semi-rigid and partial strength connections or in the development of simplified design approaches for the same class of structure.

Experimental Determination of M-φ Characteristics

Testing is typically conducted on either a cantilever specimen - intended to represent an external beam to column arrangement - or a cruciform specimen - designed to simulate the behaviour of an internal connection, see Fig. 2. Both arrangements have certain advantages and disadvantages, although it is of interest to note that relatively few tests have been conducted on internal arrangements with deliberately unbalanced loading i.e. tests that attempt to investigate the interaction between a pair of joints to the same column when one joint is more severely loaded than the other. Generally speaking simple statically determinate testing arrangements have been used, although Astaneh (1) has questioned this from the point of providing the same distribution of loads to the test specimen as would be encountered in an actual frame. This is an interesting development as the general consensus from earlier testing has been that the moment to shear ratio in the connection is of no real significance. A related problem concerns the possible influence of axial load in the column, an effect that has been absent from the vast majority of joint tests. Results from a limited number of experiments at

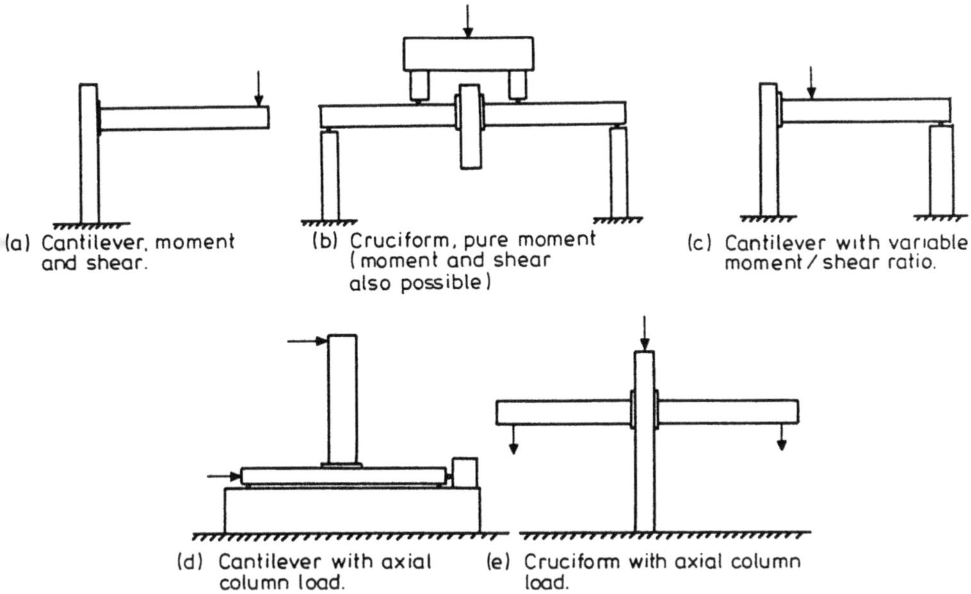

FIG. 2 JOINT TEST ARRANGEMENTS

Delft suggest that only in the event of extremely high column loads e.g. of the order of 70% of the squash load of the column, does the presence of compressive forces in the column start to have any real influence.

A point of particular importance, not really appreciated until comparatively recently, is the range and type of response for which M - f data is required if it is subsequently to be used in analyses of the behaviour of frames and subassemblages. Working back from the results of such analyses shows:

i even under proportional loading some connections will have the sign of their rotation reversed during behaviour up to attainment of the maximum load, Fig 3.

ii especially for non-sway frames, the maximum load of the frame will often be attained when most of the connections have only been required to undergo comparatively small rotations, Fig 4.

The first point makes it particularly important that testing include the unloading behaviour i.e. response of the joint under a reduction in the applied load. This is not necessarily the same as the production of full cyclic load response data of the sort necessary when considering the behaviour of structures under seismic action as the full hysterisis loop will not be required. Whilst relatively few of the test results presently available - largely those from series completed within the last five years - include the determination of unloading behaviour, it does appear that this follows a line parallel to the initial loading slope and that for the purposes of frame analysis such a representation will be

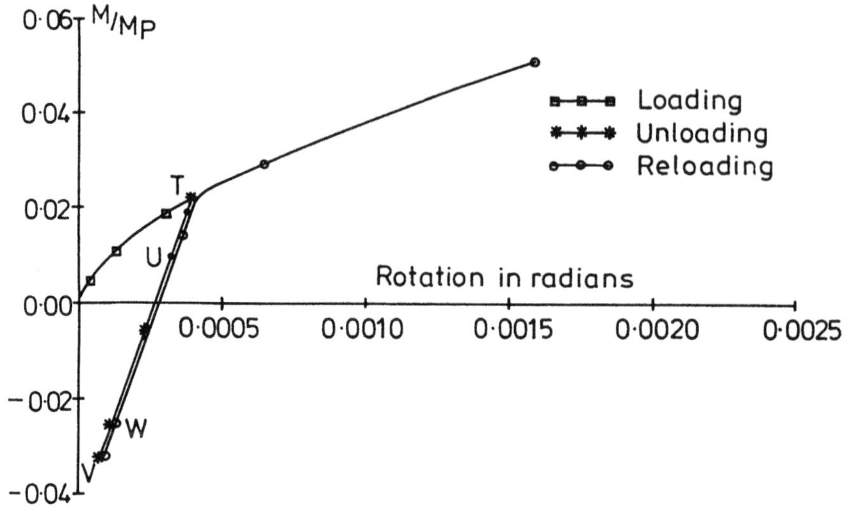

FIG. 3 BEHAVIOUR OF ONE JOINT IN A SUBASSEMBLAGE ANALYSIS UNDER PROPORTIONAL LOADING, SEE REF. 2.

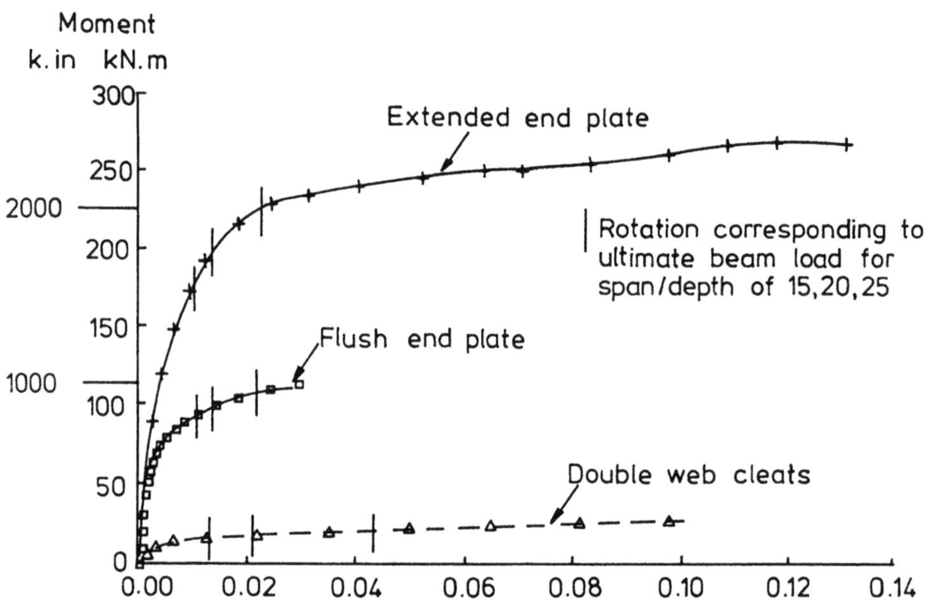

FIG. 4 BEHAVIOUR OF CONNECTIONS FOR BEAM ANALYSIS IN A NON-SWAY FRAME SHOWING ROTATIONS ATTAINED AT ULTIMATE LOAD, SEE REF. 3

acceptable. Particularly in cases of no-sway frames for which collapse is governed by the behaviour of the columns, ultimate load will be attained at very small end rotations; an example of this is given as Fig. 5. Thus connections will not be required to rotate through large angles and the most important part of the M-ϕ curve will be that located close to the origin. Once again the great majority of early tests on joints concentrated on developing the M-ϕ curve up to levels of rotation far beyond those likely to be experienced in frame structures, with the result that the measuring systems used and the sensitivity of the recording of data were such that behaviour at small rotations could only be regarded as rather approximate.

Fortunately the availability of several computer programs (5) capable of analysing the full range response of structures containing semi-rigid and partial strength joints and their use for a limited number of behavioural studies have highlighted the need for future M-ϕ joint tests to recognize the importance of the two points listed above. As a result most of those test series conducted in the last few years have paid more attention to these points.

In addition to concentrating on those aspects of behaviour that are of most significance, it is also important that tests be conducted in a fashion that makes the results usable by the research and code writing communities. This requires (6) :

i a clear definition of terms,

ii joint rotation to be clearly referenced to a point and a system of axes fixed with respect to the specimen,

ii a guarantee that measurements taken do coordinate with the definitions of the quantities being measured.

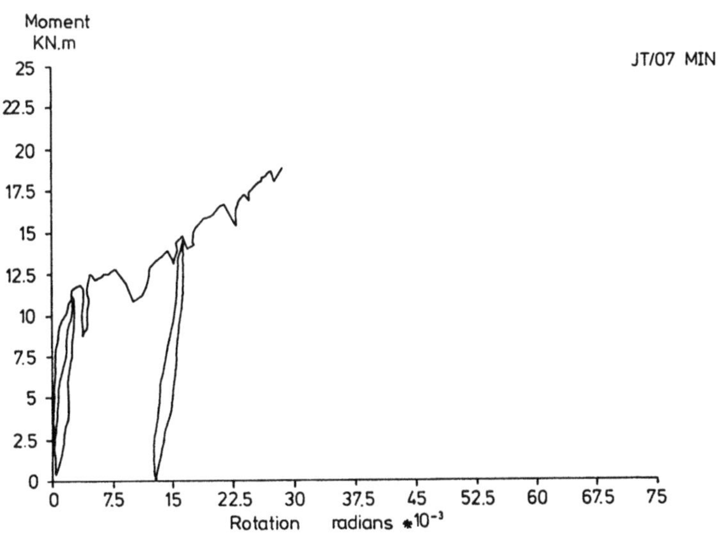

FIG. 5 EXPERIMENTAL BEHAVIOUR OF A FLANGE CLEAT JOINT SHOWING UNLOADING BEHAVIOUR, SEE REF. 4.

Whilst these requirements might sound obvious, it is disturbing, when studying test series conducted over the past few decades to note how frequently measurement systems have been used that do not permit the actual joint rotation to be determined, the definition of joint rotation given is clearly at variance with the method used to measure it, the experimental arrangement admits the possibility of effects that cannot be properly quantified etc. In addition to this the data recorded and reported should constitute a full description of the test arrangement, procedure and results obtained. Recently, a number of attempts have been made to standardize on this reporting procedure through the use of data sheets (7). These are particularly useful in focussing the minds of investigators on the need for care if the results of their work, in which they will undoubtedly have invested significant amounts of time and money, are to be properly usable.

It is not uncommon in reading the reports of test series to find that additional measurements have been taken e.g. to identify contributions to the deformations of the joint of individual components. This is particularly useful in assisting with understanding of the physical behaviour and in devising methods for the prediction of the flexibility of semi-rigid connections that rely on a summation of the flexibilities of the individual components. An example of this is the method given in Annex J of EC3.

Representation of M-ϕ Data

The exact style of representation required for M-ϕ test data depends on the intended use. For example, design methods may require only an estimate of the moment capacity and the indicative stiffness; such a method is included in EC3. On the other hand, if a complete non-linear analysis of a frame is to be undertaken the full M-ϕ curve will normally need to be available.

When using representations of M-ϕ behaviour that are based solely on curve fitting it is essential that the limits of applicability of the formulae be properly observed. For example the original study in this area by Frye and Morris (8) proposed simple polynomial expressions for a variety of joint types. However, if these are used outside the range of individual parameters e.g. cleat thickness on which the curve fitting was based, surprising and potentially unsuitable M-ϕ curves result. Behavioural models, on the other hand, offer greater promise of being generally applicable over the whole range of parameters for a particular joint type. This is because they are based on a simplified representation of key aspects of the actual physical behaviour and are likely to be valid over a much wider range. The disadvantage of such an approach at present is that it still needs to make reference to test data in order to fix the values of the 'free' parameters (9).

However, as more studies of the sensitivity of the response of complete frames to variations in the assumed joint behaviour become available, it should be possible to relax requirements on accuracy of joint representation. Results from the limited number of studies so far produced suggest that the more extensive the system under consideration the less important it is to use precise M-ϕ data. This particular issue is important for another more practical reason: design methods that recognize semi-rigid partial strength joint action need to be able to accommodate the inevitable variability in behaviour between notionally identical joints produced under site conditions. In this respect it is of interest that some studies of joint behaviour in which deliberate lack of fit effects have been included have shown that certain types e.g. end plates as shown in Fig 6, are relatively insensitive to quite substantial variations between specimens. All of this work suggests that great precision in representing M-ϕ curves may well not be necessary but confirmation of this must await the availability of a sufficiently large body of results.

Useful reviews of the current status of the representation of moment rotation characteristics are available in the book chapter by Nethercot and Zandonini (11) and in a forthcoming ECCS report.

Requirements for the Future

Although the dependence on physical testing of joints as the means of generating moment rotation characteristics is likely to reduce as prediction methods based on behavioural models become better established, it seems inevitable that some degree of testing will be required for the foreseable future. It is therefore important that such testing be properly conducted, be arranged so as to ensure that the most important aspects of behaviour are those monitored and that the results be fully and clearly reported.

This leads naturally to the question of some form of central register or collection of the results of joint moment rotation tests. Some progress has been made in this direction e.g. in the compendia of joint test data produced by Goverdham (12), Nethercot (13), Chen and Kishi (14). An extension of this has seen the establishment of a computerized data bank of joint test results at the University of Sheffield. The challenge is to expand, maintain and provide access to such a data bank so that those with a legitimate interest in moment rotation data for joints can readily be provided with the information.

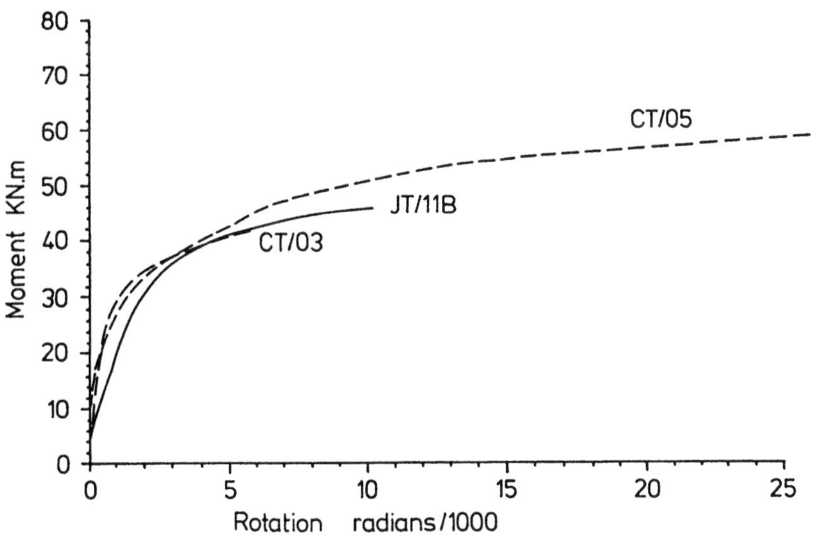

FIG. 6 M-ϕ RESPONSE OF FLAT AND DISTORTED FLUSH ENDPLATES, SEE REF. 10

Conclusions

A review has been made of the current position regarding the testing of joints to produce M-ϕ characteristics and the subsequent representation of the results in a form suitable for analysis and/or design. Suggestions for the emphasis of future testing and reporting and the necessary accuracy of representation have been made.

References

1. Astaneh, A., "Demand and Supply of Ductility in Steel Shear Connections", Journal of Constructional Steel Research, Vol. 14, No. 1, 1989, pp. 1-20.

2. Nethercot, D.A., Kirby, P.A. and Rifai, A.M., "Column behaviour in PR construction - analytical studies", ibid, also published in Canadian Journal of Civil Engineering, Vol. 14, No. 4, August 1987, pp 485-497.

3. Nethercot, D.A., Davison, J.B., Kirby, P.A., "Connection flexibility and beam design in non-sway frames" ibid, also published in American Institute of Steel Construction Engineering Journal, Vol. 25, No. 3, 1988, pp.99-108.

4. Davison, J.B., Kirby,, P.A. and Nethercot, D.A., "Rotational stiffness characteristics of some steel beam to column connections", Journal of Constructional Steel Research, Vol. 8, 1987, pp. 17-54.

5. Deierlein, G.G., Hsieh, S-H. and Shen, Y-J, "Computer-Aided Design of Steel Structures with Flexible Connections"' National Steel Construction Conference, AISC, Kansas City, March 1990, pp. 9.1-9.22,

6. Kirby, P.A., Zandonini, R. and Davison, J.B., "Experimental Determination of Moment Rotation Response of Semi-Rigid Beam to Column Joints", Proceedings Structural Stability Research Council, 1988, pp. 45-58.

7. Bjorhovde, R. and Brozzetti, J., "Data Sheet Format Presentation"' Connections in Steel Structures: Behaviour, Strength and Design, ed. R Bjorhovde, J Brozzetti and A Colson, Elsevier Applied Science Publishers, 1988, pp. 380 - 387.

8. Frye, M.J. and Morris, G.A., "Analysis of Flexibly Connected Steel Frames"' Canadian Journal of Civil Engineering, Vol. 2, 1975, pp. 280 - 291.

9. Wu, F.H., and Chen, W.F., "A Design Model for Semi-Rigid Connections", Engineering Structures, Vol. 12, No. 2, April 1990, pp. 88-97.

10. Davison, J.B., Kirby, P.A. and Nethercot, D.A., "Effect of lack of fit effects on connection flexibility", ibid, pp. 55-70.

11. Nethercot, D.A., and Zandonini, R., "Methods of Prediction of Joint Behaviour"' ch. 2 of "Structural Connections: Stability and Strength", ed. R. Narayanan, Elsevier Applied Science Publishers, 1989, pp

12. Goverdham, A.V., "A Collection of Experimental Moment-Rotation Curves and Evaluation of Predicting Equations for Semi-Rigid Connections"' Ph.D., thesis, Vanderbilt University, 1984.

13. Nethercot, D.A., "Steel Beam to Column Connections: A Review of Test Data and their Application to the Evaluation of Joint Behaviour on the Performance of Steel Frames"' CIRIA Project Record 338, London, 1985.

14. Kishi, N. and Chen, W.F., "Data Base on Steel Beam-to-Column Connections"' Rept. CE-STR-86-26, School of Civil Engineering, Purdue University, 1986.

9 DEFINITION OF A DESIGN MOMENT–ROTATION RELATIONSHIP BY TESTING

M.A. PARISI, R. ZANDONINI and P. ZANON
Department of Structural Mechanics and Design Automation, University of Trento, Italy

1 INTRODUCTION

Considerable research efforts have been recently devoted to the study of the behaviour of semi-rigid connections. From the perspective of professional practice, the knowledge of the moment-rotation (M-Φ) law of the joints is a pre-requisite for structural analysis of semi-continuous frames. At present, the knowledge of M-Φ relationships is still in a hardly usable form, even if numerous mathematical and analytical models have been proposed for the most popular forms of joints. Some of these models have been introduced in the Annexes of the present version of Eurocode 3. In Eurocodes 3 and 4 [1, 2] testing is also indicated as a viable design method. As to the determination of the joint response, however, no guidelines are given on how to perform design-oriented testing. Specifically, these should indicate what number of tests it is appropriate to perform, what quantities to measure, how to interpret test results in order to deduct an M-Φ curve that may be safely used, and, possibly, suggest a standardized form for reporting them. No generally accepted criteria exist, indeed, for defining a usable behavioural curve, once some experimental data are obtained.

All the above aspects should be considered in a reliability based framework. The problem discussed in the following concerns, then, the identification of a methodology for producing by testing M-Φ curves characterized by a defined safety level. The professional context concerned with this problem is fairly wide, including:

- design of a major structure;
- production of a standardized industrial system;
- use of a database, to be tuned to the specific context by means of a limited number of experiments;
- calibration of codes.

In the first two cases, experimentation, which is seldom carried out for direct design purposes, may be accessible as well as cost-effective.

Testing of Metals for Structures. Edited by F.M. Mazzolani.
Published by Taylor & Francis, 2 Park Square, Milton Park, Abingdon, Oxon, OX14 4RN.
ISBN 0 419 15810 3

Three possible basic approaches can be sorted out when addressing the selection of a design-oriented M-Φ curve: these correspond to pure experimentation, to a combination of testing and theoretical models, and to the sole use of theoretical models. The procedures corresponding to the first two cases will be discussed in the following sections, considering the third one as a case with a very limited practical interest at this stage of knowledge.

The example in Appendix A stresses the importance of the subject. It shows how a variation in the basic joint characteristic affects heavily the performance of the end-restrained beam. Specifically, a difference of approximately 30 percent propagates from basic data to the M-Φ relationship and, finally, to the beam load capacity. The role of standardized testing and test interpretation procedures in eliminating an important source of inaccuracy stands clearly out from the example.

2 DEFINITION OF AN M-Φ CURVE FOR DESIGN

General procedures for the definition of design-oriented M-Φ curves are proposed here and the related operational problems presented. Two principal cases are examined: the definition of an M-Φ curve by experimental analysis only and by combined use of theory and testing.

A) M-Φ design curves by experimental analysis

This case corresponds to the situation where no available theoretical model is suitable for the case under study.

The procedure may be itemized in the following steps:

i. *Identification of main parameters*: this step corresponds to the description of the connection in terms of those of its geometric and mechanical parameters that influence the M-Φ behaviour. A set of basic parameters must, then, be identified: although their variability is defined by market and/or practice, their values of interest must be further reduced to the significant range for a specific design case.

   The *Requirements* in order to fulfil the objectives of this step concern the following aspects:

   - market product data: knowledge of the kind of products to be used, as available on the market (i.e., whether European, U.S., or Japanese standards).
   - economic and architectural constraints: constraints may be fixed by the constructor either on the basis of the quality of workmanship available or on economic grounds, and by the architect for aesthetic or for usability reasons. As an example, welded or partially welded connections may be indicated as suitable in a specific case from an engineering point of view. Yet, they could be ruled out by the constructor for economic or for technical reasons, whereas a bolted connection may not please the architect.

ii. *Definition of mechanically feasible combinations of joint parameters* by elimination of non-significant cases.

   To this purpose, elements to be considered may be: the class and diameter of bolts, the angle type and thickness, the plate geometry, the type of beam section, the grade of steel, etc.

Many combinations of parameter values may be eliminated being void of structural meaning. This operation requires a knowledge base on local component behaviour, that may be derived from a variety of sources: it may consist, in the least, of knowledge embodied into codes, or be based on specific expertise if available, as well as obtained from literature data peculiar to the area under study.

Requirements:

- knowledge base on local component behaviour;
- code limitation for geometric data.

iii. *Selection of the joint configurations and of the minimum number of tests.*

The methodology to be set up for selecting the number of tests to be performed in order to obtain a certain confidence level for test results must be flexible enough to recognize a wealth of different influencing factors and situations, including the combinations of internal forces. Also, an influencing aspect is the objective for which the tests are performed: i.e., test results in view of admissible stress rather than limit state applications, or for the exploitation of design constraints as deformability -whether in service or in limit conditions- depend on different numbers of factors.

The number of joint configurations to be tested and the number of tests necessary for each configuration are of interest here.

As to the selection of specimens, on the basis of all the cases pointed out previously (see ii), different criteria may be adopted. If, for instance, the purpose of the testing program is to create a database of results, tests have to be performed on each joint configuration deemed feasible at point (ii). Otherwise, various criteria may be expressed for choosing a subset of meaningful configurations. As an example, on the basis of possible failure mechanisms, a number of configurations allowing to activate all the different mechanisms must be chosen. Different criteria could indicate configurations the behaviour of which is expected to bracket that of a series of joints. In any case, criteria should be defined allowing a significant reduction of the amount of experimental work.

For each specimen type, then, the necessary number of tests is to be indicated by means of statistical methods.

Requirements:

- identification of failure mechanisms;
- knowledge of parameter uncertainties.

iv. *Laboratory tests*

Present lack of standardization of testing procedures and of measurement criteria has hindered, to date, comparisons of results from different laboratories, making an abundance of data hardly usable beyond the specific context in which they were originated, with great dispersion of resources [3]. Steps toward standardization, however, may be expected to absorb a considerable amount of work: the definition of measurement criteria generally requires previous definition of behavioural aspects: for instance, defining what is to be intended for rotation of a connection precedes the definition of a methodology for actual rotation measuring.

Requirements:

- o standard definitions of the key response parameters;
- o testing procedures;
- o measurement criteria;
- o criteria for the evaluation of test data.

v. *Statistical determination of the mean M-Φ curve.*

Test results must be processed by statistical methods (e.g., regression analysis) to yield an M-Φ curve to be used as basis for further elaboration towards design quantities. Though the study should allow a full statistical definition of the joint response, the mean M-Φ curve seems in principle a convenient reference for the next steps of the procedure.

Requirements:

- o methods for evaluation and processing of sets of multi-variable test data.

vi. *Structural response sensitivity to joint parameters* (initial stiffness, C, ultimate moment M_u, and ultimate rotation capacity, Φ_u) both for service and ultimate limit states, with reference to the curve of the previous point, (v), and to its variability.

Subsequent steps concern the choice of a design curve compatible with specific reliability levels for the structure. In order to obtain design data, the sensitivity of the structural response to the joint parameters must first be assessed.

Requirements:

- o criteria for structural sensitivity analysis;
- o suitable methods of structural analysis. Analysis methods must be capable of including material nonlinearity and second order effects.

vii. *Definition of the level of reliability for the joint.*

Once the reliability level expected from the structure has been established and the sensitivity of the structure to the curve, as defined in (v), has been evaluated, a characteristic M-Φ curve or, at least characteristic values of the key parameters related to it must be estimated.

Requirements:

- o sensitivity to structural response (point vi.);
- o L_r, reliability level required of the structure;
- o definition criteria for resistance factors (if code supplied values are not applicable).

viii. *Determination of the design M-Φ curve*, as a function of the type of design adopted and of the limit state considered (whether at service or at ultimate).

The design curve is obtained on the basis of design constraints and of the safety level imposed by the codes (usually in terms of γ-factors).

Requirements:

- o criteria for use of (possibly different) γ-factors in the various portions of the M-Φ curve.

B) M-Φ design curves by testing and theoretical models

This case applies when a theoretical model exists, it is deemed appropriate for use, and yet it has to be calibrated to the specific kind of connection under exam. Many steps are the same as in case A. For the sake of completeness, however, all the steps are listed, though only those specific of this procedure are expanded.

i. *Identification of main parameters.*
(see A.i)

ii. *Definition of mechanically feasible parameter combinations*
(see A.ii)

iii. *Selection of the joint configurations and of the minimum number of tests.*
The main difference with the former procedure is at this point. In particular, the number of joint configurations to be examined is the same. In this case, the experiment may be carried out in a deterministic perspective. For each case it is hence possible to reduce the number of tests, possibly down to a single experiment.

iv. *Laboratory tests*
This allows experimental determination of calibration M-Φ curves (see A.iv).

v. *Estimate of the error inherent to the model*
Comparison with test results enables for each configuration an evaluation in quantitative terms of the discrepancies between the theoretical model and the actual behaviour of the connection considered. Such a comparison with experimental evidence yields, then, an error estimate for the model.
The first objective is related to the acceptance of the model on the basis of accuracy levels previously defined. A possibility is also that a limited range of application of the model is identified.
The second objective is to obtain the necessary quantities for the subsequent reliability analysis.
Requirements:
- accuracy levels for acceptance criteria (may be different for C and M)
- error estimate criteria:
  -probabilistic
  -single test

vi. *Statistical definition of the M-Φ behaviour for the joint configuration considered*
This knowledge is achieved via a suitable parametric study using the model. A mean M-Φ curve is the main final product.
Requirements:
- knowledge based analysis (as in A.i. and A.ii.);
- knowledge of parameter uncertainties;
- criteria for statistical assessment.

vii. *Structural response sensitivity to joint parameters*
(see A.vi)

viii. *Definition of the level of reliability for the joint*
(see A.vii)
Being the response determined via the model, the error inherent to the model should be taken into account in the reliability analysis. Knowledge of criteria to incorporate error in reliability analysis represent an additional requirement to those listed in A.vii.

ix. *Determination of the design M-Φ curve*
(see A.viii)

3 CONCLUSIONS

Design on the basis of experimental information is presented as a possibility also at code level. This is certainly of interest in areas as semi-rigid construction where available knowledge is not an exhaustive background for design [4].
No guidelines, however, are offered to lead the designer through the various passages from experimentation to the formulation of realistic design data. This work has attempted to point out the steps of two possible procedures aiming at the determination of a moment-rotation curve for a particular joint form. The purpose is limited to the identification of the various problems that arise and to sorting out the areas that require an effort of work and organization of sparse knowledge. The general requirements for each step in terms of data type and operational criteria have been pointed out. The final picture is that of a complex area, which is, however, worth the research and reorganization effort to make the path of design by testing effectively viable also for semi-rigid systems.

4 REFERENCES

[1] Commission of the European Communities Eurocode N. 3 "Design of Steel Structures", Draft, April 1990.

[2] Commission of the European Communities Eurocode N. 4 "Design of Composite Steel and Concrete Structures", Revised Draft, Issue 1, October 1990.

[3] P.A. Kirby, R. Zandonini, J.B. Davison, "On the Determination of Moment Rotation Characteristics of Beam-to Column Joints", to appear in Proc. "Needs in testing metals", RILEM Workshop, Naples, May 29-31, 1990.

[4] D.A.Nethercot, "Utilization of Experimentally Obtained Connection Data in assessing the Performance of Steel Frames", Proc. Session of Connection Flexibility and Steel Frames, ASCE, Detroit, 1985.

[5] R. Zandonini and P. Zanon,"Design of Beams in PR frames", AISC, National Steel Conference, Kansas City, March 1990.

APPENDIX A

An example

The example shown in the Figures from A-1 to A-4 aims at pointing out the variability in the evaluation of M_p, the ultimate plastic resistance, of W_s, the service limit load, and of the load at the ultimate limit state (W_e =elastic; W_u =inelastic).
Parameters for which variations have been considered are the yield stress for the steel, f_y, the depth of the beam, d, and the thickness of the connection angles, t.
The plastic moments of the beam obtained at the two extremes of the range of the parameter values are in a ratio of 1.31 to 1.
Limit state load resistances were obtained via the method of analysis developed by two of the Authors [5]. The ratios between the two loads obtained at the inelastic limit state, at the elastic limit state, and at the service limit are 1.33, 1.35, and 1.16, respectively. The purpose of these calculations being simply to indicate the importance of the problem, no probabilistic treatment has been applied to the values of this example: indeed, variations in the parameters are selected within the values admitted for them in various codes, Eurocodes included.

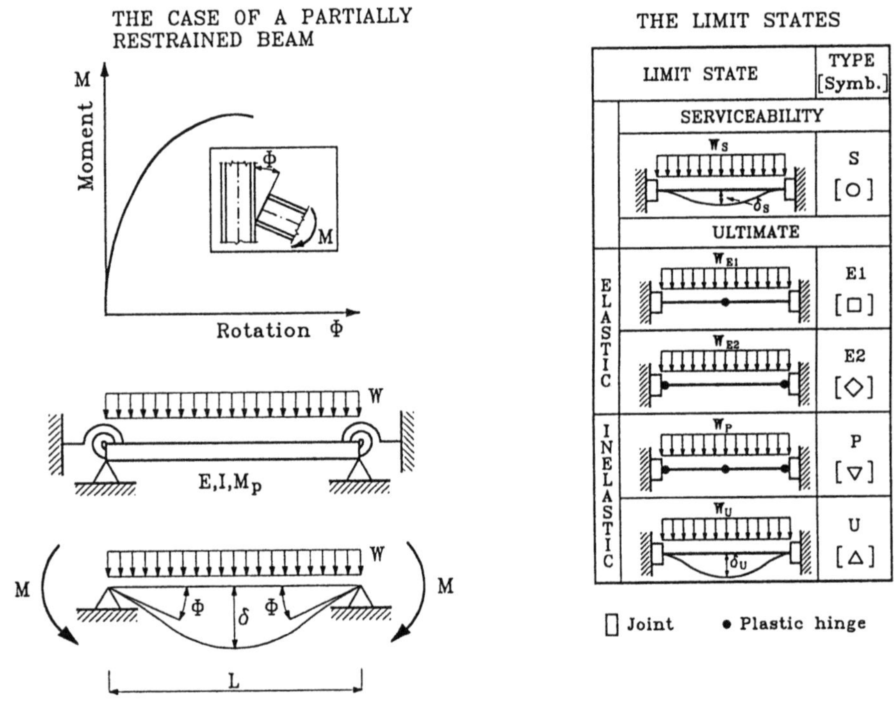

Fig. A–1 The problem

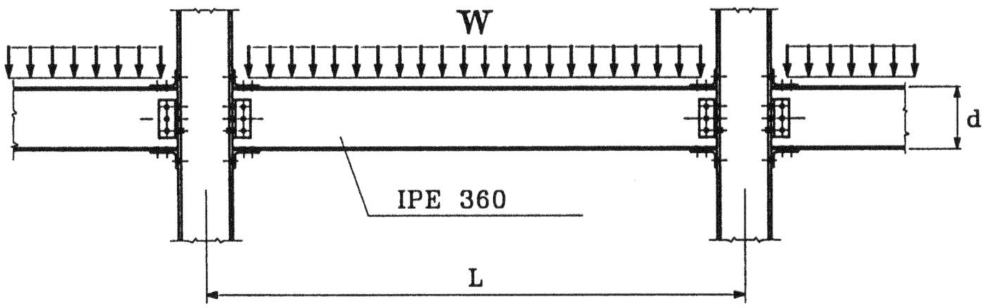

Fig. A-2 The beam

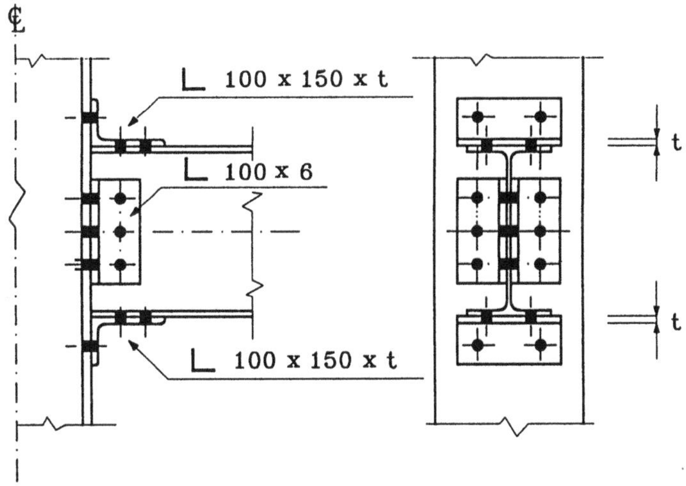

Fig. A-3 The joint

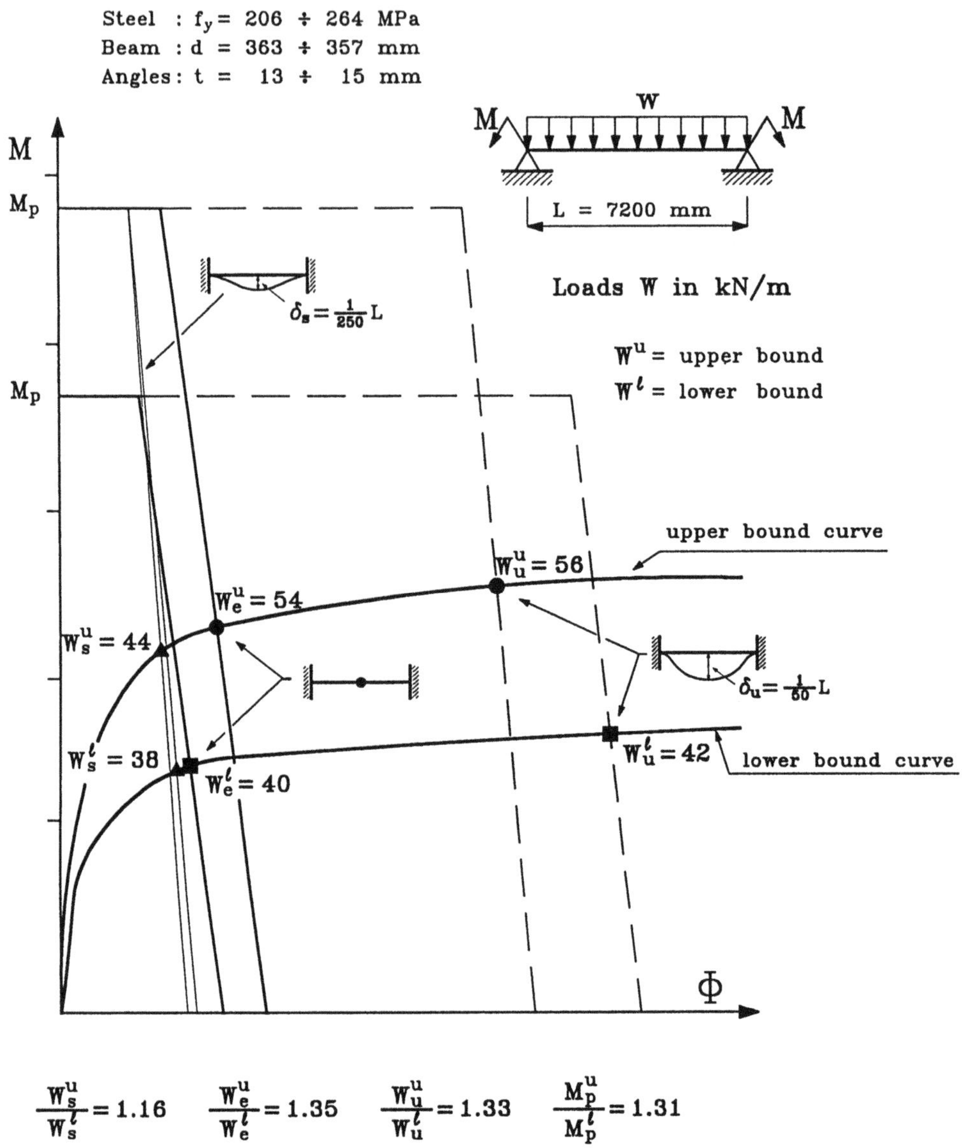

Fig. A–4 The beam performance

PART TWO
LOW CYCLE FATIGUE BEHAVIOUR OF MATERIALS, JOINTS AND STRUCTURAL COMPONENTS

10 CYCLIC BEHAVIOUR OF STEEL SUBASSEMBLAGES

L. CALADO
Department of Civil Engineering, Instituto Superior Técnico,
Technical University of Lisbon, Portugal

SUMMARY

This paper comments the behaviour of structural steel elements under cyclic load based on the ECCS [1] recommended testing procedure. Some experimental studies and numerical models on the cyclic behaviour of steel subassemblages are presented. A numerical model for damage accumulation and for predicting the failure of structural subassemblages under cyclic loading is also presented. This model is expressed as a function of the inelastic strain and the dissipated hysteretic energy. It is shown that failure of the structural element does not depend on the loading histories but depends on the accumulation of the inelastic strain.

1 INTRODUCTION

There is a wide variety of types and dimensions of steel rolled products which may be used for structural elements and connections.Steel structural elements have a high seismic resistance. However, there are a variety of conditions that may affect the performance of these structures and connections.

Experimental studies were performed over the last twenty years in universities and research centers to develop test procedures to achieve a better understanding of the seismic behaviour of structural steel elements and connections. However, the testing procedure and the interpretative parameters of the tests are different from one researcher to another, making, at times, qualitative and quantitative comparisons a difficult task.

The ECCS, European Convention for Constructional Steelwork, through its Technical Committee 1 - Structural Safety and Loadings suggests [1] a testing procedure for assessing the behaviour of structural steel subassemblages under cyclic loads. This procedure was intended as a unified way to carry out experimental testing in structural elements or even complete structures in order to make possible the comparison of results obtained by different authors. Simultaneously it was intended to make possible the assessment of the seismic behaviour of steel elements based on cyclic quasi static results using a specified loading history.

2 ECCS TESTING PROCEDURE

The procedure for assessing the behaviour of structural steel elements under cyclic loads recommended in the ECCS can be applied to plane or three dimensional tests and may

Testing of Metals for Structures. Edited by F.M. Mazzolani.
Published by Taylor & Francis, 2 Park Square, Milton Park, Abingdon, Oxon, OX14 4RN.
ISBN 0 419 15810 3

include two preliminary monotonic displacement tests (positive and negative) or obviate them. In the first case, it is called "complete testing procedure" (it needs three specimens, two for the monotonic test and one for the cyclic test) while in the second it is designated "short testing procedure" (it needs only one specimen for the cyclic test). Due to economical reasons, the second procedure is preferable.

In this type of procedure (short testing procedure) on account of the disknowledge at the beginning of the test of the yield load (F_y^+, F_y^-) and the yield displacement (δ_y^+, δ_y^-) it is necessary to perform steps of displacement sufficiently small to ensure that at least four levels of displacement are imposed before the yield displacement is reached. In this procedure the yield load (F_y^+, F_y^-) is defined as the interception between the tangent modulus (E_t^+, E_t^-) at the origin of the force-displacement curve evaluated from the first cycle and the tangent to the envelope curve of the cycles with a slope of 0.10 of the initial tangent modulus as indicated in Figure 1.

As soon as the yield limits are defined three cycles in the interval $[(2+2n)\delta_y^+;(2+2n)\delta_y^-]$ with n = 0,1,2,3... are imposed.

The end of the test is left free. However, a researcher will certainly go as far as possible, in order to obtain the maximum information possible while a design engineer will probably stop the test as soon as the code requirements are reached.

Regarding the interpretative parameters of the testing procedure, the recommendations of the ECCS look for a uniformization of these parameters which are established in ratios with which the engineer is familiar. The suggested parameters are defined taking the elastic perfectly plastic behaviour as a basis, meaning that all the ratios are equal to one for that peculiar behaviour.

Due to the possible different behaviour of the element in tension and in compression range, the parameters are evaluated in these two ranges. The quantities used in the ratios are deduced from the force-displacement curve (Figure 2) and are obtained for cycles with displacements larger than the yield displacement.

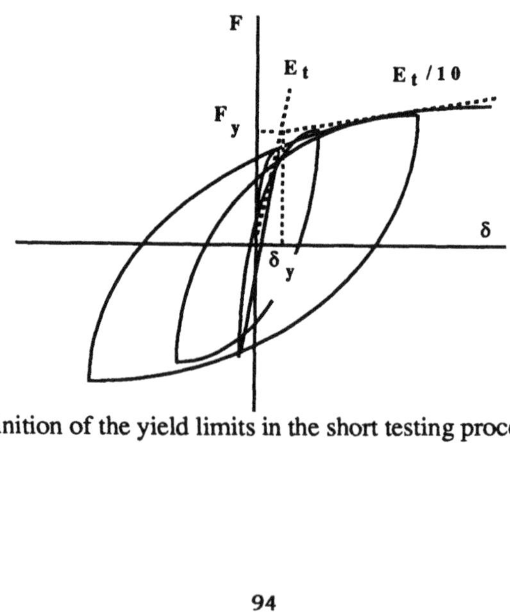

Figure 1 - Definition of the yield limits in the short testing procedure of the ECCS.

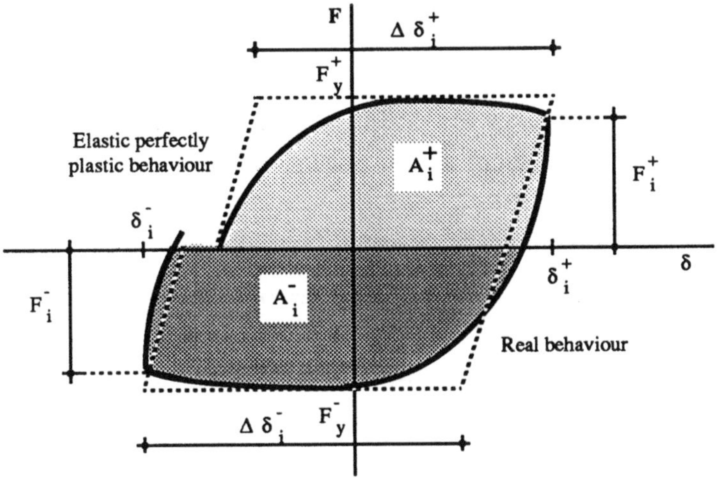

Figure 2 - Quantities to determine in a general cycle.

The proposed parameters for a typical cycle are the following:

- full ductility ratio:

$$\psi_i^+ = \frac{\Delta\delta_i^+}{\delta_i^+ + \delta_i^- - \delta_y^-} \qquad \psi_i^- = \frac{\Delta\delta_i^-}{\delta_i^- + \delta_i^+ - \delta_y^+}$$

- resistance ratio:

$$\varepsilon_i^+ = \frac{F_i^+}{F_y^+} \qquad \varepsilon_i^- = \frac{F_i^-}{F_y^-}$$

- dissipated energy ratio:

$$\eta_i^+ = \frac{A_i^+}{(\delta_i^+ + \delta_i^- - \delta_y^+ - \delta_y^-)\, F_y^+} \qquad \eta_i^- = \frac{A_i^-}{(\delta_i^- + \delta_i^+ - \delta_y^- - \delta_y^+)\, F_y^-}$$

For each group of three cycles of equal displacement, the parameters are defined as a function of the ductility (μ) (ratio of the current displacement to the yield displacement ($\mu = \delta/\delta_y$)) as follows:

- ψ (μ_i) minimum value of the three ψ_i evaluated in the group of three cycles of equal maximum displacement.

- ε (μ_i) minimum value of the three ε_i evaluated in the group of three cycles of equal maximum displacement.

- η (μ_i) average value of the three η_i evaluated in the group of three cycles of equal maximum displacement.

The test is then characterized and resumed by these values which represent discontinuous functions.

In general, the behaviour of the element is much better as its nears elastic perfectly plastic behaviour (values of these functions near one). Small values of these parameters (<< 1) may be assumed as an index of the end of the test because in this case exists a large loss of ductility, resistance and dissipated energy.

The parameters proposed by the ECCS have the advantage of being efficient in the quantitative analysis of the cyclic behaviour of structural elements and can also be considered as practical parameters to define code acceptance criteria.

3 CYCLIC BEHAVIOUR OF STEEL SUBASSEMBLAGES

Experimental tests on bracing elements and beam-columns were performed in the Dipartimento di Ingegneria Strutturale del Politecnico di Milano, which followed the ECCS recommended testing procedure. The specimens were chosen so as to model in full-size steel subassemblages likely to be met in real structures. Some basic information on the cyclic behaviour of the bracing elements and beam-columns was gained from the tests. They proved the accuracy of the developed numerical models which describe the cyclic behaviour of the element until failure occurs.

3.1 Bracing Elements

A typical plastic hysteretic behaviour of a bracing element [2] under repeated loading and its mode of failure are showed in Figure 3.

The force-displacement curve was obtained from experiment, in which a bar was subjected through pinned ends to repeated tension and compression, gradually increasing strain amplitude in each loading cycle nearly following the ECCS short testing procedure.

The initial buckling load corresponds to point A. In the subsequent force-displacement curve, the bracing element loses its resistance considerably once it buckles. This phase of the hysteretic behaviour is dominated by the plastic interaction between the bending of the brace element due to the P-Δ effect induced by the compressive force. If the sense of increment in the displacement is reversed, the force-displacement curve corresponds to the elastic recovery with a brace lengthening. In the case of bolted connections, a slip occurs in the force-displacement curve. At point B the element is fully straightened. The zone BC is characterized by the plastic elongation of the element. Reversing the direction of the displacement results in the elastic unloading of the brace.

Figure 3 shows a decrease of the ultimate compressive load with the application of the cyclic displacement, consequence of the Bauschinger effect exhibited by the steel when subjected to cyclic deformations. This decrease can also be consequence of the brace curvature. In general after an initial buckling cycle the brace develops a residual curvature which is not completely removed by the subsequent tensile yielding, and for that reason the brace behaves as an element with an initial curvature.

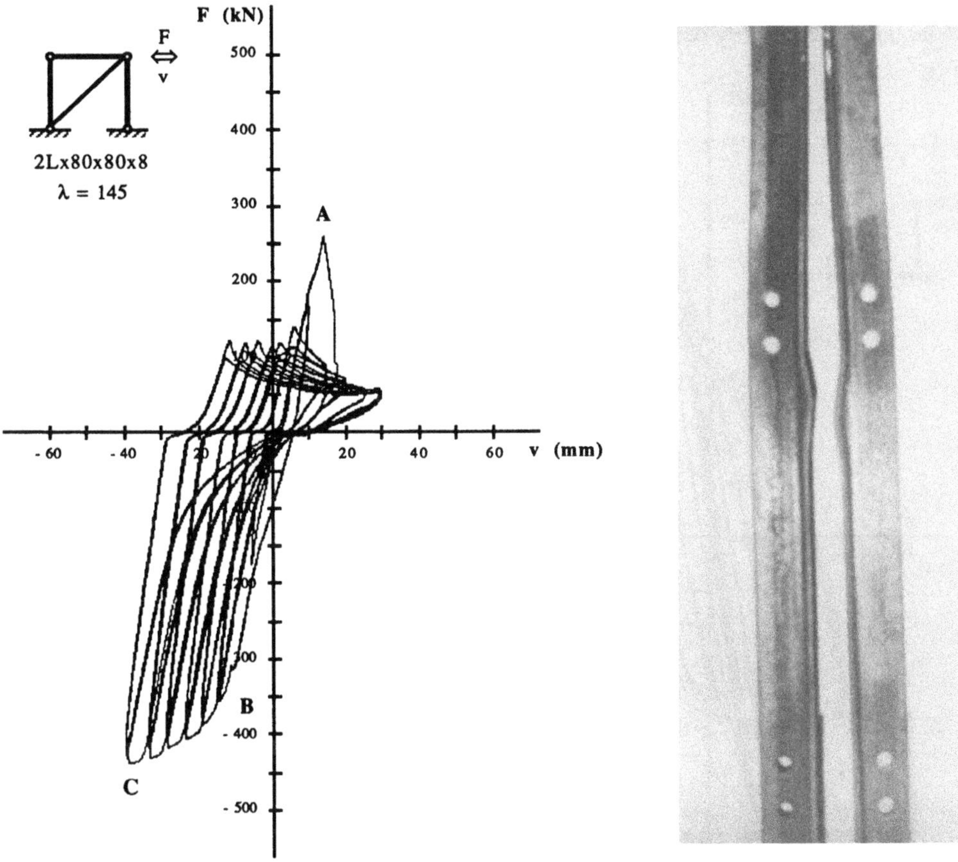

Figure 3 - Typical hysteresis loops and mode of failure of a bracing element under repeated loading.

The tests also indicated that the cross-sectional shape can affect the hysteretic behaviour of the brace. This parameter is related to the brace´s susceptibility toward lateral-torsional buckling and local buckling and for that reason with the brace compressive carrying capacity. For the cases studied (back-to-back angles and back-to-back channels) it was observed that local buckling occurred at extremely large lateral displacements.

3.2 Beam-columns

Beam-columns are typical elements of low rise building. Figure 4 shows a force-displacement hysteretic curve obtained from an experiment on a beam-column [3] subjected to repeated loading following the ECCS short testing procedure.

Experimental investigations [3] carried out on beam-columns under repeated and reversal loading showed that the development of local buckling in the flanges does not signal an immediate loss of the moment capacity. The beam-columns are able to sustain loads substantially higher than those that cause initial flange buckling. This is attributed to the considerable post-buckling strength of the plate elements. However, after the occurrence of the maximum load in the following cycles, the moment capacity deteriorates. This deterioration is higher with increasing width-thickness ratio (b/t) of the flange as a consequence of the early occurrence of local instability in the flange elements.

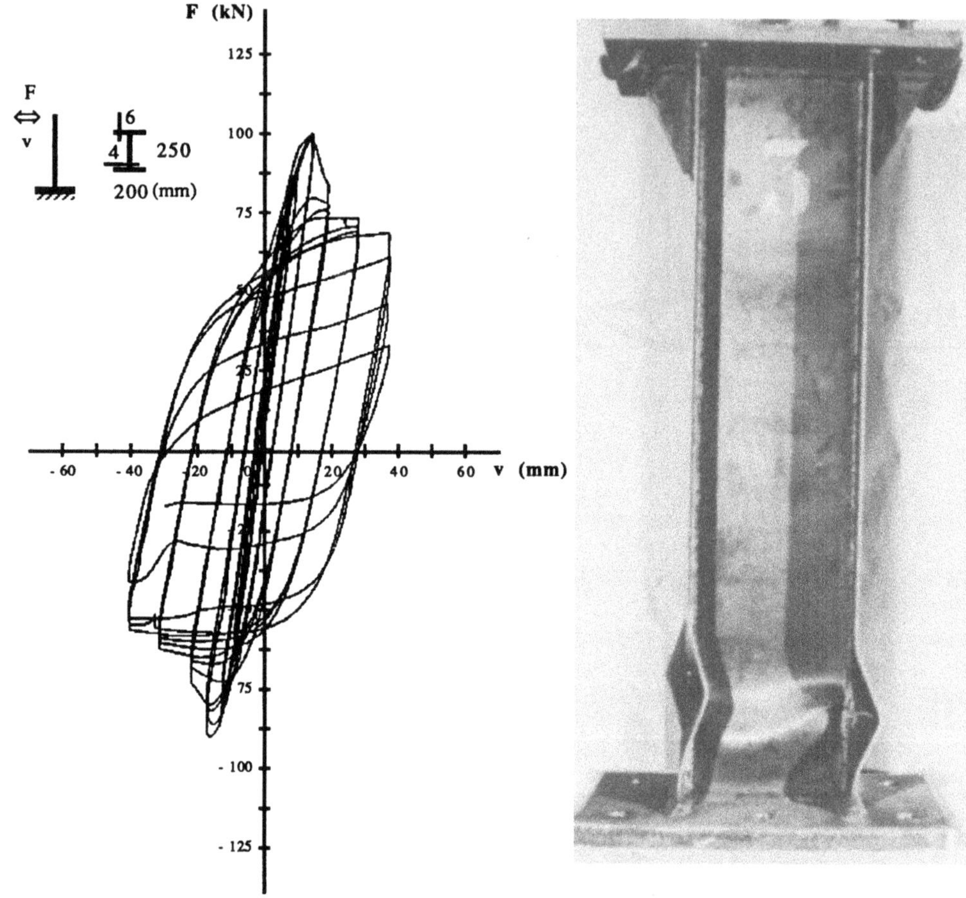

Figure 4 - Typical hysteresis loops and mode of failure of a beam-column under repeated loading.

The severe distortions of the flanges tend to induce torsional displacement of the section causing a lower load that would develop in pure flexural buckling.

This is likely to contribute to the somewhat poorer performance of H and I beam-columns as compared to box shaped cross sections. For that reason where supports in lateral direction are not provided properly, the use of a box section is better.

4 NUMERICAL MODELS FOR THE CYCLIC BEHAVIOUR OF BRACING ELEMENTS AND BEAM-COLUMNS

Several analytical models have been developed to represent the cyclic behaviour of steel subassemblages. These models can be divided into three different general types: the finite elements, the phenomenological and the physical theory models.

Finite elements provide the most realistic representation of the subassemblage behaviour, but usually demand too much computing time to be applied to practical analysis of steel elements.

Phenomenological models are based on simplified hysteretic rules which only mimic the observed force-displacement curves of the subassemblage.

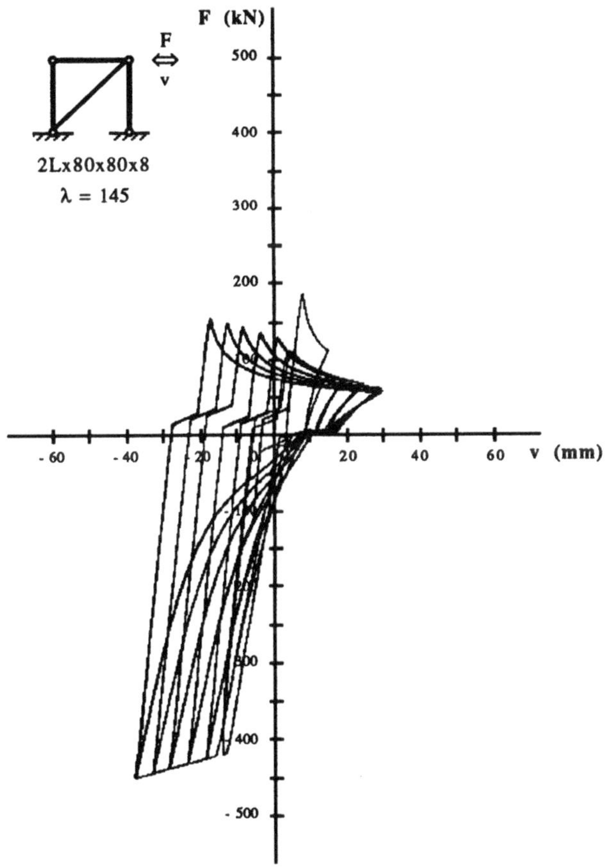

Figure 5 - Numerical hysteresis loops of a bracing element under repeated loading.

Physical theory models incorporate simplified formulations based on physical considerations that permit the cyclic inelastic behaviour to be computed. These models combine the realism of finite elements approaches with the computational simplicity of phenomenological modelling.

In this investigation the numerical models developed were of this last type. The complete formulation of the numerical model for bracing elements and beam-columns was presented in other works [2-4].

These models represent, with good accuracy, the cyclic behaviour of bracing elements and beam-columns until failure, has it can be seen in Figure 5 and 6.

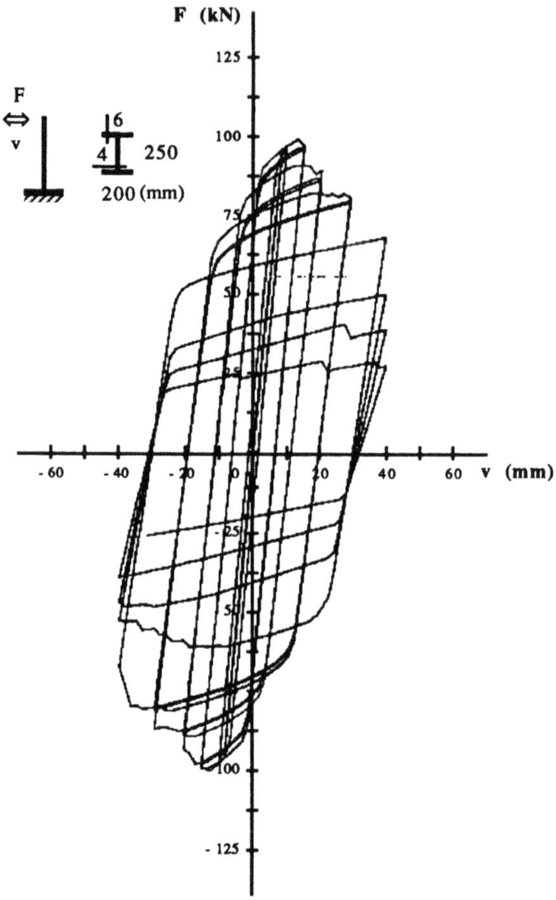

Figure 6 - Numerical hysteresis loops of a beam-column under repeated loading.

Even if the comparison was made in term of the interpretative parameters suggested by the ECCS, good accuracy was obtained.

5 DAMAGE ACCUMULATION

The studies performed on the cyclic behaviour of bracing elements and beam-columns have shown that failure was generally caused by the interaction of several physical phenomena, namely buckling, local buckling of flanges and webs and fracture. From the arising of these phenomena to the failure of the elements, a progressive deterioration of the strength and ductile capacity of the structural elements occurs.

Observation of past seismic events [5-7] has shown that structures can safely resist, without failing, even if large accumulation of damage occurs, and thus it can be assumed that the hysteretic behaviour of these elements does not necessarily lead to element or structural failure. Nevertheless, such observations lead to the conclusion that there is a threshold level of damage accumulation for which the failure is possible.

5.1 Criteria to Define Structural Failure

Due to the absence of any type of degradation for elements with elastic perfectly plastic behaviour, it can be assumed that the definition of structural failure can be referred to this standard behaviour through the use of several parameters. In the ECCS [1] recommendations, several parameters that can be used for that purpose are presented.

In the context of the cyclic behaviour of steel structural elements, it was assumed that the failure of an element can be defined based on the dissipated energy and that failure is reached when, in a given hysteretic cycle, the capacity to dissipate energy is less than a certain percentage of the capacity the same element would have if it had a elastic perfectly plastic behaviour. Failure can thus be defined by means of an inequality:

$$\eta_i = \frac{A_i}{A_i^{epp}} < \gamma$$

In this expression η_i represents the ratio between the energy dissipated in a given cycle i (A_i) and the energy that would be dissipated if the element would have a elastic perfectly plastic behaviour (A_i^{epp}), and γ is the ratio of energies for which failure occurs. It is suggested that a value of 50% be assumed for g. This assumes that failure occurs when the energy dissipated in a certain cycle (i) is less than 50% of the energy that would be dissipated if the element had a elastic perfectly plastic behaviour.

5.2 Numerical Model

Several models [8-10] have been proposed in the recent past to evaluate the damage accumulated in materials or structural elements when subjected to cyclic loading. They are all based on the the theory of metal fatigue and on the hypothesis of linear damage accumulation (Miner's rule) [8]. Based on these concepts, a model for the accumulation of structural damage [4] was developed. This model admits that the plastic deformations of the structural elements are concentrated on critic zones referred as "plastic cells" and that, after N cycles, the damage accumulated in each one of these zones is the sum of the damage due to the mean plastic deformations ($\Delta\zeta_{pi}$) caused by the several hysteretic cycles. This process of damage accumulation can be written as:

$$ID = a \sum_{i=1}^{N} (\Delta\zeta_{pi})^{c}$$

In this equation ID represents the accumulated damage after N cycles and is referred as index of accumulated damage. With respect to parameters a and c, it is assumed that a is equal to 1 and thus all the uncertainties are assumed to be represented by c. To quantify c, in order to obtain an ID value that is independent on the loading history, several numerical simulations on the cyclic behaviour of beam-columns and diagonal bracing systems were performed. It was observed that the value of c that caused a smaller variation in the ID value was 1.

5.3 Numerical Analysis

The imposed displacement histories used in this study are randomly generated and try to simulate displacement histories that are similar to the ones verified during the occurrence of seismic events. Simultaneously, the ECCS loading history was also employed.

The cases studied refer to bracing elements and beam-columns. Figure 7 shows the evolution of the accumulated damage as a function of the non-dimensional dissipated energy for each cycle, for all the loading histories up to the failure of one of the studied beam-column element.

An important conclusion that can be taken from the observation of Figure 7 is the fact that the results obtained are "independent" from the loading histories. It can be observed that, for all the loading histories, the evolution of the index of accumulated damage with the capacity to dissipate energy is remarkably similar.

At failure, the reduced value for the coefficient of variation shows that the numerical model for damage accumulation can be considered "independent" from the loading histories to which the element can be submitted, represented by different histories of imposed displacement.

This observation is very important, and allows the conclusion that the ECCS loading history can be utilized as a unified loading history to assess the cyclic behaviour of structural steel elements.

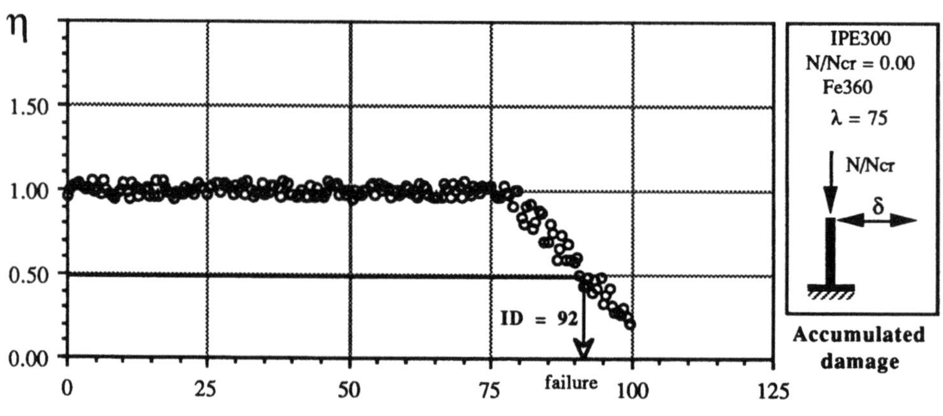

Figure 7 - Accumulated damage evolution for all the loading histories.

The studies developed had indicated that the damage index is a good indicator of the ductility of the element.

In the case of dynamic analysis [4] the value obtained for the damage index at the failure is quite similar of the damage index obtained in static tests, showing that the foresight of the dynamic failure of a structural element can be based on the damage index at failure obtained in static analysis.

6 CONCLUSIONS

The results obtained indicate that the ECCS loading history can be utilized as a unified loading history to predict the failure, the ductility and the damage index of a structural element.

The numerical model proposed for the estimation of the accumulated damage at failure through an average index of accumulated damage, besides making possible the foresight of the dynamic failure of a structural element based on the average index of accumulated damage at failure, is also a good indicator of the ductility of the element. As a matter of fact, the larger the index that characterizes the failure, the larger will be the accumulated inelastic deformations and thus the larger is the capacity of the element to dissipate energy by means of a ductile hysteretic process.

7 ACKNOWLEDGMENTS

The financial support from JNICT (Junta Nacional de Investigação Científica e Tecnológica) is gratefully acknowledged.

8 REFERENCES

[1] ECCS, Study on Design of Steel Building in Earthquake Zones, *ECCS - European Convention for Constructional Steelwork - Technical Committee 1 - Structural Safety and Loadings, Technical Working Group 1.3 - Seismic Design*, Nº 47, 1986.

[2] Ballio, G. and F. Perotti, Cyclic Behaviour of Axially Loaded Members; Numerical Simulation and Experimental Verification, *Journal of Constructional Steel Research*, Vol. 7, Nº 1, 1987.

[3] Ballio, G. and L. Calado, Sezioni Inflesse in Acciaio Sottoposte a Carichi Ciclici - Sperimentazione e Simulazione Numerica, *Costruzioni Metalliche*, Nº 1, 1986.

[4] Calado, L., Caracterização do Comportamento de Estruturas Metálicas sob Acções Sísmicas, *Phd Dissertation, Technical University of Lisbon*, Lisbon, 1989.

[5] Gere, J. and H. Shah, *Terra non Firma - Understanding and Preparing for Earthquakes*, Stanford Alumni Association, Stanford, 1984.

[6] EERI, Reducing Earthquake Hazards: Lessons Learned from Earthquakes, *EERI - Earthquake Engineering Research Institute*, Publication Nº 86-02, November, 1986.

[7] Bertero, V., B. Bresler, L. Selna, A. Chopra and A. Koretsky, Design Implications of Damages Observed in the Olive View Medical Center Buildings, *Proceedings of the Fifth World Conference on Earthquake Engineering*, Roma, 1973.

[8] Miner, M., Cumulative Damage in Fatigue, *Transactions of the American Society of Mechanical Engineers*, Vol. 67, 1945.

[9] Yao, J., F. Kozin, Y. Wen, J. Yang, G. Schueller and O. Ditievsen, Stochastic Fatigue, Fracture and Damage Analysis, *Structural Safety*, Nº 3, 1986.

[10] Park, Y., H. Ang and Y. Wen, Damage-Limiting Aseismic Design of Buildings, *Earthquake Spectra*, Vol. 3, Nº 1, 1987.

[11] Krawinkler, H., M. Zohrei, B. Irvani, N. Cofie and H. Tamjed, Recommendations for Experimental Studies on the Seismic Behaviour of Steel Components and Materials, *The John A. Blume Earthquake Engineering Center*, Report Nº 61, 1983.

11 LIFE PREDICTIONS IN FATIGUE TAKING INTO ACCOUNT MATERIAL MEMORY PROPERTIES

G. CALIGIANA
Department of Construction, Mechanical, Nuclear, Aeronautical and Metallurgical Engineering, University of Bologna, Italy

SUMMARY

Torsional fatigue life prediction algorithms are considered with particular reference to memory effect requirements. Local strain approach is employed. Stress concentration is accounted for by utilizing Neuber's rule or torque-local strain curve discretization. Mean stress influence is examined. Experimental results relative to particular sample geometries subject to typical torque-time histories of turbomachineries are compared with numerical predictions.

Symbols

M applied torque
T_a nominal shear stress amplitude
Γ_a nominal shear strain amplitude
τ_a local shear stress amplitude
τ_m mean shear stress
γ_a local shear strain amplitude
K_t theoretical stress concentration factor
N number of cycles (or blocks, if specified) to fracture
t time

INTRODUCTION

The local strain approach has been recognized as the more meaningful method to face low cycle fatigue problems in components characterized by local high stresses in the plastic range. Crucial points are: 1) the prediction of the local strain history beginning from the nominal load history and 2) the evaluation of the resulting cumulative damage.

A solution to the first problem can be the utilization of simplified methods or the use of finite element analysis [1, 2, 3].

In the second step it is very important to choose a well suited counting

Testing of Metals for Structures. Edited by F.M. Mazzolani.
Published by Taylor & Francis, 2 Park Square, Milton Park, Abingdon, Oxon, OX14 4RN.
ISBN 0 419 15810 3

method to reduce complex loading histories to a superposition of several constant amplitude cycles producing equivalent fatigue damage. Many efforts have been made in the past to choose among different counting methods at disposal [4]. Nowadays almost all the researchers agree on the use of the rainflow method.

Aim of the work is the identification and the resolutions of the problems related to the proper use of counting methods in fatigue life predictions. Attention of the author is paid to torsion fatigue life predictions of turbomachinery shafts. Test results on torsion specimens are presented to support the theoretical argumentations and to derive the conclusions.

1 FATIGUE LIFE PREDICTIONS

Two different approaches have been considered to low cycle fatigue life predictions in torsion.

A) The cyclic torque versus local shear strain curve is derived utilizing finite element analysis of the component [2, 3, 5]. For many engineering metals the "stable" hysteresis curve can be obtained by expanding the cyclic curve with a scale factor of two. A numerical discretization of this curve can be treated according to the method of the availabilities matrix introduced by Wetzel [6, 7] to derive the torque-local strain hysteresis loops starting from the torque-time history (fig. 1). A similar discretization of the shear stress-shear strain curve can be used to derive shear stress-shear strain hysteresis loops. The logical path followed by the method is evidenced as path 1 in fig. 2.

B) In the second approach the use of Neuber's rule [5, 8] permits to derive directly the shear stress-shear strain curve by starting from the nominal loading hystory (path 2 in fig. 2).

In both cases the application of Palmgren-Miner [9] rule to the closed cycles of the shear stress-shear strain curves (identified by the counting method) permits to obtain the total damage produced by the history.

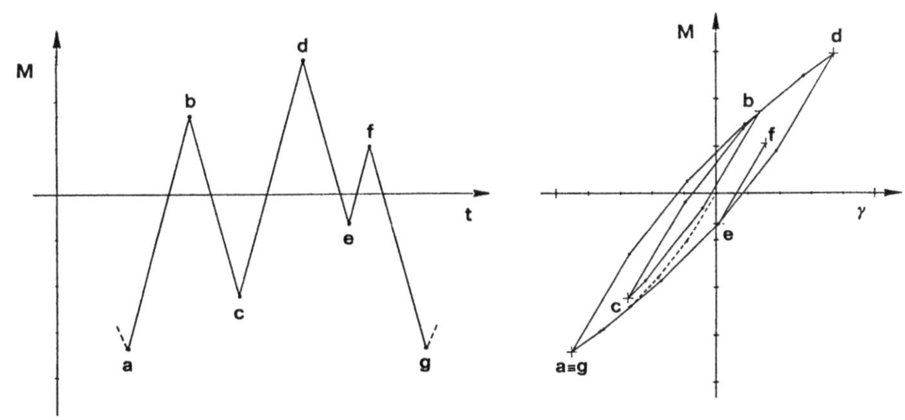

Fig. 1 Discretization of the torque-local strain hysteresis curve to predict torque-local strain hysteresis loops according to the "availability matrix" method |1|.

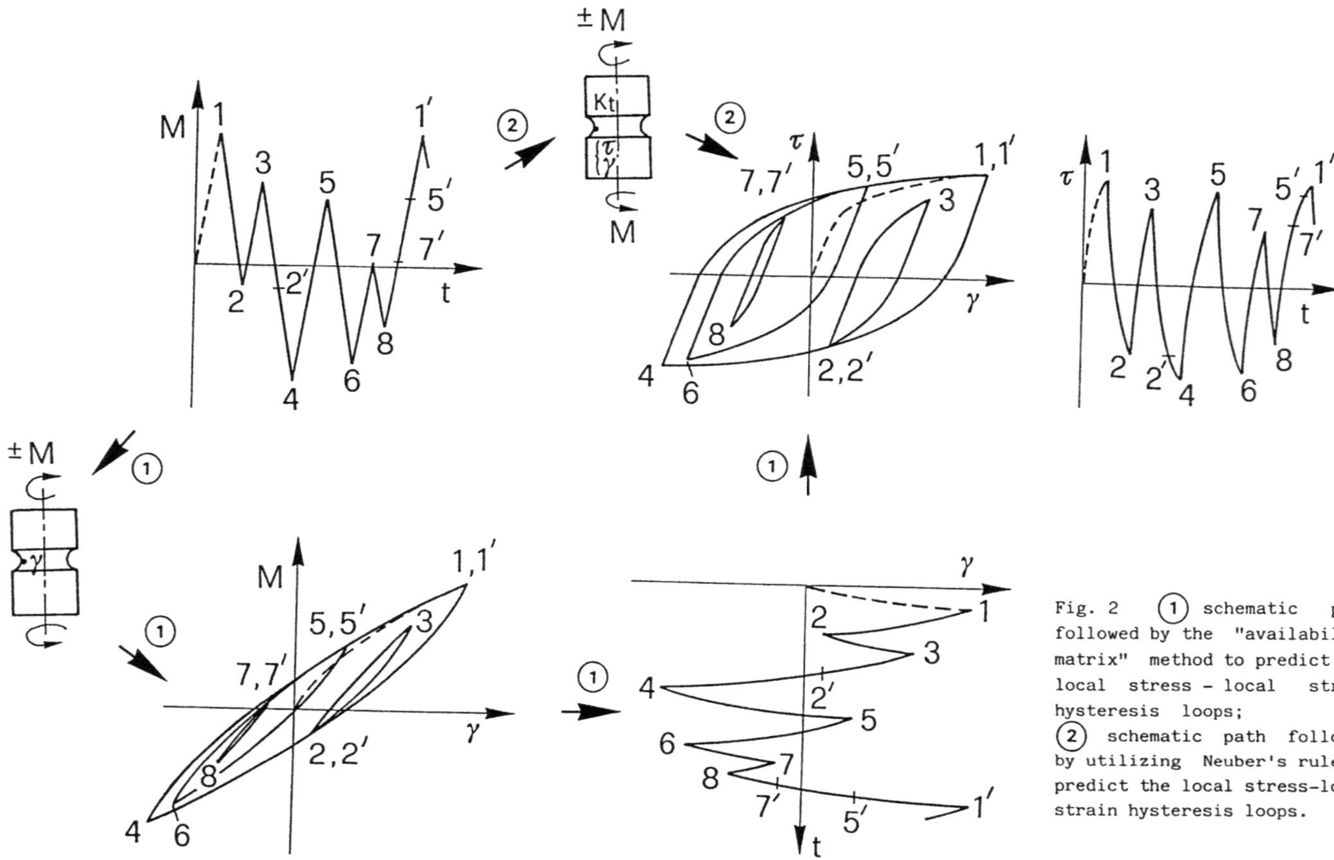

Fig. 2 (1) schematic path followed by the "availability matrix" method to predict the local stress - local strain hysteresis loops; (2) schematic path followed by utilizing Neuber's rule to predict the local stress-local strain hysteresis loops.

2 MEMORY EFFECT

Memory effect is an important aspect which must be considered in the numerical prediction of the shear stress-shear strain hysteresis loops. It derives from experimental evidence. In a periodical complex history, after the cyclic "stabilization" of the material, larger hysteresis loops always enclose smaller ones. If a small strain excursion (such as excursion 2-3 in fig. 2) occurs inside a larger strain excursion (1-4 in fig. 2), the larger hysteresis loop is not affected by the smaller one [1] and the hysteresis hystory after the closure of the smaller loop behaves as if the smaller loop had never happened. The material "remembers" the previous loading path.

A) The availability matrix method can cause the use of only fractions of elements or the skipping of a proper number of elements of the discretized curve just to rejoin the stress-strain path of the larger loop (fig. 1) and to meet the memory effect [6]. The analysis of complex histories by this method has, however, the disadvantage of long computation times and especially the need of very accurate and, sometimes, onerous finite element analysis to obtain load-local strain curve (really, this can be avoided by utilizing an approximation of the stress-local strain curve according to Neuber's rule [1, 2]).

B) On the contrary, the direct use of Neuber's rule together with an analitical modelling of the stress-strain curve is very easy and attractive. In this case the fulfilment of memory effect is not automatically guaranteed. In the followings particular attention is devoted to the right utilization of rain-flow counting method just to account for the memory effect. Reference will be made to torsional low cycle fatigue.

3 RAIN-FLOW COUNTING METHOD AND NEUBER'S RULE TO PREDICT LOCAL STRESS-STRAIN LOOPS

If the surface nominal shear stress reaches the plastic range, the usual problem of the derivation of the nominal stress history from the loading history is further complicated by the presence of the stress gradient in torsion. In any case nominal stress-time history can be predicted from torque-time history by utilizing Nadai theory [2, 5, 10, 11, 12].

The nominal stress history is rearranged to start from the highest peak or the deepest valley of the history; as a consequence, a simplified rain-flow counting algorithm [7] can be utilized (fig. 3).

While the nominal stress history is analysed by rain-flow, one peak or valley each time, a portion of the program (point 3 in fig. 3) calculates the nominal strain and the corresponding local stress and local strain according to the extension to torsional fatigue of the Neuber's rule [5, 13, 14].

The first peak or valley of the nominal stress history is read and the corresponding nominal strain is evaluated by utilizing the extension to torsion of Ramberg-Osgood model for the material curve:

$$\Gamma_a = (T_a/G) + (T_a/H')^{(1/n')} \quad (3.1)$$

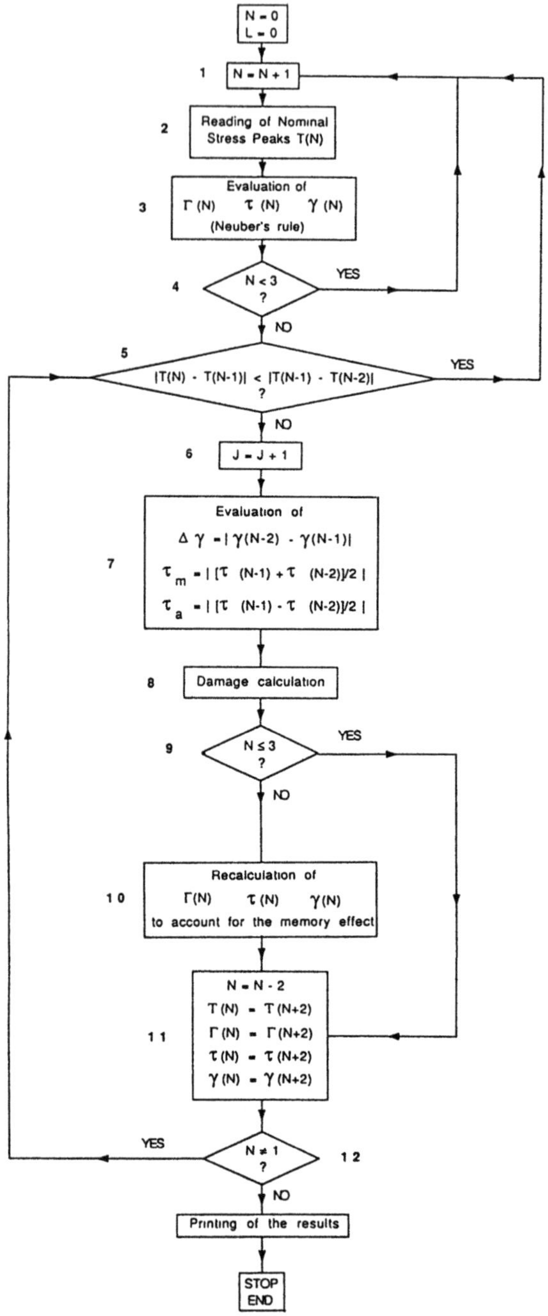

Fig. 3 Rain-flow algorithm (Neuber's rule is utilized in points 3 and 10 to obtain local stress and local strain peaks and valleys).

Local stress and strain values are obtained by solving the system of the stress-strain curve equation and the Neuber's hyperbola:

$$\gamma_a = (\tau_a/G) + (\tau_a/H')^{(1/n')}$$
$$\tau_a \gamma_a = K_t^2 T_a \Gamma_a \tag{3.2}$$

Successive nominal strains for the following peaks or valleys can be obtained directly by utilizing the equation of the hysteresis curve (paragraph 1):

$$\Delta\Gamma/2 = (\Delta T/2G) + 2\,[\Delta T/(2H')]^{(1/n')} \tag{3.3}$$

where $\Delta T = |T_i - T_{i-1}|$ and $\Delta\Gamma = |\Gamma_i - \Gamma_{i-1}|$.

Local stresses and strains can be obtained by the simultaneous solution of the equation of the hysteresis curve and of the Neuber's hyperbola:

$$\Delta\gamma/2 = (\Delta\tau/2G) + 2\,[\Delta\tau/(2H')]^{(1/n')}$$
$$\Delta\tau\ \Delta\gamma = K_t^2\ \Delta T\ \Delta\Gamma \tag{3.4}$$

where: $\Delta\tau = |\tau_i - \tau_{i-1}|$ and $\Delta\gamma = |\gamma_i - \gamma_{i-1}|$;
($\gamma_i = \gamma_{i-1} \pm \Delta\gamma$, with plus sign if $T_i = T_{i-1} + \Delta T$, minus sign if $T_i = T_{i-1} - \Delta T$).

The procedure just specified is equivalent to finding geometrically the intersection (i-th point) between Neuber's hyperbola and hysteresis curve (stress-strain curve for the first peak or valley; fig. 4b and 4c) and placing the origin of the coordinate axes, each time, at the point determined at the previous step (characterized by an index i-1).

Rain-flow identifies a closed cycle every time the nominal loading history returns to a nominal stress value it had previously reached before changing the direction of the loading path (e.g. point 2', loading path 1-2-3-2'-4; fig. 4).

To satisfy the memory requirements in this latter case, it is necessary, after the closure of the cycle, to "bring back" the index of the current point by three (instead of one) to find the origin (i-3) of the coordinate axes (fig. 4d and 4e). If there are more cycles one inside the other it is necessary to "bring back" the index of the current point by (2L+1), where L is the number of nested closed loops identified by rain-flow. The algorithm to perform the rigth choice for the origin of the axes is schematized in fig. 5.

During damage calculation mean stress can be accounted for by utilizing the extension to torsion of the model proposed by Smith et al. [15]:

$$(\tau_a + |\tau_m|)\,\gamma_a = A^2\,G\,(2N)^{2\alpha} + A\,B\,(2N)^{(\alpha+\beta)}$$

where A, B, α, β, are the constants of the torsion fatigue curve (Manson-Coffin-Basquin model),

$$\gamma_a = A\,(2N)^{\alpha} + B\,(2N)^{\beta}$$

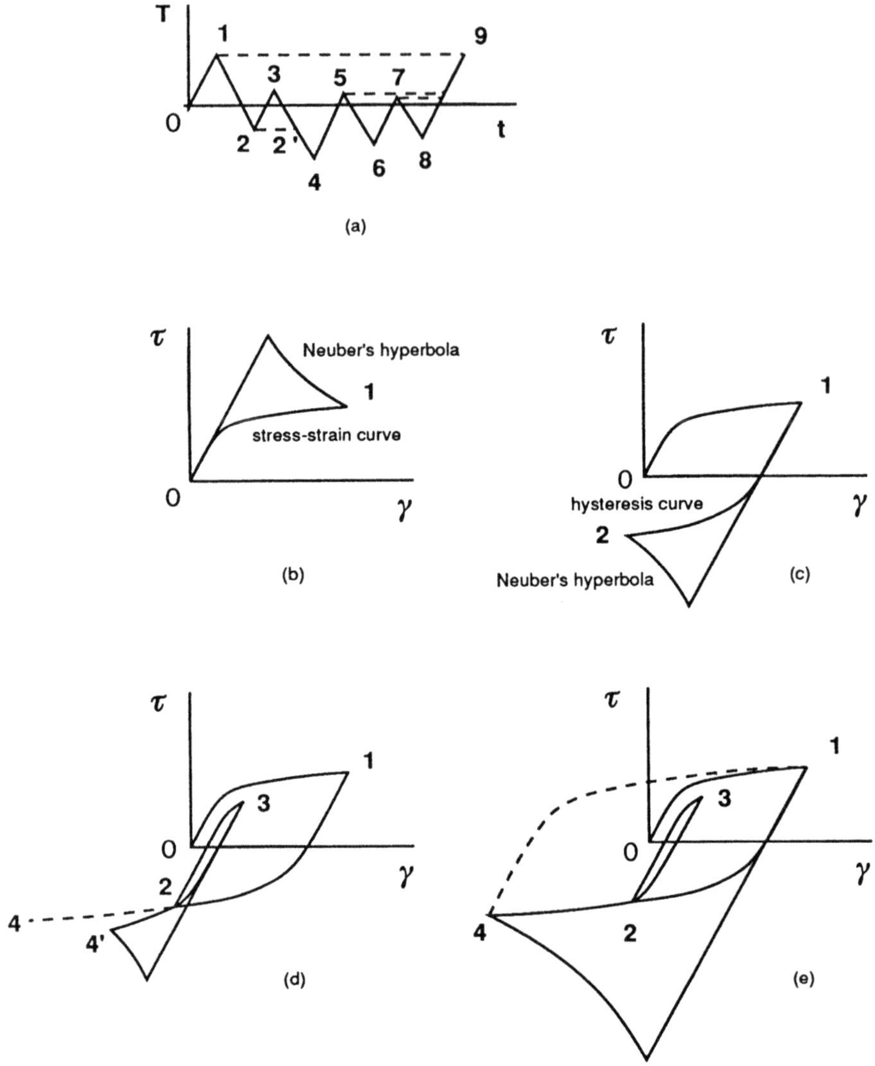

Fig. 4 a) torque-time history; b) identification of the first point by utilizing Neuber's hyperbola and cyclic stress-strain curve; c) identification of the successive peaks and valleys by utilizing Neuber's hyperbola and hysteresis curve; d) wrong choice of the origin for the coordinate axes (point 3) and wrong identification of point 4' (without considering memory properties); e) rigth choice of the origin for the coordinate axes (point 1) and rigth identification of point 4 (by including memory effect).
b), c), d), e) local stress-local strain hysteresis loops.

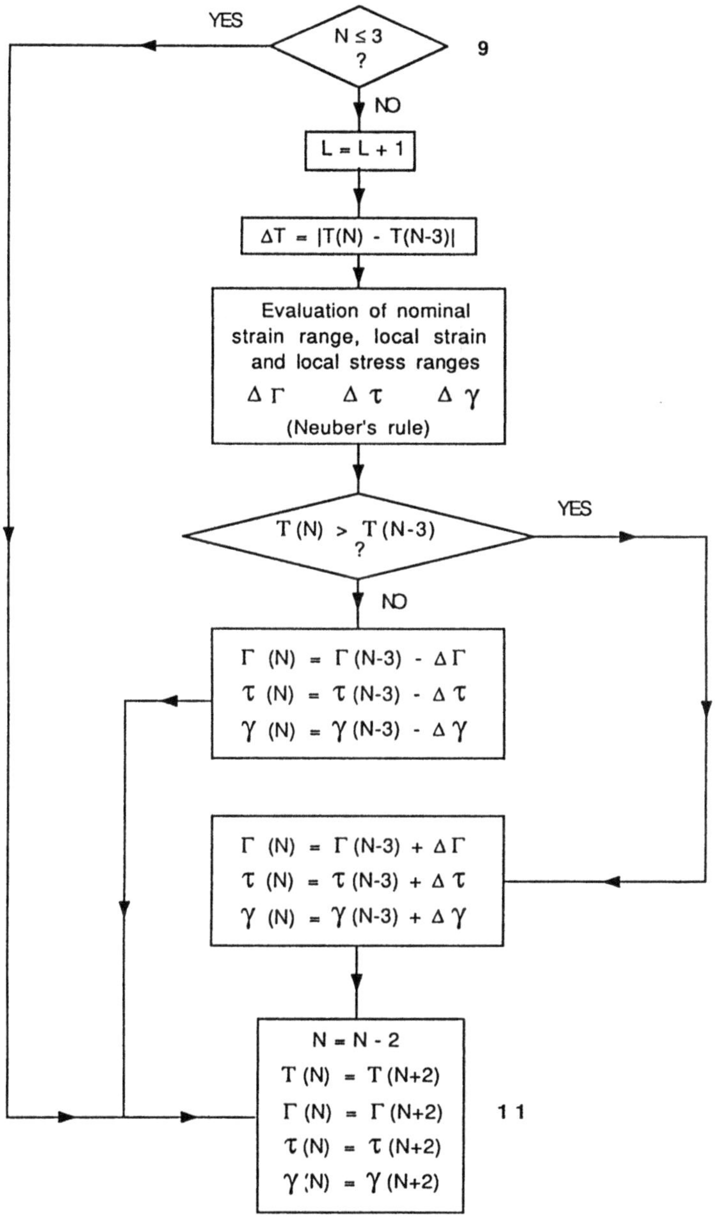

Fig. 5 Detailed description of point 10 of fig. 3. Algorithm to account for memory properties in the prediction of local stress-local strain history.

4 EXPERIMENTAL RESULTS

Many low cycle fatigue tests in torsion have been performed to verify the correlation between experimental results and numerical predictions on specimens simulating stress concentrations of real components [16].

In particular two typical torque-time histories have been applied to samples of rotor steel whose mechanical characteristics and chemical composition are reported in [16] or [17]. These histories are representative of field situations in turbomachineries: a) 120 degree out-of-phase synchronization (identified as "SYNC" in fig. 6) and b) short circuits (identified as "SC" in fig. 6). This latter normalized history is multiplied by two different scale factors to obtain two histories: SC1 and SC2. The specimens utilized, of 24 millimeter diameter, have a circumferential groove and are characterized by a theoretical stress concentration factor of 1.47 (fig. 7).

Comparison between numerical predictions and experimental results are reported in fig. 6c (on terms of number of repetitions of the whole history to fracture). The following cases are considered in the numerical predictions:

A) Neuber's rule and rain-flow counting method with and without considering memory effect (m.e. = memory effect; w.m.e. = without memory effect);

B) Neuber's rule and rain-flow counting method including memory properties and mean stress effect (m.s. = mean stress);

C) "availability matrix" method with a torque-local strain curve, as input, derived by finite element analysis [2, 3, 5].

5 CONCLUSIONS

In the fatigue life evaluation is very important to account for memory effect. This is expecially true, in variable amplitude histories, if the strain ranges of consecutive closed cycles are very different in amplitude. The failure to include memory properties leads to consider minor strain ranges during the prediction of stress-strain loops according to Neuber's rule and rain-flow counting method (fig. 8c). As a consequence, not conservative damage predictions results (solid symbols in fig. 6c).

The inclusion of memory effect reduces the nonconservatism (large open squares and triangles in fig. 6c).

The consideration of mean stress together with memory effect improves the life predictions in the case of short circuit histories and seems to have no effect in the case of out of phase synchronization (crosses and large open squares in fig. 6c). This is in agreement with the observation that, in the synchronization history, many very small amplitude cycles have a not to big mean stress, while the cycles of larger amplitude have about zero mean stress (fig. 6a); in the short circuit history, many cycles, having simultaneously significant amplitude and appreciable mean stress, are present (fig. 6b).

These latter results are always conservative and very similar to those obtained by utilizing the method of "availability matrix" (small symbols in fig. 6c).

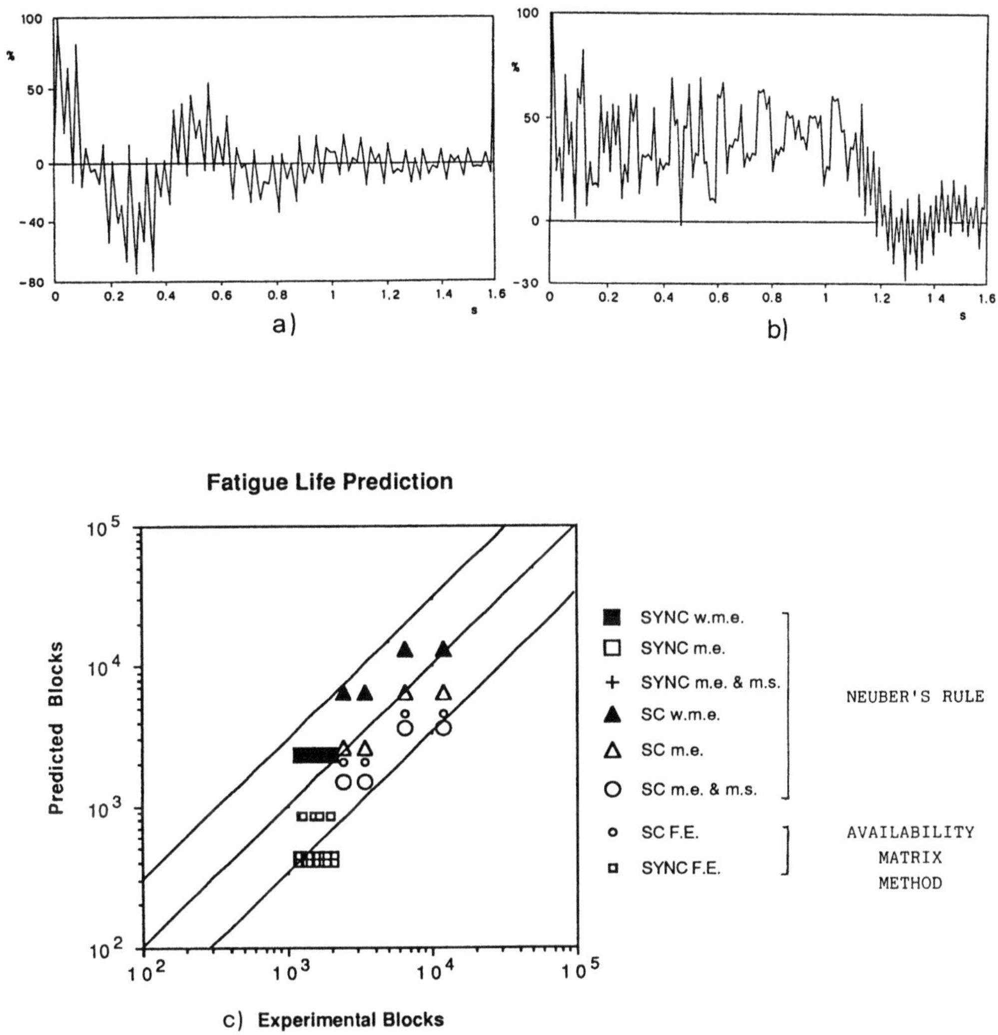

Fig. 6 a) Typical normalized torque-time history resulting from a 120° out of phase synchronization (SYNC); b) typical normalized torque-time history resulting from short circuits (SC); c) comparison between experimental results and life predictions on terms of number of repetitions of the whole history to fracture (m.e. = including memory effect; w.m.e. = without including memory effect; m.s. = including mean stress effect; F.E. = finite element prediction of torque-local strain curve).

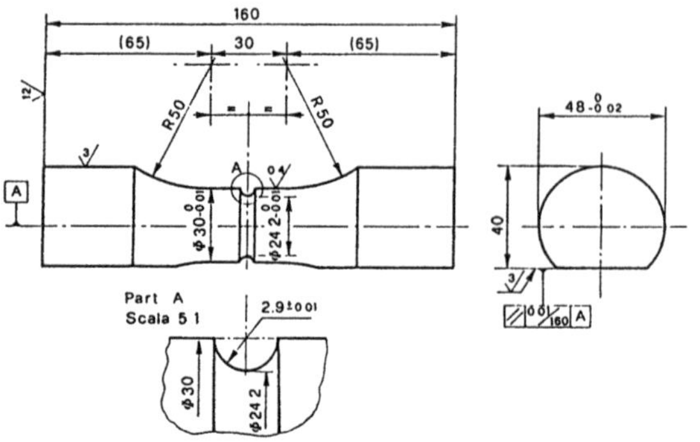

Fig. 7 Specimen utilized in the low cycle fatigue tests in torsion.

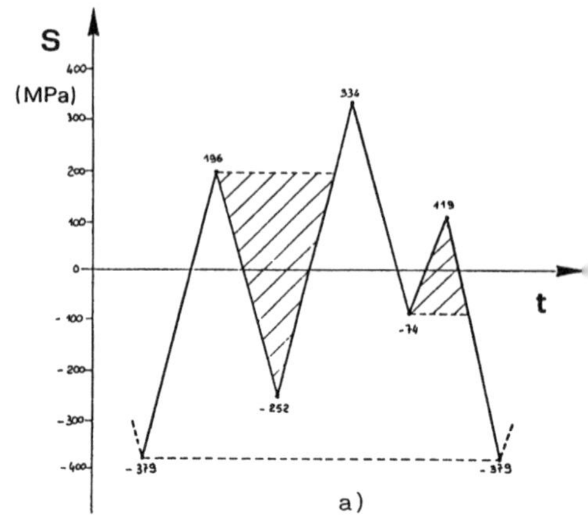

Fig. 8 a) Nominal stress history (push-pull).

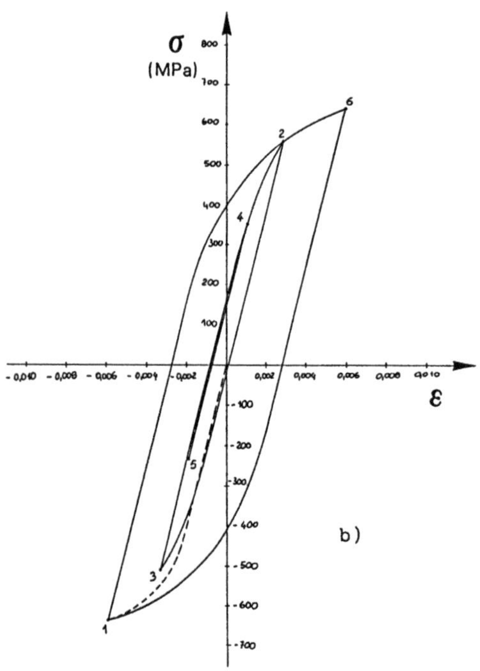

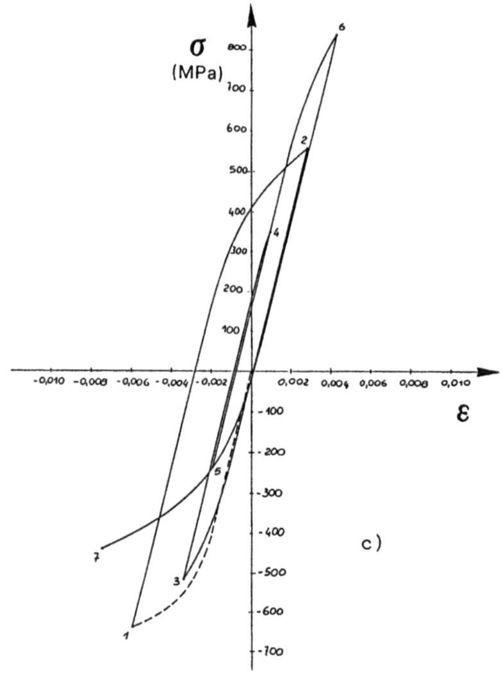

Fig. 8 b) and c) Comparison between predictions of local stress-local strain hysteresis loops, respectively, by considering and without considering memory properties of the materials (push-pull).

REFERENCES

[1] Downing, N.E., Brose, W.R., Wilson, W.K., "Notched Member Fatigue Life Predictions by the Local Strain Approach", 1977, Fatigue under Complex Loading: Analysis and Experiments, Advances in Engineering", Vol. 6, R.M. Wetzel Ed., SAE, pp. 55-84.

[2] Caligiana, G., Cesari, F., "Vita a fatica di un provino intagliato soggetto a torsione: confronto tra dati sperimentali e teorici", 1988, Atti del XVI Convegno AIAS, pp. 695-708.

[3] Caligiana, G., Cesari, F., Quieto, A., " Studio di strutture assialsimmetriche sottoposte a torsione", 1989, Atti del XVII Convegno AIAS, pp. 83- 92.

[4] Dowling, N.E., "Fatigue Failure Predictions for Complicated Stress Strain Histories", 1972, Journal of Materials, Vol. 7, No. 1, pp. 71-87.

[5] Caligiana, G., Cesari, F., "Fatigue life prediction of a notched specimen subject to torsion: theoretical and experimental results", MPA Stuttgart, 1989, Acta of the Third International Conference on Biaxial/Multiaxial Fatigue, Vol. 1, pp. 6.1-6.17.

[6] Brose, W.R., "Fatigue Life Predictions for a Notched Plate with Analysis of Mean Stress and Overstrain Effects", Fatigue under Complex Loading: Analysis and Experiments, Advances in Engineering", 1977, Vol. 6, R.M. Wetzel Ed., SAE, pp.117-135.

[7] Downing, S. D., Socie, D.F., "Simplified Rainflow Counting Algorithms", 1982, International Journal of Fatigue, Vol. 4, No. 1, pp. 31-40.

[8] Neuber, H., "Theory of stress concentration for shear-strained prismatical bodies with arbitrary nonlinear stress-strain law", 1961, Trans. ASME, Journal of Applied Mech., Vol. 28, pp. 544-550.

[9] Miner, M.A., "Cumulative damage in fatigue", 1945, Journal of Applied Mechanics, ASME, Vol. 12.

[10] Nadai, A., "Theory of Flow and Fracture of Solids", London, McGraw Hill Publ., 1950.

[11] Brown, M.W., "Torsional Stresses in Tubular Specimens", 1978, Journal of Strain Analysis, Vol. 13, No. 1, pp. 23-28.

[12] EPRI, "Determination of Torsional Fatigue Life of Large Turbine Generator Shafts", Palo Alto, U.S.A., 1984, EL-3063, Research Project 1531-1, Electric Power Research Institute.

[13] Topper, T.H., Wetzel, R.M., Morrow, J.D., "Neuber's rule applied to fatigue of notched specimens", 1969, Journal of Materials, ASTM, Vol. 4, pp. 200-209.

[14] Tanaka, K., Matsuoka, S., Kouzu, F., Nagata, K., "Fatigue Strength of a Rotor steel subjected to Torsional Loading simulating that occurring due to Circuit Breaker Reclosing in an Electrical Power Plant", 1983, Fatigue of Engineering Materials and Structures, Vol.6, No 2, pp. 103-120.

[15] Smith, K.N., Watson, P., Topper, T.H., "A Stress Strain Function for the Fatigue of Metals", 1970, "Journal of Materials, ASTM, Vol. 5, No. 4, pp. 767-778.

[16] Freddi, A., Curioni, S., Caligiana, G., " La vita a fatica torsionale negli alberi dei turbo gruppi", Roma, 1989, ENEA Ed., "La casa della Stampa" Publ., .

[17] Curioni, S., "Life Prediction analysis based on a local strain approach for shafts under torsional loading", 1987, International Journal of Materials & Product Technology, Dorgham, M.A. Ed., Inderscience Enterprises, Vol. 2, No. 1, pp.75-87.

12 LOW CYCLE FATIGUE: CHARACTERIZATION OF THE PLASTIC CYCLES DUE TO EARTHQUAKE GROUND MOTION

E. COSENZA
Istituto di Ingegneria Civile, University of Salerno, Italy
G. MANFREDI
Istituto di Tecnica delle Costruzioni, Faculty of Engineering, University of Naples, Italy

SUMMARY

The effect of destructive earthquake ground motion represents a very important case of low cycle fatigue; thus, the analysis of plastic cycles due to earthquake is the basis to characterize the inelastic behaviour of the structures in seismic areas. In this paper, with reference to some significative ground motions and to the elastic-perfectly plastic model, the distribution of the plastic cycles is analyzed in terms of number of plastic cycles, maximum plastic deformation, mean value and standard deviation of the plastic amount of the other cycles. Different probability models (Normal, Lognormal, Exponential, Beta and Gamma) are used to interpolate the real distribution of plastic cycles, and, on the basis of the statistical test of Kolmogorov and Smirnov, are compared each to the other.

INTRODUCTION

The design procedures suggested by codes in seismic areas allow to have damage to non-structural elements for medium intensity ground motion, and significative plastic damage in structures, in case of destructive ground motion. As a consequence, the strong motion represents an important case of low fatigue plastic cycles, characterized by variable number of cycles and amplitude.

The seismic verify of structures can be performed defining an adequate damage functional, and evaluating the value that it assumes for the given ground motion; values less then 1 mean that the structure doesn't collapse during the earthquake. The choice of the damage functional is a very important matter, and strictly depends on the distribution of plastic cycles, and on the behaviour of the material for cyclic actions.

The most widely used damage parameter is the "cinematic ductility" **[1]**, which represents the ratio between the maximum displacement of the structure and the yielding displacement; this parameter is significative if the distribution is characterized by only one cycle with large amount of plastic deformation. An other possibility is given by the use of the cumulative dissipated energy **[2,3,4]**: this case is important if the distribution is characterized by a large number of significative plastic deformations, and if the material (and the structural tipology), exhibits a small attitude to dissipate plastic energy.

Generally, damage functionals consider combination of the energy and of maximum plastic deformation **[5,6]**, or the extension of the linear cumulation law of fatigue that seems to offer good results **[7,8,9]** in the case of steel structures. Thus, in the general case, the choice of adequate damage functionals requires theoretical analysis (large number of numerical simulation of the behaviour of structures under cyclic action), and experimental evaluation (analysis of the experimental behaviour of the material, of the structural element and of the sub-assemblages). As consequence, the distribution of plastic cycles assume a significative importance, for reasons that we can summarize in the following way:

Testing of Metals for Structures. Edited by F.M. Mazzolani.
Published by Taylor & Francis, 2 Park Square, Milton Park, Abingdon, Oxon, OX14 4RN.
ISBN 0 419 15810 3

1. Theoretical: The exact modeling of plastic cycles allows to generate directly realistic "static" time history on structural elements, without to generate artificial ground motion and to analyze the dynamic inelastic behaviour of the structure; simplified numerical analysis are, hence, possible.

2. Experimental: The exact characterization of plastic cycles due to real ground motion allows to perform cyclic static tests which represents adequately the effective behaviour of structures under seismic actions, without to use expansive dynamic tests.

Hence, the full characterization of the distribution of the plastic cycles is a very important theoretical and experimental matter, because it allows to consider equivalent static tests, numerical and physical, which represents with sufficient approximation the real behaviour of structures under dynamic severe actions. In the case of steel structures, statistical analysis of the evaluation of damage of steel structures subjected to severe ground motion are reported in **[10,11,12,13,14]**; in particular, in **[13]**, the plastic deformation is assumed to be distributed with a Lognormal probability model. In the following, with reference to the elastic-perfectly plastic single degree of freedom model (EPP-SDOF), the optimum choice of the probability model will be performed on statistical base, by means of the Kolmogorov and Smirnov test.

EARTHQUAKE	DATE	STATION	COMP.	MAX. ACC. [g]	DURATION [s.]
Ancona	14.06.72	Rocca	NS	**0.603**	18.98
Friuli	06.05.76	Tolmezzo	NS	**0.356**	36.40
Friuli	11.05.76	Forgaria	WS	**0.303**	17.96
Friuli	11.09.76	Buia	NS	**0.229**	20.23
Campano-Lucano	23.11.80	Bagnoli Irpino	WE	**0.184**	79.15
Campano-Lucano	23.11.80	Brienza	NS	**0.220**	78.72
Campano-Lucano	23.11.80	Calitri	WE	**0.166**	86.08
Campano-Lucano	23.11.80	Sturno	WE	**0.327**	70.76
Umbria	29.01.81	Nocera Umbra	NS	**0.212**	28.26
Lazio-Abruzzo	11.05.84	Villetta Barrea	WE	**0.218**	59.80
Ixtapa, Mexico	19.09.85	Mexico City	EW	**0.171**	180.0

Tab. 1. Recorded ground motions considered in the analysis.

THE ELASTO-PLASTIC BEHAVIOUR OF THE SDOF MODEL

The parameters which characterize a EPP-SDOF are the period T, the relative damping ν and the yielding level η. In this context, the yielding level η is defined in the following way:

$$\eta(T,\nu) = \frac{F_y}{mS_a(T,\nu)} \qquad (1)$$

where F_y denotes the yielding force, m the mass and S_a the elastic acceleration spectra. For a given acceleration function $a(t)$, period T, yielding level η and damping ν (in the following the value ν=5% will be assumed), the analysis of the problem can be performed integrating numerically the equations of the motion and it provides the maximum plastic deformation Δx_{max}, the distribution of the other plastic deformations $\Delta x_i < \Delta x_{max}$, and the total number n of these plastic deformations.

The analysis has been developed considering 10 of the most significative Italian recent earthquake, further the registration of Mexico City. The characteristics of these ground motion are reported in Tab.1. It seems interesting to underline that the selected ground motion are significative of very different seismic events; i.e., in Figs. 1, 2, 3 and 4 are drawn the acceleration spectra, for different values of the required maximum cyclic ductility $\mu_c=(x_y+\Delta x_{max})/x_y$ (x_y= yielding displacement), for the recorded ground motions of Rocca, Tolmezzo, Calitri and Mexico City; in the first case, spectra with amplification only for low periods are obtained; in the second case the amplification is relative to the medium range period, and the fourth case is relative to amplifications for high periods; in the case of Calitri, a large range of periods are amplified.

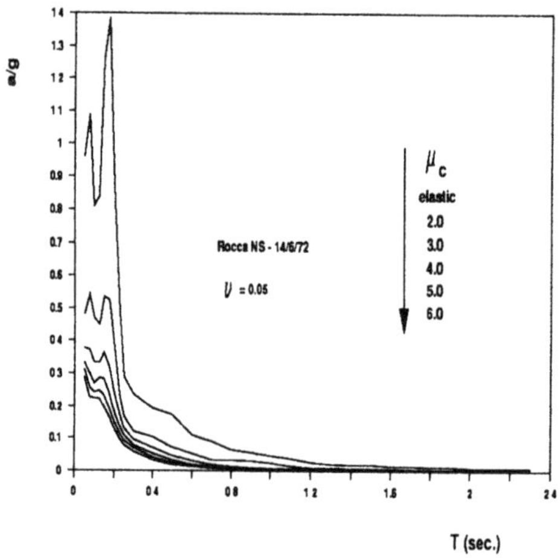

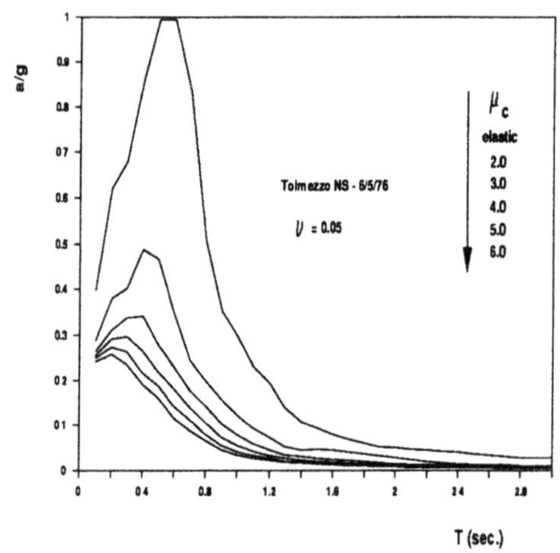

Figs. 1, 2. Inelastic acceleration spectra, with imposed ductility. Rocca and Tolmezzo.

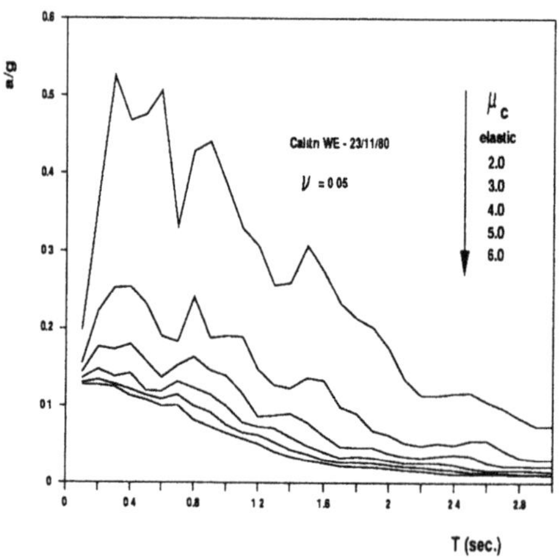

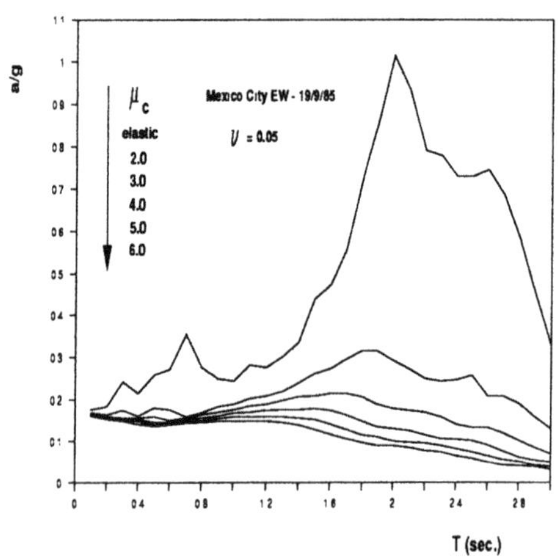

Figs. 3, 4. Inelastic acceleration spectra, with imposed ductility. Calitri and Mexico City.

Figs. 1, 2, 3 and 4 allow to obtain the yielding force which is necessary to design a SDOF that has an allowable ductility μ; in fact, for the given values of T and μ, the curve provides the maximum acceleration $a(T,\mu)$ that permits to design the SDOF for a yielding level $\eta=a/S_a$. Figs 5, 6, 7 and 8 show, for the same ground motions, the number of plastic cycles n (vertical axis) versus the period T of the SDOF (horizontal axis), for values of η=0.1, 0.2, 0.3, 0.4, 0.5, 0.6. The curves show that n generally decrease with T, and, as expected, high value of η are characterized by low value of n. It is interesting to observe that the number of plastic cycles is uncorrelated with the peak acceleration of the record; in fact, the higher values of n is relative to Calitri, which has the lower acceleration peak (0.166 g, cfr. Tab. 1). However, as shown in **[13]**, the value of n is linearly correlated to the ground motion duration; in fact, the Calitri record is the characterized by a duration of about 80 s.

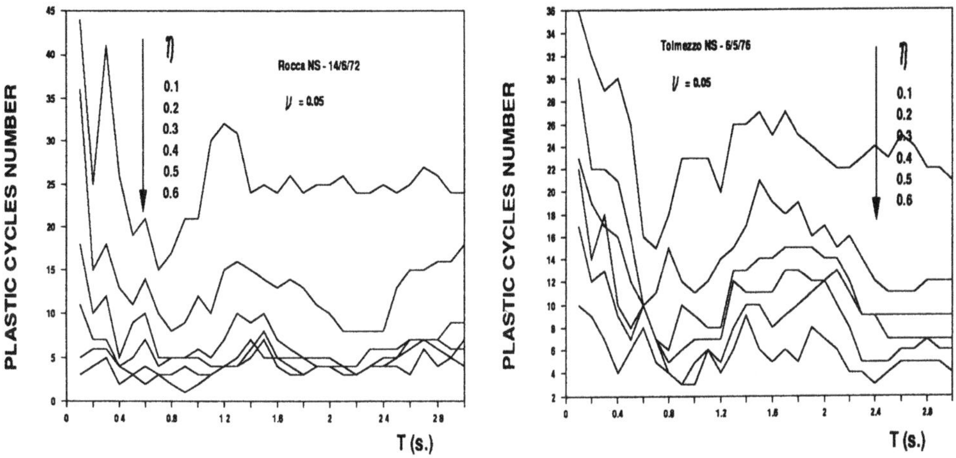

Figs. 5, 6. Plastic cycles number, with imposed yielding level. Rocca and Tolmezzo.

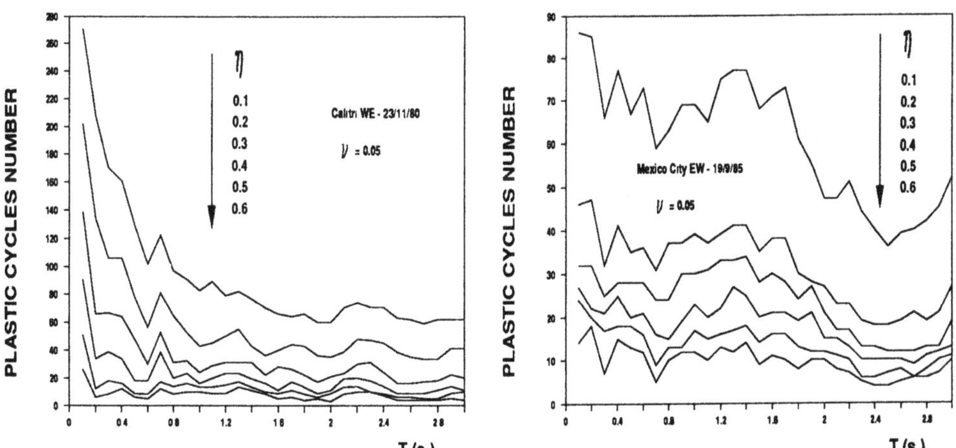

Figs. 7, 8. Plastic cycles number, with imposed yielding level. Calitri and Mexico City.

The histograms of the distribution of plastic cycles for the recorded ground motion of Calitri are drawn in Figs. 9, 10, 11 and 12; in these figures the percentage of the total number of plastic cycles *n* (vertical axis) which has a given percentage of the maximum plastic deformation, are represented; in particular, the plastic deformations of cycles are subdivided in 10 groups (0-10%, 10-20%, ..., 90-100% of the maximum plastic deformation). The other horizontal axis provides the values of the period (T=0.1, 0.5, 1.0 and 3.0 s.), and the four figures are drawn for values of η equal to 0.1, 0.2, 0.4 and 0.6.

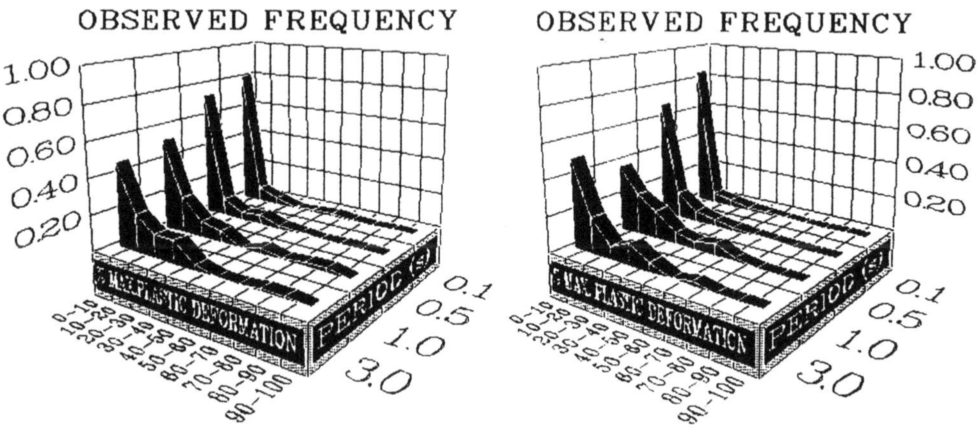

Figs. 9, 10. Observed plastic cycles frequency. Calitri record; η***=0.1, 0.2.***

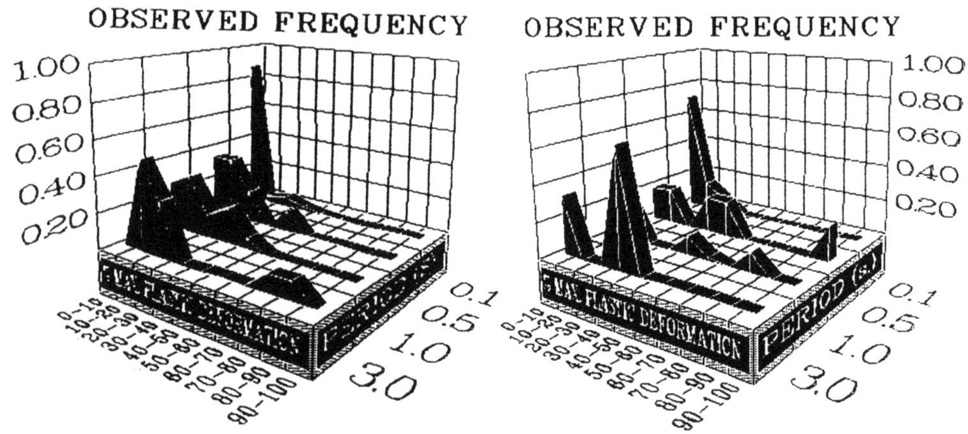

Figs. 11, 12. Observed plastic cycles frequency. Calitri record; η***=0.4, 0.6.***

The shape of the distribution depends on the value of T and η; hence, the curves for low value of T and of η are always fast decreasing, without relative maximum and with maximum values of frequencies in the range 0-10% of the maximum plastic deformation; as consequence, in these cases the plastic cycles due to the ground motion are characterized by only one cycle with large amount of plastic deformations, and many other cycles with low plastic deformation. For high values of η the frequency distribution tends to have an irregular shape;

this case is generally characterized by a low value of the number of plastic cycle n, and the statistical description tends to loss significance. In the following a comparison between different available probability models will be developed, in order to characterize in rational way the distribution of plastic cycles.

THE STATISTICAL CHARACTERIZATION OF PLASTIC CYCLES

In this paragraph, the random variable $X=\Delta x/\Delta x_{max}$ will be analyzed from the statistical and probabilistic point of view. For a given value of T and η, the seismic elasto-plastic analysis provides the maximum plastic deformation Δx_{max} , the number of plastic cycles n and a statistical sample of size n-1 $(x_1, x_2, ..., x_{n-1})$, with $x_i=\Delta x_i/\Delta x_{max}$, of the random variable X.

Firstly, the statistical characterization of the distribution of plastic cycles requires the evaluation of the sample mean m and of the sample standard deviation s; other interesting parameters **[15]** are the coefficient of "Skewness" (Skew.=0 for symmetrical distribution; positive values correspond to distribution with dominant tails on the right), and the coefficient of "Kurtosis" (flatness of the distribution; Kurt.=3 for Normal distribution). The definitions of these parameters are the following:

$$m = \frac{1}{n-1}\sum_{i=1}^{n-1} x_i \quad ; \quad s^2 = \frac{1}{n-1}\sum_{i=1}^{n-1}(x_i - m)^2 \tag{2}$$

$$Skew. = \frac{1}{s^3}\cdot\frac{1}{n-1}\sum_{i=1}^{n-1}(x_i - m)^3 \quad ; \quad Kurt. = \frac{1}{s^4}\cdot\frac{1}{n-1}\sum_{i=1}^{n-1}(x_i - m)^4 \tag{3}$$

In Figs. 13 and 14 the mean values of m and of the "coefficient of variation" s/m, evaluated for the four recorded ground motions of Rocca, Tolmezzo, Calitri and Mexico City, are drawn; the curves are given for values of η equal to 0.1-0.6. It is interesting to observe that the mean value of m is practically independent by T, and assumes the higher values for high values of η; in this case the elasto-plastic response is characterized by a low number of plastic cycles n, as evidenced by Figs. 5–8.

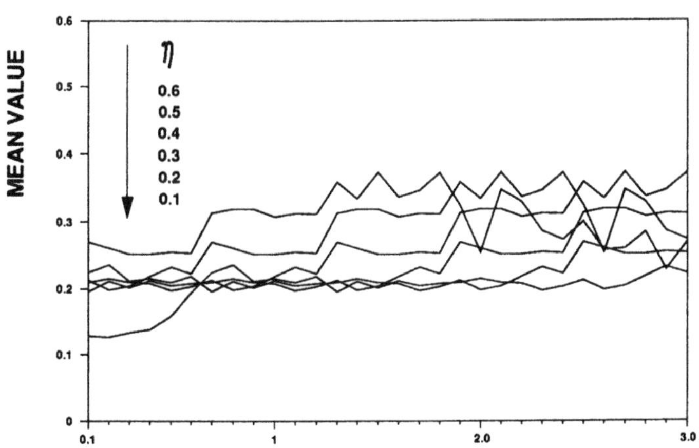

Figs. 13. Mean value spectra (average value for Rocca, Tolmezzo, Calitri and Mexico City ground motions).

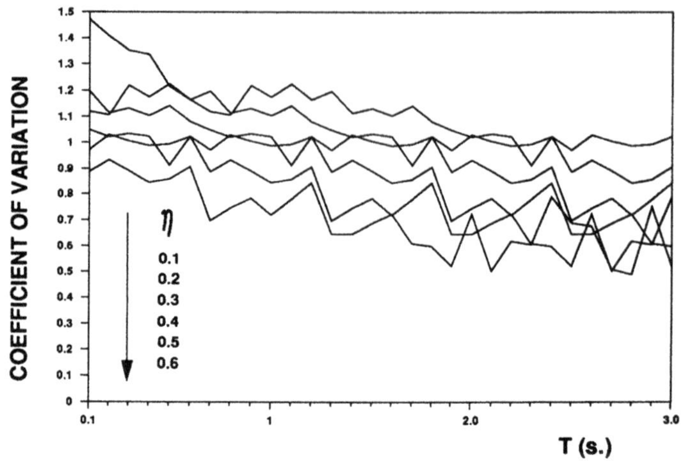

Figs. 14. Coefficient of variation spectra (average value for Rocca, Tolmezzo, Calitri and Mexico City ground motions).

The range of values of *m* is 0.20÷0.35, except rare cases. Thus, the plastic response of the model is characterized by one cycle with the maximum plastic deformation, and *n-1* cycles with an average plastic deformation of $0.20\text{-}0.35\Delta x_{max}$. However, the number of plastic cycles can be sufficient to develop low fatigue problems, as evidenced in **[5,6]**.

The value of *s/m* is decreasing with η and *T*; it is interesting to observe that the value 1 provides a good average result.

A more refined characterization of the plastic cycles can be provided interpolating the observed distribution by an appropriate probability model. In this context, the random variable *X* is defined by the density function $f_X(x)$ ($f_X(x)dx$ defines the probability that *X* assumes values in the range $x, x+dx$), and by the distribution function $F_X(x)$ ($F_X(x)$ is equal to the probability that *X* assumes values less or equal to x); f_X is equal to the derivate of F_X, and F_X is an increasing function, with values between 0 and 1.

Very interesting probability models, from the engineer point of view, are the Normal and the Lognormal; in fact, the central limit theorem affirms that, under very general conditions, as the number of variable in the sum becomes large, the distribution of the sum of random variables will approach the Normal distribution. As consequence, if the uncertainty in a physical variable may be the result of the combined effects of many contributing causes, it may be modelled as a Normal random variable. In this case we have:

$$f_X(x) = \frac{1}{\sigma\sqrt{2\pi}} e^{-1/2[(x-\mu)/\sigma]^2} \tag{4}$$

$$F_X(x) = \Phi\left(\frac{x-\mu}{\sigma}\right) \qquad with \qquad \Phi(x) = \frac{1}{2\pi}\int_{-\infty}^{x} e^{-1/2t^2} dt \tag{5}$$

where μ represents the expected value, and σ the standard deviation.

In the case of a phenomenon which arises as the result of a multiplicative mechanism acting on a number of factors, the central limit theorem applied to the logarithm of the variable, provides that the random variables has a logarithm that is distributed as a Normal; in this case the probability model is called Lognormal, and we have:

$$f_X(x) = \frac{1}{\zeta x\sqrt{2\pi}} e^{-1/2[(\ln x - \lambda)/\zeta]^2} \quad (x \geq 0) \tag{6}$$

$$F_X(x) = \Phi\left(\frac{\ln x - \lambda}{\zeta}\right) \quad (x \geq 0) \tag{7}$$

where:

$$\zeta^2 = \ln\left[1 + \left(\frac{\sigma}{\mu}\right)^2\right] \quad ; \quad \lambda = \ln\mu - \frac{1}{2}\zeta^2 \tag{8}$$

Both the Normal and Lognormal probability models needs of the estimators of mean value μ and of standard deviation σ, given by m and s. The advantage of the Lognormal, with respect to the Normal, is that the probability that $x<0$ is equal to 0; moreover, the Lognormal is "skewed", with skewness coefficient equal to $3\zeta+\zeta^3>0$; on the contrary, the normal has skewness 0: in other words, it has a symmetrical shape.

Normal and Lognormal have a finite probability that $x>1$, and this property is, for the problem considered, physically not understandable; moreover, the shape of this two probability models has in each case a maximum, and this is in contrast with the general behaviour given by Figs. 9÷12.

	μ	σ^2	Skewness
EXPONENTIAL	$\frac{1}{\nu}$	$\frac{1}{\nu^2}$	2
GAMMA	$\frac{k}{\nu}$	$\frac{k}{\nu^2}$	$\frac{2}{\sqrt{k}}$
BETA	$\frac{k}{k+\nu}$	$\frac{k\nu}{(k+\nu)^2(k+\nu+1)}$	$\frac{2(k-\nu)}{(k+\nu)(k+\nu+2)\sigma}$

Tab. 2. Characterization of the probability models.

Other interesting probability models are the Exponential, the Gamma and the Beta. The Exponential needs of the estimate of the only parameter ν (cfr. Tab. 2), by means of the estimate of μ given by m, and is defined by the following expressions, which provides a curve strictly decreasing with x:

$$f_X(x) = \nu e^{-\nu x} \quad ; \quad F_X(x) = 1 - e^{-\nu x} \quad (\nu, x \geq 0) \tag{9}$$

This model has coefficient of variation $\sigma/\mu=1$, and coefficient of skewness equal to 2. It is often used to describe the time between events **[15]**, that physically seems to be a problem similar to the one presently analyzed.

The Gamma probability model is defined in the following way:

$$f_X(x) = \frac{\nu^k}{\Gamma(k)} x^{k-1} e^{-\nu x} \quad (k, \nu, x \geq 0) \tag{10}$$

$$F_X(x) = \frac{\nu^k}{\Gamma(k)} \int_o^x t^{k-1} e^{-\nu t} dt = \frac{\Gamma(k, \nu x)}{\Gamma(k)} \tag{11}$$

where:

$$\Gamma(k)=\int_o^{\infty} t^{k-1}e^{-t}dt \quad ; \quad \Gamma(k,x)=\int_o^{x} t^{k-1}e^{-t}dt \tag{12}$$

For k=1, the Gamma is equal to the Exponential; thus, the Gamma may be considered a generalization of the Exponential.

The Beta probability model is defined by the following expressions:

$$f_X(x)=\frac{1}{B(k,\nu)}x^{k-1}(1-x)^{\nu-1} \quad (k,\nu,x \geq 0) \tag{13}$$

$$F_X(x)=\frac{1}{B(k,\nu)}\int_o^x t^{k-1}(1-t)^{\nu-1}dt=\frac{B_x(k,\nu)}{B(k,\nu)} \tag{14}$$

where:

$$B(k,\nu)=\int_o^{\infty} t^{k-1}(1-t)^{\nu-1}dt \quad ; \quad B_x(k,\nu)=\int_o^{x} t^{k-1}(1-t)^{\nu-1}dt \tag{15}$$

Table with numerical expressions of the functions of eqns. (12,15) are provided by standard integrals textbooks; for the numerical use, series expansions are available **[16]**, and were used in the present analysis.

The Beta and Gamma are defined by two parameters (cfr. Tab. 2) that can be evaluated by the estimators m and s of the parameters μ and σ; these probability models have the interesting property to have a shape depending on the values of their parameters; in fact, only if $k\leq 1$ the Gamma is decreasing, and only if $k\leq 1$ and $\nu\geq 1$, the Beta is decreasing. The Beta has also the interesting advantage to be limited; in other words, the Beta is the only examined probability model that has a null probability that $\Delta x_i > \Delta x_{max}$, that is a physically evident property.

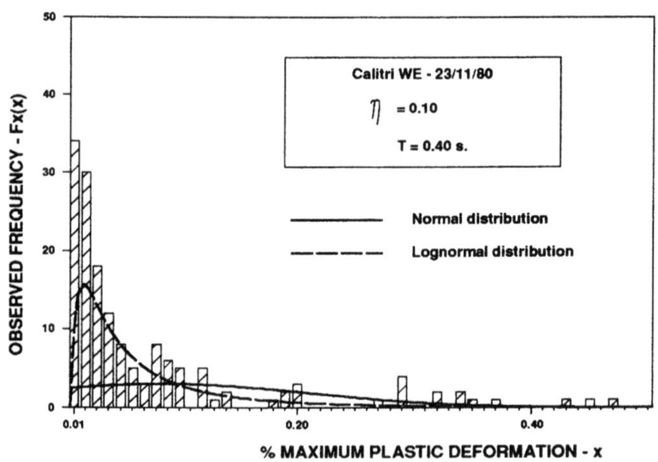

Figs. 15. Comparison between observed frequency and theoretical values (Normal and Lognormal). Calitri record; η=0.1, T=0.4 s.

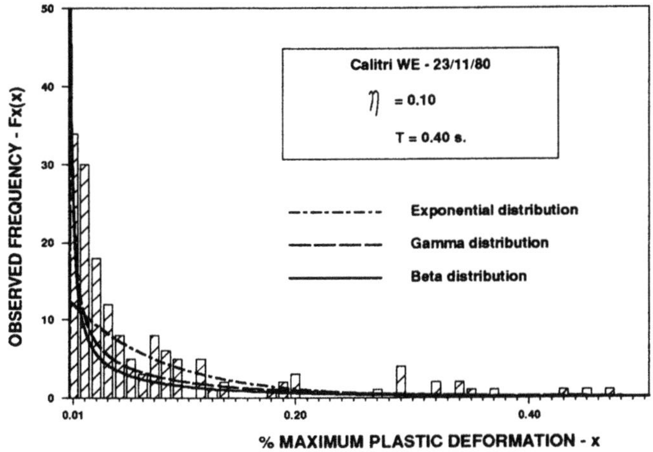

Figs. 16. Comparison between observed frequency and theoretical values (Exponential, Gamma, Beta). Calitri record; η=0.1, T=0.4 s.

A comparison between the results provided by the different probability models is given in Fig. 15 (Normal and Lognormal) and 16 (Exponential, Beta and Gamma); in the same figures are also drawn the observed frequency; the recorded ground motion analyzed is Calitri, with η=0.1 and T=0.4 s.; in this case n=161, m=0.0819 and s=0.1288 result. The curves are drawn estimating the necessary parameters by means of m and s:

Exponential: $\nu=12.47$;

Gamma: $\nu=4.836$; $k=0.3878$;

Beta: $\nu=3.171$; $k=0.2765$.

The shapes of the probability models are very different, and it is difficult to establish what is the better approximation; the only possibility of an impartial choice is to consider the maximum difference between the effective value of the cumulate frequency and the cumulate frequency given by the selected probability model; in this case, the worst approximation is given by the Normal (0.2605), and the best by the Gamma (0.1866); satisfactorily result is given, in this case, by the Lognormal (0.2409). In the next paragraph, this comparison will be developed in extensive way, by means of the statistical test of Kolmogorov & Smirnov.

THE OPTIMAL CHOICE OF A PLASTIC CYCLES PROBABILITY MODEL

The goodness-of-fit test of Kolmogorov & Smirnov **[15,17]** is based on the maximum difference D_{max} between the observed cumulate histogram and the hypothesized distribution function $F_X(x)$. This difference is a random variable, and it has been found that its distribution is independent of the hypothesized distribution of X. Its only parameter is the sample size (in this case n-1). As consequence, it is possible to evaluate the critical value D_{lim} and to perform an hypothesis test at a prescribed significance test; the value of D_{lim}, for some values of the sample size and for 5% and 20% significance level, are given in Tab. 3; more extend table are given in **[18]**. The test is passed if D_{max} is less or equal to D_{lim}.

The results of the tests for the recorded ground motions of Rocca, Tolmezzo, Calitri and Mexico City are given in Tabs. 4, 5, 6 and 7; in the same Tabs. the statistical characterization of the elasto-plastic response of the SDOF is given. Values of η= 0.1, 0.2 and 0.5 and T=0.1, 0.5, 1.0 and 3.0 s are considered; the cases with number of cycles less than 5 aren't considered because statistically insignificant.

It is interesting to observe that the average value, for the cases of Tabs. 4, 5, 6 and 7, of the coefficient of variation is 1.07, and of the skewness 1.82. These values are very close to the ones of the Exponential (respectively 1.00 and 2.00). The results of the Kolmogorow and Smirnov test, at significance level 5%, are given in the right part of the tables, where, in italic, it is given the value of D_{lim} corresponding to the number of plastic cycles less one, and the values of D_{max} for the different probability models analyzed. The shade indicate un-passed tests.

	Sample Size							
Significance Level	**5**	**10**	**15**	**20**	**25**	**30**	**35**	**Asymptotic Formula**
5%	*.563*	*.409*	*.338*	*.294*	*.264*	*.242*	*.23*	$1.36/\sqrt{n}$
20%	*.446*	*.322*	*.266*	*.231*	*.21*	*.19*	*.18*	$1.07/\sqrt{n}$

Tab. 3. Critical values D_{lim} for the Kolmogorow & Smirnov test.

The comparison between Normal and Lognormal indicates that the two probability models are practically equivalent, with results worst than Exponential, Gamma and Beta. The best results are given by the Beta and the Gamma; in particular the Beta passes all the tests in the cases of Rocca, Tolmezzo and Mexico City, and the Gamma gives bad results only in the case of Mexico City. The Exponential provides results weakly worst, but with the important simplification to have only one parameter; in other words, it requires the knowledge of the only estimated mean value m. It is interesting to observe that in about 25% of the cases the Exponential provides D_{max} less than the Gamma; in almost all these cases, the parameter estimated for the Beta are unrealistic (k>1). Thus, when in the Gamma, it results k>1, it may be useful: 1) to use the Exponential; 2) to evaluate the parameters of the Gamma by the others statistical parameters as the Skewness or the Kurtosis.

More general results are provided by Tabs. 8 (5% of significance level) and 9 (20% of significance level), where the results of the 11 recorded ground motion of Tab. 1 are summarized. In this tables the percentages of accepted Kolmogorov & Smirnov tests are provided, selecting between the 570 (30 values of T=0.1, 0.2, ...,3.0 s.; 19 values of η=0.05, 0.10, 0.15, ..., 0.95) analyzed cases only that have n>5.

The best probability model is the Gamma, that passed the 88.9% of the tests for 20% significance level, and 95.3% for 5% significance level. The Beta provides results substantially equivalent, and the Exponential gives results only weakly worst. The affidability of Normal and Lognormal is less good, probably due to the incorrect shapes, for this problem, of the densities functions. It seems very important to underline that the Exponential probability model, that needs of only the value of one parameter (this fact should be taken into account in other types of statistical tests, as the χ-square test), provides satisfactorily results; of course, the Gamma, which is a generalization of the Exponential with two parameters, provides better results.

ROCCA		STATISTICAL CHARACTERIZATION					KOLMOGOROV & SMIRNOV TEST					
η	T [s]	Cycles number	Mean value	Variar. coeffic.	Skewn.	Kurtosis	D *Limit*	Dmax Normal	Dmax Lognor.	Dmax Expon.	Dmax Gamma	Dmax Beta
0.1	0.1	44	0.135	1.04	1.79	4.00	*.207*	.197	.224	.136	.117	.121
0.1	0.5	19	0.190	0.93	1.55	2.80	*.309*	.199	.238	.113	.140	.085
0.1	1.0	21	0.245	0.92	1.52	2.73	*.294*	.318	.211	.210	.186	.163
0.1	3.0	24	0.224	1.02	1.98	4.94	*.276*	.280	.159	.153	.149	.154
0.2	0.1	36	0.150	1.19	1.63	3.10	*.230*	.253	.290	.220	.141	.102
0.2	0.5	11	0.184	0.71	1.38	2.07	*.409*	.175	.198	.235	.188	.173
0.2	1.0	12	0.346	1.11	1.22	1.67	*.391*	.317	.355	.284	.236	.204
0.2	3.0	18	0.134	1.24	2.93	10.1	*.318*	.273	.152	.137	.212	.282
0.5	0.1	<5	-	-	-	-	-	-	-	-	-	-
0.5	0.5	<5	-	-	-	-	-	-	-	-	-	-
0.5	1.0	<5	-	-	-	-	-	-	-	-	-	-
0.5	3.0	7	0.118	0.79	1.33	1.92	*.521*	.175	.304	.180	.235	.213

Tab. 4. Statistical characterization of cycles distribution for the Rocca record.

TOLMEZZO		STATISTICAL CHARACTERIZATION					KOLMOGOROV & SMIRNOV TEST					
η	T [s]	Cycles number	Mean value	Variat. coeffic.	Skewn.	Kurtosis	D *Limit*	Dmax Normal	Dmax Lognor.	Dmax Expon.	Dmax Gamma	Dmax Beta
0.1	0.1	36	0.167	1.36	1.83	4.53	*.230*	.295	.393	.343	.201	.085
0.1	0.5	26	0.170	1.06	1.47	2.60	*.264*	.240	.287	.203	.179	.132
0.1	1.0	23	0.250	0.93	1.45	2.45	*.282*	.254	.321	.241	.187	.153
0.1	3.0	21	0.256	0.94	1.26	1.80	*.294*	.210	.363	.239	.264	.172
0.2	0.1	30	0.161	1.47	2.03	5.44	*.246*	.289	.436	.394	.257	.098
0.2	0.5	16	0.246	0.89	1.53	2.88	*.338*	.211	.210	.133	.141	.103
0.2	1.0	11	0.319	0.59	1.38	2.07	*.409*	.134	.242	.250	.211	.176
0.2	3.0	12	0.357	0.63	1.36	1.93	*.391*	.151	.250	.277	.217	.177
0.5	0.1	17	0.131	0.64	2.15	5.79	*.328*	.306	.387	.391	.184	.125
0.5	0.5	7	0.285	0.55	1.27	1.76	*.521*	.337	.279	.329	.283	.286
0.5	1.0	<5	-	-	-	-	-	-	-	-	-	-
0.5	3.0	6	0.385	0.61	1.29	1.81	*.563*	.203	.247	182	.204	.158

Tab. 5. Statistical characterization of cycles distribution for the Tolmezzo record.

CALITRI		STATISTICAL CHARACTERIZATION					KOLMOGOROV & SMIRNOV TEST					
η	T [s]	Cycles number	Mean value	Variat. coeffic.	Skewn.	Kurtosis	D *Limit*	Dmax Normal	Dmax Lognor.	Dmax Expon.	Dmax Gamma	Dmax Beta
0.1	0.1	271	0.073	1.86	3.37	14.5	*.083*	.277	.411	.401	.113	.083
0.1	0.5	129	0.098	1.47	2.71	9.91	*.120*	.256	.241	.210	.181	.251
0.1	1.0	83	0.162	1.22	2.29	6.61	*.150*	.234	.188	.118	.079	.174
0.1	3.0	62	0.164	1.02	2.23	6.85	*.174*	.145	.199	.088	.078	.097
0.2	0.1	202	0.058	2.03	3.58	16.8	*.096*	.288	.426	.441	.100	.080
0.2	0.5	78	0.103	1.25	2.40	7.64	*.155*	.242	.175	.184	.120	.176
0.2	1.0	43	0.186	0.96	1.65	3.29	*.210*	.133	.267	.160	.177	.115
0.2	3.0	41	0.166	0.90	1.92	4.99	*.210*	.183	.167	.090	.104	.095
0.5	0.1	51	0.054	1.61	2.75	10.3	*.192*	.306	.377	.352	.128	.088
0.5	0.5	9	0.268	0.81	1.36	1.97	*.382*	.212	.345	.288	.272	.227
0.5	1.0	13	0.228	0.73	1.29	1.79	*.310*	.166	.193	.140	.147	.120
0.5	3.0	10	0.191	1.27	1.60	2.89	*.396*	.368	.226	.270	.206	.183

Tab. 6. Statistical characterization of cycles distribution for the Calitri record.

MEXICO CITY		STATISTICAL CHARACTERIZATION					KOLMOGOROV & SMIRNOV TEST					
η	T [s]	Cycles number	Mean value	Variat. coeffic.	Skewn.	Kurtosis	D *Limit*	Dmax Normal	Dmax Lognor.	Dmax Expon.	Dmax Gamma	Dmax Beta
0.1	0.1	86	0.140	1.60	2.18	5.71	*.148*	.288	.449	.405	.214	.053
0.1	0.5	67	0.179	1.41	1.83	4.21	*.167*	.262	.412	.370	.231	.067
0.1	1.0	69	0.203	1.27	1.76	3.86	*.165*	.251	.324	.272	.161	.163
0.1	3.0	52	0.247	1.13	1.67	3.32	*.190*	.232	.219	.152	.107	.183
0.2	0.1	46	0.189	1.31	1.76	3.96	*.203*	.254	.423	.348	.278	.134
0.2	0.5	35	0.225	1.20	1.63	3.24	*.232*	.235	.379	.318	.263	.126
0.2	1.0	39	0.286	0.98	1.51	2.76	*.221*	.180	.231	.153	.161	.103
0.2	3.0	27	0.385	0.80	1.24	1.66	*.260*	.223	.438	.381	.300	.159
0.5	0.1	24	0.162	1.47	1.85	4.13	*.276*	.282	.429	.404	.278	.118
0.5	0.5	18	0.181	1.30	1.85	4.46	*.318*	.290	.331	.273	.169	.139
0.5	1.0	17	0.232	1.16	1.73	3.60	*.328*	.243	.364	.289	.219	.132
0.5	3.0	11	0.399	0.49	1.43	2.15	*.409*	.132	.212	.289	.185	.146

Tab. 7. Statistical characterization of cycles distribution for the Mexico City record.

EARTHQUAKE	Cases with $n>5$ (Max 570)	PERCENTAGE OF ACCEPTED K & S TESTS. SIGNIFICANCE LEVEL 5%				
		NORM.	LOGN.	EXPON.	GAMMA	BETA
ROCCA	225	84.0%	90.7%	94.7%	100.0%	93.8%
TOLMEZZO	340	87.9%	78.2%	92.3%	97.9%	100.0%
FORGARIA	263	87.0%	84.4%	87.1%	94.3%	92.4%
BUIA	185	89.2%	82.2%	89.2%	97.8%	97.3%
BAGNOLI IRPINO	235	76.2%	69.4%	71.9%	91.9%	91.1%
BRIENZA	352	67.6%	70.7%	82.7%	94.6%	81.8%
CALITRI	377	70.3%	71.1%	89.4%	93.4%	88.1%
STURNO	295	78.3%	76.6%	81.4%	95.9%	84.1%
NOCERA UMBRA	325	69.2%	80.0%	86.5%	99.1%	83.1%
VILLETTA BARREA	119	63.0%	68.9%	71.4%	95.8%	80.7%
MEXICO CITY	396	78.0%	60.3%	71.7%	88.1%	93.2%
MEAN VALUES		77.3%	75.7%	83.5%	95.3%	89.6%

EARTHQUAKE	Cases with $n>5$ (Max 570)	PERCENTAGE OF ACCEPTED K & S TESTS. SIGNIFICANCE LEVEL 20%				
		NORM.	LOGN.	EXPON.	GAMMA	BETA
ROCCA	225	64.4%	75.1%	86.7%	98.7%	84.0%
TOLMEZZO	340	75.3%	60.0%	79.1%	89.7%	99.4%
FORGARIA	263	76.0%	64.6%	76.4%	84.8%	87.1%
BUIA	185	69.7%	58.9%	82.7%	93.5%	95.1%
BAGNOLI IRPINO	235	61.3%	55.7%	63.0%	83.4%	82.1%
BRIENZA	352	52.2%	55.7%	67.9%	85.2%	71.3%
CALITRI	377	55.4%	51.1%	81.4%	90.7%	81.7%
STURNO	295	67.4%	65.8%	75.3%	90.5%	80.3%
NOCERA UMBRA	325	51.0%	63.3%	75.1%	91.4%	69.2%
VILLETTA BARREA	119	50.4%	47.8%	55.5%	90.8%	71.4%
MEXICO CITY	396	54.8%	43.4%	63.1%	79.3%	87.1%
MEAN VALUES		61.6%	58.3%	73.3%	88.9%	82.6%

Tabs. 8,9. Global results of Kolmogorov & Smirnov tests; significance level 5 and 20%.

CONCLUSIONS

The characterization of the observed frequencies of plastic cycles for a EPP-SDOF subjected to strong motion allows to develop rational numerical and experimental low cycle fatigue static tests of materials, structural elements and structural sub-assemblages. These tests provide essential informations to develope adequate damage functionaly, that are the essential basis for the rational seismic verify of structures.

Based on the results of an extensive study, the following main conclusions have been obtained:

(1) The number of plastic cycles is decreasing with the period and with the yielding level, and it is strictly correlated to the duration of the ground motion.

(2) The mean value of plastic deformations cycles is variable between 0.20 and 0.35 times the maximum plastic excursion, and it is practically statistically independent of the period.

(3) The coefficient of variation of frequency of plastic cycles is decreasing with the period, and with yielding level; the value 1.0 provide a good approximation in the entire range analyzed.

(4) The observed frequency is decreasing with the amplitude of plastic cycles; in other words the plastic cycles most probable are characterized by low plastic deformations.

(5) Based on the extensive use of the Kolmogorov & Smirnov goodness-of-fit test, we can affirm that the Gamma probability model provides very good previsions of the observed frequency; the Exponential probability model, that have the decisive advantage to be defined by only one parameter and that is a simplified version of Gamma, provides sufficiently accurate results. The Beta probability model gives results weakly worst than the Gamma.

(6) The Normal and the Lognormal probability models are generally inappropriate to analyze the problem, due to their bell-shape.

REFERENCES

[1] **MAHIN S., BERTERO V.**, *An Evaluation of Inelastic Seismic Design Spectra*, Journal of the Structural Division, ASCE, September, 1981.

[2] **ZAHRAH T., HALL J.**, *Earthquake energy absorption in SDOF structures*, Journal of the Structural Division, ASCE, August, 1984.

[3] **AKIYAMA H.**, *Earthquake resistant limit-state design for buildings*, University of Tokyo Press, 1985.

[4] **UANG C.M., BERTERO V.V.**, *Use of energy as a design criterion in earthquake resistant design*, EERC Report, University of California, Berkeley, 1988.

[5] **COSENZA E., MANFREDI G., RAMASCO R.**, *An Evaluation of the use of Damage Functionals in Earthquake-Resistant Design*, IX ECEE, Moscow, September, 1990 (Submitted).

[6] **COSENZA E., MANFREDI G., RAMASCO R.**, *Un analisi della risposta sismica dell'oscillatore elasto-plastico*, Proceedings IV Congress "L'Ingegneria Sismica in Italia", Milano, October, 1989.

[7] **IEMURA H.**, *Earthquake Failure Criteria of Deteriorating Hysteretic Structures*, Vol.5, VII WCEE, Istanbul, 1980.

[8] **KRAWLINKER H., ZOHREI M.,** *Cumulative Damage in Steel Structures Subjected to Earthquake Ground Motion*, Computers & Structures, Vol. 16, n.1-4, 1983.

[9] **STEPHENS J.E., YAO J.T.P.,** *Damage Assessment Using Response Measurements*, Journal of Structural Engineering, ASCE, April, 1987.

[10] **JEONG G.D., IWAN W.D.,** *The Effect of Earthquake Duration on the Damage of Structures,* Earthquake Engineering and Structural Dynamics, Vol.16, 1988.

[11] **LASHKARI-IRVANI B., KRAWINKLER H.,** *Damage Parameters for Bilinear Single Freedom Systems*, 5-15, VII EECE, Athen, 1982.

[12] **LASHKARI-IRVANI B.,** *Cumulative Damage Parameters for Bilinear Systems Subjected to Severe Earthquake*, ATC-10-1, Critical Aspects of Earthquake Ground Motion and Building Damage Potential, 77-87, 1984.

[13] **HADIDI-TAMJED H.,** *Statistical Response of Inelastic SDOF Systems Subjected to Earthquakes,* Ph.D. Thesis, Dept. of Civil Engineering, Stamford University, December, 1987.

[14] **HADIDI-TAMJED H., LEE D.G., KRAWINKLER H.,** *Statistical Study on Seismic Response Parameters for Damage Evaluation*, paper 7-3-1, Vol. V, IX WCEE, Tokyo-Kyoto, August 2-9, 1988.

[15] **BENJAMIN J.R., CORNELL C.A.,** *Probability, Statistics and Decision for Civil Engineers*, McGraw-Hill, 1970.

[16] **GRADSHTEYN I.S., RYZHIK I.M.,** *Table of Integrals, Series and Products,* Academic Press, 1965.

[17] **ANG A.H.S., TANG W.H.,** *Probability Concepts in Engineering Planning and Design, Vol I, Basic Principles,* J. Wiley & Sons, 1975.

[18] **CRC (BEYER W.H. Editor),** *Handbook of tables for Probability and statistics*, The Chemical Rubber Co., 1968.

13 BASE ISOLATION AS PASSIVE CONTROL OF SEISMIC RESPONSE OF STRUCTURES

B. CSAK
Technical University of Budapest, Hungary

Summary

This paper deals with questions of seismic response of earthquake resistant structures and the function of design methods. We focus on the difference between traditional and modern earthquake resisting design methods.

Current earthquake resisting design concentrates primarily on the safety of structures as <u>primary systems</u>. For certain special buildings, such as hospitals, nuclear and industrial facilities etc. the purpose of minimizing damage to the buildings and protecting the functional safety of sensitive and costly equipment as <u>secondary systems</u>, is added.

To serve this purpose we give an overview of response control methods of structures.

1 Seismic response of traditionally designed buildings

The main aspects of traditional practice in the design of earthquake resistant buildings may be expressed in the following items:

- tie the structures firmly to their foundation;
- make them light;
- make them strong enough to resist the forces generated by earthquakes.

Experience has taught us that such design generates very high seismic forces in the building, which often exceed the capacity of the structural system.

In later practice to prevent buildings from very severe failures or collapsing, the design philosophy was corrected by theories of:

- ductility of structural materials and structural elements,
- energy absorbtion capacity of the whole structure.

Testing of Metals for Structures. Edited by F.M. Mazzolani. © 1992 RILEM.
Published by Taylor & Francis, 2 Park Square, Milton Park, Abingdon, Oxon, OX14 4RN.
ISBN 0 419 15810 3

Ductile structures stretch before they break, consequently the buildings undergo very large and permanent inelastic distortions during earthquakes. Buildings constructed in this way will not collapse, but will be frequently so badly damaged, that they become unsafe for occupancy and have to be demolished.

Hospitals, communication and emergency operation centres, police and fire stations, nuclear and other facilities, cannot be used when they are needed most, i.e. immediately after the earthquake.

Our latest buildings contain extremely sensitive and costly equipment, which is vital in business, commerce, education, health care and safety. The contents of these buildings are frequently more costly and valuable than the buildings themselves. Conventional seismic engineering cannot provide adequate protection for the contents of the buildings. During earthquakes the buildings become large "mixing boxes" where all the loose contents are scrambled. For this reason even our recently constructed buildings cannot be expected to remain functional after an earthquake.

The seismic response of these structures cannot be controlled, so we need to find another solution, in which we have a control system over the seismic response of structures. This is the vibration control of structures, which we use as *active*, or *passive* control.

Further we shall deal with *base isolation*, for passive seismic response control of structures.

2 Seismic response controls as vibration controls of structures

Structural vibration control has recently become an important topic of research in civil engineering. It has an essential influence on the design of modern high-rise buildings, especially with respect to their resistance to unsteady wind forces and strong earthquake motions. Passive vibration control techniques have been found more or less effective in diminishing the response of structures under hostile environments such as strong earthquakes. In recent years different base isolation systems have been developed and applied practically in buildings.

In order to limit structural dynamic response under seismic actions within acceptable ranges, two main ways of control can be considered.

(a) Active Control

This is an earthquake-proof method which does not operate by increasing the damping capacity of structures, but by measuring the input or response motion with sensors and controlling the response by adding a force to the structure or changing the dynamic characteristics of the structure.

(b) Passive Control

The passive vibration control system is almost equal to the so called base isolation, in a practical sense, though the passive system covers a wider category than the other.

In contrast to active control, in this case the main aim is to increase the damping and energy absorbtion capacity of structures. The process of energy absorbtion does not develop *inside* the structure, but within the base isolation *outside* the structure. Accordingly with this solution we can control the seismic response of whole buildings and their internal equipment as primary and secondary systems.

The isolating and energy dissipating system consists of the structure, the isolator and the energy dissipator devices, mounted between the structure and its foundation. It can effectively isolate the building from the ground motion of earthquakes, and offer high damping to dissipate the energy transmitted to the structure. So it can diminish the structural response to a very small value of the ground motion, preventing the whole structure from damage in a severe seismic action.

In contrast to conventional design, the loose contents of the buildings are also protected. Buildings therefore can be expected to remain undamaged and functional after an earthquake.

Base isolation has a new principle; instead of tying a building to its foundation, the two are separated; this solution permits the building to "float" on top of flexible pedestals during earthquakes.

In the following we present the theory and details of base isolation, with appropriate reference to published sources of information.

3 Some aspects of base isolation theory development and applications

3.1 Short historical overview: early efforts and solutions

Of the basic principle of separating the structure from seismic motions, as a short history of the development, the following is known:

According to the literature, the first base-isolated structure was proposed by Kawai in 1881. His structure had rollers at its foundation a mat of logs put on several steps lengthwise and crosswise (2).

After the San Francisco Earthquake, in 1909 an English doctor, J.A. Calantarients, patented a construction by putting a talc layer between the foundation and structure.

Proposals from Kawai and Calantarients and sliding mechanisms or real base isolation systems adopted "the shape" and "the material" respectively for easy sliding. So the mechanisms are ruled by Coulomb's equation:

$$F = \mu N$$

where F = frictional force, μ = the frictional coefficient and N = the vertical force as whole weight of the building.

In 1921 the Imperial Hotel in Tokyo, designed by Frank Lloyd Wright was completed. This building survived the 1923 Tokyo earthquake. Wright wrote later from this building: "A good cushion to survive the terrible shocks. Why not float the building on it?" He tied the building to the upper layer of good soil by closely spaced short piles that penetrated only as far as the top of the soft mud (2). The Imperial Hotel is evidence that base isolation works.

Around 1930 in the USA the idea of a flexible first storey was proposed by R.R. Martel and others. This is an idea that makes the first storey flexible, absorbs the vibration and prevents energy being transmitted to the upper stories.

The first base-isolated systems actually constructed in the world are the Fudo Band buildings in Japan.

After World War II the USA took a leading part in earthquake engineering. The 1st World Conference of earthquake engineering was convened in 1956 on the 50th anniversary of the San Francisco earthquake. Since that time the base isolation theory has been developed as follows.

The first use of rubber for earthquake isolation was in an elementary school in Skopje, after the serious earthquake in 1963. The building is a three-storey structure in reinforced concrete which rests on large blocks of natural rubber. In contrast with recent rubber bearings, these blocks are completely unreinforced, so that the weight of the building causes it to bulge at the sides (2).

3.2 Recent approaches and practice

Recently great attention has been given to this method and there is a growing concern about different systems and solutions over the world in the field of earthquake engineering. The main aim to control and minimise the seismic response of the structure as the *primary* system, and its contents, as the *secondary* system.

The isolating and energy dissipating system consists of the whole structure with energy absorbing devices mounted between the foundation and the structure, so it can control the seismic response of the structure and reduce the ground motion to a very small value, preventing the whole building from damage in a severe earthquake attack. This gives engineers a new way of designing earthquake-resistant structures. The response control and its calculation method concentrate in general on eliminating the acceleration of the building in both horizontal and vertical directions.

The Shimizu Base Isolation System of 1987 (5) has the main characteristics and details of design and test results of a modern base-isolation system. A flow chart, which outlines the design and construction of the base isolated structure is shown in Fig.1.

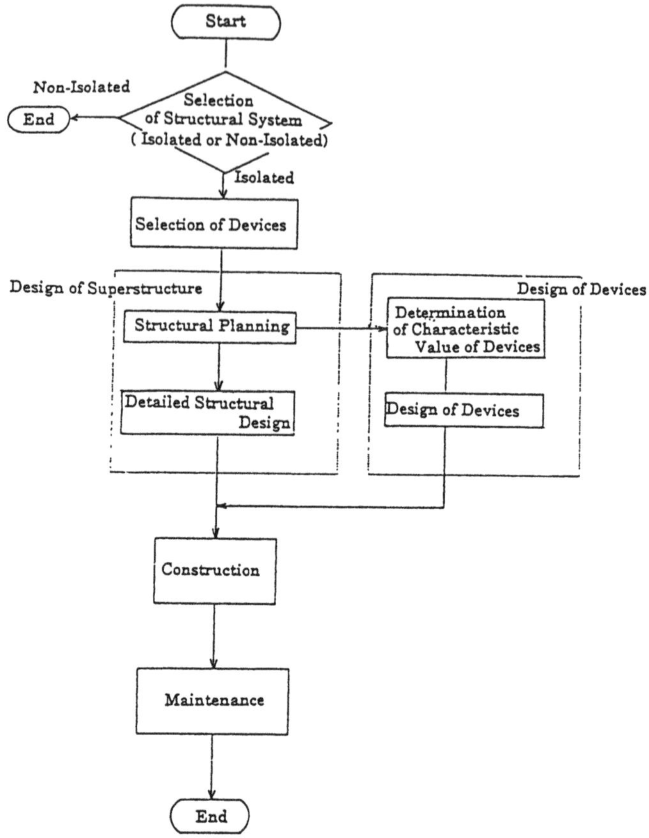

Fig. 1 Flow chart of Shimizu base isolation system.[5]

The devices used for base isolation systems have been developed in the last few decades, and therefore their durability should be fully investigated and confirmed. The system has three different combinations of devices. Fig.2 shows details of the devices, and Table 1 describes the three combinations of devices and their characteristics. Fig.3 shows the typical force-deformation relationship of the laminated rubber.

The energy absorbtion capacity of the isolation system depends first of all on the repeated low cycle fatigue behaviour of the steel damper elements. Fig.4 shows experimental results of the horizontal force deformation relationship of a steel damper. Figs. 5, 6, 7 and 8 show the response diagrams of non-base isolated and base isolated structures.

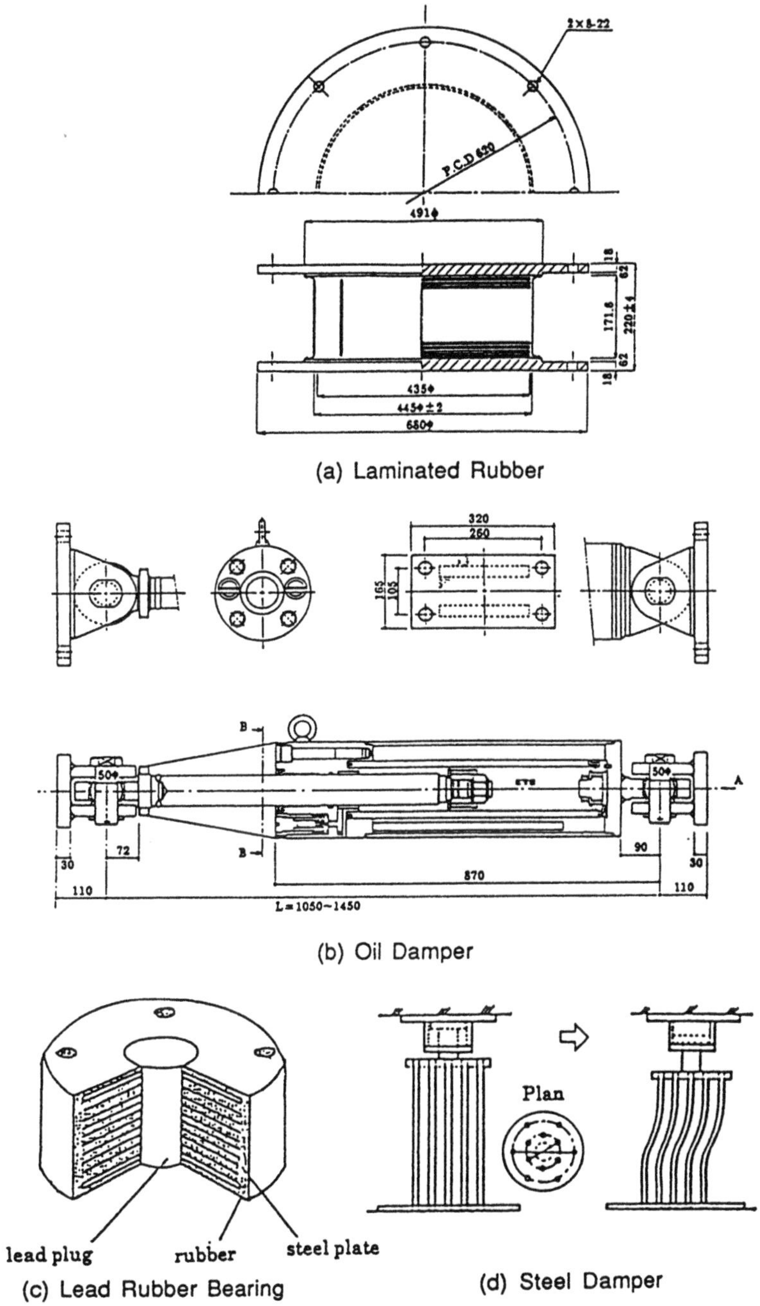

Fig. 2 Details of devices. [5]

Table 1 Three Different Combinations of Devices and Their Characteristics [5]

Laminated Rubber + Oil Damper	(1) Oil damper is effective at small amplitude, and is advantageous to small earthquake motion. (2) It is not necessary to exchange devices after earthquakes, because hysteretic energy consumption is not expected. (3) Response is linear and therefore response by large eathquake motion is clearly examined. (4) It is easy to select desirable damping characteristics.
Lead Rubber Bearing	(1) Damping is large. (2) It is hardly necessary to exchange devices , because lead will re-crystallize after plastic deformation due to relatively large earthquakes. (3) It is easy to handle the device, because support system and damping system are combined into one body. (4) Resonant phenomenon hardly appears, because vibration system is non-linear due to predominant period moving to longer when amplitude becoming large.
Laminated Rubber + Steel Damper	(1) Damping is large. (2) It is possible to minimize the amount of eccentricity when upper frame is eccentric, because support system(laminated rubber) and damping system with some stiffness(steel damper) are independent. (3) Resonant phenominon hardly appears, because vibration system is non-linear due to predominant period moving to longer when amplitude becoming large.

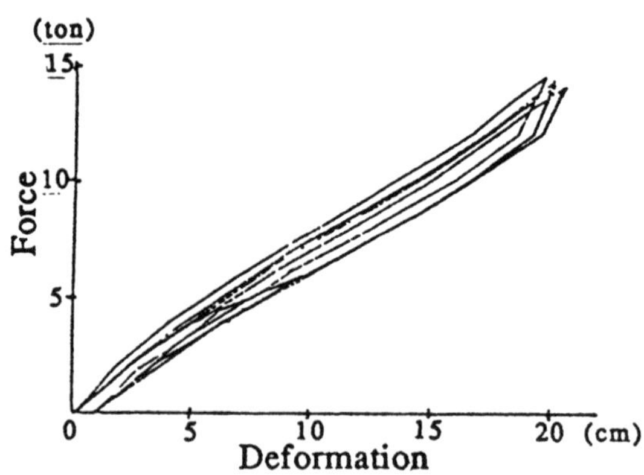

Fig. 3 Horizontal force-deformation relationship laminated rubber. [5]

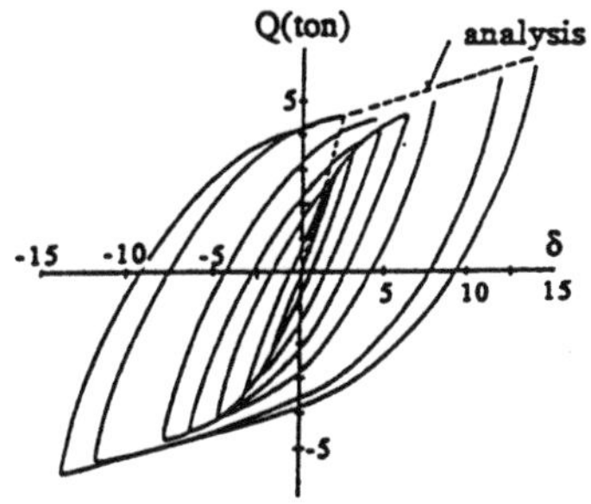

Fig. 4 Horizontal force-deformation relationship of steel damper. [5]

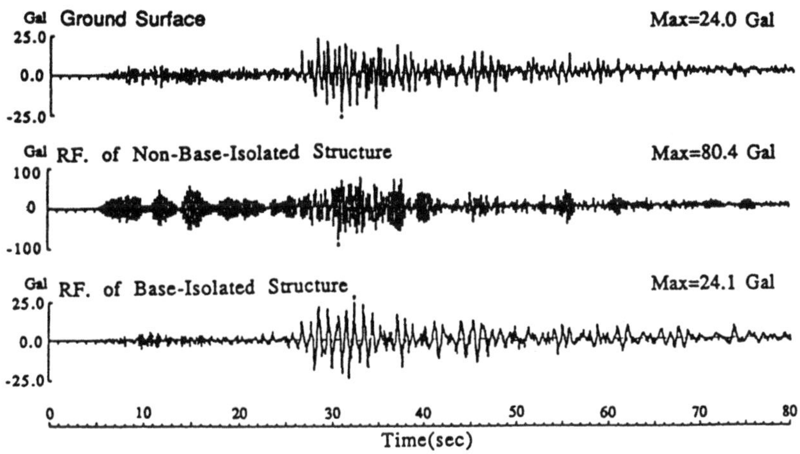

Fig. 5 Example of observed motions (X-direction) (Iwate Prefecture inland earthquake).[5]

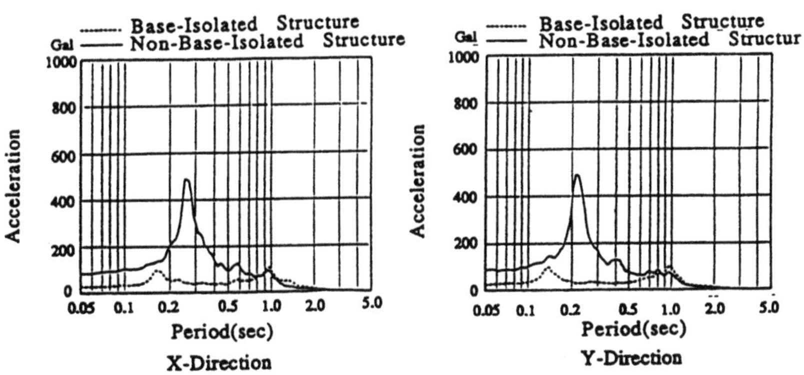

Fig. 6 Response spectra of motions at roofs (Iwate Prefecture inland earthquake). [5]

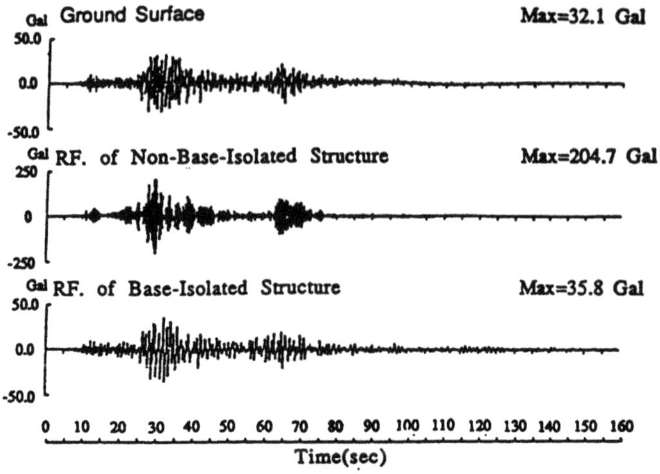

Fig. 7 Example of observed motions (X-direction) (Fukushima Prefecture offshore earthquake).[5]

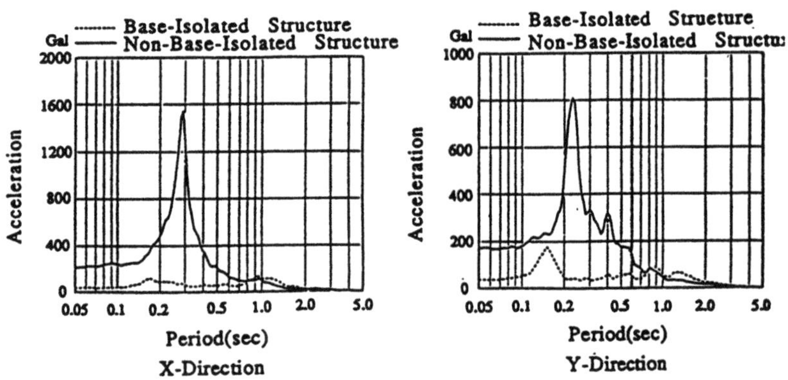

Fig. 7a Response spectra of the motions at the roofs (Fukushima prefecture offshore earthquake).[5]

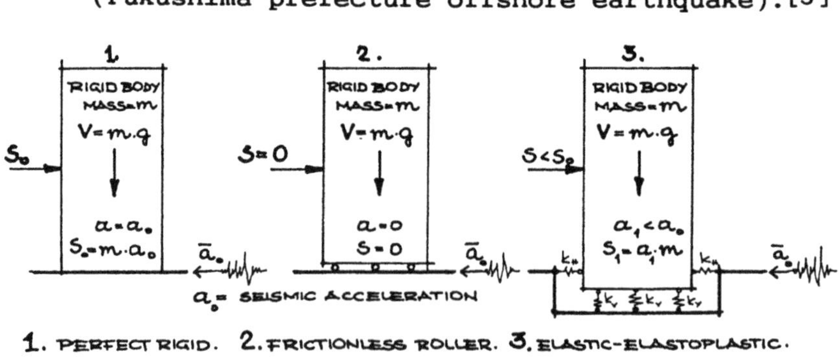

Fig. 8 Connection systems between soil and building.

4 Elastoplastic energy absorber bearing elements for passive control of seismic response of structures

A base isolation system has been developed at the Technical University of Budapest. The system is the subject of an international patent, US patent number 4,651,481, date of patent: March 24 1987.

4.1 Seismic motions, internal forces

The energies of an earthquake spread in the form of wave motions. On the inhomogenous earth surface on which the buildings are also standing, they travel in the form of irregular Love, Rayleigh etc waves. From the effect of the wave motions originate seismic forces on the building constructions. Considering the simple case of a mass $\underline{m}$, the seismic force can be expressed by the formula according to Newton's law and D'Alembert's principle

$$S(t) = m \cdot \underline{a}(t)$$

and the kinetic energy of the point of mass:

$$E_{kin}(t) = m\left[v(t)\right]^2 \cdot \frac{1}{2}$$

where $\underline{a}$ (t) = acceleration, v(t) = velocity of wave motion.

The case of buildings is not so simple. The greatness of seismic forces and energies depend mainly on three circumstances:

(a) The effective magnitude or intensity of the earthquake and the eigenvalues and geophysical characteristics of the underground. The earthquake magnitude and soil characteristics are independent from us: their manipulation is very complicated or impossible.

(b) The connections between the soil and the building: According to Fig.8, the connection between the soil and the building can be of three kinds: Perfectly rigid, frictionless roller or elastic-elastoplastic connection. The reality is between case 1 and 2 where the manipulation of connections is possible for the constructor of structures, for example by base isolation, as discussed in this paper

(c) The stiffness and dynamic characteristics (eigenvalues) of the whole building construction. According to Fig.9 the seismic acceleration that occurs on the building depends on the dynamic characteristics of the construction system, first of all on its eigenperiod time and damping coefficient.

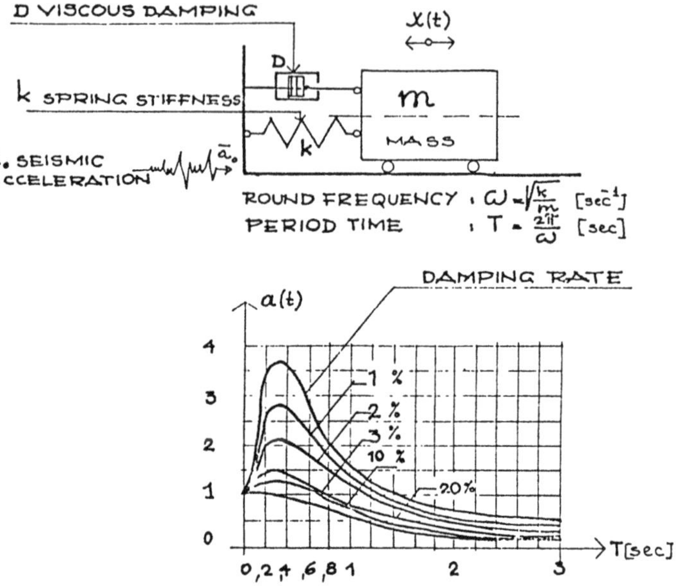

Fig. 9 Acceleration response spectra of a one degree of freedom system.

4.2 Energy absorbtion inside the structure

Kinetic energies are proportional to the velocities of motions occurring on the buildings. According to the theorem of energy law, kinetic energies are compensated, are absorbed, by internal deformation (potential) energies originated from the elastic or plastic deformations of the structure. According to the theorem of energies;

$$E_{kin}(t) + E_{pot}(t) = constant$$

$$E_{kin\,max} = E_{pot\,max}$$

To prevent the collapse of the building, m, the process of the energy absorption must exist during the whole time of the earthquake period. Thus, in this case, the process of the energy absorbtion is developing inside the structural system, and causes serious structural and other damages in the building.

Such buildings can be designed according to the principle of predeterminated collapse mechanisms (Limit design Fig. 10). Practice has proved the correctness of this theory. Buildings designed in such a way, though at the expense of severe structural and other damages, survive the earthquake they were designed for. Their building costs are 4-6% higher than the original building cost.

The subsequent repairing and strengthening of the damaged buildings can reach, however, according to experiences, 40 - 60% of the total building cost.

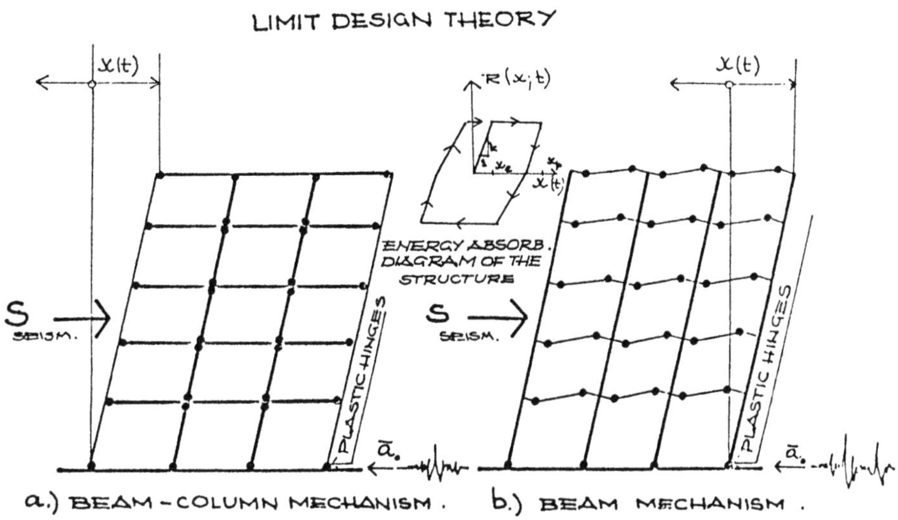

Fig. 10 Sideway mechanisms (Energy absorbtion inside of the structure.

4.3 Energy absorbtion outside of the structure. (Base isolation)

Against the previous solution, base isolation is a wholly new tendency or strategy in seismic protection of structures. Base isolation is similar in principle to the action of tires, springs and shock absorbers on automobiles which reduce, but do not completely eliminate, the impact of bumps on the highway. The reason this concept has been utilized in this important facility is to significantly reduce building damage and decrease the possibility of injury to occupants in the event of severe earthquake. The absorbtion of energies is assured not by the elastic or elastoplastic deformations of the structure, but by large elastic-elastoplastic deformations of inbuilt absorber elements between the foundation and lower structure. So the process of energy absorbtion is assured outside the structure.

The absorber elements may be built into the construction by different methods, so we can ensure that the acceleration, velocity and displacement of ground motions on the lower construction should remain under predetermined limits.

This is the demand on the one hand, and leads to the question of design and dimensioning on the other hand. The demand depends on the type of building: in the case of common buildings this can be the maximum wind load for which the structure must be designed. In special cases e.g. hospitals, important public buildings, industrial, nuclear or energy generating establishments, the seismic movements and internal forces can be limited to a degree which does not cause any breakdown.

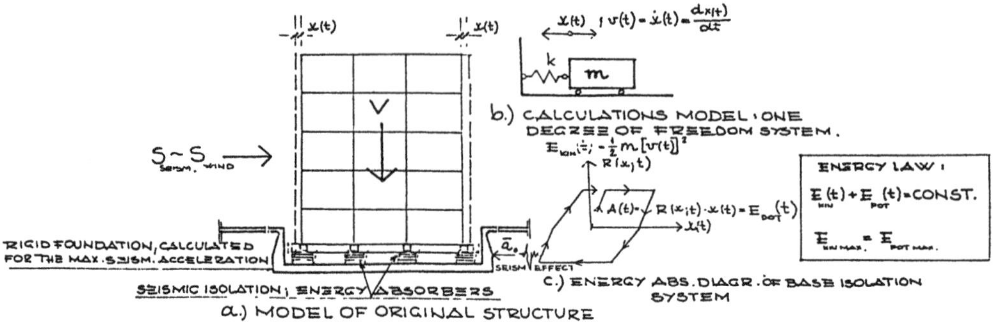

Fig. 11 Seismic isolation (Energy absorbtion outside of the structure).

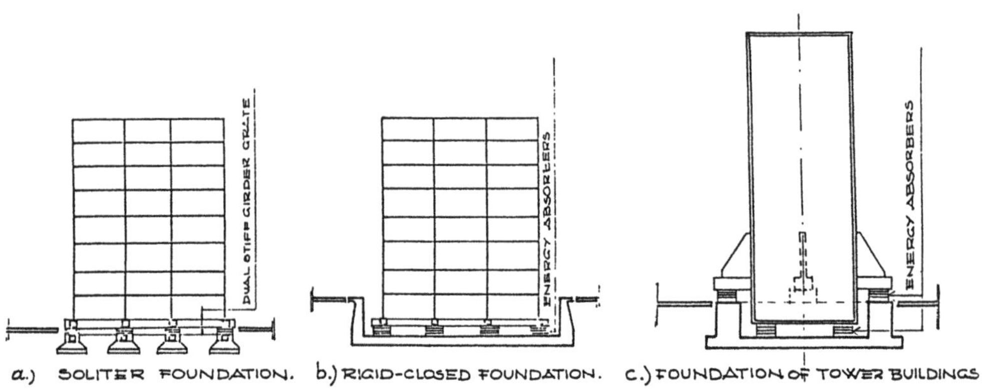

Fig. 12 Inbuilt-systems of elastic-elastoplastic energy absorbers.

4.4 Short description of the energy absorber

According to Fig.13. the system consists of two elements;

Sandwich elements: vulcanized or stacked alternate rubber and steel plates. These carry the vertical loads of the building, and give, elastic support against vertical seismic motions.

Vertical steel pins: made of special steel material, passing through bore-holes in the sandwich elements, fastened by steel tubes which are fixed into the concrete of the foundation at their lower ends and into the floor of the

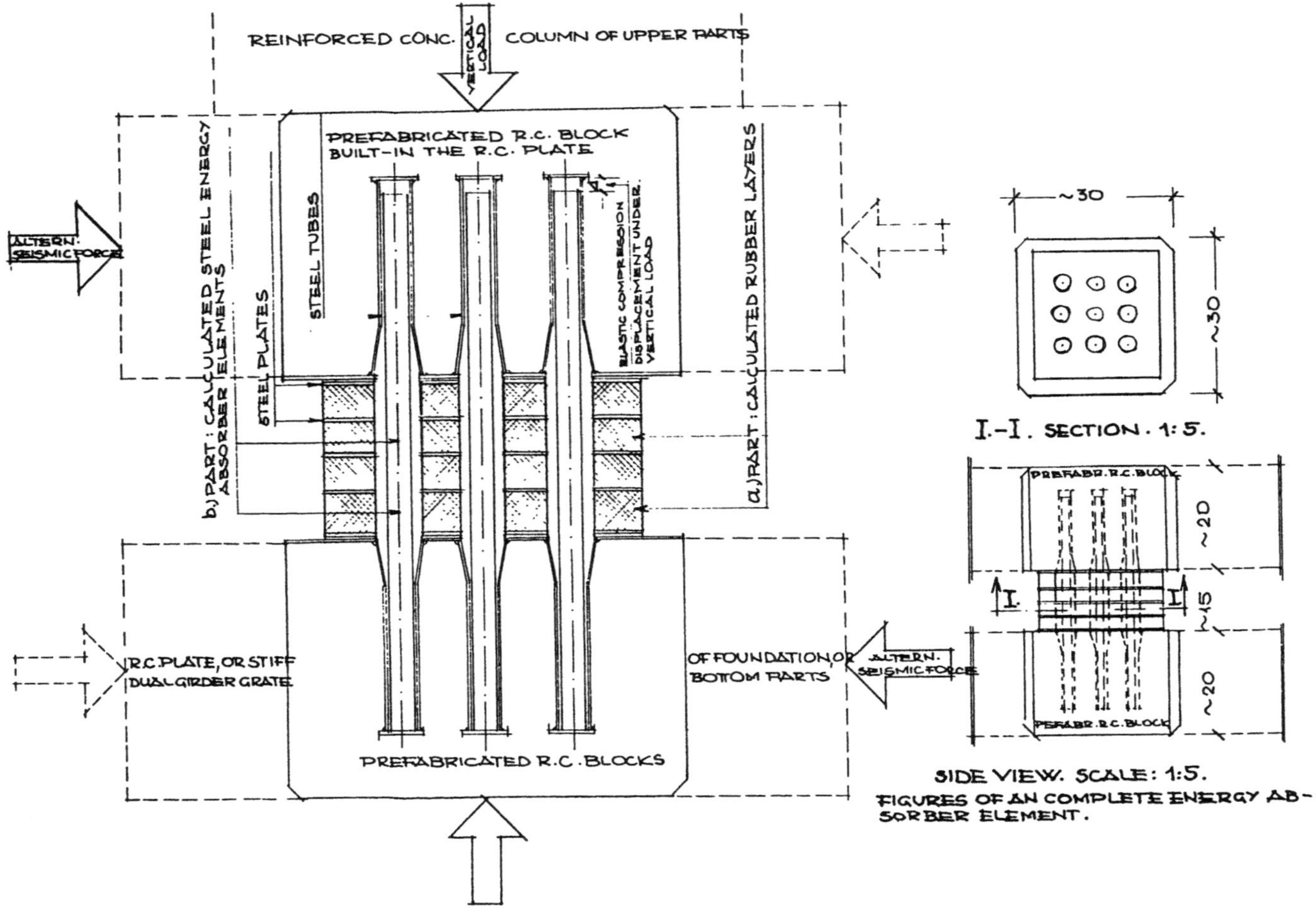

Fig. 13 Inbuilt-system and function scheme of an energy absorber element (vertical section detail) scale: 1:2.

lower structures at their upper ends. The steel pins are not loaded by vertical forces, so they ensure only the elastoplastic connection of the foundation and the upper structure against horizontal motions.

In case of repeated elasto-plastic deformations the system absorbs kinetic energies, Fig.14. If, after certain plastic horizontal displacement, according to the principle of progressive springing and following the hardening mechanical material model, a new elastic or elasto-plastic element is entering into action, this demand is also fulfilled. Consequently the motion of the building can be limited to the desired degree.

4.5 Calculation method (11, 12)

According to Fig.9, in simple cases the calculation model is a one degree of freedom system, and from this we consider the following:

- The whole building is a rigid body.
- The isolator and energy absorber system is a viscoelastic, elastoplastic system and the elastic and elastoplastic parts of them are parallel coupled.
- In the elastic regions the system is linear.
- In the elastoplastic regions the rubber elements are elastic, the steel energy absorber elements are plastic. According to Fig 16. the system is a progressive "spring system", hardening up in the different phases of displacement.
- Calculation of the eigenvalues of the swing system in the elastic and in elastoplastic range can be made according to the methods of engineering vibration theory.
- In complicated cases of multistorey and frame buildings and construction systems etc, we must calculate with a multidegree of freedom system.

4.6. Application fields

Figs. 12 to 17 show a wide range of application possiblities.

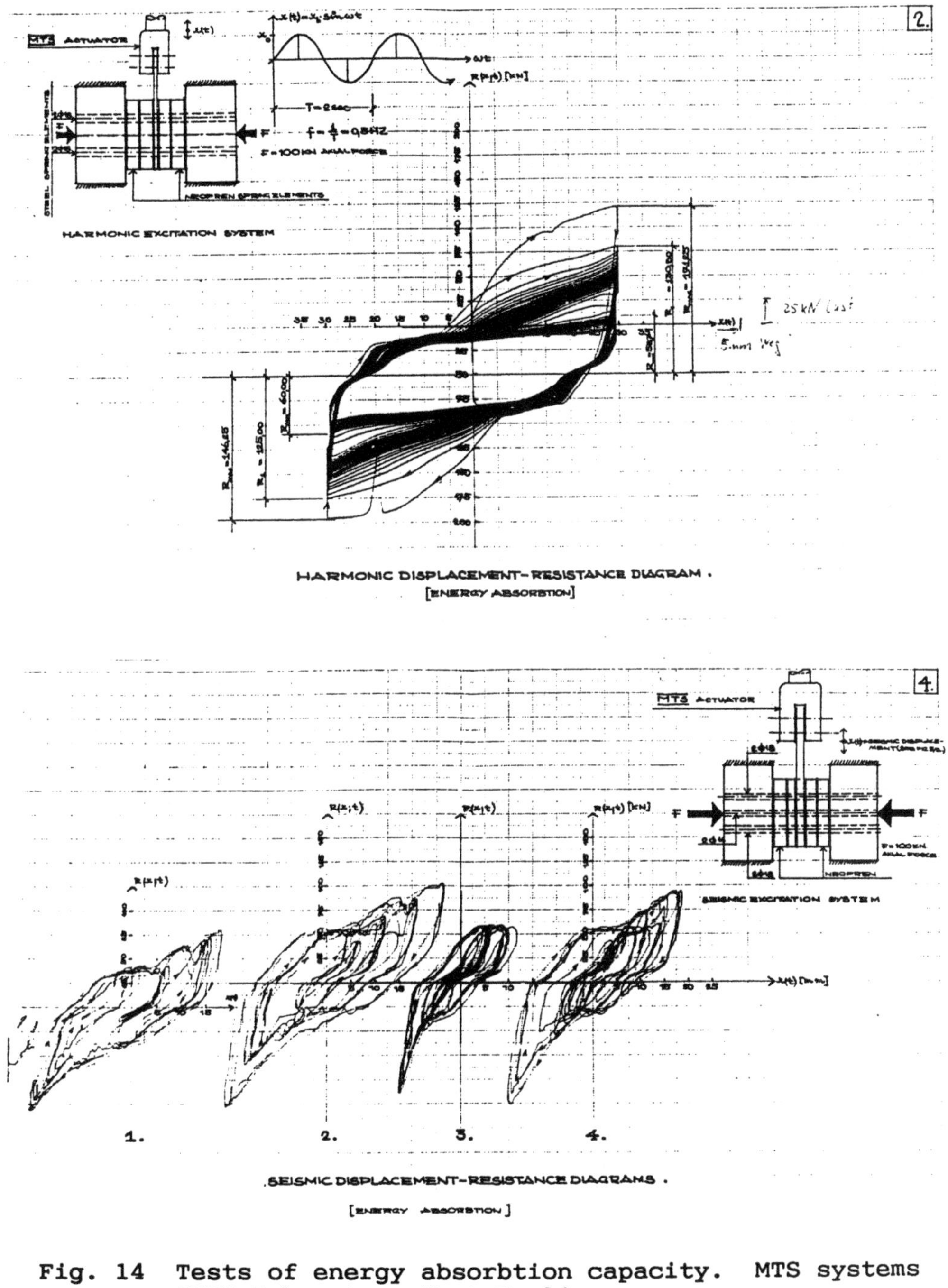

Fig. 14 Tests of energy absorbtion capacity. MTS systems GMBH Laboratory West Berlin.

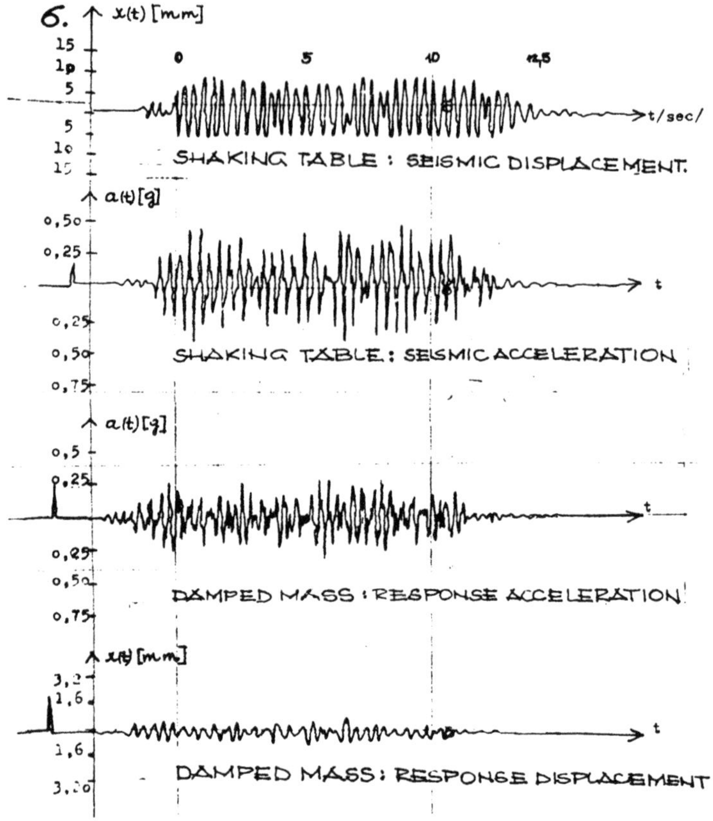

Fig. 15 **Shaking table tests. Interatom Laboratory Bensberg, West Germany.**

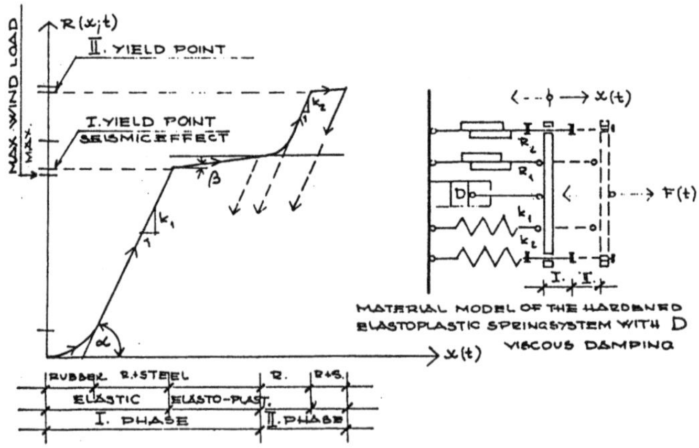

Fig. 16 **Schematic displacement resistance diagram of the hardened progressive spring system.**

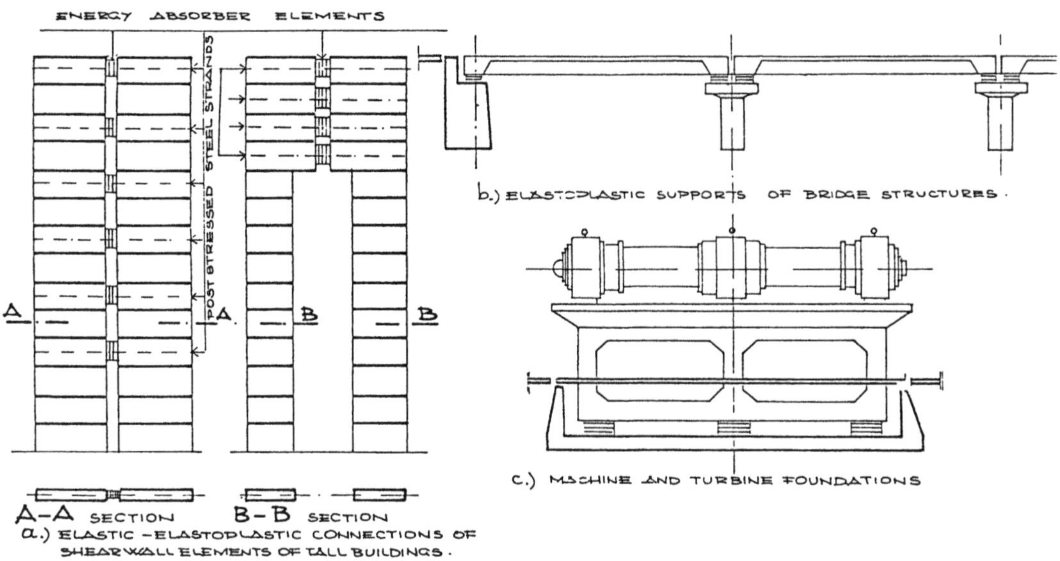

Fig. 17 Another application fields of energy absorber elements.

4.7. Advantages of the application

Technical advantages:
By the use of the system such buildings may be erected in active seismic zones in which, independently from earthquake intensity, only predetermined seismic forces, internal forces and motions will occur. In this way damage of the buildings can be entirely avoided.

Economical advantages:
According to calculations and analysis of practical examples, the use of the system increases building costs by 2-4%. There are also examples, however, where 2% building cost reduction can be demonstrated. The great economic advantage comes from the fact that buildings, designed and built by using the base isolation system, do not get damaged during an earthquake. Consequently, it is not necessary to reckon with the important costs of subsequent repairing and strengthening, and with costs resulting from the interruption of operations.

References

(1) J.M. Kelly: Progress and Protect in Seismic Isolation Proceedings of a seminar and workshop on base isolation and passive energy dissipation. March 12-14 1986, San Francisco, California. pp. 29-37.

(2) I.G.Buckle: Development and Application of Base Isolation and Passive Energy Dissipation: A World Overview. Proceedings. o.f. pp. 153-174.
(3) A.G.Tarics: The Acceptance of Base Isolation for Earthquake Protection of Buildings. Proc. 8th World Conf. on Soil Dynamics and Earthquake Engineering, June 22-24 1987, Princeton Univ., USA.
(4) Masanori Izumi: Base Isolation and Passive Response Control. Proc. 9th World Conf. on Earthquake Engineering. Aug. 2-9, 1988, Tokyo. pp. VIII. 385-396.
(5) Shimizu Co.Ltd.: Shimizu Base Isolation System. Outline of Development of Base Isolation System. At Shimizu. 1987. 6.
(6) Kajima Corp.: Dynamic Intelligent Building.
(7) Satoshi Fujita et al: Earthquake Isolation Systems for Buildings of Industrial Facilities using Various Types of Damper. Proc. 9th World Conf. on Earthquake Engineering August 2-9 1988, Tokyo.
(8) B. Csak: Gorgan Iran: Four fodder-mixer establishments, built in 1976-77.
(9) MTS-System GMB: Interatom: Laboratory and Shock-Table Tests, 1981-84
(10) B. Csak et al: Proc. 8th World Conf. on Earthquake Engineering; July 21-29 1984, San Francisco. Elasto-plastic spring elements for diminution of seismic forces and absorbtion of kinetic energies respectively seismic isolation. Vol.V.p.1063.
(11) B. Csak et al: Proc. of 8th European Conf. on Earthquake Engineering, Sept. 7-12 1983, Lisbon, Portugal. Elastic-Elastoplastic-Energy Absorber Bearing Elements for Diminution of Seismic Forces and Movements of Buildings. Base Isolation.

14 EXPERIMENTAL INVESTIGATIONS ON SHAKEDOWN: A STATE-OF-ART

S. RIZZO and A. FAILLA
Department of Structural Engineering and Geotechnics,
University of Palermo, Italy

SUMMARY

The shakedown occurrence of structures subjected to variable repeated or cyclic loads represents a safe behaviuor, for the structure is made able to adapt to such cyclic loading sequences in a purely elastic manner, although some plastic deformation of finite limited amount may have developed at initial loading phases and then stop. The paper approaches a first state-of-art on experimental reseaches on shakedown behaviour: the results obtained from various structural typologies are closely discussed and compared with the phenomenology of adaptation and the theoretical expected responses. While the importance of such a peculiar elasto-plasto-elastic behaviour is stressed by the results, the issues of further experimental research is this domain are outlined.

1 Introduction

Structures at service in almost all the engineering fields (civil, mechanical, aeronautical, nuclear, off-shore, etc.) undergo -- during their life -- to actions of various nature (loads, thermal variations, etc.) which are usually variable and repeated. Moreover the extreme values only of such actions can be estimated in the engineering practice, while the actual load time-history remains a-priori unknown.

Under such conditions plastic deformations may develop during the loading history. It has been observed however that, under appropriate circumstances, a peculiar elasto-plasto-elastic behaviour may occur, which is theoretically and experimentally ascertained as the concept of structural shakedown or adaptation: it occurs when plastic deformations of finite amount only develop during the initial loading phases, and thereafter the structure "shakes down" into a purely elastic behaviour, or also it is said "to adapt" to subsequent loadings in a purely elastic manner. The shakedown occurrance prevents therefore the failure of the structure, which could have taken place as either alternating plasticity collapse mode (when plastic deformations of opposite sign develop at some points of the structure subject to load cycling); or incremental collapse mode (when plastic deformations do not cease to develop and the structure serviceability is lost, eventually).

The theoretical concept of structural adaptation can be traced back to 1926, Gruning (1), Bleich (2) in 1932, Melan (3) in 1936, and -- since then -- it attracted the grooving interest of leading researchers: references (1) to (15) represent a very short and necessarily incomplete list of contributions on the theoretical side of the subject, the complete one being well out of this paper aims. This latter is in fact devoted to outline a state-of-art of the experimental investigations on the shakedown behaviour, carried out in order to better

Testing of Metals for Structures. Edited by F.M. Mazzolani. © 1992 RILEM.
Published by Taylor & Francis, 2 Park Square, Milton Park, Abingdon, Oxon, OX14 4RN.
ISBN 0 419 15810 3

understand the physical basis of the adaptation phenomenon and to seek closely the limits of its applicability in the engineering practice.

In the following section 2. the theoretical conditions for adaptation are resumed, along with a discussion of the main parameters and related issues. In section 3. a first list of references, (16) to (70), on experimental studies on shakedown behaviour -- although not exhaustive -- is closely discussed. The needs of further experimental research on this subject is then outlined in section 4.

2 The theoretical fundamentals of the shakedown phenomenon

Let us have a structure elastic perfectly-plastic subjected to cyclic and arbitrarily variable loadings. However loads and load combination and sequence are assumed to fall inside an assigned loading domain which takes the shape, in a multidimensional loading space, of a convex hyperpolyhedron: loading histories inside such a convex domain only are taken as the "admissible" ones (14).

Let us also assume that geometric linearity holds, i.e. to assume -- for sake of simplicity -- small strains.

The static shakedown theorem, as formulated by Melan (3), states that, once it is known the perfectly elastic response to a given loading, if it exists a distribution of time-independent self-stresses such that superimposed to the elastic stresses nowhere in the structure the plastic yield conditions are violated, then shakedown will necessarily occurs, and plastic deformations developed tends to a finite value. It is remarkable to observe that shakedown must occur at low cicle number ($N = < 10^4$), for which it falls inside the "low-cicle fatigue" field. But the actual loading sequence remains unknown, in the engineering practice: from this difficulty stems the need of mathematical formulations apt to ensure the shakedown occurrence for any "admissible" loading history.

In such a framework the actual value of the plastic deformations, although necessarily finite, will remain unknown. Such a circumstance does not ensures however that some assigned limited ductility or serviceability criteria be fulfilled, i.e. that the actual plastic deformation values be everywhere less or equal to the limit values provided by such criteria.

It follows that, in order to render meaningful the shakedown occurrence, methods for approximate assessment of the plastic deformations need to be provided in practice, methods which takes the shape of an a-priori Upper Bounding Principle, and which received a number of formulations (9, 10, 12, 14, 15).

For structures which can be discretized by finite elements connected by plastic hinges, a typical discrete formulation of shakedown conditions is as follows (7-15) (vector quantities in bold character)

$$\max_{(s,\,\mathbf{X})} s \quad \Big\| \quad s\,\mathbf{P} + \mathbf{N}^T \mathbf{R}\,\mathbf{X} - \mathbf{k} \le 0\,, \tag{2.1}$$

where: s is the loading domain multiplier; and the following vector quantities respectively collect: $\mathbf{P} = \max(\mathbf{N}^T\,\mathbf{Q}^E_j)$, the most severe perfectly elastic responses for given domain (which occurre at its j vertices); $\mathbf{R}\,\mathbf{X}$, the time-independent self-stresses; $\mathbf{k}$, the plastic yield strengths; N^T, the unit versors, outgoing with respect to the hyperplanes of the convex yield domain.

Such a formulation focuses, the first parameter to obtain in shakedown analyses, i.e. the shakedown safety factor s_{SD}, or the optimal value solution of the Linear Mathematical Programming problem (2.1).

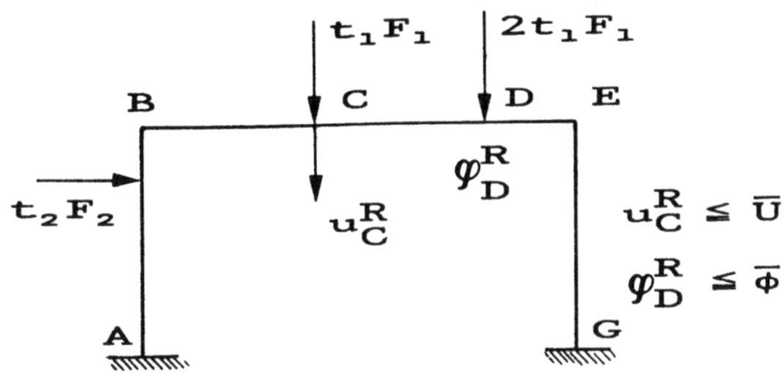

Fig. 1 Typical structure subjected to arbitrarily variable loadings

The above formulation, provided that shakedown occurred, is however unable -- as said before -- to give the finite value reached by the plastic deformation. To this purpose let U_i be some limit value assigned -- by appropriate limited ductility or serviceability criteria -- to some deformation parameter, i.e. residual displacement u^R or φ^R in Fig. 1, or residual strain or stress at some relevant node of the structure. The use of the Upper Bounding Principle allows one to re-formulate the problem by introducing a term, called perturbation vector **d** which will be appropriately defined according to the plastic deformation measure selected as relevant, and by appending a quadratic constraint which ensure that actual plastic deformations do not overcome the assigned limit. Such a quadratic constraint (so-called perturbation method in Ref. 14) constitutes the a-priori upper bounding of the actual plastic deformations developed, which are only apprimately assessed. This new formulation is as follows,

$$\max_{(s,\mathbf{X}_i,\omega_i)} s \quad \left\| \begin{array}{l} s\,\mathbf{P} + \mathbf{N}^T \mathbf{R}\,\mathbf{X}_i - \mathbf{k} + \omega_i\,\mathbf{d}_i \leq 0\,, \\ \\ \dfrac{1}{2\,\omega_i}\,\mathbf{X}_i^T\,\mathbf{a}\,\mathbf{X}_i \leq U_i\,, \quad \omega_i \geq 0\,, \\ \\ \qquad (i = 1, n) \end{array} \right. \tag{2.2}$$

which constitutes a convex Nonlinear Mathematical Programming problem. This problem carries as many quadratic constraints as the n points or nodes of the structure on which controls on plastic deformation measures are requested.

For both the problems (2.1) and (2.2) effective computation strategies have been successfully provided (16), which also leaded to the more evolute concept of "limited ductility shakedown safety factor" s_B, i.e. the maximum load domain amplifier solution of nonlinear problem (2.2), which ensures shakedown occurrennce for any admissible loading cycle inside a loading domain amplified by s_B and -- at once -- that the imposed ductility serviceability limits are not violated (or $u^R \leq U_i$). While the plastic deformations are bounded from above, the s_B amplifier bounds from below the shakedown safety factor s_{SD}, Fig. 2.

Theoretical studies provided therefore a deep interpretation of the shakedown behaviour, allowing to identify the shakedown load domain amplifier and the upper-bounds as the main parameters to obtain by analysis.

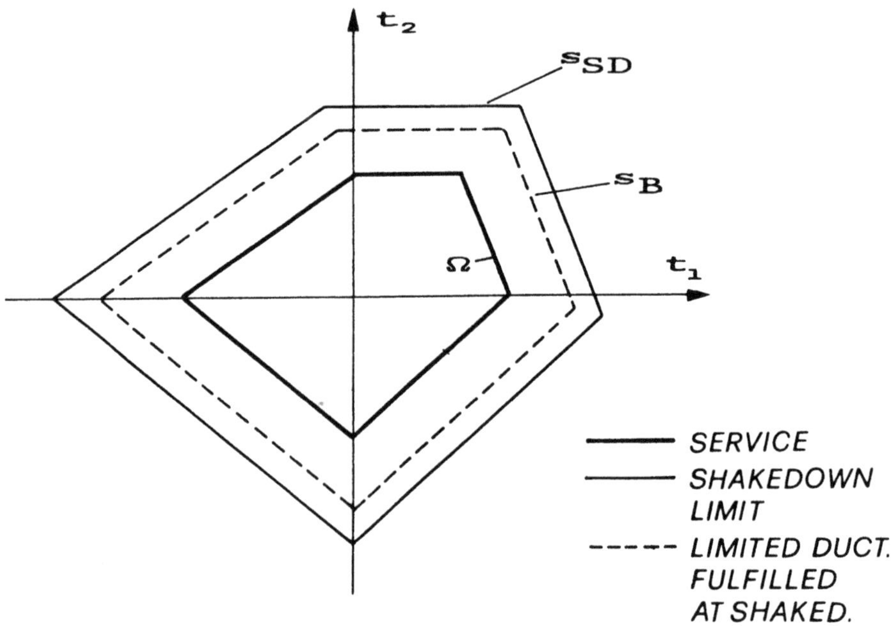

Fig. 2 Loading domains

3 An overview on experimental research on shakedown

Papers and reports devoted to shakedown experimental research trace-back to 1923, Ludwik (17), but intensive activity in this field started in the early fifties only and continued until today. References (17) to (70) are offered herein as a first trial list on this topic, with 7 papers in the fifties (18-24), 21 papers in the sixties (25-45), 14 papers in the seventies (46-59), and 11 papers in the eigthies (60-70): authors apologize to the readers for unwanted omissions or incompleteness in the above list.

Different structural typologies, suited to the civil, mechanical and nuclear engineering needs, have been tested, as well as a number of material specimens (17, 26, 37-39, 44, 65, 68) subjected to repeated loading and suitably constrained for shakedown to occur.

The first tests approached small-scale and medium scale steel propped beams, continuous beams and simple framed hyperstatic structures (one - and two - storey frames), (18-22), typologies intensively tested also subsequently (24-25, 28-30, 32-34, 40, 43, 46, 52, 57, 60-61).

Special structures or components have been also investigated such as steel arches (23, 27); space tubular beams (51); space truss nodes (62); and bridge deck or steel orthotropic grid (55).

The above studies were concerned with case-study structures mainly civil-engineering oriented, i.e. structural components for building and bridge engineering.

In latest years however the shakedown behaviour has been identified in the nuclear engineering field, as a very essential limit-state to be investigated, for safety against cyclic actions on the main structural parts of the reactors, i.e. the pressure vessels and the steam tubes. In such structures the shakedown occurrence must necessarily studied in connection with the parental "ratchetting" behaviour or the platic deformation stabilization under sustained internal pressure and loadings and repeated thermal cycles. To the shakedown and ratchetting behaviours, and to the borders between the two, are devoted the papers (31), (47-49), (53-54), (56), (58-59), (63-70): these papers are almost all in dialogue with the paper (36) by Bree, a real mile-stone paper for the safety concerns of nuclear engineering structures.

The whole of all these experimental contributions possesses a long lasting value, also in consideration of the amount of difficulties to carry out the shakedown tests:

- need of accurate and previous numerical analyses;
- identification of a set of eligible loading histories;
- benchmark design, with controls against local failures or buckling development, eventually;
- influence of residual stresses and related appropriate procedure for specimen preparations (annealing);
- time-rate effects on the loading process, i.e. the influence on test of creep and of strain-hardening;
- and last but least, time-consuming and cost.

The relevant parameters to be observed and/or measured during shakedown tests can be summarized as follows: -i) onset of plastic deformations and their development and corresponding loads levels; -ii) value of the chosen plastic deformation measures (strain, residual or total displacement at selected monitored points of specimen); -iii) no. of cycles needed for either adaptation to loads, i.e. deformation stabilization, or inadaptation, i.e. failure; -iv) time needed for cycle adaptation (for loads below the shakedown limit); -v) load intensities at shakedown occurrence, or at failure; -vi) creep occurrence; vii) local failures (local buckling).

The results obtained from the quite large number of shakedown tests reported in the literature shows the effectiveness of the shakedown theory for the good agreement betweeen predicted and measured shakedown loads, with the recorded values sligthly above the theoretical ones. Further, the shakedown load under repeated cyclic loads on fixed points and the shakedown load under repeated moving loads are identical (43), as theoretically expected.

The permanent deformations or displacements recorded were clearly of plastic and not creep nature, the latter being generally negligible in comparison with the former.

The plastic displacements or deformations measured at shakedown occurrence are markedly larger (up to five-six times) than the elastic ones, what confirms the needs for approximate estimations of them via upper-bounding methods, in practice. Also the goodness of such methods, or the stringentness of upper-bounds they provide, has not been verified for lack of comparison with the measured displacements.

Strain hardening plays non essential role in plastic deflections or deformations stabilization for load levels below the shakedown limit. It has also been observed how the finite spread of plastic hinging zones affects the shakedown deflections (60).

The number of cycles for shakedown occurrence is very low, from a few unities to less than 100 cycles, and reaches 500 cycles in the tests on shakedown-ratchetting borders, typical in the nuclear engineering field (58).

4 Issues for future experimental research in shakedown and related domains

The tests reported in the literature give results on specimen-structures subjected to cyclic action of some single assumed loading history. However theory provides analytical conditions for shakedown to occurr for any "admissible-loading-history", inside a given convex domain: the multiple or various loading-histories is therefore a main issue to be verified by future tests on shakedown behaviour.

The correctness of plastic deformations upper bounding techniques still need verifications, and comparison between recorded values and assessed ones is still waited.

The borders between shakedown and classical fatigue behaviours need also to be investigated: theoretically once shakedown has occurred the structure should respond with a purely elastic behaviour to an infinite number of subsequent load cycles, but almost all tests were stopped soon after shakedown occurrence. Does the shakedown occurrence changes the cycles number at which some "post-shakedown" fatigue-failure is to be expected?: such a question probably deserves further theoretical and experimental research.

The above issues summarize some of the main needs for further experimentally oriented research in shakedown framework, together with the study of second-order geometric effects on displacement development. On the above topic experimental researches at the author's Department is being set-up.

5 Aknowledgments

This work is part of a research on shakedown behaviour, financially supported by the Italian National Ministry of the Research and the University. The authors thank also Filippo Gianfala for manuscript preparation.

6 References

1. Gruning, M., "Die Tragfahigkeit statisch unbestimmter Tragwerke ans Stahl bei beliedig haufig wiederholtes belastung", Springer, Berlin, 1926.

2. Bleich, H., "Uber die Bemessung statisch unbestimmter Stahlwerke under der berucksichtigung des elastich-plastischen Verhaltens des Baustoffes", Bauingenieur 13, p.261-267,1932.

3. Melan, E., "Theorie statisch unbestimmter Systeme", Proc. 2nd IABSE Congress, Berlin, p. 43-64, 1936.

4. Symonds, P.S., and Prager, W., "Elastic-plastic analysis of structures subjected to loads varying arbitrarily between prescribed limits", Trans. ASME, Sez. E, J. Appl. Mech., 17, p. 315-323, 1950.

5. Neal, B.G., "The behaviour of framed structures under repeated loading", Quart. Appl. Math., 4, p. 78-94, 1951.

6. Koiter, W.T., "General theorems for elastic-plastic solids", in "Progr. in Solid Mech", North Holland, Amsterdam, p. 165-221, 1960.

7. Maier, G., "Shakedown theory in perfect elastoplasticity with associated and non-associated flow-laws", Meccanica, 6, p. 250-260, 1969.

8. Ponter, A.R.S., "On the relationship between plastic shakedown and the repeated loading of creeping structures", Trans. ASME, sez. E, J. Appl. Mech., 38, p. 437-440, 1971.

9. Sawczuk, A., "Shakedown analysis of elastic-plastic structures", Nucl. Eng. Des., 28, p. 121-136, 1974.

10. Capurso, M., "A displacement bounding principle in shakedown of structures subjected to cyclic loads", Int. J. Solids and Struct., 10, p. 77-92, 1974.

11. Martin, J.B., "Plasticity", MIT Press, Boston, 1975.

12. Corradi, L., "Mathematical programming methods for displacement bounds in elasto-plastic dynamics", Nucl. Eng. Des., 37, p. 161-177, 1976.

13. Gokhfeld, D.A., and Cherniavsky, O.F., "Limit analysis of structures at thermal cycling", Sijthoff and Noordhoff, 1980.

14. Polizzotto, C., "A unified treatment of shakedown theory and related bounding techniques", S.M. Archives, 7, p. 19-75, 1982.

15. Konig, J.A., "Shakedown of elastic-plastic structures", Elsevier, Amsterdam, 1987.

16. Rizzo, S., and Giambanco, F., "Shakedown analysis of limited-ductility structures", Meccanica, 19, p. 151-157, 1984.

17. Ludwik, P., "The behaviour of metals under repeated straining", Zeitschrift, Verein Deutscher Ing., Berlin, 67, p. 122-126, 1923.

18. Neal, B.G., "The behaviour of framed structures under repeated loading", Quart. Appl. Math., 4, p. 78-94, 1951.

19. Symonds, P.S., "Cyclic loading tests on small-scale portal frames", Proc. 4th IABSE Congr., Cambridge, p. 109-112, 1952.

20. Massonnet, C., "Plastic collapse and shake-down tests on small rolled beams", IABSE Publ., 13, p. 239-282, 1953.

21. Neal, B.G., and Symonds, P.S., "Cyclic loading tests on small-scale portal frames", Proc. 5th IABSE Congr., London, p. 109-119, 1953.

22. Baker, J., Heyman, J., and Horne, M.R., "The steel skeleton", Cambridge Univ. Press., Cambridge, 1956.

23. Stevens, L.K., "Carrying capacity of mild-steel arches", Proc. Inst. Civ. Eng., p. 493-514, 1957.

24. Neal, B.G., and Symonds, P.S., "Cyclic loading of portal frames, theory and tests", IABSE Publ., 18, p. 171-199, 1958.

25. Popov, E.P., and Mc Carthy, "Deflection stability of frames under repeated loads", J. Eng. Mech. Div., ASCE Proc., 86, p. 61-78, 1960.

26. Benham, P.P., and Ford, H., "Low endurance fatigue of a mild steel and aluminium alloy", J. Mech. Eng. Sc., 3, p. 119-132, 1961.

27. Bruzzese, E., and Augusti, G., "Incremental collapse of a steel arch bridge under repeated live and thermal loading", Proc. Conf. on Thermal Load and Creep, Naples, 1960.

28. Popov, E.P., and Becker, E., "Shakedown analysis of same typical one-storey bents", J. Struct. Div., ASCE Proc., 90, p. 1-9, 1964.

29. Bertero, V.V., and Popov, E.P., "Effects of large alternating strains of steel beams", J. Struct. Div., ASCE Proc., 91, p. 1-12, 1965.

30. Davies, J.M., "The stability of plane frameworks under static and repeated loading", Ph.D. Thesis, Manchester Univ., 1965.

31. Fessler, H., and Stanley, P., "Stresses in torispherical drumheads: a photoelastic investigation", J. Strain Anal., 1, 1965.

32. Rasmussen, B.H., "Incremental collapse of ordinary RC beams", IABSE Publ., 16, p. 439-491, 1965.

33. Davies, O.M., "Frame instability and strain hardening in the plastic theory of the structures", J. Struct. Div., Proc. ASCE, 92, p. 1-16, 1966.

34. Heyman, J., and Threfall, "The repeated loading of ductile steel structures", Proc. Inst. Civ. Eng., 34, p. 477-501, 1966.

35. Save, M., "Verification experimentale de l'analyse limite plastique de plaques et de coques en acier doux", Centre de Rech. Scient. et Techn. de l'Ind. de Fabric. Metal., Brussels, Rep. MT21, 1966.

36. Bree, J., "Elastic-plastic behaviour of thin tubes subjected to internal pressure and intermittent high-heat fluxes with applications to fast-nuclear reactor fuel elements", J. Strain Anal., 8, p. 226-238, 1967.

37. Dietrich, T., "Experiments on a type of shakedown comprising cyclic plastic deformation in alternating directions", Trans. ASME, sez. E, J. Appl. Mech., p. 780-782, 1967.

38. Pascal, K.J., and De Villiers, J.W.R., "Low cycle fatigue of steel under biaxial straining", J. Strain Anal., 2, p. 117-129, 1967.

39. Findlay, G.E., Moffat, D.G., and Stanley, P., "Elastic stresses in torospherical drumheads: experimental verification", J. Strain Anal., 3, 1968.

40. Heyman, J., Johnson, R.P., Fowler, P.P., and Gillson, I.P., "Shakedown analysis", The Structural Engin., 46, p. 97-106, 1968.

41. Pickett, A.G., and Grigory, S.C., "Cyclic pressure tests of full-size pressure vessels", Wold. Res. Council, Bull. No. 435, 1968.

42. Proctor, E., and Flinders, R.F., "Shakedown investigations on partial penetration welded nozzles in a spherical shell", Nucl. Eng. Des., 8, p. 171-185, 1968.

43. Fukumoto, Y., and Yoshida, H., "Deflection stability of beams under repeated loads", J. Struct. Div., ASCE Proc. 95, p. 1443-1458, 1969.

44. Kato, B., and Akiyama, "Inelastic bar subjected to thrust and cyclic bending", J. Struct. Div., ASCE Proc., 95, p. 33-56, 1969.

45. Sawczuk, A., "On incremental collapse of shells under cyclic loading", IUTAM Symp. on Theory of Thin shells, Copenhagen, 1967, Spinger Verlag, Berlin, p. 328-340, 1969.

46. Eyre, D.G., and Galambos, T.V., "Shakedown tests on steel bars and beams", J. Struct. Div., ASCE Proc., 96, p. 1359-1379, 1970.

47. Findlay, G.E., Moffat, D.G., and Stanley, P., "Torospherical drumheads, a limit-pressure and shakedown investigation", J. Strain Anal., 6, p. 147-166, 1971.

48. Fox, J.D., Kraus, H., and Penny, R.K., "Shakedown of pressure vessels with ellipsoidal heads", ASME Conf. on Pres. Vessels and Piping, San Francisco, 1971.

49. Meyer, J.D., and Gerstle, K.H., "Shakedown of strain-hardening structures", J. Struct. Div., ASCE Proc., 98, p. 95-110, 1972.

50. Chang, T.Y., and Chu, S.C., "Elastic-plastic deformations of cylindrical pressure vessels under cyclic loading", 2nd SMIRT Conf., Berlin, paper 17/9, 1973.

51. Del Puglia, A., and Nerli, G., "Experimental research on elasto-plastic behaviour and collapse of statically indeterminate space tubular beams", 2nd SMIRT Conf., Berlin, paper I 3/7, 1973.

52. Eyre, D.G., and Galambos, T.V., "Deflection analysis for shakedown", J. Struct. Div., ASCE Proc., 99, p. 2049-2060, 1973.

53. Anaud, S.C., and Garg, V.K., "Shakedown of rolling wheel under hub loading", J. Eng. Mech. Div., ASCE Proc., 100, p. 1237-1251, 1974.

54. Aylward, R.G., Galletly, G.D., and Moffat, D.G., "Buckling under external pressure of cylinders with toriconical or pierced torispherical ends: a comparison of experiment with theory", J. Mech. Eng. Sc., 17, 1975.

55. Ceradini, G., Gavarini, C., and Petrangeli, " Steel orthotropic plates under alternate loads", J. Struct. Div., ASCE Proc., 101, p. 2015-2026, 1975.

56. Moreton, D.N., Moffat, D.G., "Shakedown of a torispherical drumhead made from a strain-hardening material", 3rd SMIRT Conf., London, paper G 2/7, 1975.

57. Megahed, M.M., "Behaviour of simple structures under cyclic loading conditions", Engrg. Dept., Leicester Univ., U.K., Reports No. 77-12, 77-13, 77-14, 1977.

58. Moreton, D.N., and Moffat, D.G., "Shakedown of three stainless steel pressure vessel components", 3rd Int. Conf. on Press. Vessel Techn., Tokyo, p. 233-245, 1977.

59. Moreton, D.N., Moffat, D.G., and Hornby, R.P., "Techniques for investigating the yield surface behaviour of pressure vessel materials", J. Strain Anal., 13, p. 185-191, 1978.

60. Dorosz, S., Konig, J.A., Sawczuk, A., Biegus, A., Kowal, A., and Seidel, W., "Deflections of elastic-plastic hyperstatic beams under cyclic loading", Arch. Mech., 33, p. 611-624, Warszawa, 1981.

61. Ali, M.H., Konig, J.A., and Mahrenholtz, O., "Experimental investigations on shakedown of portal frames", Eng. Trans., Polska Akademia Nauk, 32, p. 342-359, 1984.

62. Karczewski, J.A., Konig, J.A., "Experimental shakedown analysis of space truss nodes", Eng. Trans., Polska Akademia Nauk, 32, p. 545-561, 1984.

63. Mahrenholtz, O., Leers, K, and Konig, J.A., "Shakedown of tubes: a theoretical analysis and experimental investigations", in Metal Forming and Impact Mech., W. Johnson Comm. Vol., Pergamon Press, p. 155-172, 1984.

64. Ng, H.W., Brookfield, D.J., and Moreton, D.N., "A technique for conducting shakedown tests at elevated temperatures", Strain, p. 7-14, 1984.

65. Cocks, A.C.F., and Ponter, A.R.S., "The plastic behaviour of components subjected to constant primary stresses and cyclic secondary strains", J. Strain Anal., 20, p. 7-14, 1985.

66. Leers, K., "Experimentelle und theoretische shakedown untersuchung an Rohren", VDI Verlag, Dusseldorf, 1985.

67. Lers, K., Klie, W., Konig, J.A., and Mahrenholtz, O., "Experimental investigations of shakedown of tubes", in Plasticity Today, Sawczuck and Bianchi Eds., Elsevier Publ., London, p. 259-275, 1985.

68. Ponter, A.R.S., Jakeman, R.R., and Morrison, C.J., "An experimental study of simplified methods for the prediction of the deformation of structures subject to severe cyclic thermal loading", J. Strain Anal., 20, p. 225-240, 1985.

69. Brookfield, D.J., and Moreton, D.N., "The ratchetting behaviour of stainless steel type 316 at ambient and elevated temperatures", 9th SMIRT, Lausanne, vol. 1, p. 427-432, 1987.

70. Brookfield, D.J., and Moreton, D.N., "The use of a Bree simulation to investigate strain accumulation in 316 stainless steel at temperatures between room temperature and 500° C", Strain Anal., 24, p. 95-102, 1989.

15 LIFE PREDICTION METHODS FOR LOW-CYCLE TORSIONAL FATIGUE

S. CURIONI, A. FREDDI and G. CALIGIANA
Department of Construction, Mechanical, Nuclear, Aeronautical and Metallurgical Engineering, University of Bologna, Italy

SUMMARY

Fatigue life predictions of components locally stressed in the plastic field is a very important task for engineers employed in the automotive and in the electric utility industries both for design and troubleshooting of service failures. Aim of the paper is the description of the utilization of the local strain approach to fatigue life estimates of samples simulating the behaviour of turbine-generator shafts subject to torsional oscillations. An outline of the experimental and theoretical job done by the authors is given.

Symbols

τ_a, γ_a shear stress, shear strain amplitudes
τ_{ott}, γ_{ott} octahedral shear stress, octahedral shear strain amplitudes
$\Delta\tau$, $\Delta\gamma$ shear stress, shear strain ranges
T_a, Γ_a nominal shear stress, nominal shear strain amplitudes
K_t theoretical stress concentration factor
N number of cycles to failure
M_a torque amplitude
θ_a angle of twist per unit length of the cylindrical smooth specimen
G modulus of elasticity for shear stress
I_p polar moment of inertia

INTRODUCTION

Turbine-generator shafts are subject to severe torsional loading histories due to vibrations produced by several different occurrences (short circuits, out of phase synchronization, etc.).

- It has been recognized [1] that only few high transient loads cause almost all the fatigue damage. The need results of the accurate consideration of the material behaviour in the low cycle fatigue range.

Testing of Metals for Structures. Edited by F.M. Mazzolani.
Published by Taylor & Francis, 2 Park Square, Milton Park, Abingdon, Oxon, OX14 4RN.
ISBN 0 419 15810 3

- The presence of geometrical discontinuities is unavoidable in such components (fig. 1) leading to yielding in some restricted areas; owing to the elastic region surrounding the plastic zone, the material behaves locally in strain control; local strain approach and strain-life methods, up to the initiation of the crack, are utilized [2, 3, 4, 5, 6].
- Crack growth (fracture mechanics approach) is not considered here.

Local strain approach needs the development of the following steps:

- analytical modelling of the material fatigue behaviour; the fatigue properties are derived by strain controlled tests (cyclic stress-strain and fatigue curves);
- analysis of the component to account for the stress raisers and to correlate the local strain to the applied nominal load;
- counting methods to identify the more damaging events in complex loading histories accounting for the "sequence" and "memory" effects [4, 7];
- cumulative damage criteria [4, 7].

Several numerical programs were developed according to the local strain approach.

Testing rigs were designed.

Experimental tests were carried out on smooth specimens, to obtain the basic data [8, 9], and on notched specimens subject to typical torque hystories (short circuit and 120° out of phase synchronization [7, 9]), to compare numerical predictions and experimental results.

Fig. 1 - Rotor of a generator before the assemblage.

1 MATERIAL MODELLING

Several simple equations must be considered to simulate analytically the behaviour of the material. Strain controlled tests on smooth cylindrical specimens can be utilized to obtain the values of the constants in the following equations.

Cyclic curve and hysteresis loops

For the torsional cyclic curve, an extension to torsion of the Ramberg-Osgood equation [10] is considered:

$$\gamma_a = (\tau_a/G) + (\tau_a/H')^{(1/n')} \tag{1}$$

This formula can be interpreted as a particular form of the more general multiaxial relation between octahedral shear stress and octahedral shear strain [11]:

$$\gamma_{ott} = (\tau_{ott}/G) + (\tau_{ott}/C)^{(1/n')}$$

Nadai theory [12] can be utilized to derive surface shear stress amplitude of the samples from the applied torque amplitude of the "stabilized" cycle. This theory accounts for the distibution of the stress as a function of the radius of the specimen in the elastic-plastic field, leading to the equation [13]:

$$\tau_a = (M_a/2\pi r_o^3)\{3 + \theta_a/[M_a/GI_p + (1/n_o)(M_a/H_o)^{(1/n_o)}]\} \tag{2}$$

The following relation is considered in Eq. 2 between the torque and the angle of twist of the bar per unit length:

$$\theta_a = M_a/GI_p + (M_a/H_o)^{(1/n_o)}$$

The assumption of the Masing hypothesis [14], according to which the hysteresis curve has the same shape of the ciclic one with both values of the coordinates doubled, leads to the expression:

$$\Delta\gamma = (\Delta\tau/G) + 2\,(\Delta\tau/2H')^{(1/n')}$$

Fatigue curve

Extension to torsional fatigue of the low cycle fatigue strain life approach due to Manson, Coffin, Basquin [5] leads to the equation:

$$\gamma_a = A\,(2N)^{\alpha} + B\,(2N)^{\beta} \tag{3}$$

which can be, eventually, modified to account for mean stress effect [7].

2 STRESS RAISERS

Several methods can be utilized to account for the local discontinuities. The following approaches have been considered [7]:

- Neuber's rule [5, 15], which relates the local stresses and strains to the nominal stresses by means of a simple equation based on the theoretical stress concentration factor, K_t (or, otherwise, on the fatigue stress concentration factor, K_f).
- the use of a calibration curve, for the particular component utilized, relating the nominal load to the local strain [4].

Neuber's rule

The extension to fatigue in torsion of Neuber's rule [1, 16] have been considered:

$$\tau_a \gamma_a = K_t^2 \, T_a \, \Gamma_a$$

Nominal strains (Γ_a) can be related to nominal stresses (T_a), in the more general way, by utilizing Eq. 1.

Torque-local strain curve

The curve relating the torque amplitude (M_a) to the local shear strain (γ_a) is defined (like all the ciclic curves) as the locus of the tips of the stable hysteresis loops at different levels of strains. Generally this curve cannot be derived experimentally for the component examined owing to the difficulties in the measurements of the strains relative to local discontinuities. A finite element analysis of the component is the best way to obtain it. In fig. 2 the M_a-γ_a curves obtained by approximate methods (linear and Neuber's rules [5]) and by finite element methodologies are compared for a specimen of 50 mm of diameter with sharp fillets between gage length and bearing ends ($K_t \approx 1.8$, fig. 7g). Different versions of the finite element programs based on Henky's total deformation theory or on the more sophisticated Prandtl-Reuss one [9, 13] were utilized. An example of the comparison of the results is shown in fig. 3.

The use of Neuber's rule or of the discretized torque-local strain curve in numerical programs allows the prediction of the fatigue life of the notched components. Deeper details on the use of both methodologies can be found in [7] together with the description of the rain-flow counting method and the cumulative damage summation technique.

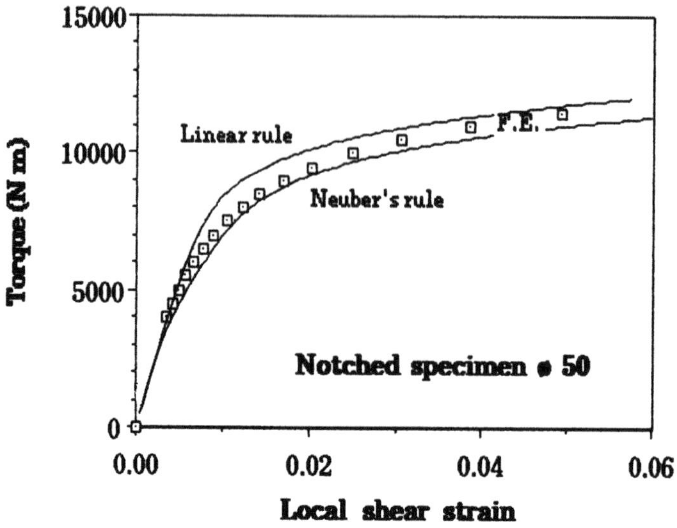

Fig. 2 - Comparison among torque-local shear strain curves obtained by approximate methods and finite element analysis for a specimen of 50 millimeter diameter.

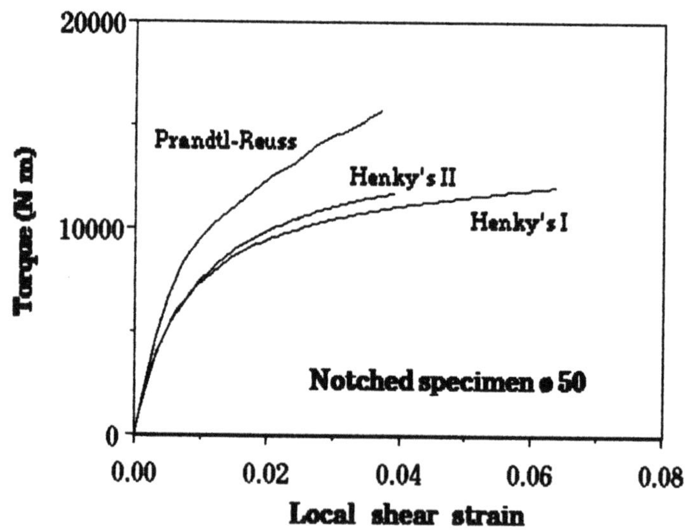

Fig. 3 - Comparison among torque-local shear strain curves obtained by different finite element programs.

3 TESTING RIGS

Push-pull tests were performed on 12 millimeter specimens by utilizing two hydraulic testing machines INSTRON series 8000.

Torsion tests were made on 18 and 24 millimeter diameter specimens by means of a testing rig which converts the linear up-down motion of the actuator of the INSTRON machine in alternate torsion [17].

A machine, able to apply torque up to 30 kN m, was designed and produced to perform tests on specimens of 50 millimeter diameter (fig. 4) [18].

Fig. 4 - Testing rig for fatigue tests in torsion (torque amplitude up to 30 kN m).

4 EXPERIMENTAL RESULTS AND FATIGUE LIFE PREDICTIONS

Basic data were derived from torsion tests performed on smooth cylindrical samples of four NiCrMoV rotor steels whose mechanical characteristics and chemical composition are reported in [9, 11]. Fig. 5 and Fig. 6 show the ciclic (Eq.1) and fatigue (Eq. 2) curves relative to the steel C1 [9]; the geometry of the specimens utilized is that of fig. 7b. For the fatigue curves in strain control, the cycles to failure were defined as those corresponding to a specified percentage loading drop in the specimen stiffness before complete fracture [17].

Tests have been performed on notched specimens (shown in fig. 7 too) subject to different loading histories. Experimental lives (expressed in cycles) are compared with fatigue predictions obtained by numerical programs [7, 9]. Fig. 8 shows an example of the data relative to the life prediction program, named FATICA 4, and to the experimental results on notched 24 millimeter specimens ($K_t \approx 1.47$, fig. 7f). The applied loading histories are shown in the figure.

5 CONCLUSIONS

Several testing rigs have been designed and produced to perform fatigue tests in torsion.

Theoretical modelling are considered to utilize the basic data obtained from smooth specimens to predict fatigue lives of notched samples subject to complex loading histories.

The use of Neuber's rule offers a conservative mean to obtain the torque-local strain curve (fig. 2).

Numerical programs based on the local strain approach allow, generally, a conservative prediction of the fatigue lives (fig. 8) of the notched specimens tested in the authors' Department [7, 9].

REFERENCES

[1] Tanaka, K., Matsuoka, S., Kouzu, F., Nagata, K., "Fatigue Strength of a Rotor Steel subjected to Torsional Loading simulating that occurring due to Circuit Breaker Reclosing in an Electric Power Plant", Fatigue of Engineering Materials and Structures, 1983, Vol. 6, No. 2, pp. 103-120.

[2] EPRI, "Determination of Torsional Fatigue Life of Large Turbine Generator Shafts", EL-3063, Research Project 1531-1, Elect. Power Res. Inst., Palo Alto, U.S.A., 1984.

[3] Dowling, N.E., "Fatigue Failure Predictions for Complicated Stress Strain Histories", Journal of Materials, 1972, Vol. 7, No. 1, pp. 71-87.

[4] SAE, "Fatigue Under Complex Loading: Analysis And Experiments", Soc. Aut. Eng., R.M. Wetzel Ed., Warrendale 1977, Vol. 6.

[5] Fuchs, H.O., Stephens, R.I., "Metal Fatigue in Engineering", Wiley & Sons Ed., 1980 (book).

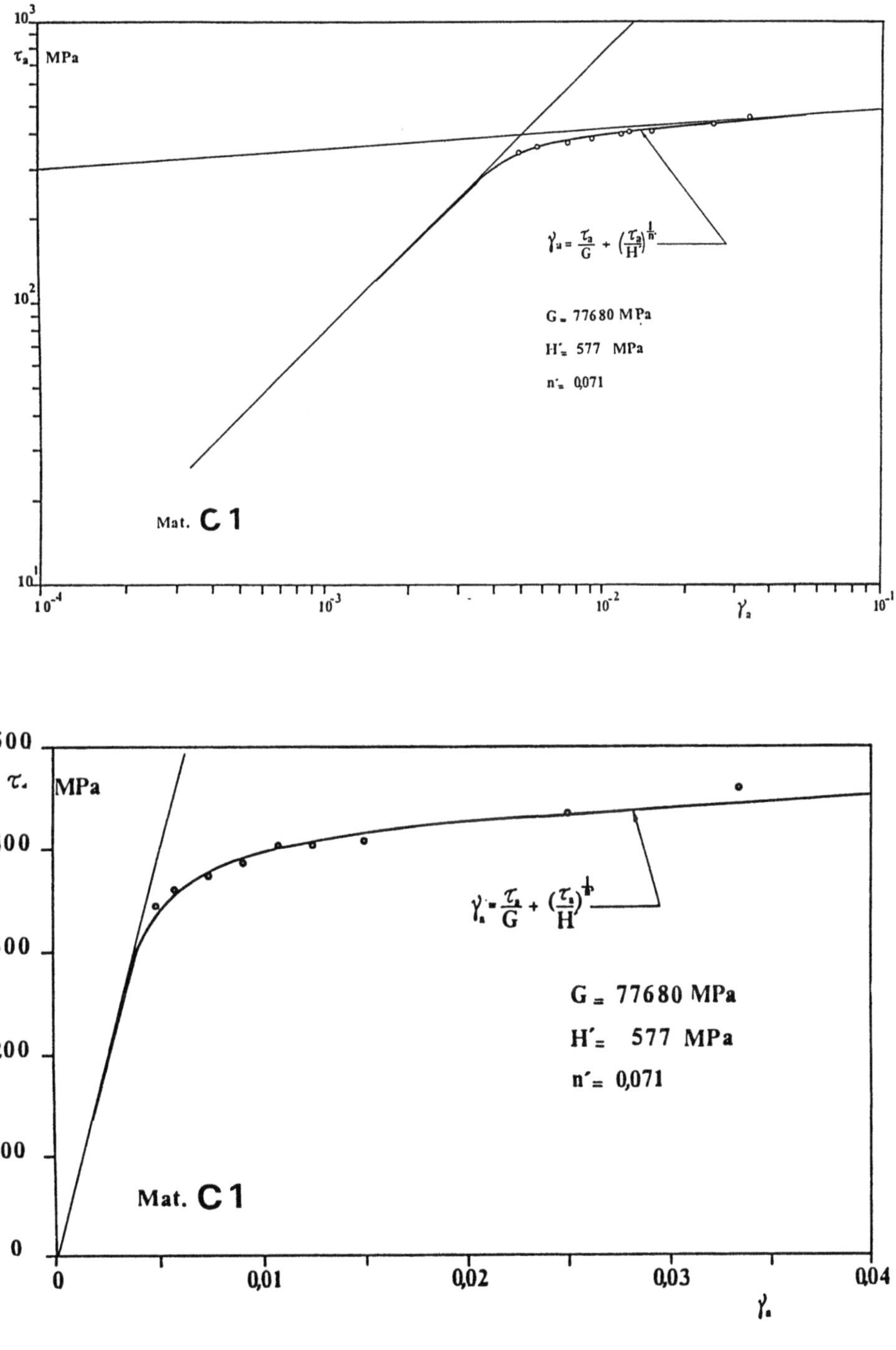

Fig. 5 - Shear stress-shear strain curve relative to C1 rotor steel |9, 11|.

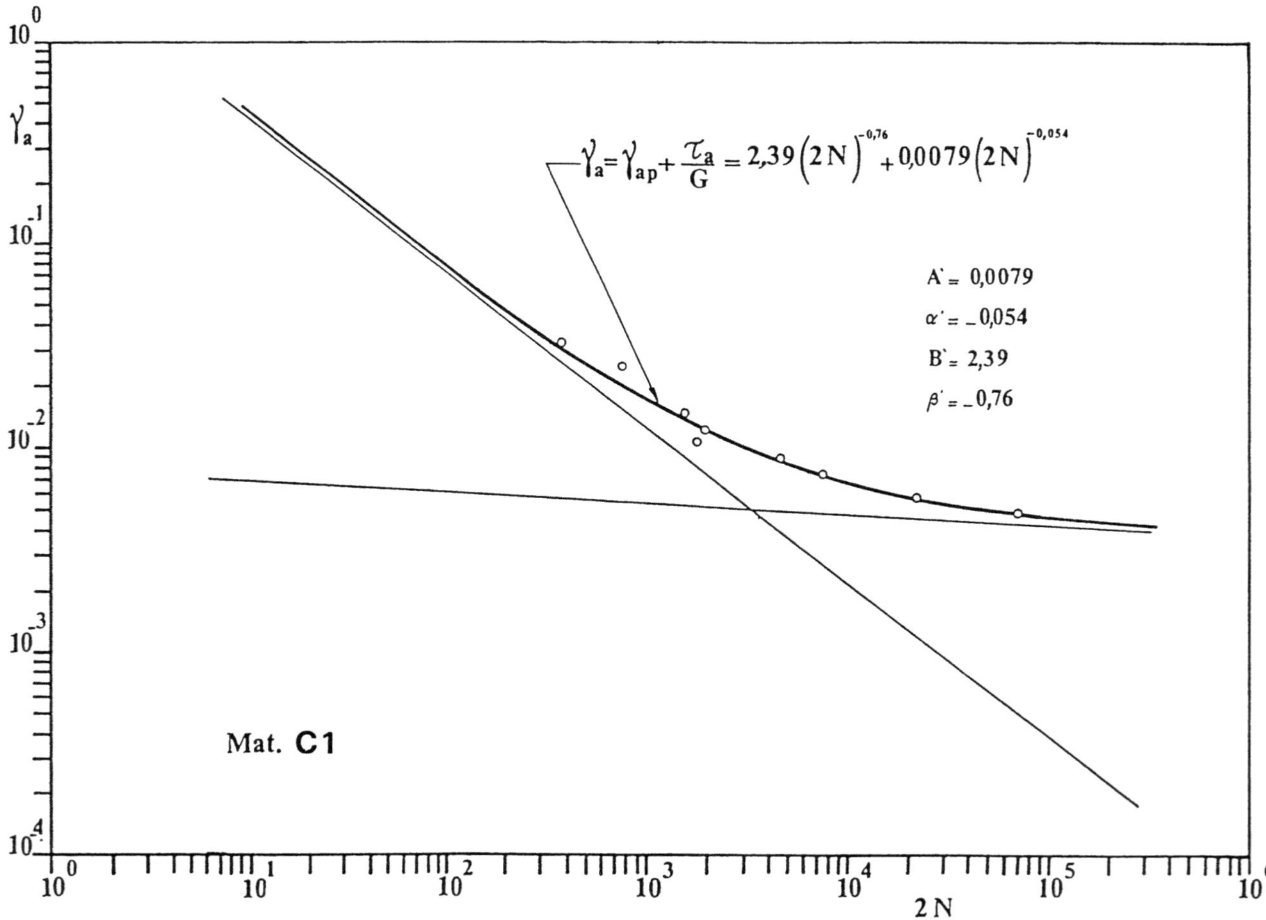

Fig. 6 - Torsional fatigue curve relative to C1 rotor steel |9, 11|.

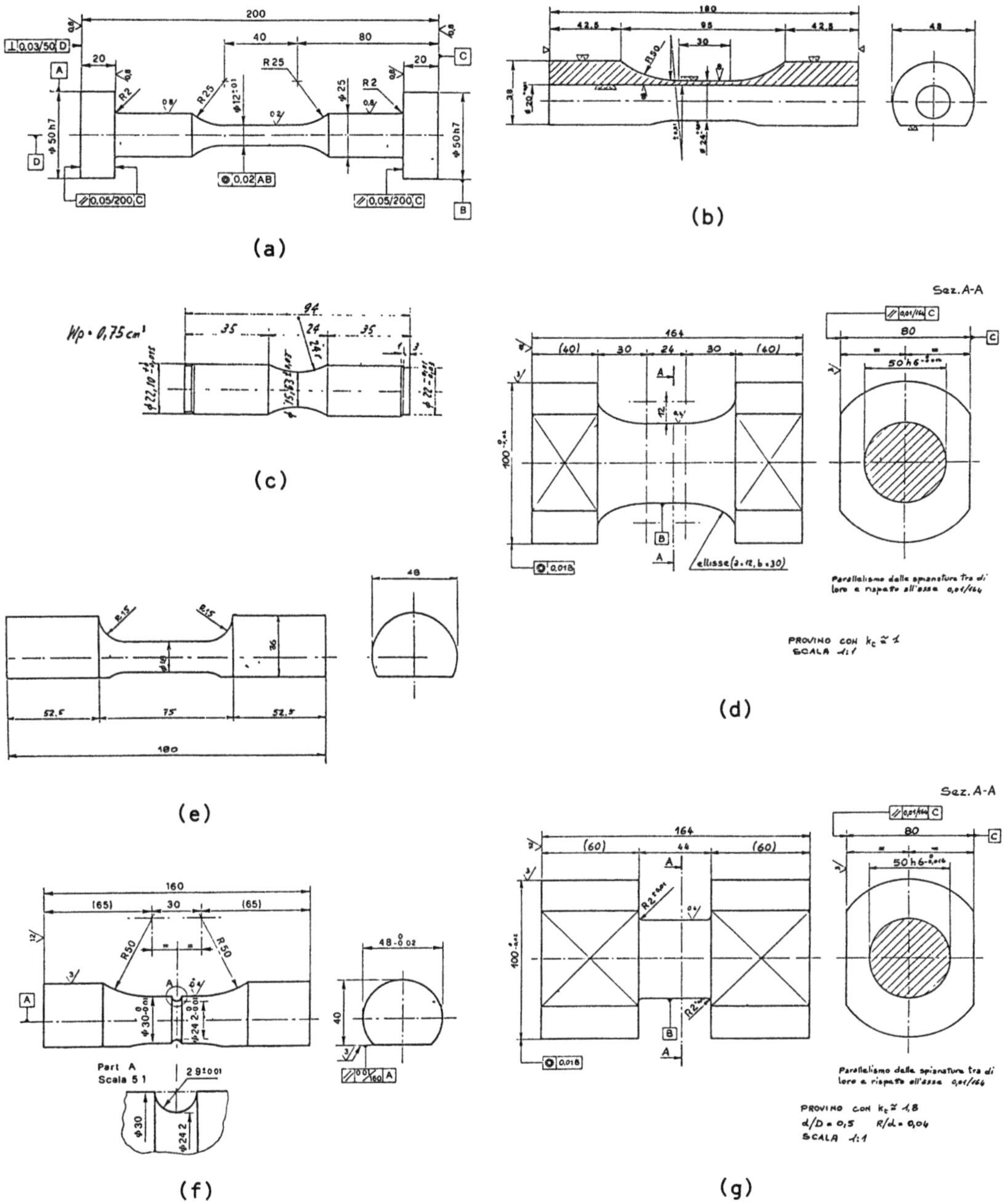

Fig. 7 - Specimens utilized in the fatigue tests. a) Low cycle fatigue push-pull tests; b), d), e) constant amplitude low cycle fatigue in torsion; f), g) constant and variable amplitude low cycle fatigue in torsion; c) high cycle fatigue in torsion.

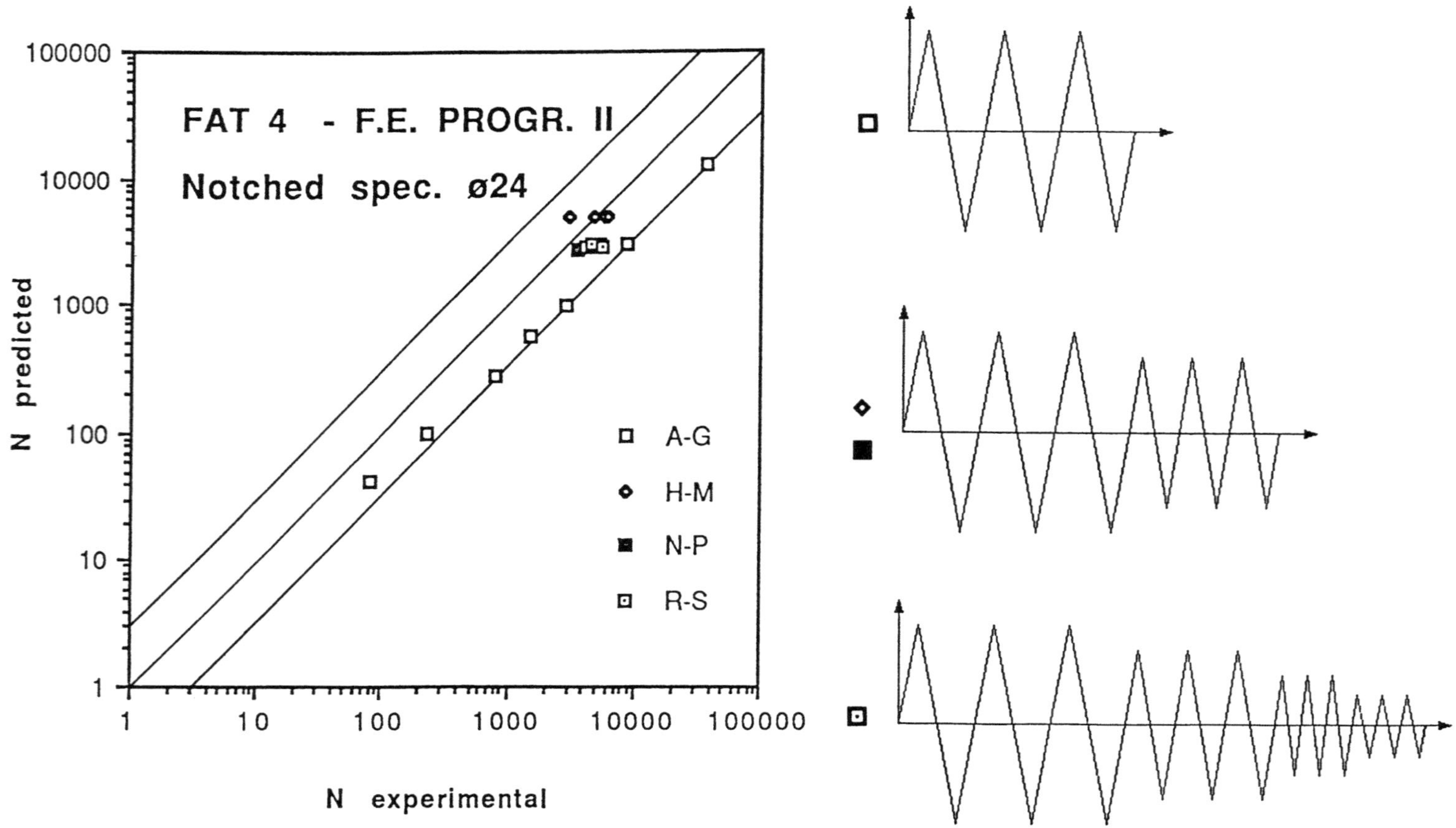

Fig. 8 - Comparison between numerical predictions of the fatigue life and experimental data.

[6] Dowling, N.E., "Fatigue Life Prediction for Complex Load Versus Time Histories", J. of Eng. Mat. and Tech., 1983, Vol. 105, pp. 206-214.
[7] Caligiana, G., "Life Predictions in Fatigue taking into account Material Memory Properties", RILEM workshop companion paper.
[8] Curioni, S., "Life prediction analysis based on a local strain approach for shaft under torsional loading", Int. J. of Materials and Product Technology, Inderscience Enterprises Limited, Geneva, 1987.Vol. 2, No. 1, pp. 75-87.
[9] Freddi, A., Curioni, S., Caligiana, G., "La vita a fatica torsionale negli alberi dei turbo gruppi", ENEA Roma, 1989 (book).
[10] Ramberg, W, Osgood, W.E., "Description of Stress-Strain Curves by Three Parameters", NACA TN-902, 1943.
[11] Freddi, F., Curioni, S., "Metodi di previsione di vita di organi soggetti a fatica oligociclica", Atti del Convegno in onore di Antongiulio Dornig, Milano, 1990, pp. 217-234.
[12] Nadai, A., "Theory of Flow and Fracture of Solids", London, Mc Graw Hill, 1950 (book).
[13] Caligiana, G., Cesari, F., "Fatigue Life Prediction of a Notched Specimen subject to Torsion: Theoretical and Experimental Results", Third Int. Conf. on Biaxial/Multiaxial Fatigue (Acta), 1989, Vol. 1, pp. 6.1- 6.17.
[14] Morrow, JoDean, "Cyclic Plastic Strain Energy and Fatigue of Metals", Internal Friction, Damping and Cyclic Plasticity, ASTM STP 378, American Society for Testing and Materials, 1965, pp. 45-87.
[15] Neuber, H., "Theory of stress concentration for shear-strained prismatical bodies with arbitrary nonlinear stress-strain law", Trans. ASME J. Appl. Mech., 1961, Vol. 28, pp. 544-550.
[16] Topper, T.H., Wetzel, R.M., Morrow, J.D., "Neuber's rule applied to fatigue of notched specimens", Journal of Materials, JMLSA, 1969, Vol. 4, pp. 200-209.
[17] Caligiana, G., "Low cycle fatigue in push-pull and torsion tests: numerical prediction and experimental results", Int. J. of Materials and Product Technology, Inderscience Enterprises Limited, Geneva, 1987, Vol. 2, No. 1, pp. 88-104.
[18] Caligiana, G., "Attrezzatura per le prove di fatica torsionale", Tecnica Italiana, Cormons, Gorizia, 1988, No. 4, pp. 357-376.

16 EXPERIENCE AND NEEDS IN INELASTIC CYCLIC TESTS OF MATERIALS, JOINTS AND TUBULAR BRACES

L.-W. LU
Center for Advanced Technology for Large Structural Systems,
Lehigh University, Pennsylvania, USA

SUMMARY

Much experimental investigations have been carried out to study the behavior of steel, structural components, joints, assemblages and frames under monotonic and cyclic loading. The paper summarizes some recent experience in testing steel material, semi-rigid joints and thin-walled tubular braces and discusses future research needs.

INTRODUCTION

Since the 1950's, experimental research has been conducted to study the mechanical behavior of steel subjected to inelastic load cycles. This work provided information for developing theories to describe cyclic behavior of steel as well as experimental verification of the proposed cyclic plasticity models. The theoretical material models are necessary in developing predictions for the response of structures or structural elements under cyclic loading.

Starting from the early 1960's, laboratory experiments on cyclically loaded beams, beam-columns, braces, shear links and joints have been successfully carried out. From these experiments a basic understanding of the inelastic cyclic behavior has emerged. The experimental data base and the theoretical development allow the formulation of recommendations and rules for designing structural components to achieve desired behavior in a seismic-resistant structure. Much of this effort has been led by researchers from the U.S., Japan and European countries.

The overall behavior of two- and three-dimensional steel frames under cyclic loading has been studied experimentally in the recent years. Some of the tests had bare steel girders; others had reinforced concrete slabs acting compositely with the girders. Both moment-resisting moments and concentrically and eccentrically braced frames were tested to their ultimate load or deformation capacities. Most of the tests were performed with a predetermined displacement program; a few were conducted using the "pseudodynamic method."

This paper views the recent work on (1) material behavior characterization, (2) joints, and (3) welded tubular braces. Needed research in these areas is also discussed.

Testing of Metals for Structures. Edited by F.M. Mazzolani. © 1992 RILEM.
Published by Taylor & Francis, 2 Park Square, Milton Park, Abingdon, Oxon, OX14 4RN.
ISBN 0 419 15810 3

MATERIAL TESTS

Limited repeated and reversed axial load tests have been conducted on specially prepared specimens to study the hysteretical behavior of steel (Ref. 1, for example). Unfortunately, most of these tests were not performed on steels commonly used in construction. The results of these tests have been used in developing cyclic plasticity models for steels (2, 3). The computational models proposed by Mosaddad and Powell (4) has been applied by Dunai and the author (5) to study the cyclic behavior of top-and-seat angle connections. Carefully performed tests on steel under cyclically applied biaxial stresses are available and have been reported by Ellyin (6), who developed an anisotropic hardening rule for elastoplastic material. These tests show that the stable uniaxial cyclic curve can be used as the basis to predict the biaxial stable response.

There is the need to conduct (1) repeated and reversed uniaxial tests on structural grade steels and (2) repeated and reversed biaxial tests on structural steels under both proportional and non-proportional loading.

JOINTS

Much of the past research on joints have been on full moment beam-to-column joints and papers summarizing this work has appeared in a special issue of the Journal of Construction Steel Research (7). The recent work has emphasized semi-rigid or flexible joints and joints between composite girder (steel girder with composite deck) and steel column. The effect of joint flexibility on the behavior of individual members and overall frames has also been extensively researched.

A program of testing was conducted on a series of extended end-plate connections and top-and-seat-angle connections. The major objective of the tests was to evaluate the behavior and failure modes of these connections, with an emphasis on connections stiffness, maximum moment capacity, and energy dissipation characteristics under reversal loading. Thirteen large-size connections were tested to failure, mostly under cyclic loading. Each connection consisted of a W27 x 94 beam attached to a W14 x 193 column, both made of 50 ksi steel. The end plates and the angles also had the same nominal yield stress.

High strength, one-inch diameter A325 bolts were used in most of these connections. Some connections included snug-tightened bolts, i.e. bolts pre-tensioned to approximately 50% of AISC recommended value (70% of the ultimate stress of the bolt material). Other connections had bolts fully pre-tensioned (referred to as "tight" in the subsequent discussions) to the recommended value.

The parameters that were varied in the end-plate connection tests included the plate thickness, bolt pitch and the amount of bolt pre-tension. The connections were designed to yield at approximately 70% (1-inch end plate) and 55% (3/4 inch end plate) of the beam plastic moment Mp. These values are reasonable for many circumstances, as beam design is often controlled by deflection and drift constraints. The parameters examined in the top-and-seat-angle connections were the number of bolt rows (one or two rows of bolts connecting the angles to the column flange), bolt pitch and the amount of pre-tension in bolts. In order to achieve moment resistance, large-size angles, two rows of bolts, and equally large top-and-seat angles, 7/8 in. thick, were used in most of the tests. These deviations from common practice are part of an experimental effort to explore the potential of the top-and-seat-angle connection as a moment carrying and energy-dissipating connection.

The following is a general description of the behavior of the end-plate connections tested. The response to external loading of an end-plate connection depends on a complex interaction of column flange distortion, end-plate bending and bolt elongation. (For simplicity, column behavior is not included in this discussion.) At a certain load, the end plate and column flange contact forces due to bolt pre-tension were relieved. The end plate became separated from the column at the beam tension flange and the connection's stiffness began to change. The contact surface then receded toward the bolts, and plate bending became substantial. The connection's configuration was such that the largest moment in the end plate occurred at the beam flange, where the end plate first experienced yielding. Depending on the restraining capacity of the bolts, a second position of yielding might occur at the bolt line. As the bolts' pre-tension was relieved and the end plate began to yield, the connection's stiffness gradually decreased in a non-linear manner. Finally, the connection failed due to: 1) end-plate shear, 2) end-plate excessive deformation, 3) bolt fracture, o, or 4) weld fracture.

Results from the end-plate connection tests indicate a very significant effect of prying action of the plate. The bolt group of the tension region was subjected to forces approximately 1.2 - 1.3 times the forces required to maintain moment equilibrium. Figure 1 compares the prying action of a typical bolt for the 1-in. snug and fully pre-tensioned (tight) cases. The solid lines represent the value of bolt force that would be present if prying action did not exist. The contribution of prying action is denoted by Q. For the snug case, prying effects became substantial at a value of 20% Mp, while for the tight case substantial prying effects began at 40% Mp. The prying forces at loads near the ultimate, however, were the same for both cases.

The major observations that can be made from the results of the end-plate connection investigation are:

- Snug tightened bolts have minimal effects on connection behavior.
- Prying effects should be considered in design procedures, since prying caused up to a 30% increase in the bolt forces.
- In order to avoid plate shear failure, end-plate connections should be designed to allow for redistribution of end-plate moments by placing a minimum requirement on the distance between the beam flange and bolt line.

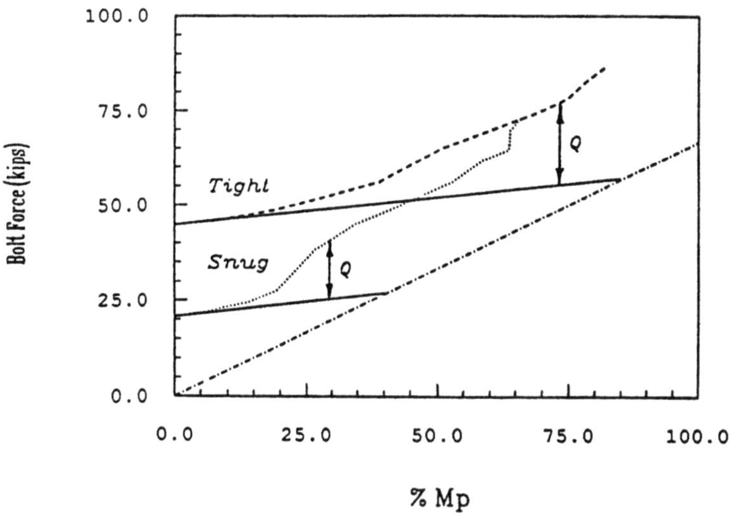

Figure 1: Effect of Prying Action on Bolt Force, End-Plate Connection.

- Reversal loading decreases the connection's stiffness characteristics on successive cycles; however, connection stiffness may increase if repeated loading is applied after reversal loads have yielded the connection.
- For connections that require energy absorption capabilities, thinner end-plate or increased bolt pitch may respond in a more ductile manner and exhibit similar strength.

Other conclusions and the detailed results of the investigation have been presented in several recently published papers (8, 9, 10). The research has led to the development of a design procedure that takes into account the effects of prying and shear (11), or new model for inelastic analysis, and an understanding of the energy absorption characteristics of the connecting plates.

The observed behavior of the top-and-seat-angle connections is summarized below. The connections having two bolt rows under full pretension acted quite differently than those under snug-tightened condition. Initially, in a fully pre-tensioned connecting angle, a perfect contact surface perhaps existed between the angle leg and the column flange. When the pre-tension of the bolts in the inner row was overcome, the contact surface receded toward the outer row. Because the full pre-tension value was near the proportional limit of the bolt, the inner row of bolts yielded before the contact forces were relieved at the second bolt row (that is, before the second bolt row began to assist in carrying the tensile force). Thus, the prevalent action in a two-bolt-row connection was one in which the inner bolt row deformed excessively, while the outer bolt row did not participate. This resulted in a more flexible connection and the tendency of early yielding in the angle leg between the bolt lines. In a snug-tightened connecting angle leg, the situation was different. Under low loads, the snug-tight pre-tension was overcome, but as the contact surface receded toward the second bolt row, the response of the bolts in the first row was well within the linear elastic portion of its load-deformation relationship. This allowed the first bolt line to act elastically, while the small amount of pre-tension was also overcome in the second row. The second bolt row became functional at a low load, so a larger force was required to put any bolt into yield as compared with the full pre-tension case. Figure 2 shows a comparison of the moment-rotation curves of two identical connections, one with snug bolts and the other fully pre-tensioned bolts.

The major findings of the top-and-seat-angle connections studies are:

- The connections can resist a very significant amount of bending moment. The moment capacity of the two-bolt-row connections was about 40 to 50% of the plastic moment of the W27 x 94 beam. The single-bolt-row connections developed a maximum moment equal to 25% of the beam plastic moment. Part of the difference is due to the longer bolt pitch in the single-bolt-row connections.

- The connections showed good ductility in the inelastic range. However, in order to achieve good ductility, the bolts must not fail by fracture.

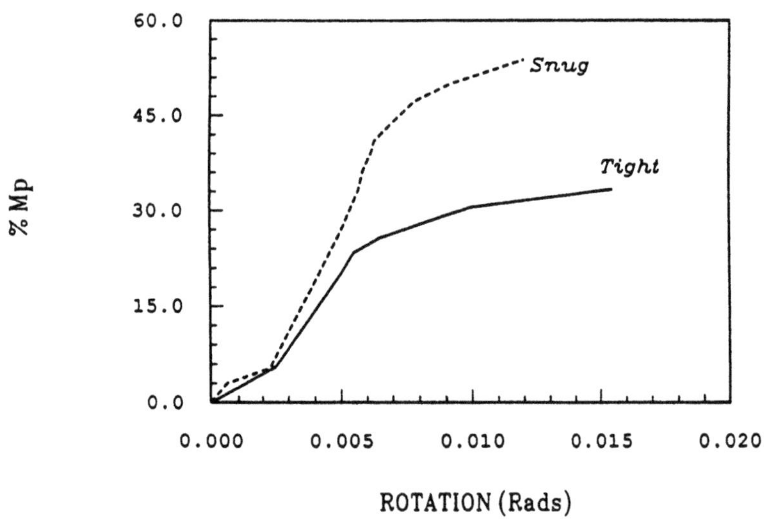

Figure 2: Moment-Rotation Curves of Top-and-Seat-Angle Connections with Two Bolt Rows.

- Prying action was significant and should be considered in bolt design.
- The behavior of the connections depended primarily on the bolts of the critical line in the tension angle.
- The two-bolt-row connections exhibited linear behavior well above the service load of the connected beam.

The detailed results of the study and other findings can be found in Refs. 8, 9 and 10.

Connections between composite girder and steel column has received recent attention because of the wide spread use of steel girders with stud connected metal deck floor slab. The concrete slab is subjected to compression and is fully effective when a positive bending moment is applied to such a connection. The composite action that develops at the beam-and-column interface increases substantially the moment capacity of the connection. This increase in moment capacity in turn increases the lateral load resistance of the structural frame.

Figure 3 shows the hysteresis loops of a composite joint assemblage test conducted at Lehigh University (12). The positive load P produced positive bending moment on the connection. This test demonstrated deformation characteristics of a composite joint in that: (1) There were remarkable increases in stiffness and strength due to composite action when subjected to positive loading; (2) the hysteresis loops were very stable but with noticeable pinching due to opening and closing of the concrete cracks; and (3) good ductility existed even after complete crushing of the concrete.

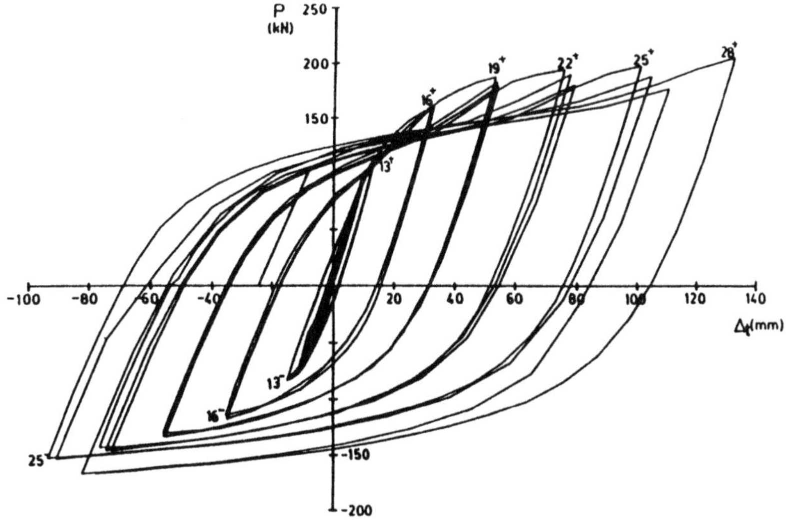

Figure 3: Hysteretic Behavior of Composite Joint Assemblage.

There is a lack of experimental data on semi-rigid, beam-to-column-web connections subjected to cyclic loading. Additional tests, coupled with detailed theoretical analysis, of extended end-plate connections designed for energy absorption are needed. No study has been carried out on composite connections and joint assemblages subjected to bi-directional bending due to gravity as well as lateral loading.

TUBULAR BRACES

Experimental and analytical studies of inelastic behavior of tubular members fabricated by welding two identical angles, either continuously or intermittently, have been made. Such members are used as braces in concentrically or eccentrically braced frames or web or diagonal elements in truss girders. The behavior of the member is affected by combined local and overall buckling (known as interactive buckling), which eventually results in fracture occurring at the severely deformed region of the member. The problem of fracture becomes especially critical when cyclic tension and compression are applied to the member. If the angles are continuously welded and if the width-to-thickness ratios of the angle legs satisfy the "compactness" requirement (13), the initial failure is usually due to overall buckling. However, when intermittent welds are used, local buckling could play a major role in affecting the ductility of the member even if the compactness requirement is met. Knowledge about inelastic cyclic behavior of tubular members fabricated with intermittent welding is extremely limited and rational design procedure for such members is not available.

The experimental study included tests on 2-1/2" by 2-1/2" square section members fabricated from 2-1/2" x 2-1/2" x 3/16" angles made of structural carbon steel. The variables examined were the member length and the length of the intermittent welds. The dimensions of the angle section meet the compactness requirement for an axially compressed member. The tests were carried out with cyclically applied axial displacements in both tension and compression. Figure 4, taken from Ref. 14, shows the hysteretic behavior of one of the test specimens, which had an overall length of 60 cm and a weld length of 10 cm. The load-carrying capacity of the member in compression (negative load in Fig. 4) decreased considerably after each cycle of load application. At the end of test, the compression strength was reduced to about 50% of the strength achieved in the first cycle.

The behavior of some of the test specimen has been analyzed by using the finite element program ADINA (15). The analysis emphasized the interaction between local and overall buckling and included both the initial out-of-flatness of the angle legs and the out-of-straightness of the member. The results indicate that in the case of monotonically applied axial compression, the load-deformation response of the member, after attainment of the maximum load, has significantly affected local buckling.

Research needs in welded tubular or box members include studies on (1) the size and length requirements of intermittent welds in axially loaded members which are designed to achieve both ductility and strength, (2) welding requirements (size of welds, fillet weld vs. groove weld, etc.) for box columns fabricated from plates, and (3) the interaction

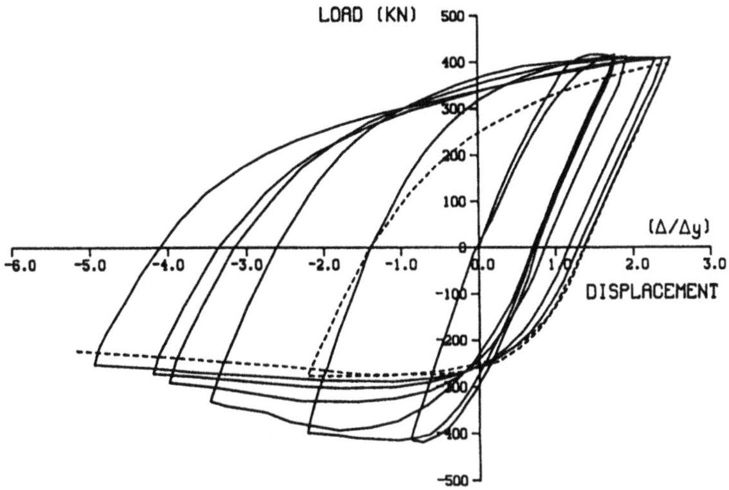

Figure 4: Hysteretic Behavior of Tubular Member with Intermittent Welds.

between local and overall buckling and its effect on strength and ductility of tubular members subjected to combined bending and axial force.

CONCLUSIONS

Considerable amount of experimental research has been conducted to study the behavior of steel material, structural elements and members, joints and frames. The results have been used to establish analytical models for predicting the behavior of the test specimens as well as to develop design provisions or procedure for the types of members and joints investigated. This paper has presented a brief overview of the work on 1) cyclic behavior of steel, 2) semi-rigid and composite joints under repeated and reversed bending moment, and 3) tubular braces subjected to cyclic axial tension and compression. For each topic, the future research needs are discussed.

Much additional research is required on problems related to inelastic cyclic response of structures and structural components. It is important that the current worldwide research effort be continued and expanded so that seismic-resistant building structures can be designed in the future to satisfy the requirements of safety, reliability and economy.

REFERENCES

1. Morrow, J. D., "Cyclic Plastic Strain Energy and Fatigue Metals," in "Internal Friction, Damping, and Cyclic Plasticity," Special Technical Publication No. 378, American Society for Testing and Materials, 1965, pp. 45-87.

2. Petersson, H. and Popov, E. P., "Constitutive Relations for Generalized Loading," Journal of Engineering Mechanics Division, ASCE, Vol. 103, No. EM4, 1977, pp. 611-627.

3. Popov, E. P. and Petersson, H., "Cyclic Metal Plasticity: Experiments and Theory," Journal of Engineering Mechanics Division, ASCE, Vol. 104, 1978, pp. 1371-1388.

4. Mosaddad, B. and Powell, G. H., "Computational Models for Cyclic Plasticity, Rate Dependence, and Creep in Finite Element Analysis," Report No. UCB/EERC 82/26, University of California, Berkeley, 1982.

5. Dunai, L. and Lu, L. W., "Analytical Studies of Monotonic and Cyclic Behavior of Flexible Connections with Bolted Angles," Report No. 90-06, Center for Advanced Technology for Large Structural Systems, Lehigh University, 1990.

6. Ellyin, F., "An Anisotropic Hardening Rule for Elastoplastic Solids Based on Experimental Observations," Journal of Applied Mechanics, ASME, Vol. 56, September, 1989, pp. 499-507.

7. Chen, W. F., editor, "Steel Beam-to-Column Building Connections," special issue of Journal of Construction Steel Research, Vol. 10, 1988.

8. Chasten, C. P., Fleischman, R. B., Driscoll, G. C. and Lu, L. W., "Top-and-Seat-Angle Connections and End-Plate Connections: Behavior and Strength under Monotonic and Cyclic Loading," Proceedings, 1990 National Engineering Conference, AISC, Nashville, Tennessee, pp. 6-1 to 6-32.

9. Fleischman, R. B., Chasten, C. P., Lu, L. W. and Driscoll, G. C., "Top-and-Seat-Angle Connections and End-Plate Connections: Snug vs. Fully Pre-Tensioned Bolts," Proceedings, 1990 National Engineering Conference, AISC, Nashville, Tennessee, pp. 4-1 to 4-29.

10. Fleischman, R. B., Chasten, C. P., Lu, L. W. and Driscoll, G. C., "Cyclic Behavior of Large-Size Flexible Beam-to-Column Connections," Proceedings, Fourth National Conference on Earthquake Engineering, 1990, Palm Springs, California, Vol. 2, pp. 687-696.

11. Chasten, C. P., Lu, L. W. and Driscoll, G. C., "Prying and Shear in End-Plate Connection Design," Report No. 90-08, Center for Advanced Technology for Large Structural Systems, Lehigh University, 1990.

12. Lee, S. J. and Lu, L. W., "Cyclic Tests of Full-Scale Composite Joint Assemblages," Journal of Structural Engineering, ASCE, Vol. 115, No. 8, August, 1989, pp. 1977-1998.

13. AISC, "Manual of Steel Construction: Load and Resistance Factor Design," American Institute of Steel Construction, Chicago, 1986.

14. Chen, C. C., "Approaches to Improve Seismic Resistance of Steel Building Frames," Ph.D. dissertation, Lehigh University, Bethlehem, Pennsylvania, U.S.A., 1991.

15. Ma, J. L., "Simulation of Monotonic Behavior of Intermittently Welded Box Columns," CE 481 Special Problem report, Lehigh University, Bethlehem, Pennsylvania, U.S.A., 1990.

17 LOW CYCLE FATIGUE FRACTURE LIMITS OF STRUCTURAL MATERIALS AND STRUCTURAL ELEMENTS

M. YAMADA
Faculty of Engineering, Kobe University, Japan

1 Introduction

Low cycle fatigue fracture limits of structural materials, elements, members as well as joints are the bases of ultimate state theory and design of structures. This Report for the RILEM Workshop on Needs in Testing Metals, Naples May 1990, presents the experimental data assembled by the author over 30 years on this problem in order to make clear the needs in testing metals to develop the ultimate state theory of structures and to establish the international standard testing methodology at RILEM.

2 Objectives and scope

In order to make clear the safety of structures against severe external loads like earthquake excitation, wind loads etc, the ultimate state of structures, i.e. fracture behaviour and fracture conditions of structures, must be clarified with regard to their external load characteristics. External loads on structures like earthquake excitation or wind load are usually random but their effects upon structures are simplified and abstracted into some simple and pure states. Practical structural design must be based upon such simple and universal bases. The author has already proposed [1, 2] that these simplified effects of external loads upon structures are as follows:
Earthquake excitation as two extreme conditions;

(1) ultimate excitation of vibration: resonance state,
(2) ultimate excitation of one way loading; impulsive state.

The effect of vibration excitation and its ultimate states correspond to the fatigue behaviour and fatigue fracture of structural materials, members and structures. Conversely, impulsive excitation and its effect on structures correspond to one way impulsive loading. Between these two opposite extreme conditions, structures will behave at their ultimate state under earthquake excitation. Especially under

Testing of Metals for Structures. Edited by F.M. Mazzolani. © 1992 RILEM.
Published by Taylor & Francis, 2 Park Square, Milton Park, Abingdon, Oxon, OX14 4RN.
ISBN 0 419 15810 3

earthquake excitation structures vibrate for very few cycles and seem to select the same frequency wave for themselves from the random earthquake waves and to resonate, and consequently to amplify their effect [3].

For the comprehensive evaluation of the ultimate fracture limit of such fatigue fracture, the low cycle fatigue fracture limit gives a very good universal standard. Therefore the author has already proposed the low cycle fatigue fracture limit of the structural members as the evaluation basis or design criteria of aseismic design of structures [1, 2].

One-way impulsive loadings are simulated by one-way loading of structural members with regard to loading rate. Such one-way loading of structural members is not so simple, because the fracture limit of ductility under one-way loading is always too difficult to define by practical experiment [1]. To avoid such uncertain definition of fracture limits of structural members under one-way loading, the author has already proposed the use of half cycle or quarter cycle value of low cycle fatigue fracture limit of structural members as the fracture limit under one-way loading [1].

As the universal evaluation base of the ultimate states of vibrated structures, the fatigue fracture limit is very convenient as standard condition. For earthquake loading the low cycle fatigue fracture limit and for wind loading the high cycle fatigue fracture limit are therefore very convenient [3].

Low cycle fatigue fracture limits of structural materials and members are usually obtained by constant strain amplitude tests for materials and constant deformation amplitude tests. These two tests are fundamentally different from each other. Test results are usually illustrated as the relationship between (plastic) strain amplitude versus number of cycles to fracture for materials and (plastic) deformation amplitude versus number of cycles to fracture for structural members. There exists again however, the difficulty of defining the fracture of structural members. The author has also proposed the definition of fatigue fracture of structural members as the loss of axial resistance to sustain at the prescribed constant value or the loss of resistance of the member against load [4].

It must be noted that the establishment of the relationships between the deformation amplitude and number of cycles to fracture of structural members under the action of the prescribed constant axial load is the most important task. For this purpose, establishment of simple standard testing rules for low cycle fatigue of structural materials, members and joints is desirable. This is the main proposal of the author as the most important task of our RILEM Task Group, Needs in Testing Metals for the RILEM Technical Advisory Committee.

3 Hypothesis on low cycle fatigue fracture limits

Researchers on low cycle fatigue fracture limits of metals were initiated at the enquiry into the cause of the failure of the first jet passenger airplane, the Comet, in England in 1954. Many experimental investigations were carried out not only by the simple constant strain amplitude method but also by multistep strain amplitude tests.

Hypotheses on low cycle fatigue fracture limits were also proposed by many researchers on the basis of such experimental results. The first important fact clarified by experimental research was that it is not the total dissipated energy but the cumulative damage energy that causes low cycle fatigue fracture of metals. Manson and Coffin [9, 10] proposed their Cumulative Damage Energy Hypothesis of metals as follows:

$$D = N \left(\frac{\Delta\varepsilon_p}{\varepsilon_p} \right)^2 = 1 \tag{1}$$

where N = number of cycles to fracture,
$\Delta\varepsilon_p$ = plastic strain amplitude (range).
ε_p = ($0.88\ \varepsilon_{Frac.}$) for ductile fracture,
ε_p = ($0.67\ \varepsilon_{Frac.}$) for macrocracking

as illustrated in Figs 1 and 2.

The application of such cumulative damage hypothesis on structural members was proposed by the author [1, 2, 13] for earthquake resistant designs to evaluate their aseismic capacity. For structural design it must be discussed on structural members, therefore investigations must be carried out mainly on the deformation level and not directly on the strain level. Strain level occurs mainly at the fracture of cross sections of the member.

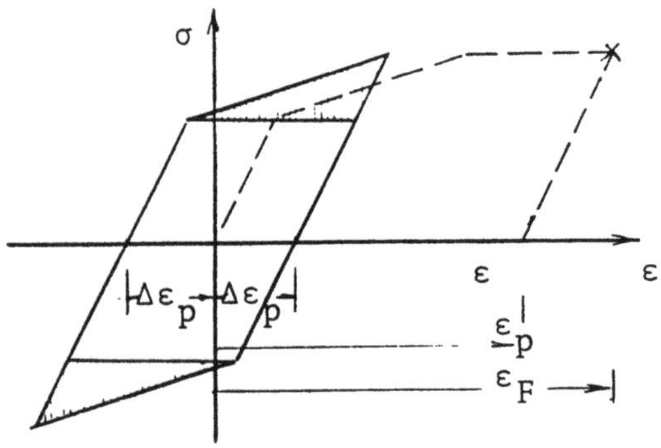

Fig. 1 Damage energy

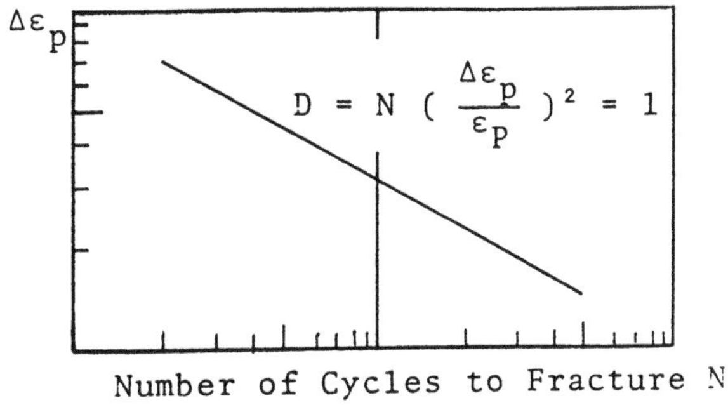

Fig. 2 Cumulative damage energy hypothesis by Manson and Coffin [9, 10]

4 Low cycle fatigue fracture limits of structural materials - metals (steel)

There are already many test data on the low cycle fatigue fracture limits of metals. Testing methodology was established already in the fields of metallurgy and mechanical engineering. However, it must be noted and checked on the definition of strain in the testing method.

In the field of structural engineering, low cycle fatigue fracture test data must be prepared on every applied practical structural steel as the base of research. Fundamental tests on the low cycle fatigue fracture limits of structural carbon steels are desirable and recommended to summarize these test results.

5 Low cycle fatigue fracture limits of structural elements

Steel structural members are usually composed of several panel elements such as flange panels and web panels. These panel elements are stressed in their panel plane. The low cycle fatigue fracture limits of such structural steel panels have already been tested experimentally by the author and reported [3, 5] such as those illustrated in Fig. 3, as well as analytical results [5].

6 Low cycle fatigue fracture limits of structural members

Low cycle fatigue fracture limits of steel structural members such as wide flange and box cross sections were presented and discussed already by the author as the evaluation basis of the aseismic capacity of steel structures under various loading conditions [1, 2]. These members are tested mainly as beam-columns by deformation amplitude controlled cyclic bending under the action of a prescribed constant axial load

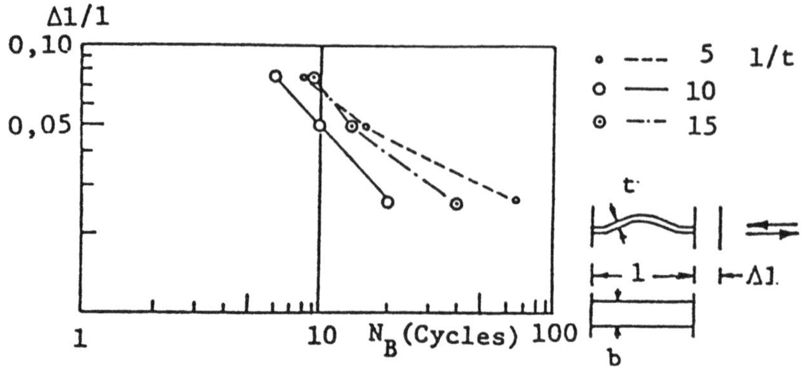

Fig. 3 Low cycle fatigue fracture limits of steel panels by Yamada et al. [3]

level [4]. There are two types of low cycle fatigue fracture modes of these structural member beam-columns, i.e. loss of the prescribed constant axial resistance and loss of the transversal resistance. The former fracture mode occurs under a smaller deflection amplitude and shows moderate cycles to fatigue fracture (N) in Fig. 5 and the latter occurs under larger deflection amplitude and shows lower cycles to fracture (P) in Fig. 5.

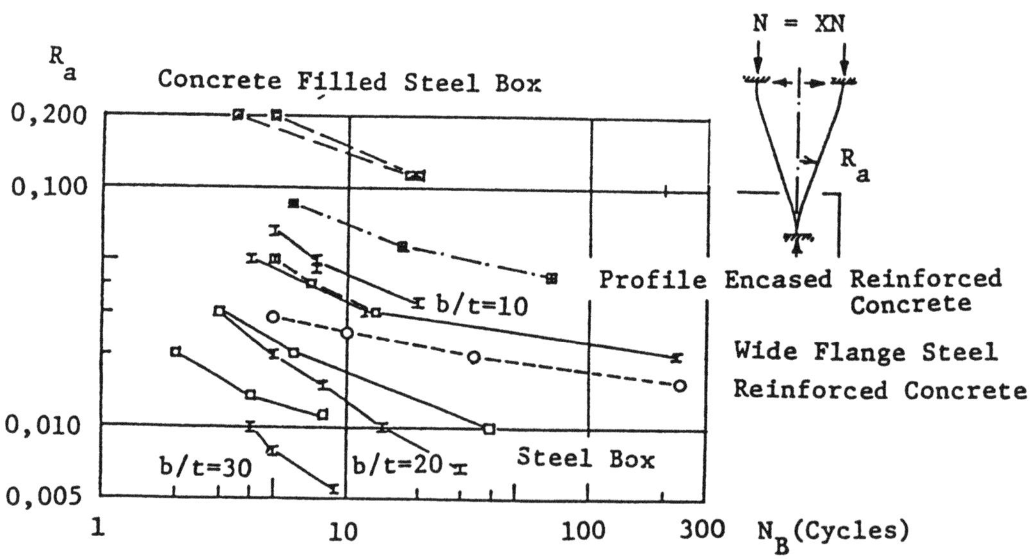

Fig. 4 Low cycle fatigue fracture limits of various kinds of beam-columns by Yamada et al. [3]

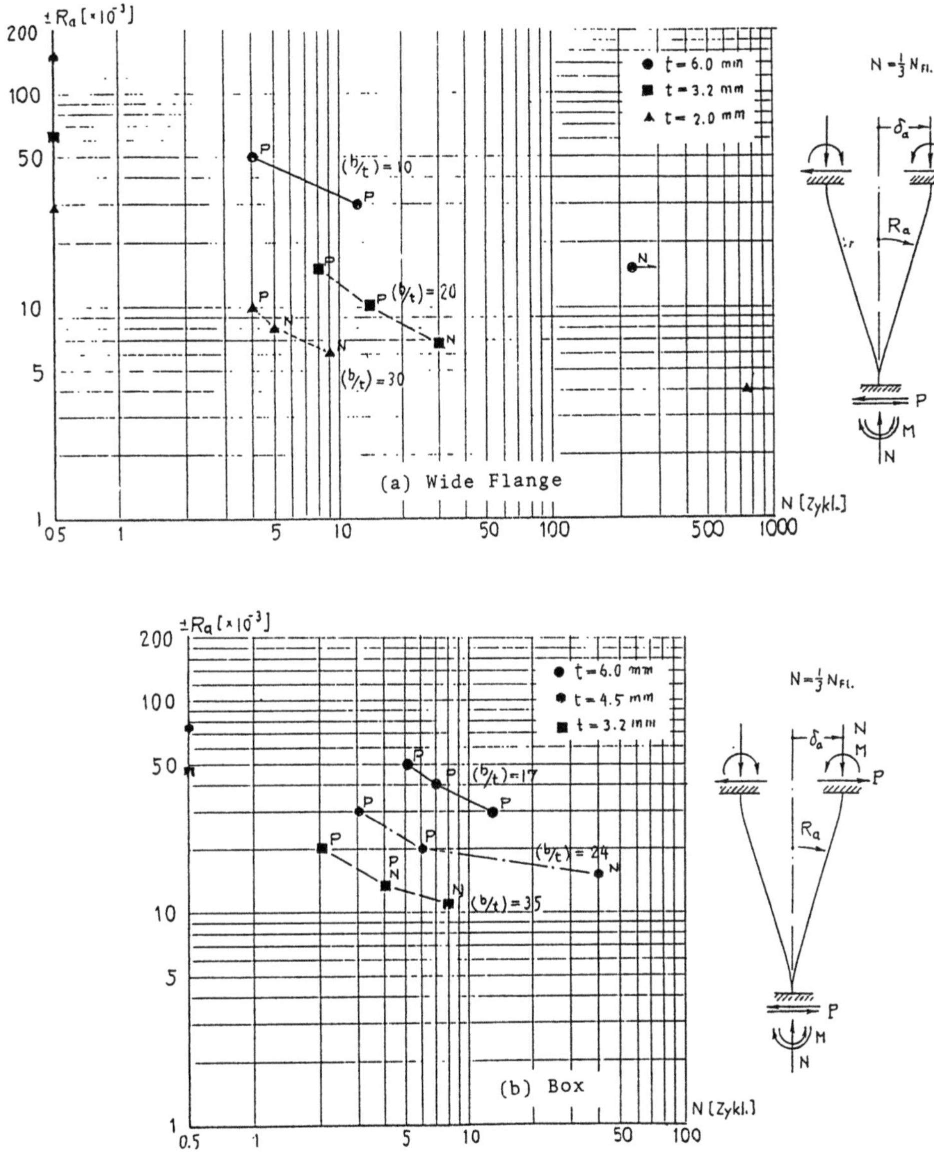

Fig. 5 Low cycle fatigue fracture limits of steel beam columns by Yamada, Kawabata, Yamanaka [4]

7 Low cycle fatigue fracture limits of joints

Joints in steel structures are usually composed of fasteners or weldings. Such joints are usually designed with far higher or larger safety factors than jointed members because of the strain concentration at joints in ordinary structural design principles. In structural design no fractures at joints or at connections between members is assumed to occur.

However, the low cycle fatigue fracture limits of fastener joints must be clarified in order to determine the safety factor against fatigue fracture of such joints. Fig. 6 shows the test results of such fastener joints by the author [7] and the cumulative damage of such fastener joints leading to low cycle fatigue fracture by typical slipping type behaviour.

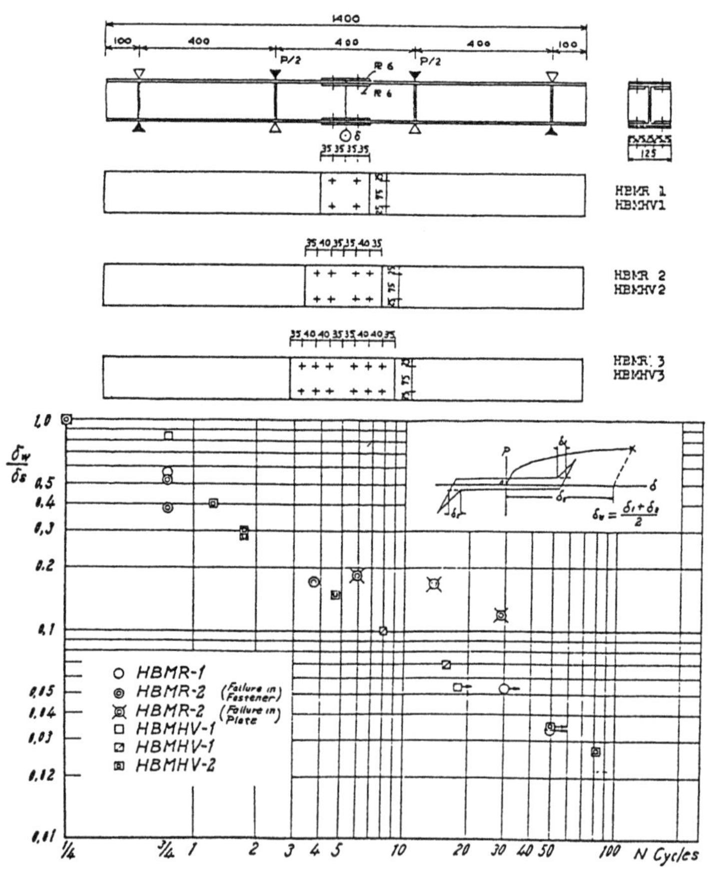

Fig. 6 Low cycle fatigue fracture limits of steel fastener joints by Yamada, Nakaza [6]

8 Low cycle fatigue fracture limits of steel rigid frames

Low cycle fatigue fracture limits of rectangular unit steel rigid frames are presented by the authors [7]. In this case the fracture of structures is also discussed by constant sway amplitude tests of frames under the action of a prescribed constant axial load. Results are clarified by the low cycle fracture condition of steel elements shown in Fig. 7.

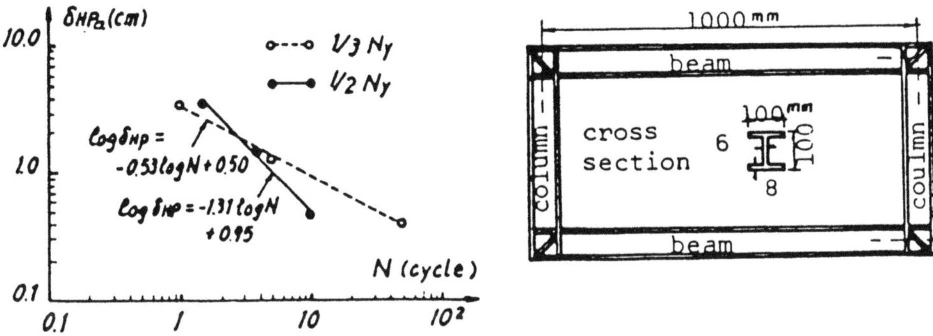

Fig. 7 Low cycle fatigue fracture limits of steel rigid frames by Yamada, Tsuji, Murazumi [7]

9 Concluding remarks

Low cycle fatigue fracture limits of structural materials, elements and members as well as joints are presented and discussed as the basis of evaluation of the aseismic capacity of structures. The needs of the discussion on the testing methodologies are empasized for not only for structural materials but also and more especially for structural members for the purpose of giving a basis for investigation of the ultimate states of structures and to establish structural design methods based upon the ultimate states.

References

[1] Yamada,M. Low Cycle Fatigue Fracture Limits of Various Kinds of Structural Members subjected to Alternately Repeated Plastic Bending under Axial Compression as an Evaluation Basis or Design Criterial for Aseismic Capacity, Proc., 4 WCEE, Santiago, Chile, Vol. 1, 1969, B-2, pp.137-151.

[2] Yamada,M. Effect of Cyclic Loading on Buildings, Proc., International Conference on Planning and Design of Tall Buildings, Vol.II, Structural Design of Tall Steel Buildings, ASCE-IABSE, Leigh, 1972, pp. 725-739.

[3] Yamada, M. et. al. Fracture Ductility of Structural Elements and of Structures, Proc., 9 WCEE, Tokyo/Kyoto, Vol.IV, 1988, pp. 219-224.
[4] Yamada, M., Kawabata, T., Yamanaka, K. Biege-Ermudungsbruch von Stahlstutzen mit I - und Kastenquerschnitt, I Versuche, Stahlbau, 58, H. 12, Dec. 1989, pp.361-364.
[5] Yamada, M., Hashimoto, Y. Ermudungsbruch von Blechstreifen unter wechselseitig wiederholtem Druck und Zug, Stahlbau, (the press).
[6] Yamada, M., Nakaza, I. Low Cycle Fatigue of Riveted and High Strength Bolted Joints in Steel Beams, RILEM-International Symposium on the Effects of Repeated Loading of Materials and Structural Elements, Mexico, Vol.VI, 1966, pp.1-14.
[7] Yamada, M., Tsuji, B., Murazumi, Y. Elasto-plastic Cyclic Horizontal Sway Behaviour of Wide Flange Unit Rigid Frames subjected to Constant Vertical Loads, IABSE-Symposium, Lisboa, 1973, IABSE-WC Rep. Vol. 12, pp.151-156.
[8] Fujitani, H., Yamada, M., Kawamura, H., Tani, A. Ultimate Earthquake Response and Damage Analysis of Structures by Pulse and Finite Resonance Method, Proc., 9 WCEE, Tokyo/Kyoto, Vol.V, 1988, pp.159-164.
[9] Manson, S. S. Fatigue; A Complex Subject - Some Simple Approximations, Experimental Mechanics, Vol.5,7, Jul. 1965, pp. 193-226.
[10] Coffin, L. F., Tavernelli, J. F. The Cyclic Straining and Fatigue of Metals, Trans., Metallug. Soc., AIME, Vol.215, Oct., 1959, pp. 794-806.
[11] Kasiraj, I., Yao, J. T. P. Fatigue Damage in Seismic Structures, ASCE, Vol. 95, No.ST 8, Aug. 1969, pp.1673-1692.
[12] Rosenbluth, E., ed. Design of Earthquake Resistant Structures, Pentech Press, London, 1980, p.116.
[13] Yamada, M. Verhalten plastischer Gelenke in Stahlbetonbalken, Vorbericht, 7. Kongress, IVBH, Rio de Janeiro, 1964, pp.963-970.

PART THREE

HIGH CYCLE FATIGUE BEHAVIOUR OF MATERIALS, JOINTS AND STRUCTIONAL COMPONENTS

18 ITALIAN CODES ON FATIGUE TESTING

B. ATZORI
Department of Mechanical Engineering, University of Padua, Italy

SUMMARY

The present note describes the various Italian Codes (or Instructions) dealing with the subject of fatigue. The aim of the description is to focalize what in Italy has already been done in the field of high cycle fatigue testing of materials, joints and structural components and what, instead, needs to be done. That could be useful as a basis both for a comparison with the existing Codes in different Countries and for the definition of possible working areas for a Technical Committee in the field of testing metals for high cycle fatigue behaviour of materials, joints and structural components.

The analysis and the comparison of the various documents considered, performed taking into account the state of the art about fatigue design, shows clearly the need to improve the existing procedures and the need of assessing new procedures to cover existing lacks in testing materials, joints and structures.

LIST OF ITALIAN CODES ON FATIGUE

Italian Codes on fatigue can be divided into two main fields:

- Codes on fatigue testing
- Codes on design of structures or structural components.

Only the documents pertaining to the first group are of direct interest for this note, given the objectives of the analysis. The documents belonging to the second group contain usually only a small part devoted to fatigue; moreover they give instructions on how to design a fatigue-prone structure,

Testing of Metals for Structures. Edited by F.M. Mazzolani.
Published by Taylor & Francis, 2 Park Square, Milton Park, Abingdon, Oxon, OX14 4RN.
ISBN 0 419 15810 3

not on the way the structure or the material or a critical component should be tested to assess its fatigue strength.Nevertheless also the documents of the second group are here considered and discussed because they are useful to point out the real problems that could arise today in considering fatigue for structural design and therefore the more concrete and urgent needs to be fulfilled by the Codes of the first group.

The esisting Italian Codes on fatigue testing are not all of the same kind and can be moreover classified into three different classes:

- group 1a: general principles and operations, as the symbols to be used, the ways for representing and analysing the results, the necessary calibrations for the fatigue testing machines;
- group 1b: evaluation of the fatigue strength of the base material, to be performed on smooth specimens;
- group 1c: evaluation of the fatigue strength of typical structural components.

The existing Italian Codes known by the Author which deal with the subject of fatigue are the following:

Group 1a: General Principles

1) UNI 3964: Mechanical testing of metallic materials, fatigue testing at room temperature, general principles.
Last edition: May 1985
Elaborated by: Subcommission 4 "Mechanical and technological testing" of the Technical Commission of UNSIDER.

2) UNIMET-UNSIDER: Calibration of fatigue testing machines.
In preparation by: Subcommission 5 " Fatigue testing" of the Technical Commission of UNSIDER.

Group 1b : Fatigue strength of base material

1) UNIMET-UNSIDER Project E12.14.009.0: Mechanical testing of metallic materials, axial load fatigue testing.

Last edition: October 1988 (tentative proposal).
Elaborated by: Subcommission 5 "Fatigue testing" of the Technical Commission of UNSIDER.

2) UNIMET-UNSIDER: Mechanical testing of metallic materials, rotating bending fatigue testing.
In preparation by: Subcommission 5 "Fatigue testing" of the Technical Commission of UNSIDER.

Group 1c : Fatigue strength of structural components

1) UNI-ISO 3800 part 1; Threaded fasteners, axial load fatigue testing, test methods.
Last edition: Dicember 1978.
Elaborated by: ISO and adopted without variations by Subcommission 4 "Threads and bolts" of the Commission "Mechanical parts" of UNI.

Group 2 : Desing of structures or structural parts.

1) CNR study document : Shaft design, construction and testing.
Last edition: May 1985(pending publication)
Elaborated by: Commission "Mechanical constructions" of CNR.

2) UNI 8634 : Aluminium alloy structures, instructions for design and construction.
Last edition : Dicember 1985
Elaborated by : Subcommission 13 "Aluminium structures" of the Technical Commission of UNIMET.

3) CNR 10011-85: Steel structures, instructions for design, construction, testing and maintenance.
Last edition : June 1986
Elaborated by : Commission "Steel structures" of CNR.

4) CNR 10021-85: Steel structures for lifting appliances instructions for design, constuction, testing and maintenance.

Last edition : May 1988.
Elaborated by : Commission "steel structures" of CNR.

5) CNR 10022-85: Cold formed steel structural shapes, instructions for structural application.
Last edition : May 1988
Elaborated by : Commission "Steel structures" of CNR.

6) CNR 10028-85: Aluminium structures for lifting appliances, instructions for design and construction.
Last edition : May 1988.
Elaborated by : Commission "Steel structures" of CNR.

7) UNI 7670: Mechanisms for lifting appliances, instructions for design.
Last edition: March 1988
Elaborated by: Commission "Lifting appliances and their fittings" of UNI.

CODES SHORT DESCRIPTION

In the following a very short description of the principal contents of the quoted codes will be given, taking into account, for each document, only the part devoted to fatigue.

Group 1a: only the first of the two documents (UNI 3964) will be considered here, because the second one is not yet available, not even as a tentative proposal. UNI 3964 is substantially correspondent to ISO/R 373. It deals with specimens both smooth and notched, for the evaluation of fatigue strength in high cycles fatigue. It considers only room temperature and force controlled tests with very simple load histories (constant amplitude). The principal definitions and the symbols necessary to deal with the very simple situations considered are given and the procedure to report the obtained results are indicated. Also the principal diagrams to plot the dependency of fatigue strength on loading conditions (static stress or stress ratio) are described in detail. As far as the statistical analysis of the results is concerned, only the evaluation of the fatigue limit applying the staircase method is specified and explained.

Group 1b: also for this group only the first document is available at the moment and will be considered here. Project E 12.14.009.0 deals only with smooth specimens (of circular or rectangular cross section) and axial loading conditions. Istructions on dimensions of specimens, their manufacture and storage are given. Also the fundamental requirements for a correct preparation and execution of the tests are indicated, together with the procedures to report the obtained results.

Group 1c: This code represents the only example of the Italian Codes, known by the Author of this note, which give detailed instructions to assess the fatigue strength of a mechanical component. It deals with threaded fasteners subjected to axial load at room temperature. Only simple loading conditions are considered (constant amplitude). The symbology to be used and the fundamental requirements for a correct preparation and execution of the tests are given, together with some instructions on how to represent the obtained results and the data to be recorded and indicated on the test report.

Group 2: all the documents of this group are related with design of structures and not with their fatigue testing, but each of them contains a part, usually not very extensive, devoted to fatigue design. Although they are not concerned with fatigue testing, a short description of these parts is believed to be of importance here because only the analysis of the needs of the designer for the real applications could clearly indicate the lacks of existing Codes on fatigue testing which ask for a more urgent and concrete solution.

The first point to be noted, which is common to all the documents on structural design, is that, as far as fatigue problem is concerned, they are related mainly with structural components, and not with base material. This is obviously very natural, as fatigue is a local phenomenon, which more likely occurs where stress concentrations arise. The most typical structural components considered in the Codes are welded joints, bolted or riveted joints and a few notches, very usual expecially in mechanical constructions.

As a second point, it is interesting to note that all the documents on structural design refer to nominal stress, without considering the possibility of a fatigue design based on local stress or local strain, measured or evaluated by numerical analysis.

As far as the loading conditions are concerned, third point, only mono-axial loading conditions are usually considered, without distinction for axial or bending. For more complex loading conditions (bi- or tri-axial) the possibility of giving detailed instructions is often stated to be difficult and premature for the state of the art. The fatigue strength is given usually as indipendent on the static stress and on the stress ratio, with a few exceptions.

The load history will be considered as point four. In all cases the possibility of constant amplitude load history and the one, more realistic, of variable amplitude, are both included in the codes on structural design.

Another point (5) relevant to the aims of this note concerns the assessment of the fatigue reliability of a structure. Usually it is based on a finite-life approach, not on the fatigue-limit concept. Moreover, when considering real time histories, the fatigue limit appears to loose almost completly his signification, due to the application of a modified Miner's rule.

Last, but quite important in the present contest, several documents explicitly suggest the opportunity of executing "specific and suitable tests" for assessing the fatigue strength of particular components, without specifying how to perform the tests and how to use the results. Experimental fatigue tests are suggested in a general form for not well known geometries, for which fatigue data are not included in the Codes (as.in CNR 10011) or, more specifically, for some particular cases, as in CNR 10022, where fatigue tests are requested for steel plates less than 3mm thick and for spot welded joints.

CONCLUSIVE REMARKS

The short description of the points of the Italian Codes on fatigue mainly consistent with the purpose of the present note, which has been just accomplished, on one side clearly shows the possibility to improve and unify the existing procedures for the experimental evaluation of high cycles fatigue behaviour of metals, on the other side points out how many of the needs which arise in applying the Design Codes are not covered by the Codes related to fatigue testing, and shows the need of assessing new procedures to cover the existing lacks.

As far as the possibilities of an improvement of the Codes are concerned, the most urgent points to be developed, which will be important also for the development of new procedures, seem to be the following:

1) the improvement and the extension of the symbology which is necessary to deal with fatigue problems, much more extended than the one given by UNI 3964, as shown by Design Codes. The present situation is likely to cause confusion and misunderstanding in several instances;
2) the definition of unified procedures for the statistical analysis of fatigue data also for the finite-life range, not only for the fatigue-limit evaluation;
3) the inclusion of more realistic time histories, with variable amplitude loading, instead of considering only constant amplitude loading.

As far as the possibilities of an extension of the Codes are concerned, the most urgent points to be developed seem to be the following:

1) the definition of procedures to assess the fatigue strength of typical structural components, to be selected between the most usual in practical applications. The first example could be that of structural joints (welded, bolted, riveted or glued) or that of notches typical in many mechanical constructions;
2) the definition of procedures to evaluate the fatigue

reliability of a component applying a local stress or local strain concept, both for the evaluation of the fatigue strength of the material and for the assessment of the load history.

The above mentioned needs, if fulfilled, will contribute to improve the reliability of fatigue-prone structures even in the short time. Beside that they will make possible for the future to have fatigue data more complete and easier to analyse and to compare between them, giving at the end more reliable fatigue data for design Codes.

Obviously the knowledge connected to the various needs which have been enucleated in this note is at present at different levels from case to case: in some cases much can be already done, in some cases more research is needed.

Other possibilities than the ones pointed out in this note could also be evidenced. A preliminary extension of this analysis, which will compare the situation existing in several Countries, could improve the picture, arriving at a better definition of the real needs and of the concrete possibilities for their fulfilment.

19 ANALYSIS OF FATIGUE TEST RESULTS ON FIELD-WELDED RIB SPLICES IN ORTHOTROPIC STEEL BRIDGE DECKS

S. CARAMELLI, P. CROCE, M. FROLI
and L. SANPAOLESI
Institute of Construction Science, University of Pisa, Italy

SUMMARY

In this paper are collected and analysed the fatigue test results of several research programmes carried on field-welded rib joints in orthotropic steel bridge decks.

The comparison puts into evidence how the fatigue behaviour of the various types of connections can be related to many parameters such as form of the stiffeners, root gap and residual stresses.

In spite of the relatively large number of test results, some aspects are not yet completely investigated and consequently are indicated needs for further research work.

1 INTRODUCTION

Almost all orthotropic steel bridge decks of some importance are made by shop executed transportable tiles spliced on field both in the transversal and in the longitudinal direction.

In modern bridges closed ribs with trapezoidal or triangular cross section are mostly used [1] and it is generally agreed that welding is the most appropriate way of field-splicing.

But, because of the closed section, it is quite difficult to perform good quality welds since the operator must work in an unfavourable position.

Testing of Metals for Structures. Edited by F.M. Mazzolani.
Published by Taylor & Francis, 2 Park Square, Milton Park, Abingdon, Oxon, OX14 4RN.
ISBN 0 419 15810 3

As the transverse jointing occurs frequently in a steel bridge deck and fatigue cracks have been discovered in many field-welded rib joints [2], [3], research programmes have been carried out in several countries.

2 FATIGUE TESTS ON REAL SCALE RIB SPECIMENS

2.1 Dutch results

In 1974 Tromp [4] investigated the fatigue behaviour of field-welded joints between trapezoidal ribs. The specimens were full scale single trapezoidal stiffeners, steelgrade Fe 510, jointed at middle span by a connecting rib butt spliced in overhead position on tack welded backing strips, root gap 4 mm (connection type A-1, figure 1).

The tests were executed by pure bending on a four points test scheme as illustrated in figure 2.

In 1986 an ECSC collective research programme on the fatigue behaviour of orthotropic steel bridge decks started in six European Countries [5].

Kolstein and De Back [6] performed constant amplitude tests with the same modalities and on the same type of trapezoidal test specimens used by Tromp.

The influence of the root gap of the welds was examinated by introducing a gap between the stiffener and the connecting rib of 4, 2 and 0 mm, respectively in types A-1, A-2, A-3 of figure 1.

Beside that, tests were performed on specimens with improved edges preparation (type D, figure 1).

The fatigue test results are plotted in the S-N diagram of figure 2 on a log-log scale together with the mean resistance line NL related to type A-1 joints.

From their analysis it may be deduced that the fatigue strength increases with the enlargement of the root gap, whereas the difference due to limb preparation does not show to have a great influence, even if it would seem to be evident that the type D geometry facilitates the execution of the welds.

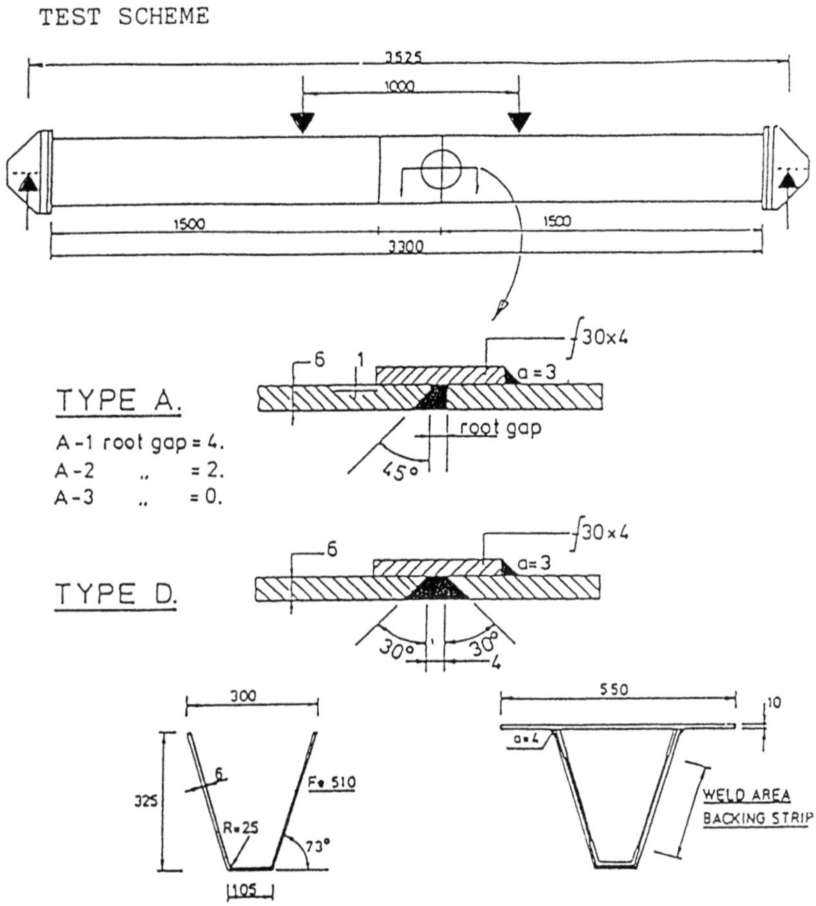

Fig. 1: Dutch test specimens geometry, test scheme and welds geometry.

2.2 Japanese results

In 1988 Yamada, Kendo, Aoki and Kikuchi [7] tested trapezoidal rib connections, similar to those tested in the Netherlands, in order to assess the influence of weld sequences and contact defects upon the fatigue life of the joints.

The tests were carried out on a four points test scheme as reported in figure 3 where the geometries of the welds and of the defects are also illustrated.

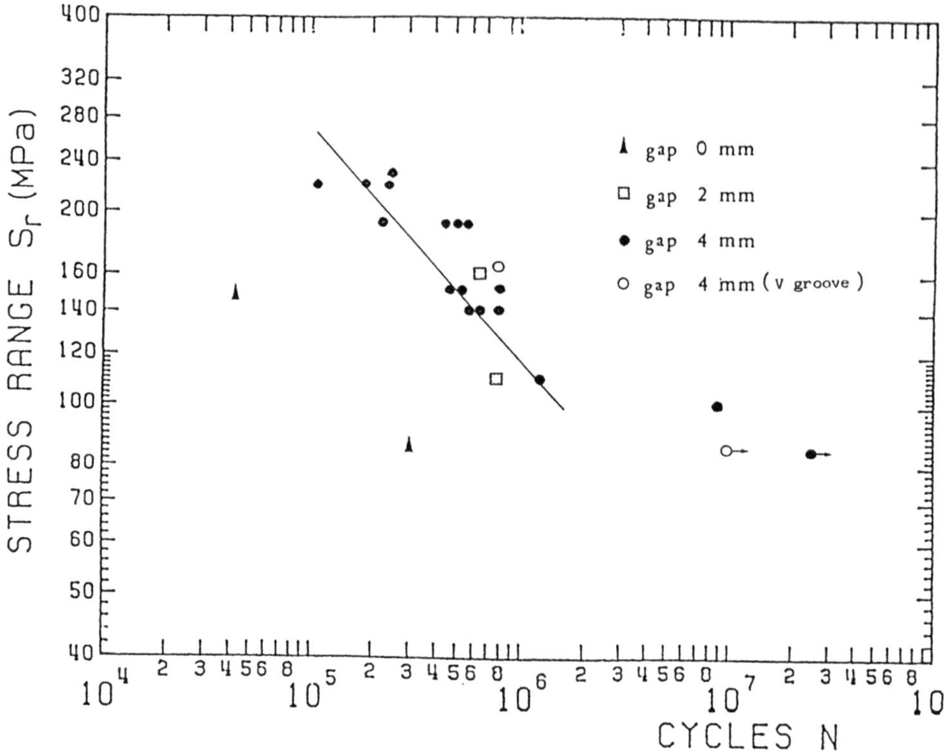

Fig. 2: Dutch fatigue test results (trapezoidal ribs).

In type L and type M specimens the webs of the ribs were welded before the lower plate in two passes whereas in the other specimens the sequence was inverted.

It was found that distribution and intensity of residual stresses depended on the welding sequence: at the lower limb of type L and type M specimens were measured tension residual stresses of around 5 KN/cm² while for the others there was a compression stress of about 10 KN/cm².

The results of these tests are reported in the S-N diagram of figure 4 in which N1 indicates the mean resistance line for type F and type G' joints, N2 is the line related to type H, N3 to type L and N4 is the line of type G specimens.

Type H and type G specimens are very similar;

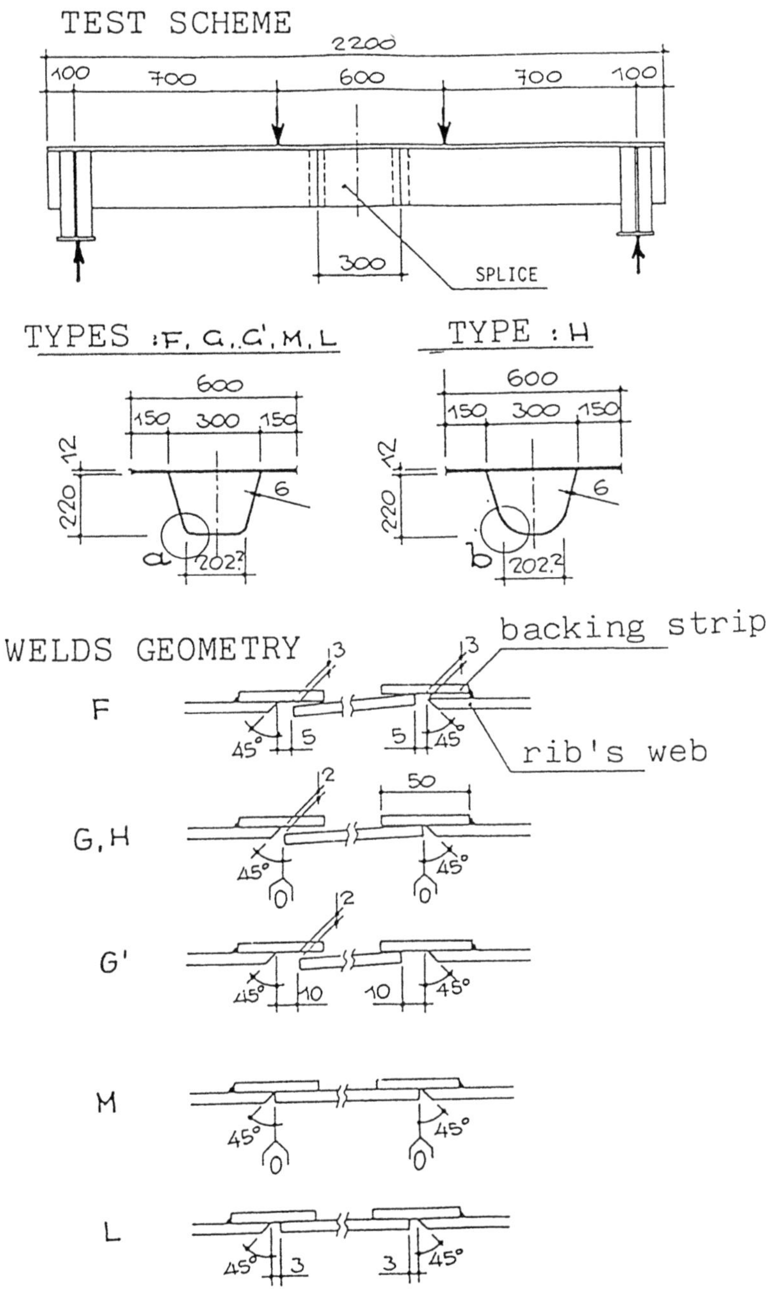

Fig. 3: Japanese specimens geometry, test scheme and welds geometry.

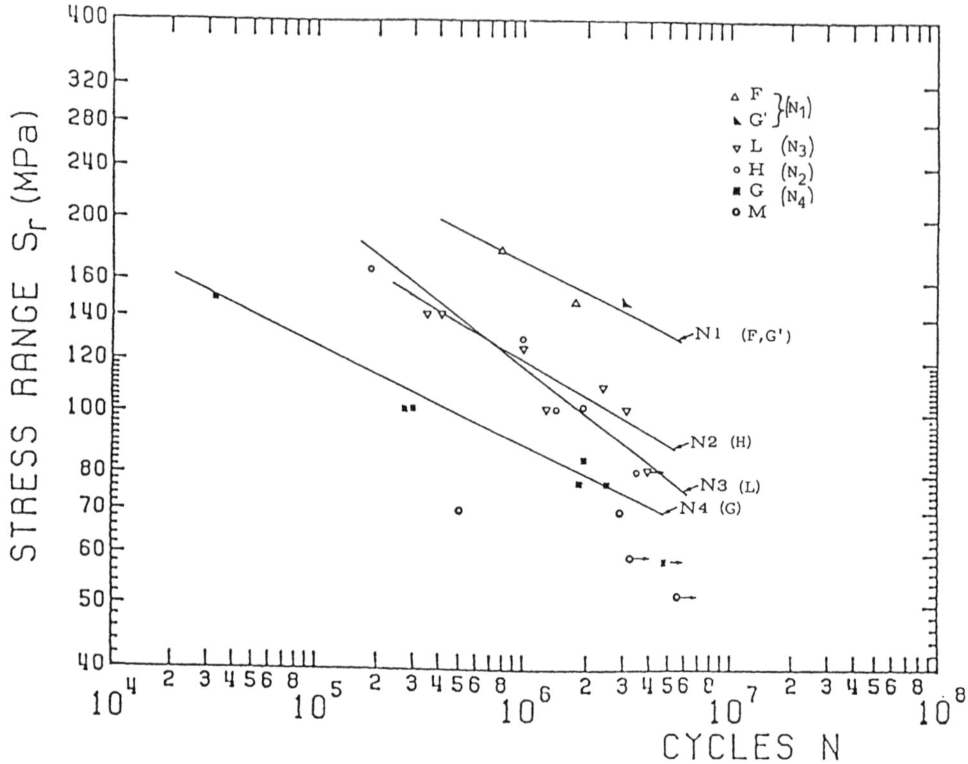

Fig. 4: Japanese fatigue test results (trapezoidal ribs).

nevertheless, line N2, related to type H specimens, is much above line N4, related to type G specimens. That was due to the presence in these latters (and in type M specimens too) of penetration defects of 1.8 -2.8 mm caused by bad execution of the welds.

The same Japanese Authors had previously already obtained other results at the Osaka University and at the Public Works Research Institute (see [7]) by testing specimens analogous to those previously examined.

In type O specimens, tested at Osaka, the root gap was 4 mm, one of the edges prepared with a 50 angle and the welds were carried out in two passes: during the first pass, the webs were welded before the lower plate; in the second the sequence was inverted.

In both type X and type Y specimens, tested at PWRI, the root gap was 6 mm and a edge prepared with a 35 angle, but while the contact defect in X specimens was 1-2 mm, in Y specimens it was 3-5 mm.

These results are reported in figure 5 where their mean resistance lines are indicated with O for the Osaka tests and respectively with J1 and with J2 for type X for type Y joints.

The results show the influence of both the root gap and the weld sequence: insufficient gaps, in fact, cause the weld's incomplete penetration and the fatigue strength decreases when traction eigenstresses are present in the lower plate (specimens J1 and J2). On the contrary, the extent of the contact defect between the rib and the backing strip does not seem to influence, in any

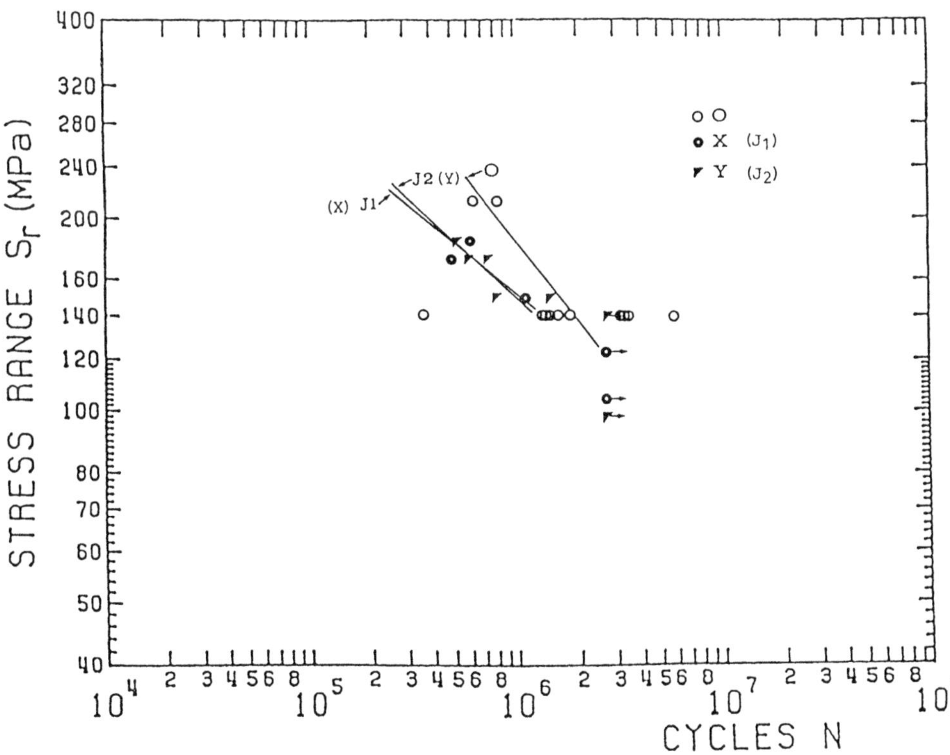

Fig. 5: PRWI fatigue test results (trapezoidal ribs).

appreciable way, the fatigue behaviour of the splices as one can see comparing type J1 to type J2 results.

2.3 British results

The fatigue behaviour of field executed joints between stiffeners of triangular shape was investigated in 1982 at the Transport and Road Research Laboratory by Cunninghame [8] who tested two groups of specimens on a four points test scheme (see figure 6).

In the first group a 19 mm thick plate was inserted as

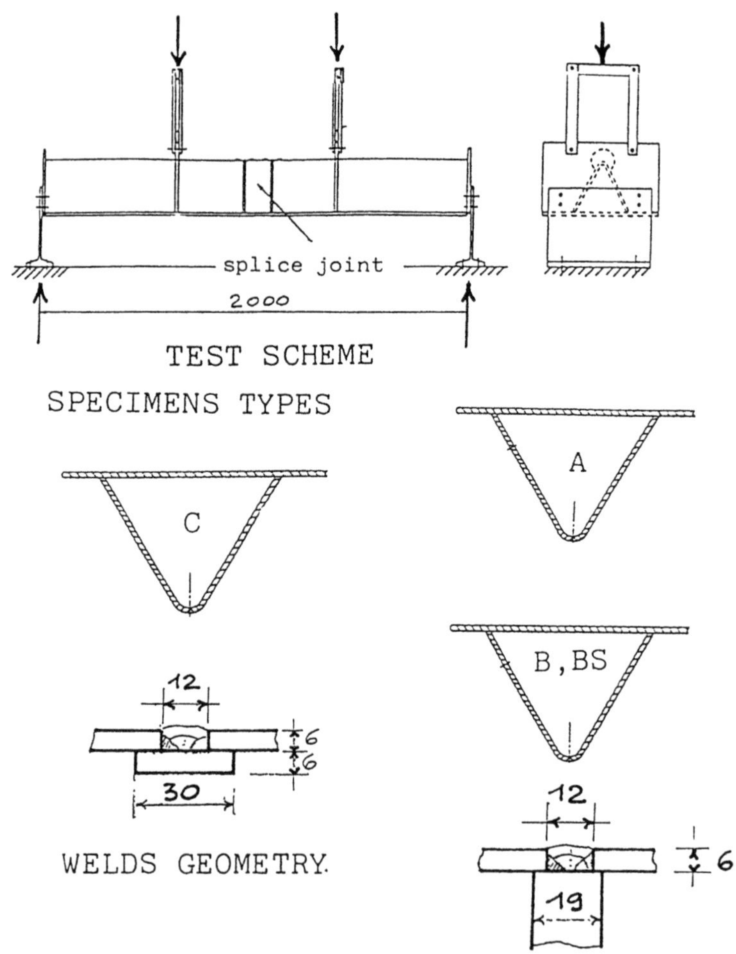

Fig. 6: British specimens geometry, test scheme and welds geometry.

a water tighting diaphragm at both ends of the ribs to be spliced and then used also to back the connecting rib (see figure 6). The first group was then further subdivided in three sets of specimens: types A, B and BS.

Types A and B differ from each other only for the weld sequences, while the type BS consisted of type B specimens previously submitted to mechanical stress relieving.

The execution modalities of the second group of specimens (type C) are illustrated in figure 6. For these specimens, the connecting rib was butt welded without edge preparation on a backing strip with a root gap of 12 mm.

The results, all very similar, are diagrammed in figure 7 in which I1 indicates the mean resistance line related to type A specimens, I2 that of type BS specimens, I3 that of type B specimens.

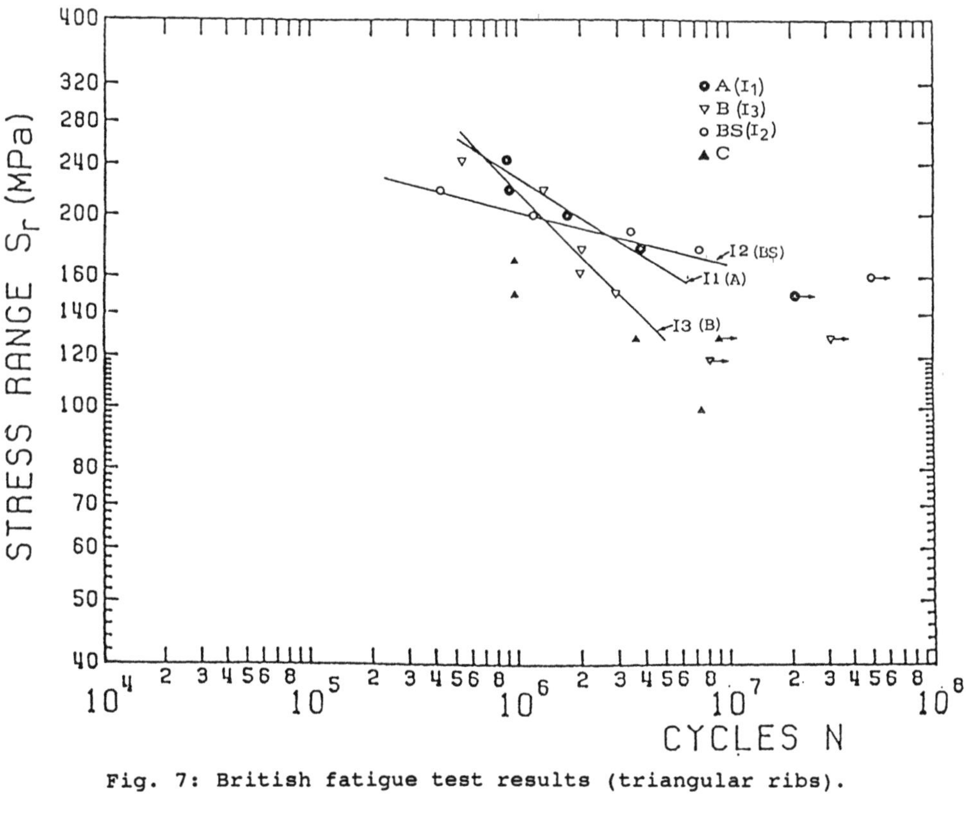

Fig. 7: British fatigue test results (triangular ribs).

2.4 Italian results

Partner of the ECSC collective research programme before mentioned at point 2.1., the Istituto di Scienza delle Costruzioni of the University of Pisa carried out constant amplitude tests on triangular shaped ribs connected at middle span by two types of field-welded splices [9].

In the first type of joint (type I) the ribs were butt welded to the connecting rib on a backing strip, one edge prepared with a 45 angle V-groove and the root gap was 6 mm.

In type II joints the stiffener webs were butt welded with complete penetration, both edges prepared with a 60 X-groove, root gap 1-2 mm. The weld is executed at first from inside, then grooved on the external part and completed from outside in overhead position. Due to the supposed superiority of the welds, this type of joint was thought to behave better than type I.

All the weldings have been checked by means of visual and magnetic controls. The butt weldings have been also 100% X-raied and repaired (only one time) when found not acceptable according to the UNI 7278 Italian Standard.

Type I and type II joints geometries are illustrated in figure 8 together with the test scheme, while in figure 9 are collected the fatigue test results.

The comparison shows that, contrary to all expectation, the fatigue behaviour of type I joints is better than that of type II joints.

Further investigations carried out on narrow strips, butt welded with the same modalities of type I and type II welds, put into evidence that the fatigue behaviour of the two sets of strips, free from residual stresses, was the same while, on the other hand, it was found that, by submitting to mechanical stress relieving rib specimens of both types, the level of residual stresses was much higher

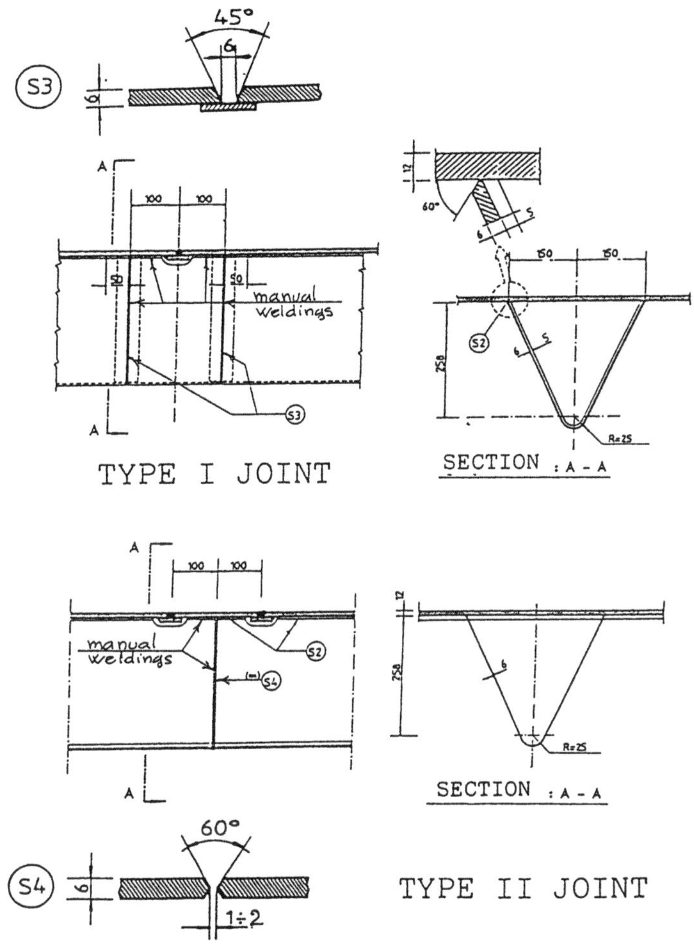

Fig.8: Italian specimen geometry, test scheme and welds geometry.

in type II specimens than in type I specimens.

That allowed to conclude that the fatigue life of the two types of joint depends mainly on residual stresses which are probably caused, according to Yamada and Oth., by the welding sequence.

3 CONCLUSIONS

The mean resistance lines of the available results on real size specimens are drawn in the diagram of figure 10,

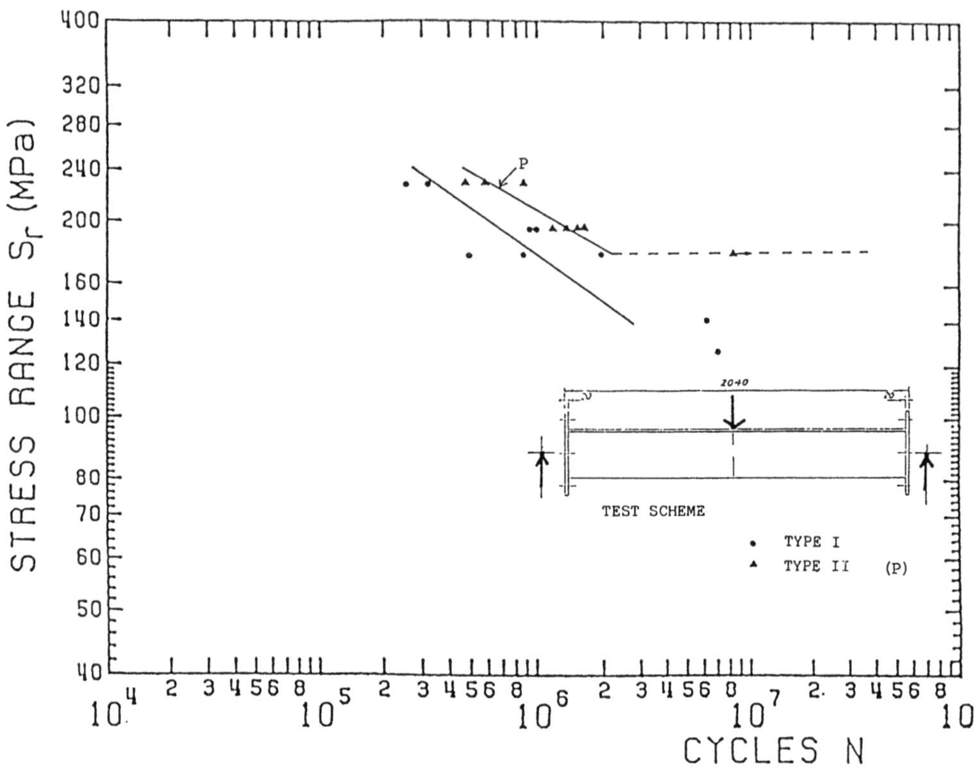

Fig. 9: Italian fatigue results (triangular ribs).

in which P stands for the curve obtained in Pisa on type I joints.

In the same figure the 5% fractile ECCS line for backing strip welds (class 71) is also reported.

At first sight, the results seem to be quite scattered, and yet if only those related to the specimens with a root gap larger than 4 mm are considered (curves I, P, O, J, NL and N1), then the scatter band reduces greatly and the lines are well above that estimated by the ECCS.

A further grouping together of the results on the basis of the shape of the rib's cross section, trapezoidal for lines O, J, N1 and NL and triangular for lines P and I1, reveals that the data obtained by the different Authors does to a large extent agree. In particular the superior

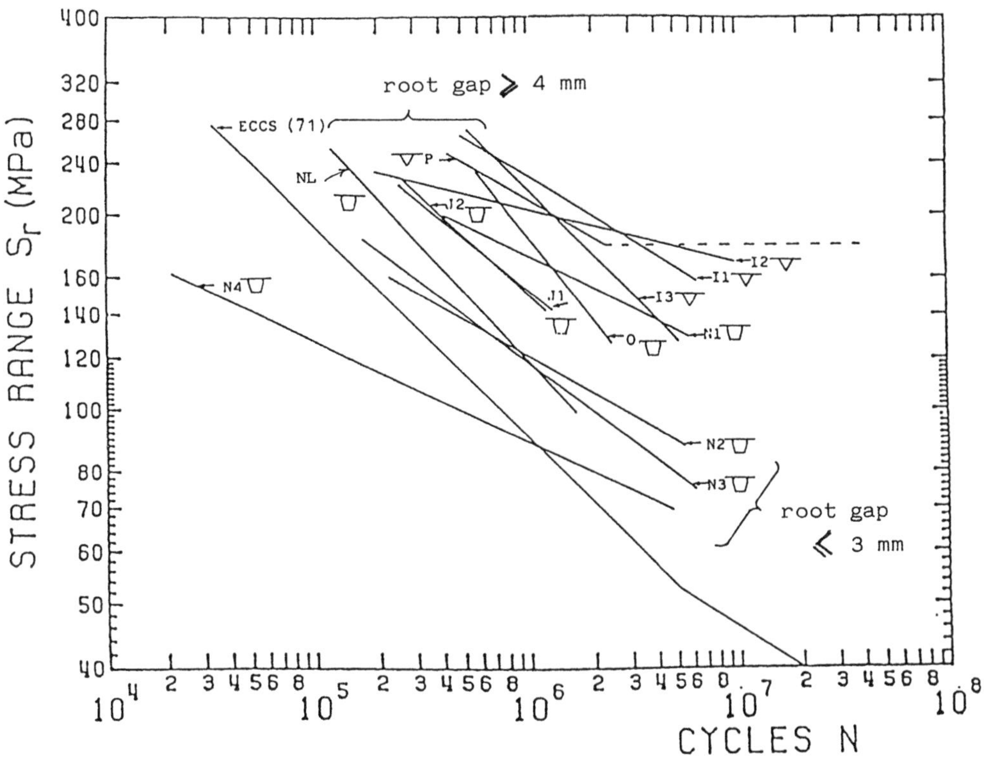

Fig. 10: Comparison between fatigue data on field-welded rib joints.

behaviour of the triangular stiffeners would suggest that the fatigue life is dependent also upon the rib shape.

It can be concluded that, while the agreement on the dependance of the fatigue behaviour of the joints on the root gap of the welds seems to be general, the influence of the shape of the ribs should be further checked.

On the other hand, the residual stresses clearly influence the fatigue strength of the splices but it is not yet completely clear the way they are influenced by the welding sequences.

At the end, it can be noted that low stress amplitude data are presently very few and therefore there is a need for further reaserch work in this direction.

REFERENCES

[1] R.BARTELLETTI, S.CARAMELLI, P.CROCE, L.SANPAOLESI, Tipologie degli impalcati in lamiera irrigidita per ponti stradali, Proceedings XII C.T.A. Congress: Giornate Italiane della Costruzione in Acciaio; Isola di Capri (1989).

[2] R.WOLCHUK, Lessons from weld cracks in orthotropic decks on three european bridges, ASCE Journal of Structural Engineering, (1990).

[3] K.YAMADA, Fatigue assessment of orthotropic steel decks of girder bridges, Proceedings of IABSE Workshop: Remaining fatigue life of steel structures; Lausanne (1990).

[4] W.A.J.TROMP, Fatigue of field splices in ribs of orthotropic steel bridge decks, Report 6-74-15, Stevin Laboratory, Delft university of Technology; Delft (1974).

[5] S.CARAMELLI, P.CROCE, S.DANIELI, M.FROLI, L.SANPAOLESI, Indagine sperimentale sul comportamento a fatica degli impalcati da ponte in acciaio a piastra ortotropa, Costruzioni Metalliche n.2/3, (1988).

[6] M.H.KOLSTEIN, J.DE BACK, Measurement and interpretation of dynamic loads on bridges - Phase 3: Fatigue Strength of Orthotropic Steel Decks, ECSC Convention 7210-KD/609, Final Report, Delft University of Technology, Stevin Laboratory; Delft (1989).

[7] K.YAMADA, A.KONDO, H.AOKI,Y.KIKUCHI, Fatigue strength of field-welded rib joints of orthotropic steel decks. IIW Doc. XIII-1282-88, Department of Civil Engineering, Nagoya University; Nagoya (1988).

[8] J.R.CUNNINGHAME, Steel bridge decks: fatigue performance of joints between longitudinal stiffeners, TRRL Laboratory Report 1066, Transport and Road Research Laboratory; Crowthorne (1982).

[9] S.CARAMELLI, P.CROCE, M.FROLI, L.SANPAOLESI, Fatigue behaviour of orthotropic steel bridge decks, Proceedings of IABSE workshop: Remaining fatigue life of steel structures; Lausanne (1990).

20 A RE-ANALYSIS OF FATIGUE TEST RESULTS ON WELDED STEEL MEMBERS: KNOWLEDGE GAPS, RESEARCH NEEDS AND STANDARDIZATION REQUIREMENTS

C.A. CASTIGLIONI
Department of Structural Engineering, Milan Polytechnic, Italy

SUMMARY

Existing data base of fatigue data are briefly discussed in this paper, together with further research needs in the field of long life fatigue testing of welded structural steel details.
As jet no internationally recognized recommendation has been proposed for fatigue testing methods and design of experiments on full scale members, aim of this paper is that of discussing some of the topics that such a recommendation should cover.
It is hope of the author that discussions which might arise about the contents of this paper be fruitful and may lead to the formation of a Drafting Panel for such a Document.

INTRODUCTION

In 1972 and 1974, two reports were published [1,2], summarizing the results of the NCHRP Project 12-7. This two reports represented a keystone for following developments of studies on fatigue behavior of welded structural steel details.

Prior to the National Cooperative Highway Research Program (NCHRP) Project 12-7, only approximate fatigue design relationships were developed because of the limitations of the test data available at that time. Furthermore. often in previous studies many variables were introduced into an experiment, with too limited a number of specimens, and this made it impossible to clearly establish the statistical significance of stress variables. types of details and type of steel, as well as of quality of fabrication on fatigue strength.

What was required was a comprehensive test program to be performed under controlled conditions so that analysis of the resulting data could lead to define the actual influence that the various parameters had on the

Testing of Metals for Structures. Edited by F.M. Mazzolani.
Published by Taylor & Francis, 2 Park Square, Milton Park, Abingdon, Oxon, OX14 4RN.
ISBN 0 419 15810 3

fatigue behavior of welded details. The NCHRP Project 12-7 [1,2] provided such a test program. It was developed to provide a statistically designed experimental program under controlled conditions, and involved fatigue testing of some 530 beams and girders with two or more details each. Large-size specimens were used to overcome some of the limitations of smaller, simulated specimens (such as residual stress fields and shear lag). The specimens were fabricated with various details that are commonly used in the design of bridges, including coverplates, web and flange attachments, and stiffeners. Three different types of steels were used to study the influence of yield stress on fatigue life. All specimens were tested under constant cycle amplitude loading. The order of testing of the specimens for each beam type was randomized so that the effect of uncontrolled variables (i.e. temperature, humidity, laboratory, human factor...) would also be random. The assignment of specimens to locations within the design was also done randomly so as to distribute any variation inherent in the fabrication. The fabricator was instructed to use usual fabrication techniques, workmanship and inspection procedures. The randomization of the uncontrolled variables allowed the effects of these variables to be included as a random error in the analysis of the experiment, and also prevented any systematic bias due to these effects on the controlled variables.

The data of the NCHRP research Project were used to develop the AASHTO Specifications [3] for fatigue design of steel bridges.

Since 1974, several major fatigue studies were carried out on beam-type specimens, all over the world. Tests were performed in East Germany, Japan, Switzerland, Office of Research and Experiments (ORE) of the International Union of Railways (West Germany, Poland, England and Holland), as well as in the U.S.. The additional studies evaluated the applicability of the findings of the original NCHRP test program to fabrication conditions elsewhere in the world, and were used to develop fatigue design codes. The additional tests augmented the NCHRP findings and often defined the fatigue strength of details that were not tested under the original NCHRP program, like for example the Japanese data, that were obtained in order to develop fatigue specifications for the design of long-span bridges for the Honshu-Shikoku Bridge Authority, and contained details typical of welded box members [4-6].

Practically, over 1,500 additional test data points were added to the existing database of approximately 800 fatigue test results [7,8]. The original database contained a limited number of welded steel details. This resulted from the need for adequate replication of the data at identical stress conditions so that the influence of stress parameters could be examined. In the new test programs it was generally accepted that stress range and detail type are the two important parameters that determine fatigue strength. Therefore it was possible to perform fatigue tests on a wider variety of details and geometries, many of which had not been previously studied.

Recently ECCS adopted a set of fatigue design curves [9], representing a major effort to reevaluate the fatigue design specifications. The analisys was based on the original NCHRP test data, limited data obtained from a specially designed test program as well as other sources. Later, Eurocode-3 [10], adopted the same fatigue design approach proposed by ECCS, and the ECCS database, eventually enlarged by more recent data, was published in tabular form in the Background Document 9.01 [11].

Practically at the same time, Fisher and Keating [8], published an extensive database, aimed to the revision of the AASHTO Specifications [3].

The data review contained in [8] is aimed only to welded bridge details, while ECCS/EC-3 [11] database is somewhat wider, in the sense that it encompasses also non welded details, like riveted or bolted connections, because is more generally aimed to structural steel details, but is probably not as comprehensive on welded details as is database [8].

In this paper, data presented in both documents [8] and [11] are briefly discussed and critically reviewed, with the aim of identifying further research needs and knowledge gaps. From comparison between the two documents and from their discussion, it clearly appears and is also well stated by the authors, that problems arose when combining togather test data from different research programs mainly because each test program was developed independently, and was based on a different set of test specifications procedures. In fact, despite the various national and international Provisions for fatigue tesing of materials (i.e. [12-22]), no internationally recognized Recommendation has been jet proposed for fatigue testing methods and design of experiments on large scale structural members and details. Thus, in the concluding section of this paper, some of the topics that such a Recommendation should cover are presented and discussed in the hope that this might represent the basis for a fruitful discussion leading to the formation of a Drafting Panel for tentative recommendations for fatigue testing procedures of full scale structural steel members and details.

RESEARCH NEEDS

At present, EC-3 [10] main provisions are based on the adoption of a set of fatigue resistance curves which define the strength of various classes of details. The proposed fatigue curves consist of 14 equally spaced curves in a log-log plot. The slopes of all curves are set to a constant slope of -3.0 up to $5*10^6$ cycles, where a constant amplitude fatigue limit is defined. At $5*10^6$ cycles two options are provided. One option changes the slope to -5.0 up to $100*10^6$ cycles, where a cut off is provided. The intercept at $100*10^6$ cycles establishes a fatigue limit, regardless of the type of loading. The second option correspond to a straight line extension of the -3.0 slope S-N curves.

The European approach to fatigue design is different from that of other countries (e.g. [3], [23]). In fact, despite the great activity that in the past ten years was spent in the development and international harmonization of recommendations for the fatigue design of steel structures, there are still some areas of disagreement.

A lack of test data obliges those committees developing design recommendations to estimate behavior in some areas. This is particularly true for long fatigue lives ($N > 2*10^6$ cycles). Furthermore, some recommendations are formulated on the basis of data obtained from tests of small specimens. Extension of these recommendations to large structures is not possible since there are few results from large-scale tests. As a result disagreements originate from different estimates of long-life fatigue strength and how these estimates should be represented within the design recommendations [24].

In order to resolve these disagreements, testing activity has increased, particularly long life tests of large scale elements. Recently, new fatigue test results have been collected and evaluated for proposed revisions to design recommendations in the USA [8]. Also, testing is planned and in progress in several locations, for example [25-27]. Differences between results from small-scale specimen testing and large-scale beam testing have been observed for both constant and variable amplitude loading. Small-scale specimen test data can give unconservative estimates of fatigue strength even for detail types which traditionally have been thought [28] to require no correction [8].

In both documents [8] and [11], however, it is clearly stated by the authors that big difficulties were encountered when trying to incorporate in the database data from different research programs. In fact, a great care was adopted in both documents to select the most comprehensive fatigue programs and those for which the clearest informations was farwarded in the test reports, and it was found that each fatigue test program was carried out independently, according to each particular testing procedures. This led, even on the same category detail, to great discrepancies in the results that can be attributed not only to the fatigue design practice, proper to each laboratory, but also from the detailed fabrication procedure and quality achieved in the preparation of the specimens to be tested. This is particularly true for the welded specimens. Unfortunately, these very important aspects and informations often do not appear in the test reports. Practically, each group of fatigue test specimens is fabricated with certain discontinuities which are not fully known or even properly identified in the experimental reports. These factors play a major role when different sets of data are grouped together. Of particular importance is the defect size, as was noticed in the Japanese studies [29], where each series of tests could be interpolated by an S-N curve with a slope near -3.0, but when combined togather showed that major deviations existed between series, resulting in a wide variation of the test data and also causing the slope provided by the regression analysis to change.

Furthermore, when variable amplitude fatigue tests are involved, very seldom in the reports are fully reported data regarding the loading history; often only the type of distribution function adopted for the loading cycles is described, but few informations are given obout loading blocks and cycle sequence.

Also failure criteria adopted in the different testing program are not homogeneous. This means that the "end of test" conditions differed from one program to the other, generating further confusion and scattering the test data. Some data were also biased by retrofitting procedures, performed during the test, which complitely spoilt the significance of the data, from the point of view of the assessment of the fatigue strength of the detail.

Practically, it can be concluded that each group of fatigue test data, coming from different research programs, stands on its own, and that the existing fatigue test data found in the literature are for one thing or the other somewhat non homogeneous [11].

By comparing and re-analyzing Fatigue Recommendations adopted by various Countries [30], it clearly appears that, under constant amplitude loading cycles, design documents model fatigue strength vs. fatigue life to be a straight line when plotted in a log-log diagram. Typically this line

TYPE OF DETAIL	CYCLES TO FAILURE						STRESS RANGE (MPa)
	$2.10^6<N<5.10^6$		$5.10^6<N<10^7$		$N>10^7$		
	C.A.	V.A.	C.A.	V.A.	C.A.	V.A.	
Plain rolled beams	4		6		5		200-300
	2						300-400
			1				> 400
Longitudinal welds						3	60-80
	1		2	1		2	80-100
	20	5	6	3	5		100-200
	14			2			200-300
	5			1			> 300
" " in box girders	13						> 100
Flange splices					3		60-80
	7				7		100-200
	2						> 200
Web stiffeners						2	60-80
	13		7		1		80-100
	12		1				> 100
Web attachments						7	10-20
	2			1	1	19	20-40
	3		2		1		40-60
	1						> 60
Web gussets					2		40-60
	7		2				60-80
	4						> 80
Flange tip attach.			2		2		40-60
	12		3				60-80
	8						80-100
	12						> 100
Flange surf. attach.	4		5		1		40-60
	6		1				60-80
	9				2		80-100
	8						> 100
Coverplated beams						8	10-20
th<20mm			1	1	10	9	20-40
	20	1	4	2	4		40-60
	8						60-80
	6						> 80
Coverplated beams				1	9	5	20-40
th>20mm	10		3		4		> 40
total data	213	6	46	11	58	55	

TABLE 1.

breaks and becomes orizontal at the crack-growth threshold stress range, called Constant Amplitude Fatigue Limit (CAFL). Despite the many test data available in the literature with regard to small specimens, the majority of data for large-scale structures are taken from tests at stress ranges which are well above the CAFL [8,11].

Most structures are subjected to variable-amplitude loading. Above the CAFL, Miner's summation, rainflow analysis and the fatigue-strength relationship lead to the definition of an effective stress range. Such a definition allows use of constant-amplitude relationships for variable-amplitude loading conditions [8,11,24,31]. Interest in variable amplitude fatigue strength has recently concentrated on fatigue lives most representative for performances of civil engineering structures. For example, bridges are often expected to withstand stress cycles of $100*10^6$ cycles, or more at low stress ranges (10-40 MPa). Stress ranges below the CAFL influence fatigue life if there is at least one cycle in the variable amplitude stress spectrum above the CAFL. This influence is due to a gradual exceedence of the threshold stress range by smaller cycles as crack length, driven by those cycles above the CAFL, increases.

Detailed informations on the CAFL and of the fatigue behavior at long endurance ($N > 10*10^6$ cycles) are then required in order to develop reliable design criteria. Unfortunately, despite the many test data available in the literature with regard to small specimens [28], the majority of data for large-scale structures are taken from tests at stress ranges which are well above the CAFL [8,11]. Table 1 summarizes, for different structural details, the number of test data presented in [8], that were carried out over $2*10^6$ cycles, at both constant and variable amplitude. It can be noticed that, for fatigue lives $2*10^6 < N < 5*10^6$ cycles, a total of 213 test data were obtained under constant amplitude stress ranges, while only 6 test data are available. For fatigue lives $5*10^6 < N < 10*10^6$ cycles, 46 test data are available for constant amplitude loading, and only 11 for variable amplitude loading. For fatigue lives $N > 10*10^6$ cycles, 58 data are available for constant amplitude loading and 55 for variable amplitude loading. It is also important to notice that most of the data of this last group, are relative to few details only. In fact, for constant amplitude loading, 27 data points are related to coverplated beams (for both plates thickness > or < 20 mm.), while for variable amplitude loading, out of 55 total data, 22 relate to coverplated beams and 26 to web attachments.

These considerations lead to the following conclusions:

a) The quantity of data presently available at stress ranges near constant-amplitude fatigue limits does not correspond to their importance to fatigue life and/or damage assessments. Additional work, both experimental and theoretical is needed to define constant amplitude fatigue limits for most categories of details.

b) Additional work is needed in the extreme life region of most categories of joints, in order to obtain more thorough informations about variable amplitude loading fatigue strength and how it is influenced by the loading spectrum.

c) A great variability exists between test data from different programs. No internationally recognized methods for fatigue testing and design of experiments are in fact being agreed. This means that each test program risks to be carried out independently, according to each particular testing procedures. In order to avoid to vanificate the future efforts of many researchers, an agreement should be found and some international testing procedures should be published, to be used as a reference for carrying out and interpreting tests, like International Organizations and Committees working in different research areas did already (for example T.C. 1., T.W.G. 1.3 (Seismic Design) of ECCS [31]).

STANDARDIZED TESTING PROCEDURES REQUIREMENTS

Recommendations for standard testing procedures should be intended as a reference, to produce an adequate and, as much as possible, an unified way to carry out tests in order to characterize the fatigue behavior of structural joints and components, as well as to interpret the results of these tests.

The following topics should be addressed in the document:

1) TESTING CONDITIONS: tests can be performed either under load controlled or displacement controlled conditions;

2) LOADING HISTORY: testing can be performed either under constant or under variable amplitude loading cycles. In this second case, indications on load spectrum type, loading blocks, cycles sequence and probability of exceeding a given stress range (e.g. the CAFL) should be given. Presently tests have been performed with

loading spectra of the Reyleigh type [32-34], but numerical studies [32] showed that the type of loading spectrum influences the fatigue life.

3) STRESS RATIO: in welded structures, because of the presence of residual stress fields, effective stress ratios differ from nominal ones, usually showing limited range of variations, and rather high values [32].

4) SPECIMEN SIZE: as well recognized by many authors, only full scale beam and girder specimens should be used in fatigue testing of welded members and joints. In fact, residual stress fields do not fully develop in most type of small scale specimens.

5) FAILURE CRITERIA: in laboratory, several failure conditions may be recognized for a fatigue test, and on the adopted failure criterium, depends the "end of test" condition. Mainly the following criteria have been adopted in the literature:
a) First appearance of a crack visually detected or physically measured;
b) Through thickness crack: a crack initiated on the surface of a plate and propagated through its thickness, reaching the back surface;
c) Complete fracture ($K_I = K_C$) or large displacements which become so large that the applied load cannot be mantained at the jacks.
d) Other engineering definition based, for example, on an increase in a global deformation parameter for the beam under test (e.g. mid span deflection f becomining 1.5-2-3 times the elastic nominal mid-span deflection for the undamaged beam f_I).

6) RETROFITTING PROCEDURES: sometimes, test programs were performed in order to evaluate the effectiveness of retrofitting procedures. If these retrofitting procedures are performed before a "failure condition" is reached, the test data is biased, and cannot any more be used in fatigue strength assessment analyses for details in as-welded conditions.

7) CRACK DETECTION METHODS AND PROCEDURES: it should be stated which procedures are to be adopted for detecting and measuring cracks: visual inspection (with or without magnifying glasses), die penetrant, NDT (e.g. X-rays, Ultrasonic testing, Magnetic particle testing), or others.

8) SPECIMEN MANUFACTURING PROCEDURES: manufacturing and design procedures for the specimens should be specified to correspond

to common design and manufacturing practice, with reference to major internationally recognized design codes (e.g. EC-3, AISC,...).

9) INITIAL CRACK SIZE: initial defects should be detected as thoroughly as possible, reporting their locations within the specimen and their size. Of course these factors are strictly connected with the design and manufaturing procedures, previously discussed.

10) MATERIALS AND MATERIAL PROPERTIES : should be reported for all performed tests, and should at least include yield strength and fracture toughness.

11) TEST REPLICATION FOR STATISTICAL SIGNIFICANCE: a minimum number of test results at each stress range level should be required in order to reach a certain significance level, and to lead statistically to a confident interpretation of the results.

12) STATISTICAL ANALYSIS OF TEST DATA: usually, a linear regression analysis has been performed when interpreting fatigue test data. This is due to the nearly linear plot of the S-N data in a log-log scale. The fractile α to be adopted in the determination of the lower confidence limit should be defined (usually is assumed α=5%).

13) CUMULATIVE DAMAGE MODELS: it should be stated which damage model to adopt in damage assessment. Miner's rule has been widely adopted, but it does not always lead to safe results, and alternative methods have been proposed (e.g. [35]). Furthermore, as damage model are usually based on an "allowable damage" parameter (e.g. allowable number of cycles at a given stress range), it should be clearly stated how this parameter is to be computed.

REFERENCES

[1] Fisher, J.W., Frank, K.H., Hirt, M.A. and McNamee, B.M., "Effects of weldments on the fatigue strength of steel beams", NCHRP Rept. 102, 1970.

[2] Fisher, J.W., Albrecht, P.A., Yen, B.T., Klingerman, D.J. and McNamee, B.M., "Fatigue strength of steel beams with welded stiffeners and attachments", NCHRP Rept. 147, 1974.

[3] AASHTO Standard Specifications for Highway Bridges, 13th Ed., American Association of State Highway and Transportation Officials, 1983.

[4] Miki, C., Nishino, F., Tajima, J., and Kishimoto, Y., "Initiation and propagation of fatigue cracks in partially penetrated longitudinal welds", Proc. JSCE, vol 312, Aug. 1981, pp 129-140.

[5] Tajima, J., Asama, T., Miki, C., and Takenouchi, H., "Fatigue of nodal joints in box-section members in a bridge truss", Proc. IABSE Colloquium "Fatigue of Steel and Concrete structures, Lausanne, 1982, pp. 353-359.

[6] Shimokawa, H., Takena, K., Fukazawa, M., and Miki, C., "A fatigue test on the full size truss chord", Proc. JSCE, vol 344, Apr. 1984, pp.57-64

[7] Keating, P.B., Halley, S.A., and Fisher, J.W., "Fatigue test database for welded steel bridge details", Fritz Engineering Laboratory, Rept. 488.2, Lehigh University, May 1986.

[8] Keating, P.B., and Fisher, J.W., "Evaluation of fatigue design criteria on welded details", NCHRP Rept. 286, 1986.

[9] ECCS "Recommendations for the fatigue design of steel structures", Publication n. 43, European Convention for Constructional Steelwork, Brussels, 1985.

[10] "Eurocode-3", Chapter 9, "Fatigue", 1989.

[11] "Background informations on fatigue design rules. Statistical evaluation", Background Document n.9.01 to EC-3, 1st draft, 1989.

[12] "Standard Terminology Relating to Fracture Testing", ASTM, E 616-82.

[13] "Standard Definitions od Terms Relating to Fatigue Loading", ASTM, E 912-83a.

[14] "Standard Definitions of Terms Relating to Fatigue Testing and the Statistical Analysis of Fatigue Data", ASTM, E 206-72/79.

[15] "Standard Practice for conducting Constant Amplitude Axial Fatigue Tests of Metallic Materials", E 466-82.

[16] "Standard Practice for Fracture Testing with Surface-crack Tension Specimens", ASTM, E 740-80.

[17] "Standard Test Method for Constant-Load-Amplitude Fatigue Crack Growth Rates above 10^{-8} m/cycle, ASTM, E 647-83.

[18] "Standard Practice for Presentation of Constant Amplitude Fatigue Test results for Metallic Materials", ASTM, E 468-82.

[19] "Prove meccaniche dei materiali metallici: Prove di fatica a temperatura ambiente. Principi Generali", UNI, 3964/73.

[20] "Determinazione della tenacita' di frattura in condizioni di deformazione piana", UNI, 7969, 1979.

[21] "Prova di resilienza su provetta a due appoggi con intaglio a V, per l'acciaio", UNI, 4713, 1979.

[22] "Prove di resilienza secondo Charpy per l'acciaio. Prova di rottura a

flessione per urto su provetta con intaglio ad U od a buco di chiave", UNI 4431/60.

[23] Swiss Standards SIA 161, "Steel Structures", Zurich 1985.

[24] Smith, I.F.C., Castiglioni, C.A., Keating, P.B., "Analysis of Fatigue Recommendations Considering New Data", IABSE Proceedings P-137/89, Zurich, 1989.

[25] Klippstein, K.H., Variable amplitude load fatigue", Detailed work Plan, Project No. DTFH61-86-C-00036-I, University of Pittsburgh, PA, 1988.

[26] ICOM. Unpublished test results, Swiss Federal Institute of Technology, Lausanne, 1987-1989.

[27] Keating, P.B., and Fisher, J.W., "Full scale details under variable amplitude loading", The Welding Institute, Proceedings of the Int. Conf. on "Fatigue of Welded Construction", Brighton, 1987.

[28] Gurney, T.R. and Maddox, S.J., "A re-analysis of fatigue data for welded joints in steel", Rept. E/44/72, The Welding Institute, Cambridge, 1972

[29] Miki, C., Nishino, F., Harabayashi, Y. and Ohga, H., "Fatigue Strength of Longitudinal Welded Joints Containing Blowholes", Proc., JSCE, vol. 325, sept. 1982, pp.155-165.

[30] Castiglioni, C.A., "Considerations on some recent Recommendations for the fatigue design of steel structures", Costruzioni Metalliche, n.1, 1988, pp. 9-30.

[31] ECCS, "Recommended testing Procedure for Assessing the Behavior of Structural Steel Elements under Cyclic Loads", Publication n. 45, European Convention for Constructional Steelwork, Brussels, 1986.

[32] Keating, P.B., "High Cycle Fatigue behavior of Welded Details Under Variable Amplitude Loading", Ph.D. Thesis, Lehigh University, Sept. 1987.

[33] Shilling, C.G., Klippstein, K.H., Barsom, J.M., and Blake, G.T., "Fatigue of Welded Steel Bridge Members under Variable-Amplitude Loading", NCHRP Rept. 188, 1978.

[34] Fisher, J.W., Mertz, D.R., and Zhong, A., "Steel Bridge Members under Variable Amplitude, Long Life Fatigue Loading", NCHRP Rept. 267, 1983.

[35] Gurney, T.R, and Maddox, S.J., "An Alternative to Miner's Rule for Cumulative Damage Calculations", Proc. of the IABSE Workshop "Remaining Fatigue Life of Steel Structures", Lausanne, 1990, pp. 189-198.

21 FATIGUE BEHAVIOUR OF AA6005A-T6 (AlMgSi0.7) ALUMINIUM ALLOYS – EVALUATION OF THE EARLY FATIGUE CRACK PROPAGATION STAGE BY USING SEM FRACTOGRAPHY, FCPR CURVES AND LEFM MODELS

R. JACCARD
Alusuisse-Lonza-Services Ltd, Zurich, Switzerland

SUMMARY

The early stage of fatigue crack propagation, where fatigue cracks initiate at defects and propagate to various crack geometries, can be evaluated by using LEFM models and FCPR data acquired under Kmax constant testing condition. Depending on the physical appearance of the initial defect (a single loose grain, a series of loose grains, inclusions, single and multiple initiation sites etc.), the fatigue crack growth has to be studied in at least two to three phases. This is necessary to apply appropriate LEFM models individually during a specific phase. As the evaluation demonstrates, most fatigue life is consumed in the first phase.

Kmax 10 const. FCPR should be used for fine grain material. In case of coarse grain material or coarse grain surface layers thicker than about 0.1mm, FCPR data of Kmax 6.7 const. have to be applied.

1 Introduction

The successful application of LEFM for evaluating the fatigue behavior of metallic structures depends on three important informations, i.e. the FCPR (Fatigue Crack Propagation Rate), the description of the defect condition (crack geometry) and the accurate calculation of the stress intensity along the fatigue crack path. As shown in (1), if the fracture mechanics approach is combined with SN data information, the lack of accuracy either of the stress intensity or the crack geometry can be compensated by a fatigue life calibration, provided correct FCPR are used. In chapter 2 FCP data of the aluminium alloy AA6005A-T6 (AlMgSi0.7) are discussed and the sensitivity of the fatigue prediction with respect to the type of FCPR curve demonstrated. The evaluation of typical defects occurring in the AA6005A-T6 alloy based on SEM fractography and LEFM fatigue life prediction is summarized in the third chapter.

Testing of Metals for Structures. Edited by F.M. Mazzolani. © 1992 RILEM.
Published by Taylor & Francis, 2 Park Square, Milton Park, Abingdon, Oxon, OX14 4RN.
ISBN 0 419 15810 3

2 Fatigue Crack Propagation Rates of AA6005A-T6

During the past decade, Alusuisse (2,3,4) has acquired in house and from specialized laboratories a considerable amount of FCPR data of different aluminium alloys. Some of these data have been published (5,6,7). All available FCPR data of AA6005A-T6 are included in fig. 1a to 1f. Fig. 2 is a blow up of fig. 1f. It is used to document the different FCPR used for the fatigue life predictions in this paper. The basic principles of FCPR testing are explained in fig. 3. FCPR used to be measured under a constant R ratio condition, where the ratio of the minimum stress intensity to the maximum stress intensity (R value or R ratio) is kept constant during the test. This leads to R sensitive FCPR curves. The curves of low R ratios deviate considerably from the curves of high R ratios, as can be seen in the first two figures 1a and 1b. This deviation is primarily caused by the fatigue crack closure phenomenon (5,6). In the Kmax constant laboratory approach as proposed by Herman (6), the maximum stress intensity is maintained at a constant level; while the stress intensity range is gradually reduced, the R values are rising continuously. Since Kmax is constant, the plastic zone at the crack tip remains also unchanged. This prevents fatigue crack closure. The FCPR data of Kmax constant are always free of closure. In extruded aluminium alloys closure is expected to occur at low R ratios and becomes neglectable at R ratios higher than about R=0.6. More information can be obtained from (5,6).

Since fig. 2 contains four different FCPR curves which are free of closure and therefore suitable for the evaluation of the early stage of fatigue crack propagation, it is necessary to check their effect on fatigue life. This was done by a fatigue life prediction as documented in tab. 1 and 2. For the fatigue life prediction, the computer software FAGRO (8) and two simple LEFM models, an edge crack in a semi-infinite plate subjected to a uniform tensile stress and a similar crack geometry but subjected to a stress gradient, were used. The stress gradient was defined by a function which gradually decreased the surface stress concentration Kt from 3.0 down to unity over a depth of 2mm below the surface according to fig. 4. Three stress range levels, 50, 20 and 15 MPa, as indicated in col.1, tab.1 and 2, were selected within the range of fatigue strength of welded structural details. The assumed initial crack lengths (col.2) are also in good agreement with the size of observed defects occurring in such structures. In the third column of tab.1 and 2, the stress intensity range at the initial crack length is also listed so that the begin of FCP can be followed on the FCPR polygon. Three FCPR were used in the fatigue life evaluation based on the assumptions mentioned before and the corresponding number of K-cycles listed in column 4 to 6 in tab. 1 and 2. Comparing the results of both tables and fig.(2), the following conclusions can be drawn:

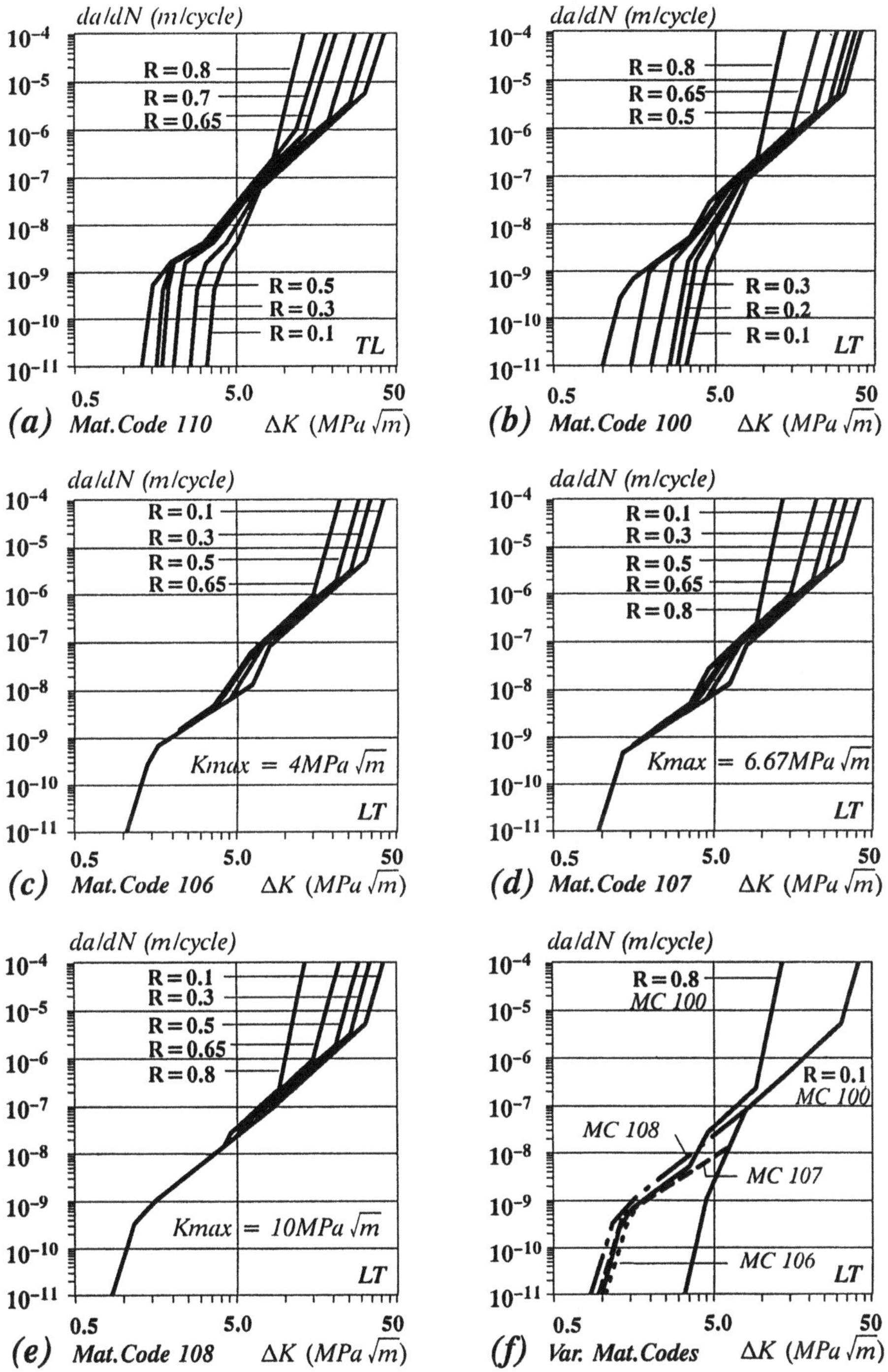

Fig. 1a–f : FCPR of AA6005A–T6 (AlMgSi0.7), different testing conditions

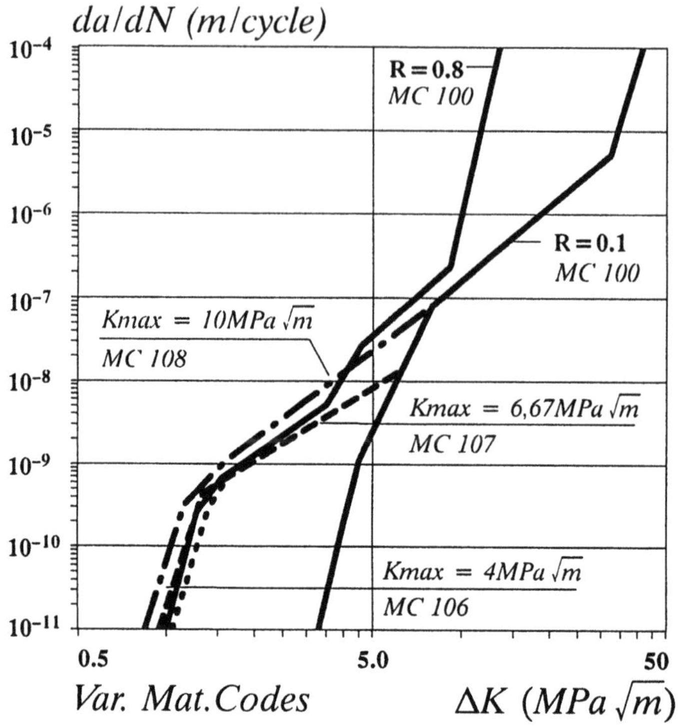

Fig. 2: Blow up of fig. 1f, various FCPR of AA6005A-T6 (AlMgSi0.7)

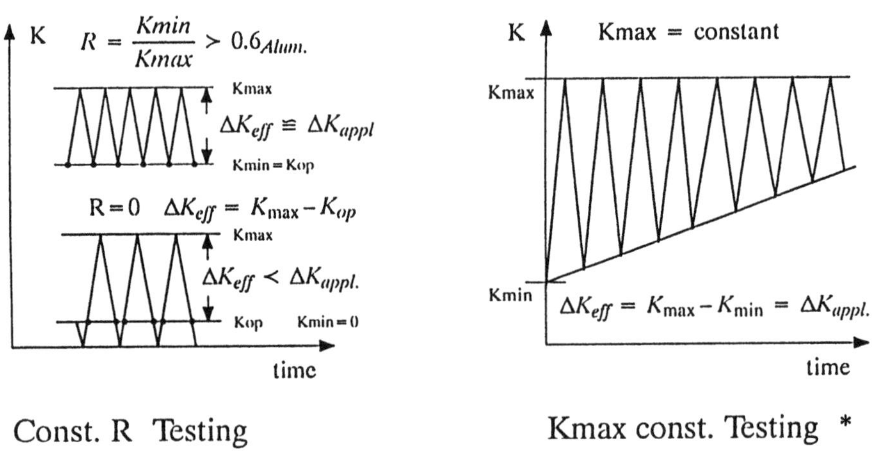

Fig. 3: Basic principles of FCPR testing: "constant R" & "Kmax constant" (6)

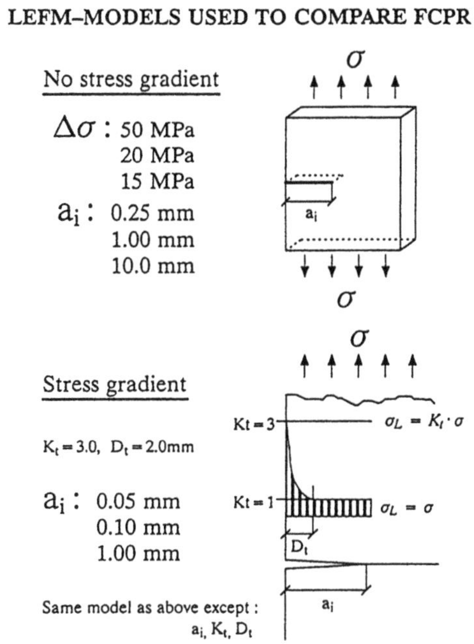

Fig. 4: LEFM models used in the fatigue life prediction in Tab. 1 and Tab. 2

- FCPR of Kmax 10 const. are quite conservative compared to constant R=0.8 data.
- The fatigue lives of the three predictions fit the expected scatter of SN data
- Using FCPR of Kmax 6.7 constant rather than Kmax 10 constant yields a two to three times longer fatigue life.

The fatigue life depends not only on the initial stress intensity range or the initial FCPR but also on the initial crack length. The relative growth of the fatigue crack is quite important since it controls the increase of the stress intensity range and the FCPR. This observation can be confirmed by comparing the second line to the last line in tab. 1 and by checking the first and the last line in tab. 2. Both cases indicate that the fatigue life is about proportional to the ratio of the initial crack length if the initial stress intensity is the same. **The fatigue life of two structures cannot be estimated by comparing the stress intensity ranges or the FCPR. The fatigue life estimation must always be based on the integration of the fatigue equation (Paris Power Law).**

Table 1: Fatigue Life Prediction based on different FCPR of AA6006A-T6
(final crack length Ae=50mm, uniform tensile stress)

1	2	3	4	5	6
			R=0.8 c.	Kmax10 c	Kmax6.7c
SR	Ai	DKi	Fat.Life	Fat.Life	Fat.Life
N/mm2	mm	MpaSm	K-Cycles	K-Cycles	K-Cycles
50	.25	1.57	660	458	1'091
50	1.00	3.14	223	184	566
50	10.00	9.92	2	2	2
20	.25	0.63	1'969'508	162'364	515'023
20	1.00	1.26	5'474	3'632	7'794
20	10.00	3.97	632	627	2'210
15	.25	0.47	83'658'000	3'873'435	11'824'633
15	1.00	0.94	47'547	12'847	35'429
15	10.00	2.98	2'174	1'676	5'363

Table 2: Fatigue Life Prediction based on different FCPR of AA6005-T6
(final crack length Ae=10mm, stress gradient Kt=3 Dt=2.0mm)

1	2	3	4	5	6
			R=0.8 c.	Kmax10 c	Kmax6.7c
SR	Ai	DKi	Fat.Life	Fat.Life	Fat.Life
N/mm2	mm	MpaSm	K-Cycles	K-Cycles	K-Cycles
50	.05	2.12	82	66	193
50	.10	2.96	59	52	164
50	1.00	6.70	23	22	60
20	.05	0.85	9'781	2'266	6'473
20	.10	1.18	1'770	1'130	2'672
20	1.00	2.68	959	654	1'799
15	.05	0.64	354'040	31'372	97'862
15	.10	0.89	12'987	3'730	9'518
15	1.00	2.01	2'387	1'541	3'389

3. Life Prediction by Evaluation of SEM Data

In the previous chapter it was shown that the FCPR curves of Kmax 10 MPA SQRT(m) constant yield the most conservative result in fatigue life prediction among all measured FCPR data of aluminium alloys and that the Kmax 6.7 MPaSQRT(m) const. curve improves the fatigue life by a factor of two to three. It is therefore important to investigate the performance of these data in the early stage of fatigue crack propagation and the evaluation of defects described by SEM fractography. Some of the evaluations from (9) are included in the last nine pages, p. 10 - p. 18.

The initial crack lengths Ai and Ci of surface cracks interpreted from the defects shown on SEM prints are listed in tab. 3. These initial crack lenghts were then used for fatigue life prediction. In column 6, the errors or the deviations of these calculated results from the measured number of cycles are tabulated. The agreement is excellent.

The initial crack lengths were calculated by using the fatigue life calibration technique based on the measured number of cycles and the stress ranges as described in (1). These results are documented in table 4. Two crack geometries were used: the Half Disk Surface Crack (HDSC) and the Shallow Elliptical Surface Crack (SESC). Three FCPR curves were included and the results are listed in table 4. The crack lengths which fitted the defect dimensions best are tabulated in mm. All other results are normalized with respect to these values. From tab. 4 it is obvious that constant R FCPR of low R ratios are not adequate to interprete initial defects since these defects behave like short cracks.

Table 3: Summary of SEM Evaluation of Defects occurring in AA6005A-T6

Specimen:		Ai: mm	Ci: mm	Ci/Ai:	(*) Error %	Note:
Fine Grain (Kmax 10 const.):						
7-3	"L"	0.0125	0.02	1.6	1	N = 54'000 cyc.
	"M"	0.0025	1.225	490	7	SR = 290 MPa R=0.1
	"R"	0.009	0.04	4.4	3	
7-8		0.044	0.06	1.4	1	N = 5'364'000 cyc.
						SR = 90 MPa R=0.65
Coarse Grain (Kmax 6.7 const.) (1):						
7-6		0.07	0.11	1.6	5	N = 540'000 cyc.
		0.025	0.155	6.2	4	SR = 126 MPa R=0.1
10		0.008	0.14	17.	1	N = 930'000 cyc.
		0.030	0.03	1.0	8	SR = 153 MPa R=0.1
2		0.011	0.10	9.1	3	N = 570'000 cyc.
		0.027	0.04	1.5	1	SR = 153 MPa R=0.1

(*) Deviation from measured num. of cycles; (1) Kmax 10 const. yields approx. 1/2 life

Table 4: Initial Crack Lengths as determined by Fatigue Life Calibration using different FCPR (relative crack lengths) :

	[K max c. 10.0(*)] Ai	[K max c. 6.67(*)] Ai	[R = constant] Ai
TS 7-3: SR = 290 MPa R = 0.1 N = 54'000 cycles			Fine Grain
HDSC:	0.02000 mm	5 x	10 x
SESC:	0.00279 mm	10 x	(R=0.1)
TS 7-8: SR = 90 MPa R = 0.65 N = 5'364'000 cycles			Fine Grain
HDSC:	0.05063 mm	1.3 x	4 x
SESC:	0.01423 mm	1.4 x	(R=.65)
TS 7-6: SR = 126 MPa R = 0.1 N = 540'000 cycles			Coarse Grain
HDSC:	0.5 x	0.08159 mm	10 x
SESC:	0.5 x	0.02334 mm	(R=0.1)
TS 10: SR = 153 MPa R = 0.1 N = 930'000 cycles			Coarse Grain
HDSC:	0.7 x	0.02906 mm	16 x
SESC:	0.7 x	0.00776 mm	(R=0.1)
TS 2: SR = 153 MPa R = 0.1 N = 570'000 cycles			Coarse Grain
HDSC:	0.7 x	0.03620 mm	14 x
SESC:	0.7 x	0.00955 mm	(R=0.1)

(*): (Units of Stress Intensity: MPaSQRT(m))

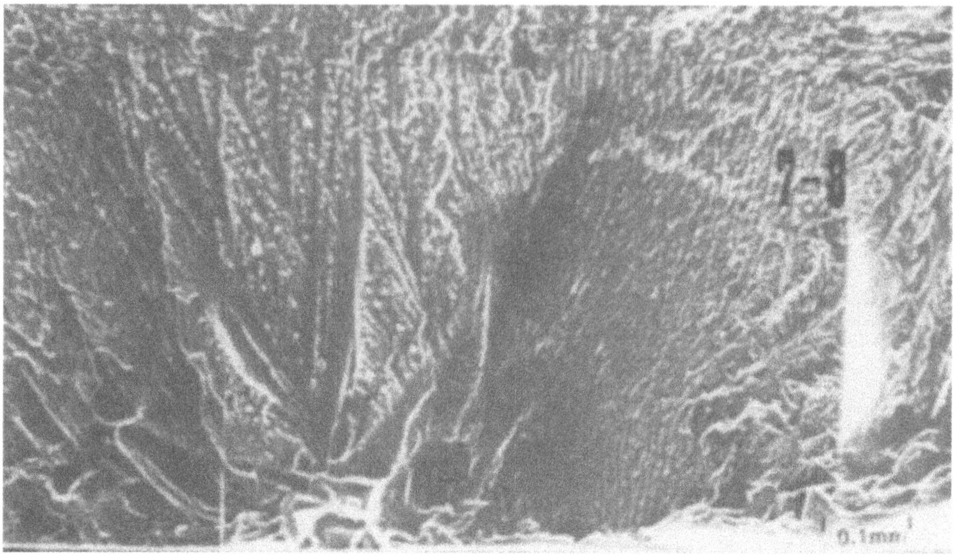

Fig. 5: Defect area of specimen 7-8, fatigue crack initiated at loose grains.

Specimen 7-8:
At the bottom of fig. 5, a little left to the middle, a 0.28mm wide and 0.092mm deep thump nail area can be observed. This area is composed of 8-10 grains ranging between 0.02-0.05mm by 0.04-0.06mm. In the center of the bottom there is a cluster of three loose grains measuring 0.044mm by 0.11mm. This cluster represents the initial surface crack Ai/Ci=0.044mm/0.055mm. In fig. 6 phase 1 of the FCP is shown. It is terminated at a crack length of 0.5mm. The predicted crack fronts are in good agreement with the fatigue appearances on the crack surface indicating the global direction of the FCP. The second phase is shown in fig. 7.

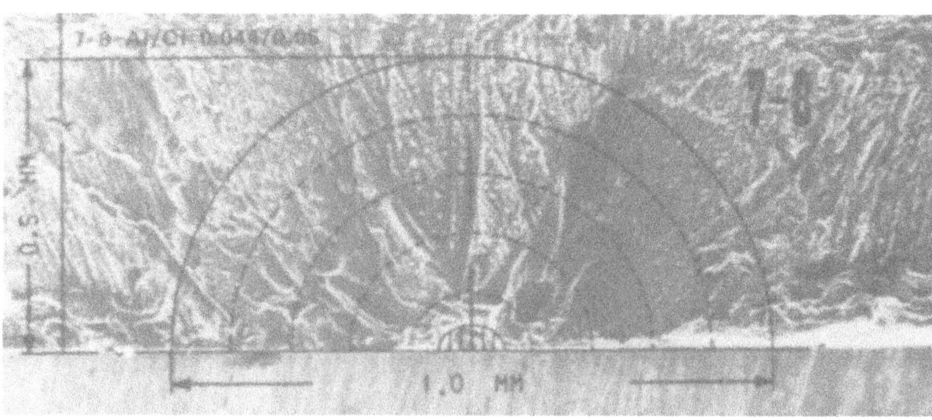

Fig. 6, specimen 7-8: Interpreted initial crack length (Ai/Ci=0.044/0.06mm)

Fig.7, spec. 7-8: Final phase of FCP, shaded area is phase 1 shown in fig.6

Specimen 7-6:

The fracture surface of this example contains a typical zone of coarse grains. According to fig. 8, the larger grains are subdivided into smaller grains near the surface, forming a 0.02 to 0.025mm thick layer over a length of about 0.31mm. These smaller grains became probably loose due to corrosion and act as the initial crack of the dimension Ai/Ci=0.025mm/0.155mm. Since some larger grains seem to become loose as well, another initial defect can be defined by a larger grain and three smaller ones. This second crack has an Ai=0.07mm and a Ci of 0.11mm. It is shown in fig. 9. The final phase is recorded in fig. 10. The fatigue life integration was performed using the FCPR data collected under a K max constant 6.7 MPaSQRT(m) condition. Both LEFM models were also reevaluated using the FCPR data of Kmax constant 10 MPaSQRT(m). The resulting fatigue lives had about half the number of cycles as computed in the first case. The last initial crack Ai/Ci=0.07mm/0.11mm was then finally subjected to the FCGR data of Kmax constant of 4.5 MPaSQRT(m). This leads to a fatigue life exceeding the first one by a factor of four.

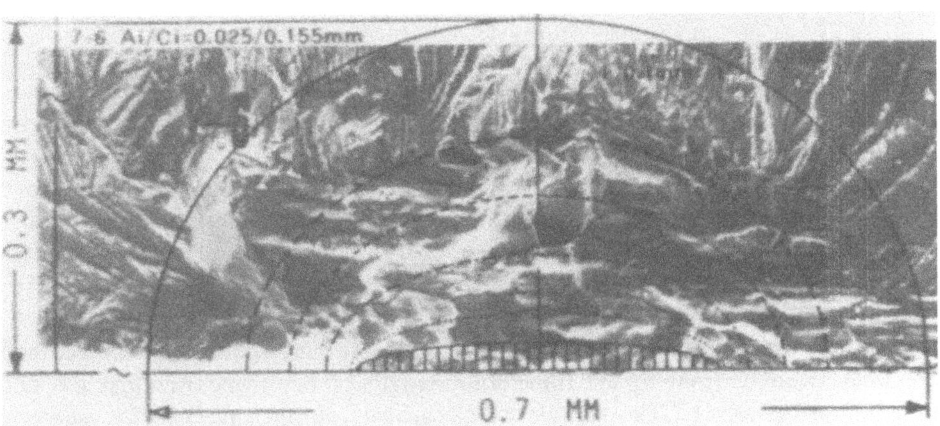

Fig. 8, specimen 7-6: Phase 1 of FCP, initial crack Ai/Ci=0.025/0.155mm (a series of loose grains)

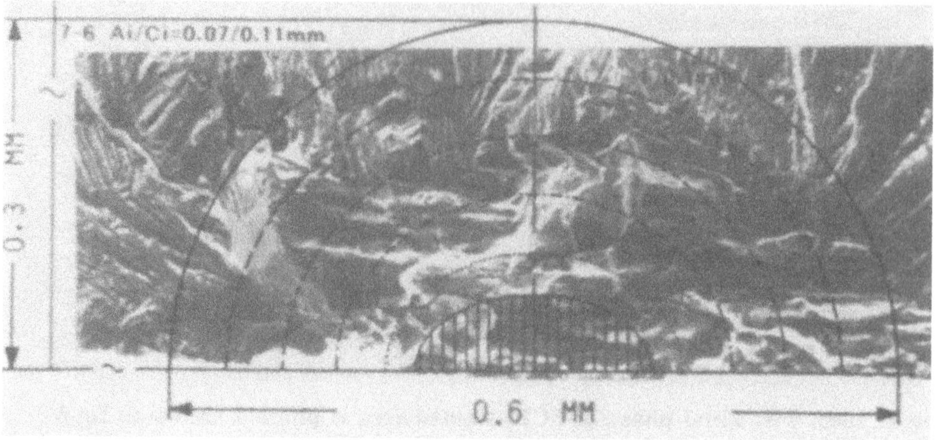

Fig. 9, specimen 7-6: Phase 1 of FCP, initial crack ai/Ci=0.07/0.11mm

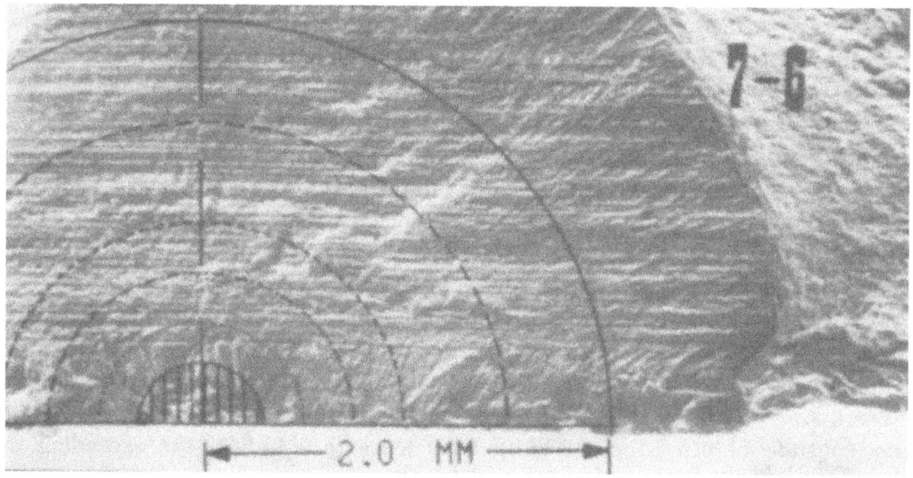

Fig 10, specimen 7-6: Final phase of FCP

Specimen 2:
This is another example of Fatigue Crack Initiation (FCI) at a coarse grain zone. Fig. 11 documents the overall fracture appearances. The half circles are crack fronts at 0.5, 1.0, 1.5, 2.0, 2.5mm from the origin of the crack initiation. At point 'A' (middle of bottom) there is a coarse grain zone. Fig. 12 reveals the details of this particular area. Two assumptions were made to evaluate the FCP: FCI at three loose coarser grains and at a series of eight to nine loose grains. The FCP was studied in two stages. Phase 1 included the growth from the initial Ai/Ci=0.027/0.04mm in the first FCI case and the Ai/Ci=0.011mm/0.10mm in the second one up to 0.5mm. The second phase covered the range from 0.5mm to the final crack length of 5mm and was just the continuation of phase 1. The fatigue crack growth during one part of phase 1, from the initial crack Ai up to 0.14mm, is depicted in fig. 13, in the case of three loose grains and in the other case of multiple loose grains in fig. 14. Fig. 15 shows the FCP during the entire phase 1 of fig. 14.

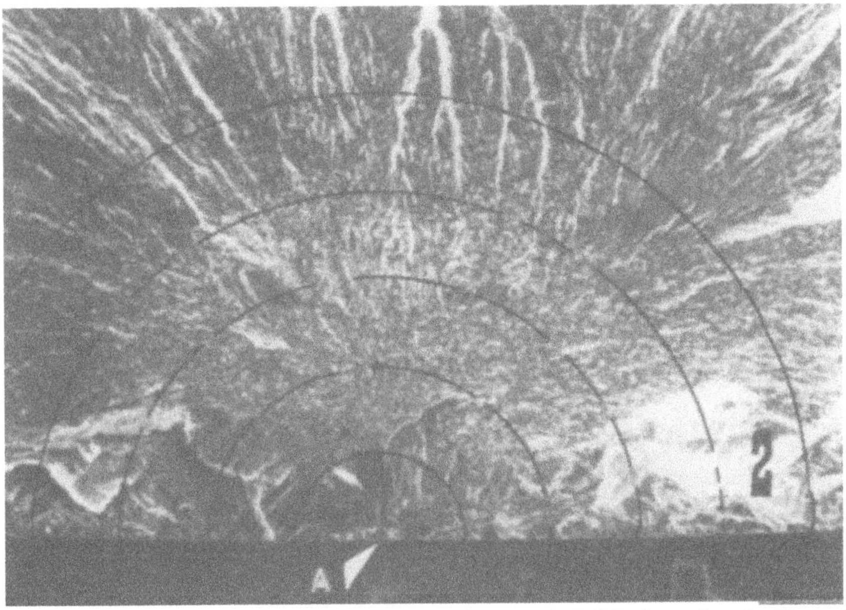

Fig. 11, specimen 2: Overall view of the crack initiation site, half circles are crack fronts at 0.5, 1.0, 1.5, 2.0, 2.5mm from the origin

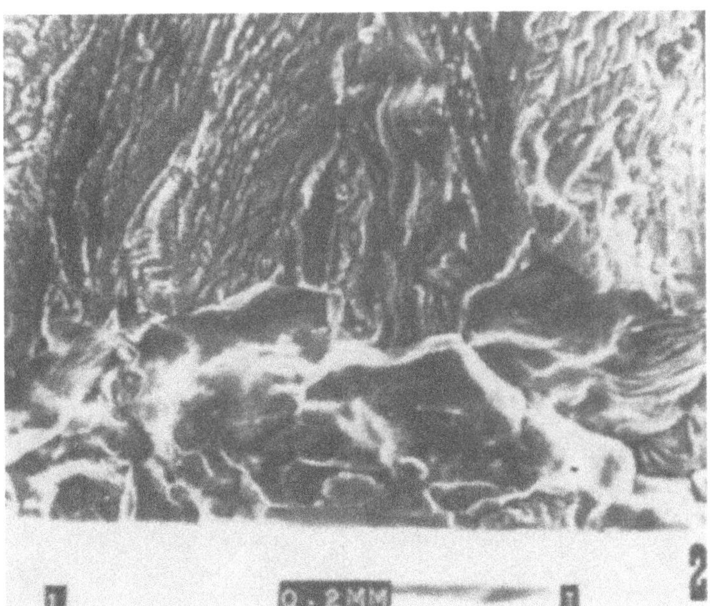

Fig. 12, specimen 2: Close up of coarse grain zone

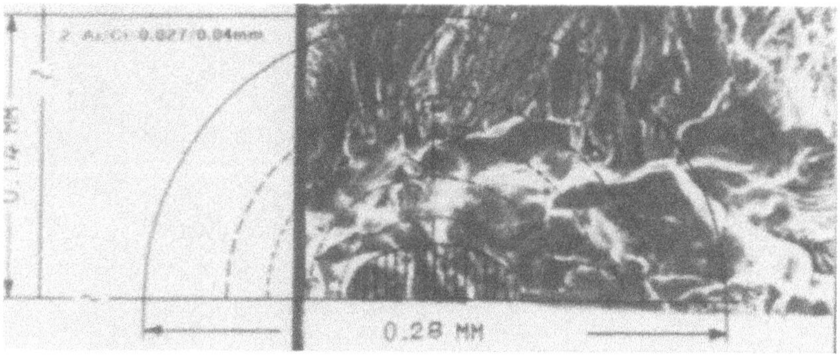

Fig. 13, spec. 2: Initial crack Ai/Ci=0.027/0.04mm (3 loose coarse grains)

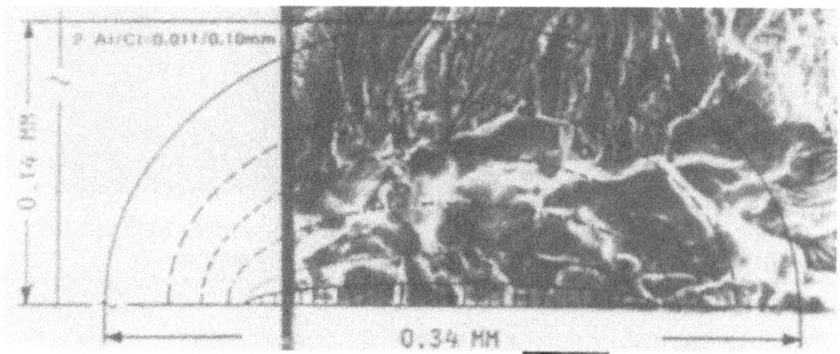

Fig. 14, spec. 2: Initial crack Ai/Ci=0.011/0.10mm
(a series of eight loose smaller grains)

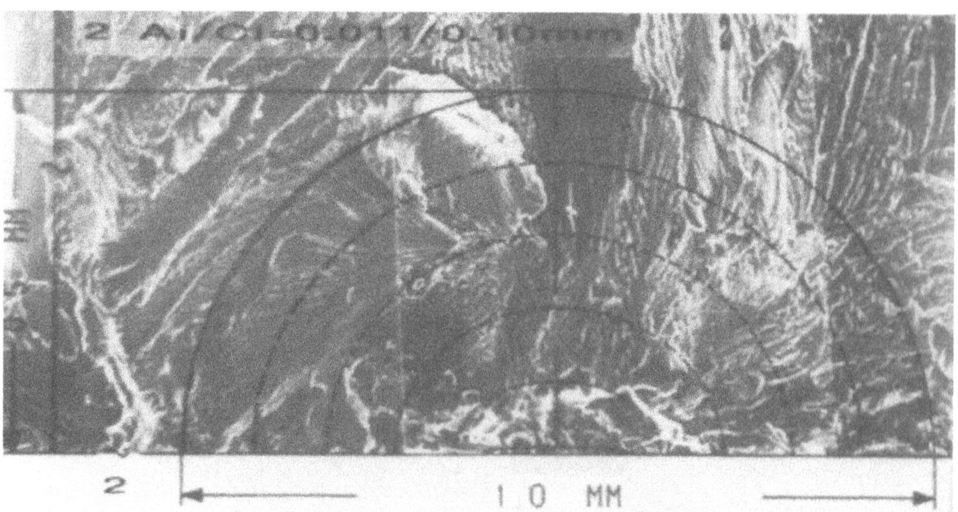

Fig. 15, specimen 2: Phase 1 of FCP as in fig. 14

Fig. 16, spec.10: Final phase of FCP from initiation site, half circles are crack fronts at 0.5, 1.0, 2.0, 3.0, 4.0, 5.0 mm from the origin. The smallest half circle is the final crack length of phase 2 from fig. 19

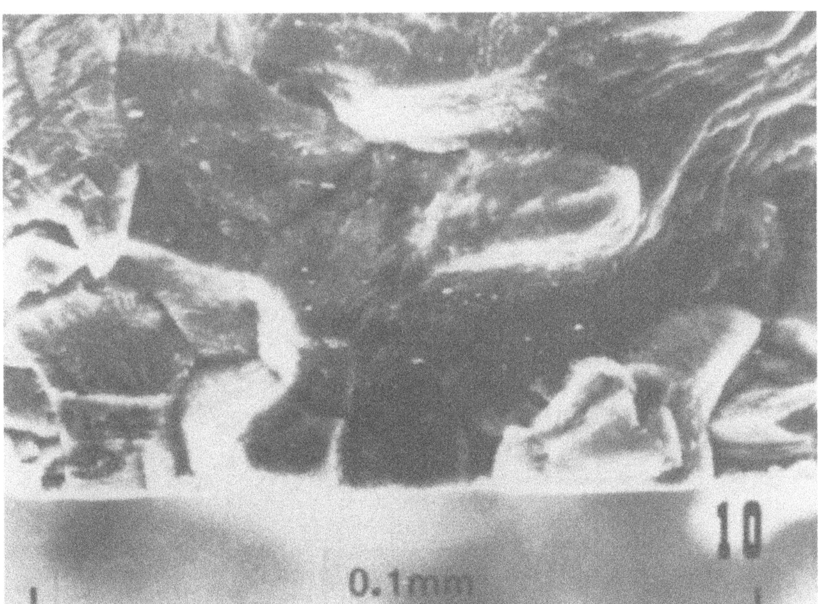

Fig. 17, specimen 10: Close up of coarse grain zone, near surface finer grains of 0.01-0.025mm diameter, on top coarser grains of 0.05-0.07mm

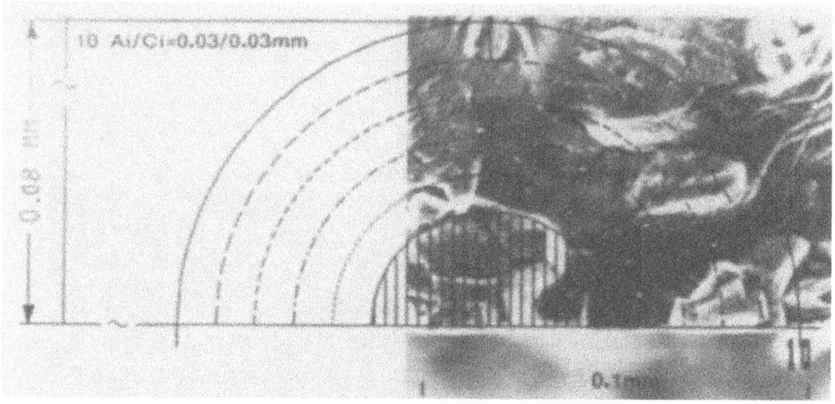

Fig. 18, spec. 10: Initial crack Ai/Ci=0.03/0.03mm, phase 1

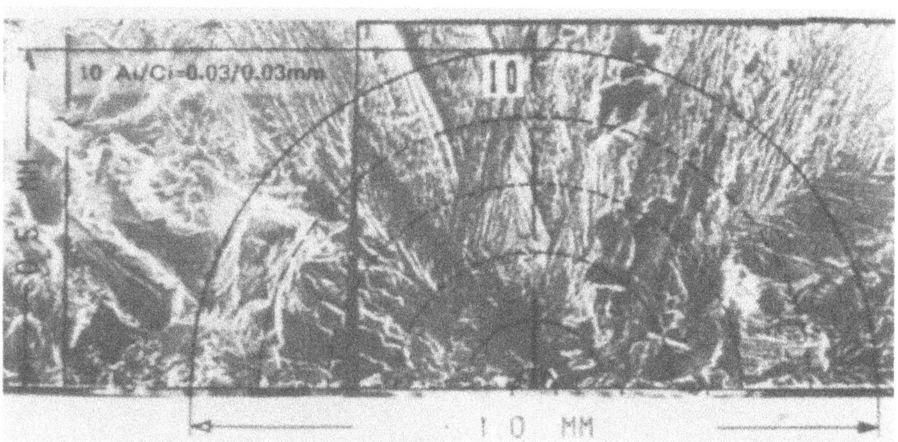

Fig. 19, spec. 10: Phase 2 of FCP, shaded area is end of phase 1 (fig. 18)

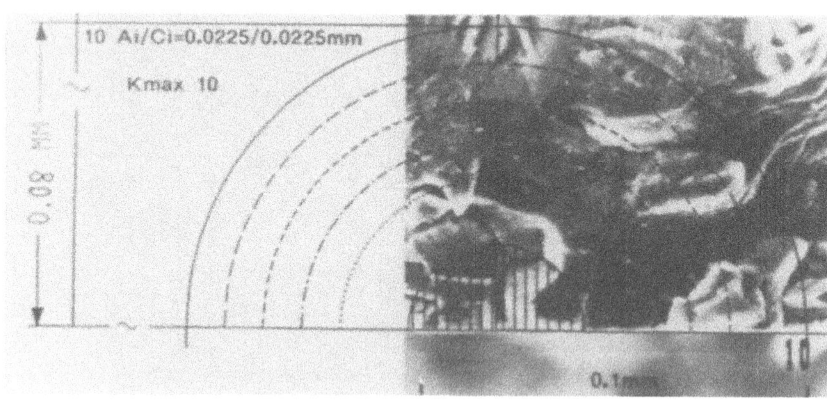

Fig. 20, spec. 10: Phase 1 of FCP, calibrated initial crack length based on Kmax 10 const. Ai/Ci=0.025/0.025mm is obviously too short.

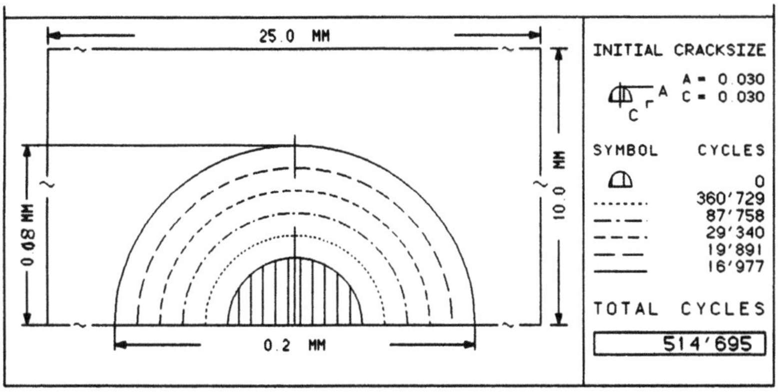

Fig. 21, spec. 10: Phase 1 of FCP as shown in fig.18, the FCP from the init. crack of 0.03mm to 0.08mm consumes 514'000 cycles or 60% of the total life

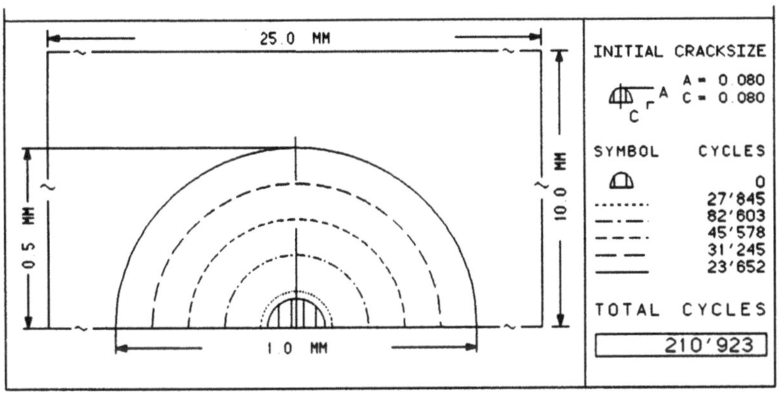

Fig. 22, spec. 10: Phase 2 of FCP as shown in fig.19, the FCP from 0.08mm to 0.5mm takes 210'923 cycles or about 25% of the total fatigue life.

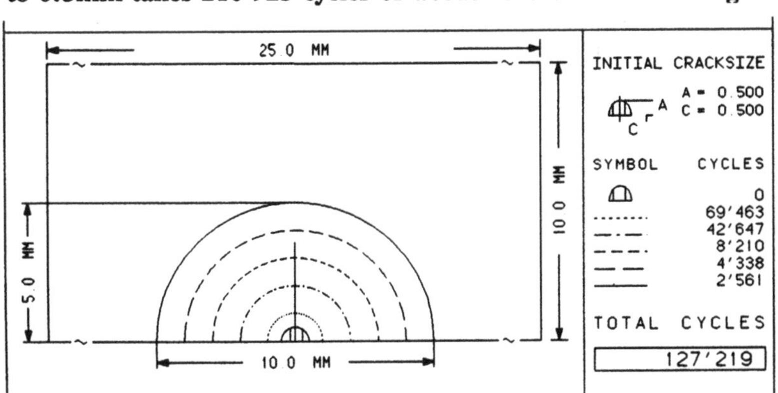

Fig. 23, spec. 10: Final phase of FCP as shown in fig. 16, less than 15% of the total fatigue life is consumed during FCP from 0.5mm to 5.0mm.

Specimen 10:
The overall view of the fatigue crack initiation site is shown in fig. 16. Specimen 10 is also an example of FCI at a layer of coarse grains located at the extruded surface. According to fig. 16, this layer is almost 1.5 mm thick. Near the surface, this layer also contains finer grains of 0.01-0.025 mm in diameter. The coarse grains on top of these measure about 0.05-0.07 mm. A close up of the fatigue crack initiation site is shown in fig. 17. The loose grains in the lower left corner are taken as the initial defect and circumferenced by a half disk crack. The radius of this initial crack is 0.03 mm. The first phase of FCP from 0.03 mm to 0.08mm is shown in fig. 18. The next stage, phase 2, where the fatigue crack propagates from 0.08 mm to 0.5 mm is documented in fig. 19. The calculated fatigue crack fronts at 0.1, 0.2, 0.3, 0.4 and 0.5 mm from the origin correspond well to the faceted appearances of the fracture surface. Looking a little closer at fig. 19, some other coarse grains and inclusions can be identified but these were obviously not the origin of the FCP. To complete the study, the initial crack length was also calculated by a fatigue life calibration based on FCPR of Kmax 10 constant. The resulting crack length of 0.025 mm is shown in fig. 20. It is too short and the assumed FCPR of Kmax 10 const. are consequently too conservative. Fatigue crack fronts were calculated by using the computer program FAGRO and FCPR of Kmax 6.7 constant. The results according to the three phases of FCP are shown in the last three figures 21 to 23. It is important to know that about 60% of the total fatigue life is consumed during the first phase of FCP, from the initial crack length of 0.03 mm to 0.08 mm, and that only 15% is used to propagate the fatigue crack from 0.5 mm to 5.0 mm in the last phase. Finally, phase 1 was reevaluated by using the same initial crack length of 0.03 mm and FCPR of R=0.1 constant. The resulting fatigue life is over four hundred million times too high. The crack fronts which correspond to fig. 21 are shown in fig. 24 below.

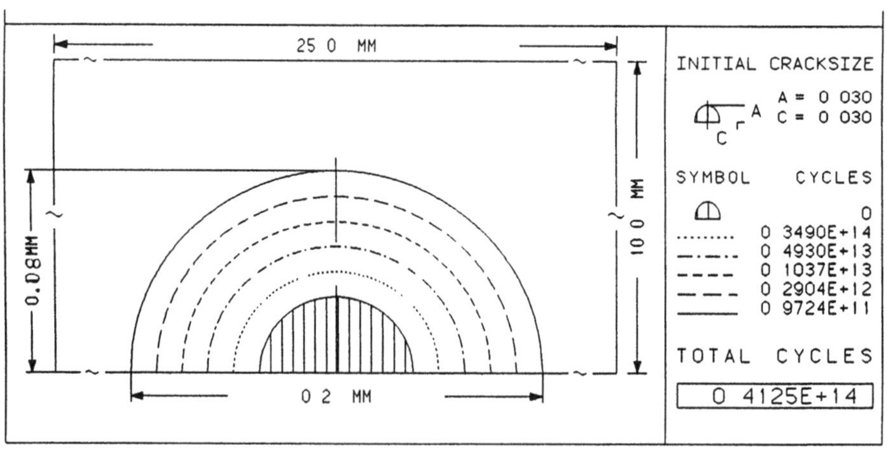

Fig. 24, spec. 10: Phase 1 of FCP, reevaluated using FCPR data of R=0.1 const. The resulting fatigue life indicates that the assumed FCPR data are not adequate

CONCLUSIONS

FCPR curves of Kmax 10 const. yield the most conservative result in fatigue life prediction compared to all other FCPR, the R=0.8 const. and the Kmax 6.7 c. All three curves can be used to analyze the FCP of welded structures and short crack problems since the resulting fatigue lives fit the expected scatter of SN data. In case of coarse grain material the Kmax 6.7 constant FCPR should be used since Kmax 10 constant curve is too conservative.

The early stage of fatigue crack propagation, where fatigue cracks initiate at defects and propagate to different crack geometries, can be evaluated by using LEFM models and FCPR data acquired under Kmax constant testing condition. Depending on the physical appearance of the initial defect (a single loose grain, a series of loose grains, inclusions, single and multiple initiation sites, etc.), the fatigue crack growth has to be studied in at least two to three phases. This is necessary to apply appropriate LEFM models individually during a specific phase. If the initial defect consists of a series of loose grains spread over multiple sites, the life evaluation becomes complex and several LEFM models, e.g. crack geometries, are applicable during the first phase. Phase 1 describes the FCP from the initial crack (loose grain, oxide skin, etc), up to about 0.5mm. As the fatigue life calculations demonstrate, most of the fatigue life is consumed during phase 1.

References:

1. Jaccard R. (1985) Fatigue life prediction of aluminum structures based on local stress fields. 3rd. Int.Conf.on Aluminium Weldments, Munich F.R.G. also IIW Doc. No. XIII–1196–86 and XV–606–86
2. Alusuisse Data Base of SN Data and FCPR of various aluminium alloys FCPR Data acquired by automated test systems (ASTM E–647) since 1980.
3. Fracture Technology Associates, Pleasant Valley, Pa.18951, USA.
4. Professional Services Group, Inc. , Hellertown, Pa. 18055, USA. (former DEL–Research Laboratory Hellertown, Pa. 18055, USA.)
5. Stofanak R.J., Hertzberg R.W., Miller G., Jaccard R., Donald K.J.(1983) On the cyclic behavior of cast and extruded aluminium alloys. Part A: F C Propagation. EFM, 17–6, 527, Part 2: Fraktograp. EFM 17–6, 541
6. Herman W.A., Hertzberg R.W., Jaccard R.(1988) A simplified labor. appr. for the prediction of short crack behavior in engineering structures. Fatigue Fract. Engng. Mater. Struct. Vol. 11, No.4, pp.303–320,1988
7. Jaccard R., FCP of Aluminium Alloys,IIW Document XIII–1376–90, 1990.
8. Jaccard R., FAGRO a computer program to evaluate fatigue crack growth, Language: Fortran IV, Hardware VAX's,(VMS)
9. Jaccard R., SEM Evaluation of AA6005A–T6 Defects, Case Studies 1989/90.

22 PROCESSING FATIGUE TEST DATA – THE ALFABET PROJECT

D. KOSTEAS
Institute of Civil Engineering III, Steel Structures, Technical University of Munich, Germany

ABSTRACT

The initial development of the Aluminium Data Bank (ALDABA) covering primarily fatigue behavior data of aluminium components has led to the concept of a comprehensive system of information sources linked to design procedures and reliability estimates for aluminium structures, the Aluminium Fatigue Behavior and Evaluation Task (ALFABET).

INTRODUCTION

The Aluminium Data Bank is a joint project of the Technical University of Munich, Germany, and the Iowa State University, Ames, Iowa, USA. It serves the harmonized documentation,analysis, evaluation and presentation of material data, especially in the area of fatigue, from small specimen earlier tests and more recent comprehensive data generated from larger welded components and full–size elements. Data from other aluminium joint types as well as substantial fracture mechanics and crack propagation data is being added.

The further development is visualized under the ALFABET Project. The Aluminium Data Bank shall be coupled through an application oriented interface to other information sources like literature, handbooks, expert knowledge and instruments such as design concepts and procedures, codes, recommendations. Of great importance is here the management of a large body of information in conjunction with the options offered, from a retrospective search of characteristic values from the original sources to the evaluating and design procedures.

While the above outline the two main goals from the user's point of view it has been necessary to deal with additional questions of why, what for, advantages and disadvantages, how, whom to address, what to documentate and process in dealing with a distinct package of information as is the case within the design concepts for aluminium structures. Some of these ideas are mentioned briefly as introductory remarks to the specific task of ALFABET since they help provide guidelines along which the project itself will have to move. But these also provide elements for further work and harmonization of procedures in testing and evaluating material and structural component behavior, i.e. tasks to be addressed within the intended work of RILEM.

Testing of Metals for Structures. Edited by F.M. Mazzolani. © 1992 RILEM.
Published by Taylor & Francis, 2 Park Square, Milton Park, Abingdon, Oxon, OX14 4RN.
ISBN 0 419 15810 3

The development of a data bank has the purpose to use computers to provide a better access to data, but also help scrutinize this data. Material data should be computerized to the same level as other computerized engineering tools, such as design procedures and recommendations. These steps help broaden the range of materials choices and share materials experiences more effectively. The developed materials data system, and in our specific task the structural design system itself, should exhibit easy access, accuracy, availability for further computerized use, timelessness, completeness and uniformity, data comparison and user-friendly presentation.

Looking into the types of databases that may be developed we recognize that these may be built some times merely as published data. The user of a database wishes to have a reliable distributor of the database, most times he will need multiple (compatible ?) databases, he will have his own computer and he will not want to learn many systems, he needs user-oriented on-line help.

Developing a public available database will be associated with some disadvantages though. There may be a weak or diffuse objective, the product itself not user-friendly. Also inconsistency in the materials (fatigue in our case) parameters, stale data, and data acquisition will depend on volunteer effort. End user does not control activities of database manager. Further, there will generally not be enough experts to validate data and funding of the database activities will be variable and insufficient.

Being aware or even having experienced some of the above disadvantages we try to avoid them as far as possible in the development of ALFABET by: (a) emphasizing the collecting and reporting of new materials and components behavior information, (b) focusing on high priority industry fatigue data requirements (the fact should be stressed here that the AlDaBa and ALFABET projects are design-oriented and not so much material selection tools), (c) standardizing and automating fatigue data collection and storage formats, (d) periodically purging the active system of stale data, (e) supplementing the expertise of validation experts with knowledge base systems, and (f) adequate and consistent funding of activities. The last point is, admittedly, a point of major concern.

I would like to close these introductory remarks by adding some general information on the procedural part of the data bank development to be followed in the special case of a fatigue data bank, which in my opinion serves also one of the first RILEM tasks on requirements on data acquisition and computerization. Thus the general materials data flow in a data bank development could be as follows:

- data generation by testing or in-service measurement
- data analysis by modelling and statistical procedures
- data dissemination by publication and databases themselves
- data combination and comparison, re-analysis and "standardization" of values
- distribution of handbooks, publication of databases

Finally, types of fatigue data required and suggestions on the data format are summarized in the following Fig. 1 and 2. As will be seen in the following several of these suggestions have already been taken into account and have been incorporated in the format of the Aluminium Data Bank.

Standard Material Coupon Tests
* Material Data
* Stress-Life
* Strain-Life
* Failure Location
* Crack Growth Rate

Material/Structural Coupon Design Verification Tests
* Duplicate local geometrical details
* Duplicate fabrication and service preparation details
* Simulate expected service history

Component and Full-Scale Design Verification Tests
* Duplicate complete fabrication history
* Dulpicate joining and set-up conditions
* Simulate expected service history
* Compare to or perform in-service measurements

Fig. 1: Required fatigue data

Material Identification
reference source, data base or other

Test Identification
type, standard or other, date

Description of specimens
identification
preparation
joining procedure
inspection and results
specimen layout reference
specimen orientation
specimen geometry

Test Parameters and Procedures
date
test facility and engineer
test environment
testing machine and fixture
loading parameters
loading rate
strain instrumentation
data collection method
sample rate

Test Results and Analysis
no. of specimens
rate of loading
loading parameters
loading statistical data
mechanical parameters
physical parameters
srength or/and strain
  nominal, average
  stand. dev., cov
Strength or/and strain
  at failure
  nominal, average
  stand. dev., cov
failure location
failure mode
test validation, data quality
raw test data source
remarks, if any such as
  significant deviation
  from standard

Fig. 2: Fatigue test data format

EARLY DEVELOPMENT

First decisive steps in the computerization of fatigue test results in the form of a data bank for aluminium can be traced back to initial work by Munse followed by efforts by Sanders around 1970 who started the Welding Research Council Aluminium Data Bank at Iowa State University [1]. Data were initially stored on punched cards.

Statistical and regressional analysis software was developed in the late sixties by the author at the University of Karlsruhe [2] and applied thereafter on consecutive comprehensive evaluations on aluminium fatigue test data. A cooperation was initiated in the early eighties under the Committee on Aluminum Fatigue Data Exchange and Evaluation (CAFDEE) [3] resulting since 1985 in a simultaneous installation and operation of the Aluminium Data Bank at the Iowa State University and the Technical University of Munich.

One of the primary tasks of the last years has been the user-oriented adaptation of available aluminium test data and software for its evaluation, whereby the former main frame computer operation has been abandoned in favor of the now broadly available personal computers.

The logical extension and applicability in engineering design led to the concept of a management system for such material, technological, manufacturing, quality control and reliability information and processes as needed during the design procedure of aluminium constructions, Fig. 3.

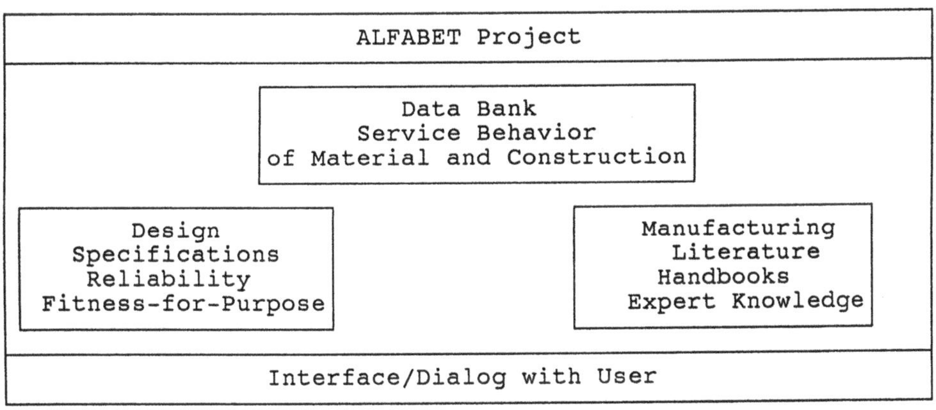

Fig. 3: Outline of the ALFABET project concept

The first task is almost complete by now and consists of what is understood as the Aluminium Data Bank (ALDABA). The bank is itself a part of the second more comprehensive task, the Aluminium Fatigue Behavior Evaluation Task (ALFABET) Project currently under development. These two are described in more detail in the following chapters.

THE ALUMINIUM DATA BANK

The original PC version

In the first stage of the data bank, from 1987 to 1989, the use of commercially available software was favoured since development and employment had to be performed in parallel. Extensive analyses, significant to drafting the ECCS European Recommendations for fatigue design in aluminium, were carried out [4] – this work was invaluable in testing procedures and providing decisions for future applications.

The data bank management was based on the dBaseIII+ software package. A positive fact was the rather powerful structured query language for data retrieval and selection, but concerning the user menu, graphics, colors or hardware programming we were confronted by serious handicaps. In order to circumvent licence problems with directly applicable programs, i.e. *.EXE or *.COM files, dBase program run-versions were constructed by means of the compiler system "Clipper". Available data was stored in three parts: (a) the literature data bank, (b) the detail and data description data bank, and (c) the fatigue test data points themselves and their mathematical processing.

Parts (a) and (b) were written in dBaseIII+, whereby the rather complex statistical formulas of part (c) were executed in Pascal – all parts integrated to one program by means of "Clipper". Part (b) is essential to the user since it allows data set selection and detail classification. A comprehensive parameter field has been covered by the formats of the literature and detail description data banks, see Fig. 4.

The data bank covers primarily and for the time being fatigue test data of various alloys in structural engineering applications (alloy groups 5xxx, 6xxx and 7xxx), base material and numerous welded joint types. The original CAFDEE data bank included approx. 15000 data points on small specimens, which have been rechecked and validated to a large extent up to now. Though interesting in many ways these older data on small specimens present problems regarding the capability to lead to reliable conclusions [5]. There are serious gaps in the documentation even of geometrical or welding and at times testing parameters. Beyond this small specimen data will not reflect actual component behavior in service. These facts have been widely acknowledged and led to similar decisions on the area of aluminium constructions in the beginning of the eighties for experimental research in fatigue and eventually for evaluating carefully performed observations, the latter based mainly on large components.

In the course of the comprehensive fatigue test series at the Technical University of Munich between 1982 and 1990 a nucleus of approx. 1000 extensively documented and published data points on large structural components, beams, have been generated. Further fatigue tests performed by industry on components and made available during the recommendations drafting process [6] in the last two years we hope will become publicly available in the near future. These together with recent work performed by Fisher and Menzemer will increase the contingent to more than 2500 data points. It remains to be mentioned that in the course of the last year's evaluations a considerable amount of almost another 20000 data points on small specimens again provided by industry [6] has served in drafting the European Recommendations [7].

A. General
1. Data Set No.
2. Literature No.
3. Page No., Table No., etc.
4. Test Year (19__)
5. Confidential (Y/N)
6. Last Check (dd/mm/yy)
7. Remarks

B. Joint Description
1. Basic Weld Type
2. Edge Preparation
3. Weld Geometry

Joint Dimensions
4. R (mm)
5. F (mm)
6. p (degrees)
7. H (mm)
8. D (mm)

C. Specimen Description
1. Specimen Description
2. Thickness (mm)
3. Width (mm)
4. Tested Yield Strength (MPa)
5. Yield Point Off-Set (%)
6. Tested Ultimate Strength (MPa)
7. Ultimate Strain (%)
8. Gage Length (mm)

D. Welding Process and Procedure
1. Welding Process
2. Welding Procedure
3. Welding Position
4. Filler Material (4 Letter AA designation)
5. Filler Material (Product Name or National Designation)
6. Welding Wire Diameter (mm)
7. Shielding Gas
8. Shielding Gas Flow Rate (l/min)
9. No. of Passes
10. Welding Speed (cm/min)
11. Welding Current (Ampere)
12. Welding Voltage (Volts)
13. Non-destructive Examination Technique
14. Discontinuities
15. Pre-Weld Treatment
16. Post-weld Treatment I (Notch Shape Modification Methods)
17. Post-weld Treatment II (Mechanical Treatments)
18. Post-weld Treatment III (Thermal Methods)

E. Base Material
1. Alloy Designation (4 Letter AA Designation)
2. Temper or Heat Treatment
3. Product Name or National Designation
4. Fabricator
5. Product Form
6. Chemical Composition
7. Tested Yield Strength (MPa)
8. Yield Point Off-Set (%)
9. Tested Ultimate Strength (MPa)
10. Ultimate Strain (%)

F.Test Conditions
1. Test Machine
2. Mode of Loading
3. Orientation of Weld and Load
4. Environmental Conditions
5. Type of Loading
6. Temperature (°C)
7. Frequency (Hz)
8. Failure Criteria
9. Basis of Recorded Stress or Strain
10. R-Ratio
11. Number of Stress Levels
12. Number of Datapoints

Fig. 4. Test data description parameters

The new version

A user-friendly data bank, allowing ready access to information and compatibility to all other steps in the structural design procedures should enable a simple, logical and reliable dialog with the user. This demands a menu-driven program with controls against unintended input errors.

Consequently the decision was made by the end of 1989 to choose Turbo-Pascal 5.5

(modified just recently by the new version 6.0) as the new language – this being at the time the only one allowing an object-oriented-programming. Another advantage for the user was reached by using the much more rapid and reliable BIOS system to control input/output functions and not any more DOS, avoiding in this way any harmonization problems with other programs.

A main feature in this new version is the menu guiding the user through the whole program. An object-oriented window and menu management system had to be developed with simple mouse- or key-driven selection of program functions. In a dialog with the user further information may have to be supplied, its plausibility being checked immediately. Common compilers on the market do not offer such language capacities, so all these tools had to be written in assembler language, guaranteeing maximum control of the hardware.

Outline of the current data bank

In the last year the statistical/regressional part of the bank was ready for mouse-supported operation. The main menu appears on the monitor starting the program with "ALFABET.EXE", outlining the procedure of an actual evaluation, Fig. 2. Data must be read in and analyzed. Results can be presented in table form or in graphics, S-N diagrams, either on the screen or on Epson compatible 24-pin printers and on HP-GL compatible plotters.

Following further the first menu point "Load Data Set" Fig. 5 demonstrates the options available. The user can either directly input a single data set or select one from the table of contents of the bank ("Overview"). If he wishes he may ask for a simultaneous analysis of more data sets ("Combine Sets") and form or edit a respective list ("View List"). He also has the possibility to transfer and save and later retrieve such a list in ASCII format on a diskette.

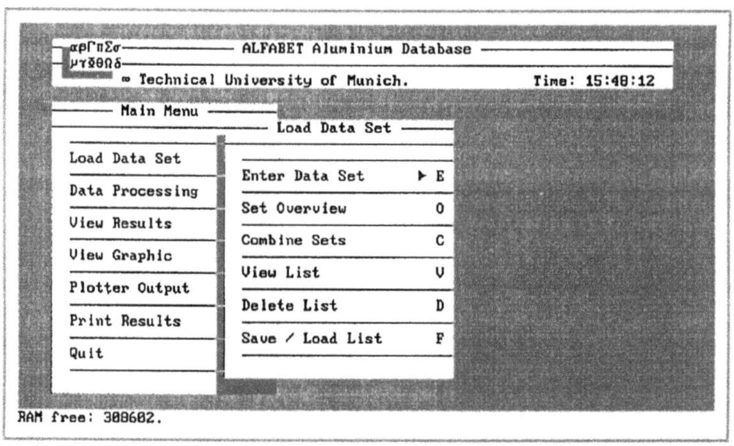

Fig. 5: System menu outlining data set options

The window shown on Fig. 6 gives the two main statistical/regressional analyses available, the linear regression and the maximum-likelihood analysis, the latter accounting appropriately for run-outs. In the case of the linear regression a version assuming run-outs as failed test results or excluding these from the analysis altogether is provided. Also a version performing an outlier test and excluding respective data from subsequent analysis. Further statistical comparisons and significance tests, which had originally been developed and were available for the main-frame operation [8] will eventually be taken over in this version as well.

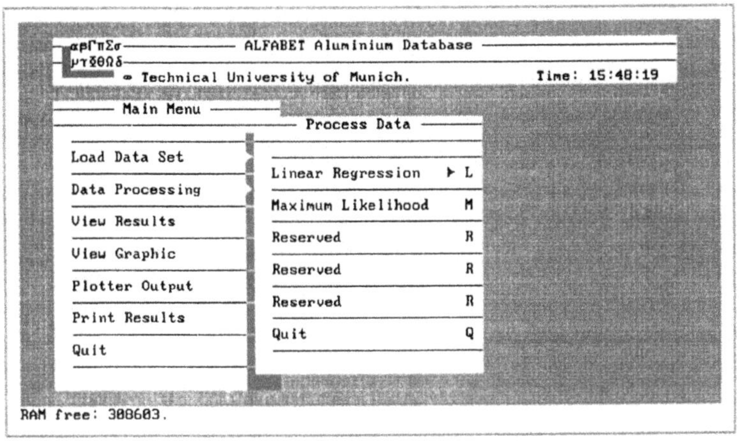

Fig. 6 System menu outlining analysis options

Through "View Results" main analysis results appear giving all values significant in subsequent design operations. "View Graphic" on the other hand presents a log-log S-N diagram with or without the data points, mean and mean minus two standard deviations regression lines – again from all these options a hardcopy may be produced by printer or plotter and results can be saved on a file for further applications.

In the last months work has commenced on the other two parts of the data bank, i.e. the literature and the structural detail description data bases, to transform these as well from dBaseIII+ to Pascal. In order to manage necessary information in a similar way as for the mathematical part described above the appropriate tools for selecting and sorting of data sets on the basis of several characteristic qualities have to be developed again.

Closing the description of the Aluminium Data Bank I would like to underline its two main features:

1) availability of well documented fatigue strength data but also fracture development and crack propagation data on large structural components, and
2) concentration on design and reliability analysis as the two characteristics distinguishing this data bank from others covering only material selection.

THE ALFABET PROJECT

Linking the Aluminium Data Bank as one information source to further such sources and decision-forming tools (literature, handbooks, manuals, expert knowledge, the latter being naturally limited in its availability by time or age) and then to design recommendations, together with appropriate reliability and fitness-for-purpose estimates offers a complete system for the design of aluminium structures. Of prime interest will be the development of an interface between these information sources and the user, which will provide guidance to the user allowing for design levels of varying sophistication degree according to the desired application. Since respective efforts on the development of design recommendations have demonstrated the need for further clarification of manufacturing and quality control parameters to allow for reliable classification of structural details, a close cooperation with the industrial side providing "manufacturing and welding technology" to supplement the above efforts on the "design technology" is strongly recommended and desired.

Work associated with the development of the data bank as stated above is understood as Phase I of this project. Phase II deals with the link between data description and data analysis. As already pointed out this is of essential importance to the detail classification and evaluation in the design procedure and reliability analysis. Extensive structural detail description and fatigue data, see Fig. 1, supported by graphics and cross-reference to detail design parameters, manufacturing options, quality criteria and strength behavior must be provided. In this step occurs also the input of corresponding welding technology information to an extent appropriate and necessary to the corresponding design and reliability level.

Several add-on packages occupy a position between Phases II and III, such as finite element modelling and analysis, reliability analysis, fracture mechanics analysis and life estimations. These are to be linked to respective material and component values as given through the data bank. Work has already commenced in independent projects recording such data.

In Phase III the actual integration of the above steps to the whole system has to be undertaken, Fig. 7. It will provide the interface and feedback system for the user. An essential part of this work is the verification, operability and close adaptation to application needs of the system. It is self-evident that such an ambitious task can only be carried out as a joint venture.

Continuing efforts to enhance the volume of the data bank are independent of the above phases and will have to be performed parallel to other efforts and on a continuous basis as new information from research or from further sources in literature keeps coming in. Besides the large contingent of data as mentioned previously the following data input tasks are being considered for the immediate future:

- further component test data in currently performed fatigue tests in Europe and USA,
- data as given in japanese literature,
- validation of component test data as outlined in [9],
- comprehensive fracture mechanics properties data for aluminium alloys currently being recorder at the TUM, and

extensive crack propagation test data generated by Jaccard [10] or Graf and Kosteas [11,12] at the TUM in the early eighties for base material, weld metal and HAZ material of different aluminium alloys and weldments under variation of stress ratio and plate thickness.

A.Rough goal definition
  Expert System for static and fatigue design
  formulation of application
B.Initial Information Accumulation
  (largely unstructured in this phase)
  1. Literatur survey and compilation
  2. Information from practice
  3. Definitions
  4. Documentation of standard recommendations etc.
     cross-references to other documents
     design and manufacturing rules
  5. Quality criteria
     Fitness for purpose, NDT, Detail classification
  6. Available software
     Drafting of software (design rules)
C.Knowledge Presentation
  Accumulated documentation material to be structured
  ordered and presented in an abstract and universal
  format
D.Main Information Accumulation
  Application oriented, enhanced documentation
E.Pilot or Prototype study
  Performed for a number of case-studies
  Check ES format, alter, amend, re-check (Iteration)
F.Detailed Study
  ES integration in actual working format
G.System Development
  ES adaption after D
H.Check and Validation Phase
  Actual design cases compared to ES solution
I.Installation and Acceptance Phase
  Efficiency in industrial application
J.Maintenance and Updating

Fig. 7. Development stages of the expert design system

CLOSING REMARKS

The essential elements in using the computer within the ALFABET project to record, evaluate and apply data are that it helps us to provide better access to better data, to reach the same degree of computerization as in other engineering tools,enhance the range of manufacturing options and apply materials and procedures more effectively. Emphasizing on the collection and reporting of new information also promotes the standardization and consequent automation of fatigue data. A better overview of available data is reached, allowing also for removal of obsolete or inappropriate data.

REFERENCES

1. Sanders,W.W.,Jr. and Day,R.H., Fatigue Behavior of Aluminum Alloy Weldments. WRC Bulletin 286, Welding Research Council, New York, Aug. 1983.
2. Steinhardt,O. und Kosteas,D., Die Schwingfestigkeit geschweißter Aluminiumverbindungen aus der Sicht optimierter Lebensdauerfunktionen. Mitteilungen der versuchsanstalt Stahl, Holz und Steine der Universität Karlsruhe, Reihe 3, Heft 5, 1971
3. Kosteas,D.,Kirou,I. and Sanders,W.W.,Jr.,Fatigue Behavior of Aluminium Alloy Weldments - A CAFDEE Report, vol. 1 to 4, Eng.Res.Inst., Iowa State Univ., Ames, 1986.
4. Kosteas,D. and Uhry,A., Fatigue Behaviour of Aluminium Structures - European Full-Scale Tests. CEC-COST 506 - EAA, Parts 1 to 3, Munich/Voreppe 1989.
5. Kosteas,D., Structural Details in Aluminium Classification and Design Parameters. Proceed. ICSAS 91,Singapore, 22-24 May 1991.
6. Unpublished internal reports and materials provided by Alusuisse, Zürich/Switzerland for the EAA-ECCS Fatigue Workshops Munich and Zürich 1990 (edited by Dr. R. Jaccard/Alusuisse).
7. European Recommendations for Aluminium Alloy Structures -Fatigue Design. ECCS TC2-TG4, Final Draft 9.0, 1990.
8. Sanders, W. W., jr., Maddox, S. J. and Kosteas, D., Aluminium Fatigue Data - Description and Principles for Evaluation. Proceedings 2nd INALCO Munich, May 1982, p. II.2.1-5
9. Ondra,R. and Kosteas,D., Evaluation of Welded Attachments Under Combined Fatigue Loads. Proceed. ICSAS 91, Singapore, 22-24 May 1991.
10. Jaccard, R., Fatigue Crack Propagation in Aluminium. IIW Doc. XIII-1377-90.
11. Graf,U. and Kosteas,D., Predicting fracture in Aluminium weldments. in: Failure Analysis - Theory and Practice, Proceedings 7th European Conference on Fracture. Ed.: E. Czoboly, Budapest, 1988, Vol. II, p. 1276-1288.
12. Graf,U., Zum Ermüdungsrißfortschrittsverhalten der mittelfesten Aluminiumlegierungen und ihrer Schweißverbindungen. Manuscript of a dissertation 1990, not yet published.

23 TREATMENT OF R-RATIO EFFECTS IN THE FATIGUE DESIGN OF ALUMINIUM COMPONENTS

D. KOSTEAS
Institute of Civil Engineering III, Steel Structures, Technical University of Munich, Germany

ABSTRACT

Scant and incomplete experimental evidence, especially on full-size structural components, has inhibited a widely acceptable and quantitative statement on the effect of R-ratio variations on fatigue strength. Before the background of accumulating experimental data with the aluminium beam fatigue testing program at TUM and new comprehensive data on small specimens made available to evaluation lately, an attempt has been undertaken to account for such effects in the European Recommendations for fatigue design of aluminium structures.

The paper describes data from four different sources, compares the respective results and follows the procedure up to the final proposal for design, which is based on whether there is knowledge or not about the stress situation – and especially about residual stresses – in the component. In the first case higher allowable fatigue design stresses may be used.

INTRODUCTION

Assumptions and statements made in the following were based on information and experimental test data available prior to the RILEM workshop in the spring of 1990. This holds especially for the test results and evaluations in connection with the ECCS-EAA Fatigue Workshops in Munich and Zürich in February and March 1990 respectively. These intermediate results helped formulate the outline on R-ratio effects presented. Due to the discussions following on the ECCS Committee 2 in relation to the European Recommendations for Aluminium Alloy Structures in Fatigue Design (ERAAS Fatigue) on meetings in the summer and winter of last year, as well as further intensive evaluations and the enforcement of general principles governing the fatigue design procedure format, the final R-ratio dependency of fatigue strength design values is not identical to the first actual results. This published version of the original oral presentation at the RILEM Workshop allows to include, for the sake of completeness, the respective decisions as given in the now finalized ERAAS document.

There is unfortunately only scant and incomplete experimental information to allow a quantitative statement on the effect of R-ratio on the fatigue strength of aluminium structural elements. Accordingly the treatment of this possible effect has varied in the past with the data available or the viewpoint of those drafting the respective recommendation, Fig. 1. Roughly, earlier recommendations accept R-dependent fatigue strength values, since they seem to be based on small specimen test

Testing of Metals for Structures. Edited by F.M. Mazzolani.
Published by Taylor & Francis, 2 Park Square, Milton Park, Abingdon, Oxon, OX14 4RN.
ISBN 0 419 15810 3

results, or to apply to rather simple elements or even only to base material with no significant residual stresses. More recent fatigue design rules,

Aluminium Fatigue Design Recommendations

R-independent	R-dependent
	DVS 1608 (1969) "LdV" (1970)
	ALCAN Handbook (19
	UNI 8634 (12/1985)
	Austrian Recomm. (3/1988)
Ontario Highway Bridge Code (1979/1983) Recommended Specifications Aluminium Association (10/1985)	Assoc. American RR Freight Cars (3/1980)
BS 8118 : (1985) (1989) light/heavy struct. el.	PC 118 : (1969)
ERAAS Fatigue Design up to Draft 8:1985	ERAAS Fatigue Draft 8.6:1990 and final document

RI 1-1

Fig. 1: Accounting for R-dependency in recommendations

in steel as well as in aluminium, attempt to emphasize on the behavior of larger, full-size components and welded constructions rather and thus assume the presence of more or less significant residual stresses. Consequently, the latter are based on a R-ratio-free design concept and do not make use of any possible beneficial effect for R values near or below zero. Although not explicitly stated it can be derived from respective experimental evidence that fatigue strength values at approximately R=+0.5 or higher have formed the basis of the proposed design values. The fact that the ERAAS are given both as R-independent and R-dependent reflects the growing knowledge level during the recommendations' drafting and the wish to cover possible beneficial structural cases.

RESIDUAL STRESS MEASUREMENTS AT TUM

Within the comprehensive welded beam fatigue testing programs at the Technical University of Munich (TUM) it was possible to undertake initial residual stress measurements on full-size welded components of 7020 and 5083 alloys, Fig. 2 [1,2,3].

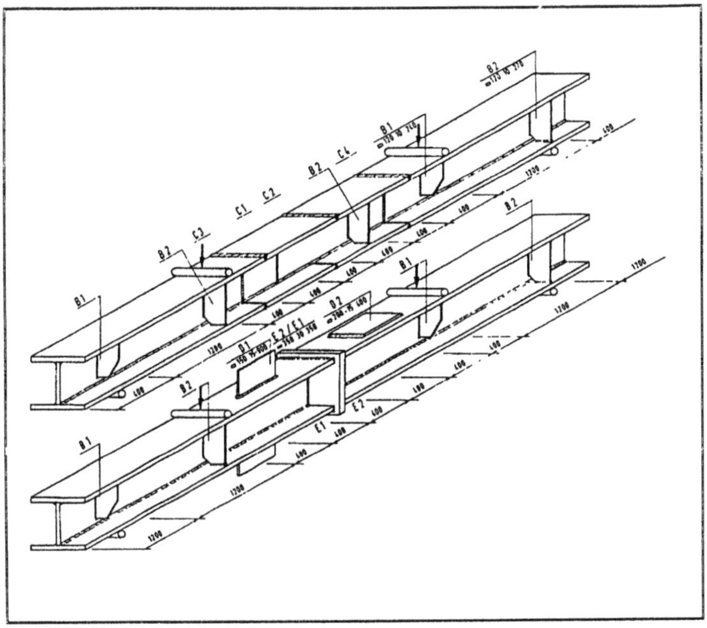

Fig. 2 : TUM welded aluminium beams for fatigue tests

The measurements themselves and the difficulties associated with the interpretation of these limited results or the sometimes somewhat contradictory information have been described elsewhere [4,5] and allow presently only the following general conclusions:

The presence of residual stresses in welded components of considerable magnitude, reaching values up to the 0.2–yield–limit of the HAZ, was documentated, Fig. 3.
Scatter is an inherent feature in measurements made by the hole–drilling method.Efforts were undertaken to make measurements sufficiently close to the welds, approximately 1 mm off the weld toe, by using rosettes with the strain gages arranged on one side only.
There were some but no significant differences in residual stresses in the structural details of the 7020 and 5083 alloy.

As seen in Fig. 3 the highest values of residual stresses were recorded at fillet welded longitudinal or transverse attachments on the beam flanges. Residual stresses in these details attained values of over 180 MPa. Residual stress measurements at butt welded splices and cruciform joints made with butt–like full penetration welds attained values of between 120 and 140 MPa.

No answer can be offered for the time being to the question of whether initially measured residual stresses are maintained, and at what magnitude, during subsequent load cycling of the components. Indirect measurements at strain gage monitored crack initiation sites showed values up to 120 MPa. These gages may not have been sufficiently close to the maximum residual stress site, though. Residual stresses, therefore, may have been higher.

The general lower location and steeper slope of S-N curves giving the behavior of components versus small specimen data agree with similar findings in steel weldments (see DIN 15015 and Eurocode 3 or ECCS TC6).

There are still difficulties in interpreting differences between S-N curves with stress ratios of -1 and +0.1. The first show at times approximately 40% higher fatigue strength values of stress range plotted in a log-log diagram at two million cycles. The effect is more pronounced in the case of 7020 alloy details. It does not seem prudent, though, to allow for a bonus in design values in such cases of large components, since sufficiently reliable data is still missing.

DATA BACKGROUND

The first comprehensive data evaluation attempt in 1985-1986 was confined to analysis of fatigue tests with small specimens only, the so-called CAFDEE test data [6].

During the second data evaluation stage interest focused on larger components, the welded and base material beam fatigue tests performed by the TUM and earlier respective tests made available by the aluminium industry. These results were summarized 1989 in [7]. Again comparative analysis of fatigue tests at two different R-values were possible with the beam test results from TUM only. A set of small specimen data at different R levels was analyzed in parallel though, allowing for a comparison along the lines of this homogeneous data a comparison with earlier assumptions in analyses and recommendations. The decision was taken at this stage to refrain from attempting a quantification of the R-ratio effect on fatigue strength and operate along the assumption of larger structural elements with respective significant residual stresses and consequently adopt an R-independent design procedure.

Discussions that followed on the next ECCS TC2 meeting in Grenoble in November 1989 and different reactions to the above proposal prior to and during the third comprehensive data evaluations at the EAA-ECCS Fatigue Workshops in Munich, February 1990, and Zürich, March 1990, now encompassing the more recent test results of the second TUM beam fatigue test program and a new large contingent of further test data on small specimens made available by the aluminium industry [8], expressed the wish for a better differentiation of cases possible in practice. Information on R-dependency was summarized once more, opening the possibility to the design engineer to distinguish between more cases in practice and utilize beneficial effects where this would seem appropriate.

Information on the data analyzed on the three mentioned evaluation stages is summarized:

Stage I: Test data and its statistical regressional analysis can be found under [6], but due to the inhomogeneous character and the unreliability in the documentation of this rather old material it was decided not to use it further for purposes of the European recommendations.

Stage II: (a)
Type: ALS (Alusuisse/Zürich bzw. Neuhausen) small specimens, 6005A alloy base metal, butt weld transverse
Sample size: 424 and 407 respectively typical data set sample size: 41
Life range: 10^4 to $2*10^6$ cycles
R-ratio: values between -3 and +0.7
Information: see Fig. 4 and Fig. 5 EAA-Report March 1989 [7]

(b)
Type: TUM beams, 7020 alloy
butt weld transverse, longitudinal and transverse fillet weld attachment on beam flange
Sample size: total/at R=0.1/at R=-1
butt weld 43/23/15
long. fillet attach. 53/19/34
transv. fill./cover plate 43/15/28
Life range: $6*10^4$ to $2*10^6$ cycles
R-ratio: values for -1 and +0.1
Information: EAA Report 3/89 [7]

Stage III: (c)
Type: TUM beams in 7020 and 6005A alloys, due to the still running tests of the 1987-1990 test series at the time of these evaluations and the resulting small sample sizes only a limited number of cases give valid statements
Sample size: web-to-flange fillet welds 87
cover plate 26
Life range: 10^4 to $5*10^6$ cycles
R-ratio: -1, +0.1, +0.6
Information: [3] and materials Fatigue Workshop Munich [8]

(d)
Type: small specimens, different structural details and alloys of the 5000/6000/7000 series
Sample size: base metal 7020 : 2209
5000 : 1719
6000 : 5035
notches, holes : 851
butt weld : 6027
web stiffener or non-load-carr. transv. fillet : 1586
cruciform or load-carrying transv. fillet : 1262
Life range: 10^4 to 10^7 cycles
R-ratio: -1 and +0.1
Information: Fatigue Workshop Documents [8] Alusuisse Zürich data

STAGE II EVALUATION

The relationship in Fig. 4 is based on mean fatigue strength values at one million cycles estimated by linear regression analysis, excluding run-outs.

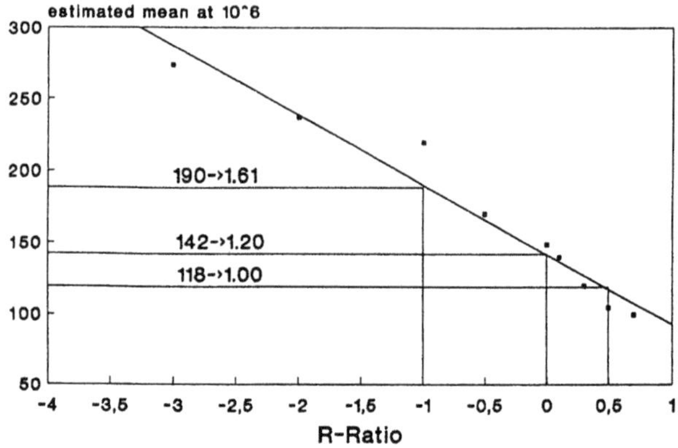

Fig. 4: R–ratio vs. fatigue strength for base metal small specimen test data

During the Munich Workshop though, a second relationship, with somewhat different values, was established, this one based on lower boundary curves for the test data at two million cycles:

R	-3	-2	-1	-0.5	0/+0.1	+0.3	+0.5	+0.7
Sr [MPa]	166	166	166	129	113	96	86	79
$S_r/S_{0.5}$	1.93	1.93	1.93	1.50	1.31	1.12	1.00	0.92

This was considered more representative and was used (at this stage of development) to establish the fatigue strength enhancement factor f(R) in the case of base material, see further on Fig. 6. It is mentioned at this point already that the respective value stated on the first column of Fig. 8 has been derived directly from the figures of the above table.

Concerning information on the transverse butt weld specimens again as mean stress values at one million cycles the relationship of Fig. 5 is established.

Or, again, the following relationship if strength values at two million cycles are used:

R	-3	-2	-1	-0.5	0/+0.1	+0.3	+0.5	+0.7
Sr [MPa]	131	131	76	(95)	54	54	50	46

S-1/S+0.5	= 1.52
S-1/S0	= 1.41

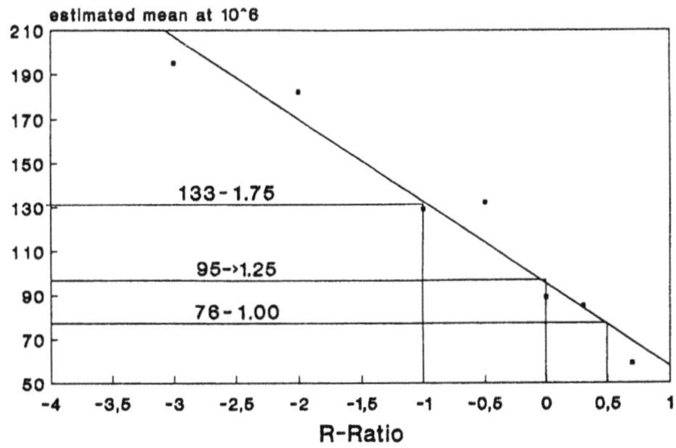

Fig. 5: R–ratio vs. fatigue strength for butt weld transverse small specimen fatigue data

Initially it was thought that both base material and (butt) welded joints could be treated by a common assumption due to the factors originating from Fig. 4 and 5 with data from small specimen tests. Comparison of different data exhibited different fatigue strength ratios at respective R values and there was no clear overall picture. Some of this information will be given in the following.

The proposal thus for welded details, according to Fig. 6, assumes a relationship parallel to the one for base material but on a lower level. The characteristic f(R) value at R=–1 equals 1.6 and reaches 1 for R=0, being thus on the safe side.

The above mentioned result of S–1/S0=1.41, established on the basis of a rather large and homogeneous sample of small specimens, together with the almost identical value of S–1/S0=34.0/23.6=1.44 as calculated for the large welded components of the TUM aluminium beams, Fig. 7, lie in accordance with further minimum values as will be shown, and indicate thus the limit to be attained with some degree of reliability for welded details.

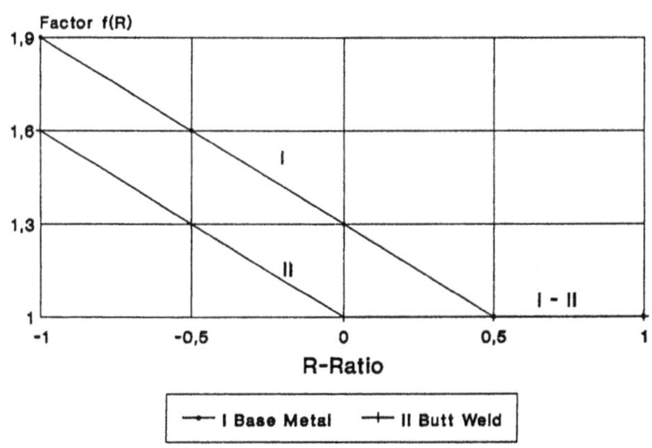

Fig. 6: Factor f(R) from source II–(a) base material and (b) butt weld transverse

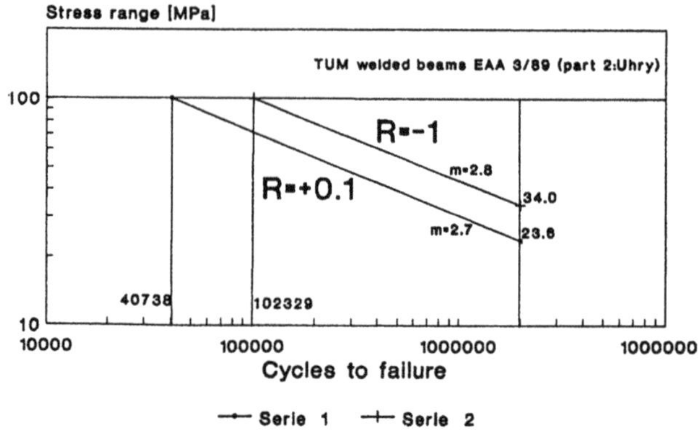

Fig. 7:Effect of R–ratio on TUM beam tests of the 1982–1986 research program [1,7]

STAGE III EVALUATION

Additional support in the case of welded beams came from test results of the current TUM beam test program 1987–1990. Valid statements based on a reliable number of test points for 7020 beams were possible in the case of longitudinal non–load–carrying fillet welds (web–to–flange welds with start and stop positions or tack welds) giving at $2*10^6$ cycles

f(R)=1.14 for R=+0.1 to −1 and
f(R)=1.57 for R=+0.6 to −1

and similarly in the case of transverse load–carrying fillet welds of flange attachments, i.e. cover plates

f(R)=1.24 for R=+0.1 to −1 and
f(R)=1.60 for R=+0.6 to −1.

The fourth information source added significant weight to the evaluations by the considerable amount of small specimen data, almost 19000 data points, on base material and various weldments in the standard 5000, 6000 and 7000 alloy series.

The following three Fig. 8, 9 and 10 summarize minimum/mean/maximum enhancement factors f(R) at $2*10^6$ cycles and between the R values 0 and −1. Especially for welded details, i.e. joints with assumed significant residual stresses, it was decided to accept an enhancement factor between 0 and −1 only. For values R>0 the factor should be equal to one.

The Fatigue Workshop evaluations were completed by information on the treatment of the R–ratio effect on fatigue strength as these had been included in a recommendation and a standard for the design of aluminium structures. Fig. 11 and 12 give information from the proposal for the Aluminum Association in the U.S. and Fig. 13 and 14 attempt to reconstruct from rather limited information the scatter band of the factor f(R) for several welded details analyzed and providing a background for the italian standard on aluminium design.

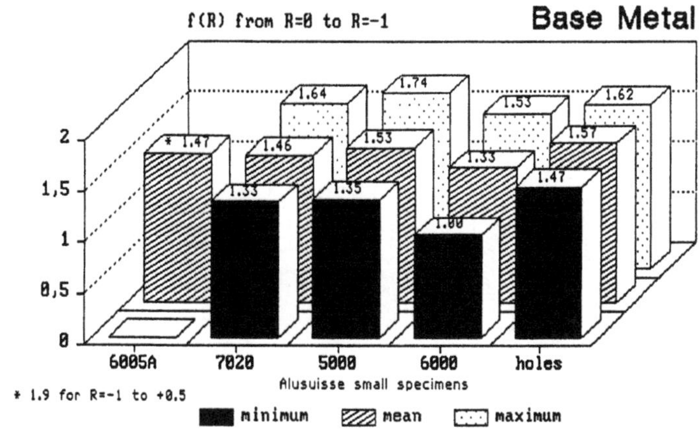

Fig. 8: Enhancement factor f(R) at $2*10^6$ cycles for base material

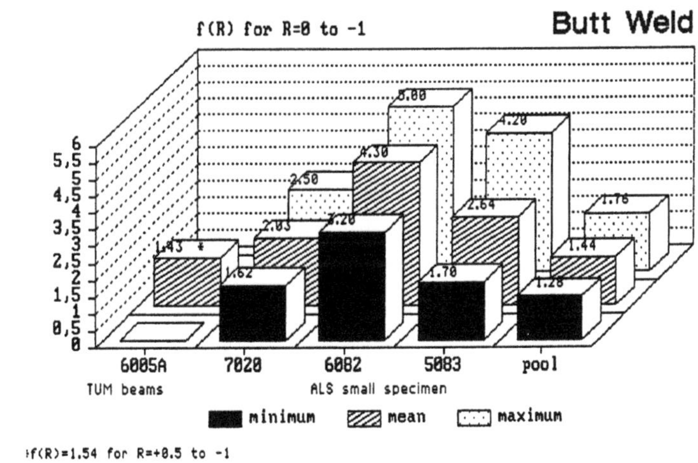

Fig. 9: Enhancement factor f(R) at $2*10^6$ cycles for transverse butt weld

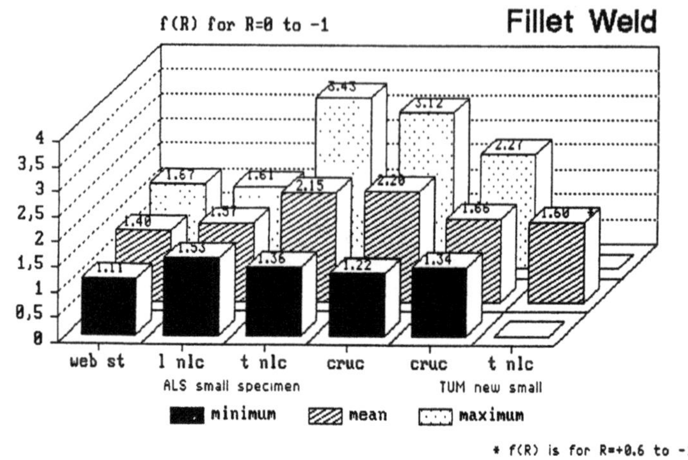

Fig. 10: Enhancement factor f(R) at $2*10^6$ cycles for longitudinal or transverse, load-carrying or non-load-carrying fillet weld

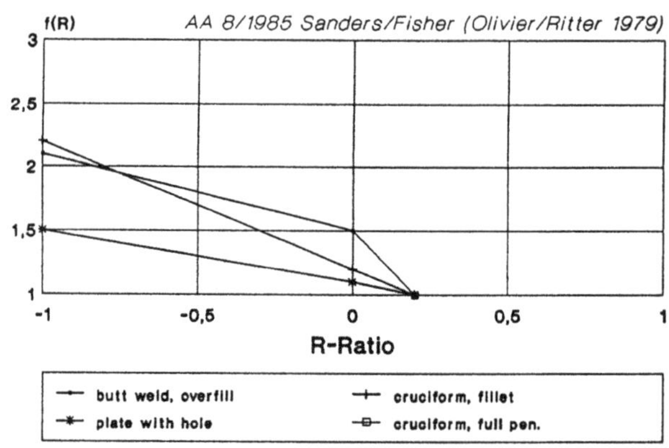

Fig. 11: Enhancement factor f(R) in design recommendations draft of the Aluminum Association – (a)

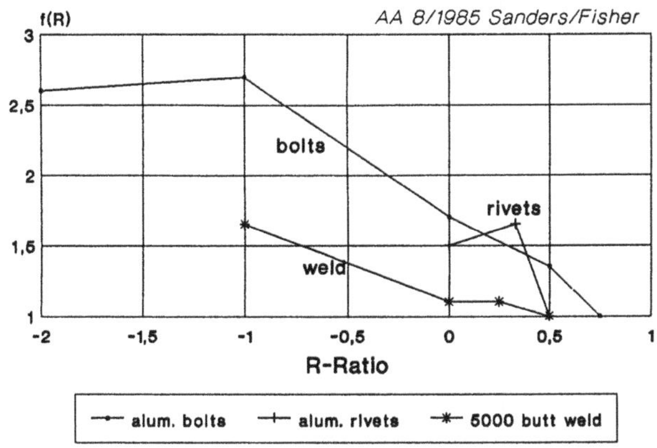

Fig. 12: Enhancement factor f(R) in design recommendations draft of the Aluminum Association – (b)

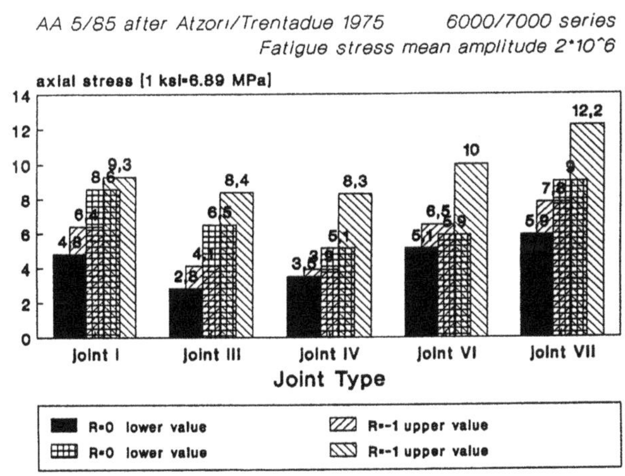

Fig. 13: Mean stress amplitude values at $2*10^6$ cycles for several welded joint types in 6000 and 7000 alloys

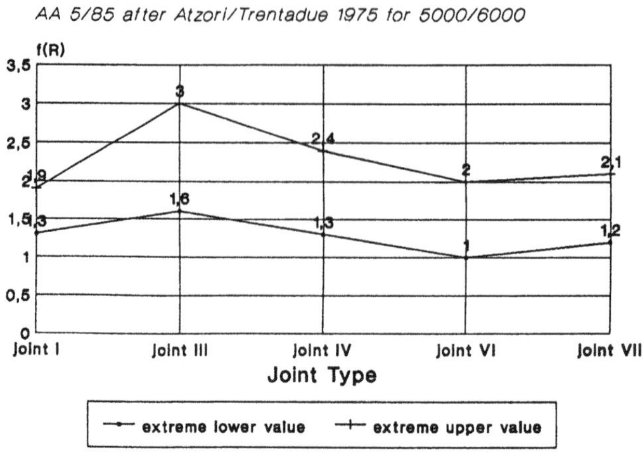

Fig. 14: Extreme limit values for the enhancement factor f(R) for several welded joint types in 6000 and 7000 alloys

DRAFTING A PROPOSAL FOR THE FACTOR f(R)

Using in general the mean values observed for the enhancement factor f(R) in all the cases outlined above a design proposal was drafted for the European Recommendations following the evaluations of the Workshops. The original proposal, as first suggested in Fig. 6, was now extended to include two practical cases for both base material elements and weldments, Fig. 15. In the one case knowledge of residual stresses in the structural element is available and if taken into account in the design procedure allows for higher design stresses. Should no information on the stress situation be available or not considered in design then a lower line gives the respective values of f(R) vs. R for

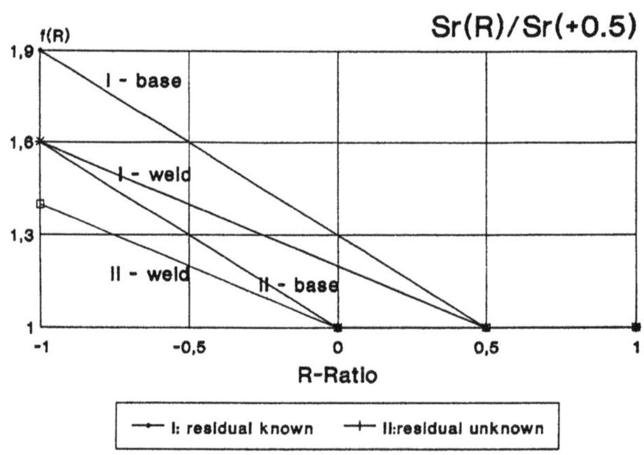

Fig. 15: Proposed design values for the enhancement factor f(R) after ERAAS Draft 8.6 (summer 1990)

base material and weldments. The concept consists thus of two pairs of parallel lines each. The line for base material and residual stresses not known is not covered by actual data but based on an assumption following discussion on the Fatigue Workshop and a respective proposal by Dr. M. Ogle, The Welding Institute.

FINAL ERAAS DESIGN PROPOSAL FOR f(R) AND CLOSING REMARKS

The above proposal was discussed intensively, arguments concentrating on variability and scatter of enhancement factors f(R), the reliability of data sources, also the their comparability, but above all on how accurate and realistic a definition of residual stress patterns can be in practice. Trying to cover these uncertainties it was decided to lower further the design proposal for the f(R) relationship, practically assuming now values justified by the majority of lower limits in the analyzed data (as depicted for instance, but not exclusively, in Fig. 8, 9 and 10). Thus the proposal after Fig. 16 was adopted as the final design proposal for the European Recommendations [9].

Line I is applicable only to base material and wrought products where there are no residual stresses acting on the structural elements or where such stresses can be fully taken into consideration in the design concept. Line II is to be used generally for all connections, also welded ones, where knowledge on the residual stress situation is available and is being considered. Line III covers those cases where significant global residual stresses are present but cannot be considered in the design analysis.This last case with no allowance for an enhancement factor includes thus all welded details in general practice.

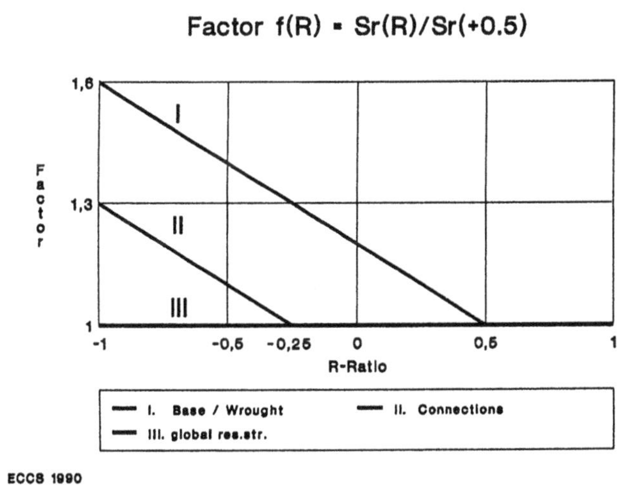

Fig. 16: Design values for the enhancement factor f(R) after the final ERAAS Fatigue Design document (December 1990)

In order to give a full picture of the final proposal and its rules as it was included in the ERAAS document the respective clauses are given in the following Appendix.

This may seem as a rather severe set of rules, departing also from the original position of the first evaluations. The introductory statements on residual stress measurement results should be considered very carefully, though. The existence and behavior of residual stresses on larger elements under fluctuating loads is by no means quantifiable at the moment and an extrapolation of these results to every possible case in practice does not seem prudent.

It must be born in mind that in the case of base material there was satisfactory experimental evidence available only in the form of small specimen tests, i.e. cases with "known" residual stress or equal to zero.

The similar magnitude of strength ratios at a certain R value between larger components and small specimens seem to indicate that a fair amount of residual stresses may still be locked in the latter (this has also been shown by test results by Maddox and Webber a few years ago). Information about the specimen sizes, welding procedures and parameters would be necessary to clarify this matter, but were not available at the time of the above evaluations.

To correctly read and apply values of the factor f(R) it should be mentioned once more that the diagram states only values relative to the fatigue strength at R=+0.5, the latter being the basic design value at $2*10^6$ cycles of a specific structural detail relating to the classification or design curve value given in the data sheet of the Recommendations.

APPENDIX

The following clauses from the ERAAS Fatigue Design, final Version, December 1990 cover:

B.3 Influence of R–Ratio

For stress ratio values less than R=+0.5 a fatigue strength enhancement factor f(R) may be considered and is given in Fig. 16. This factor is stated for a fatigue life $N=2\cdot10^6$ cycles. For other fatigue life values see Clause B.3.2.

B3.1 The following cases shall be distinguished (see Fig. 16):

I Base material and wrought products in structural elements, assuming that any residual stresses can be neglected for such elements

$f(R) = 1.6$	for $R < -1$
$f(R) = -0.4\cdot R + 1.2$	for $-1 \leq R \leq +0.5$
$f(R) = 1$	for $R > +0.5$

II Connections (welds or fasteners) in simple structural elements.

$f(R) = 1.3$	for $R < -1$
$f(R) = -0.4\cdot R + 0.9$	for $-1 \leq R \leq -0.25$
$f(R) = 1$	for $R > -0.25$

III For complex two– or three–dimensional structures with residual stresses a constant factor shall be used

$f(R)=1$

For connections in structures which have been adequately stress–relieved curve I shall be used

B3.2 For the calculation of fatigue strength at values R other than +0.5 and N between $1\cdot10^4$ and $5\cdot10^6$ cycles the following procedure is given, assuming that the basic design line as given in the Data Sheets is rotated around its value at $1\cdot10^4$ cycles.
The resulting slope is:

$$m(R) = \frac{-\,2.301}{\log f(R) + \log \Delta\sigma(2\cdot10^6) - \log \Delta\sigma(1\cdot10^4)}$$

where

f(R): from Fig. 16
$\Delta\sigma(2\cdot10^6)$: the corresponding fatigue stress range of the basic design curve at $2\cdot10^6$ cycles (for R=+0.5).
$\Delta\sigma(1\cdot10^4)$: the corresponding fatigue stress range of the basic design curve at $1\cdot10^4$ cycles (for R=+0.5).

For fatigue life values $N > 5\cdot10^6$ cycles the fatigue strength values of design classes for welded details B, C, D, E, and F are defined assuming that the slope of the design curve is $m'(R)=m(R)+2$. For base material designclass A use $m'(R)=m(R)$.

REFERENCES

[1] Poalas, K., Graf, U. und Kosteas, D.:
Ermüdungsverhalten geschweißter Aluminiumbauteile (Fatigue behavior of welded aluminium components)
Schweißen und Schneiden, 40(1988), 3, S.130–133

[2] Kosteas, D.:
Summarizing results of the aluminium welded beam fatigue programme
IIW Doc. XIII–1240–87

[3] Ondra, R. und Kosteas, D.:
Ermüdungsverhalten von Aluminiumgroßbauteilen – Ergebnisse des dritten TUM-Forschungsprogramms Fatigue behavior of aluminium full-size components – Results of the third TUM research program
To be published in Schweissen und Schneiden in 1991

[4] Kosteas, D.:
Influence of residual stresses on S–N curve parameters of aluminium weldments
IIW Doc. XIII–1239–87

[5] Poalas, K., Graf, U. und Kosteas, D.:
Ermüdungsverhalten geschweißter Aluminiumbauteile (Fatigue behavior of welded aluminium components)
Schweißen und Schneiden, 40(1988), 3, S.130–133

[6] Kosteas, D., Kirou, I. and Sanders, W.W., Jr.:
Fatigue behavior of aluminium alloy weldments. Analysis of data from tests
Part 1: Aims. Scope. Explanation of methods
Part 2a: Alloy 5000 (pages 1–670)
Part 2b: Alloy 5000 (pages 671–968)
Part 3: Alloy 6000 (pages 1–377)
Part 4: Alloy 7000 (pages 1–633)
A CAFDEE report as a joint study from the Engineering Research Institute of Iowa State University and the Chair of Steel Structures of the Technical University of Munich. Published by ISU, Ames, Iowa, USA, Aug. 1986

[7] Kosteas, D.:
Fatigue Behaviour of welded aluminium structures – European full-scale tests
Part 1: Analysis, Results and Proposals (March 1989)
Part 3: Data Base (March 1989)
Part 2: Statistical Analysis (April 1989 by Mme Uhry)
Commission of the European Communities, European Aluminium Association, Project COST 506 Phase I, Munich, March 1989/Voreppe April 1989

[8] Jaccard, R. (editor):
Unpublished internal reports and materials provided by Alusuisse-Lonza Services, Zürich/Switzerland for the EAA-ECCS Fatigue Workshops in Munich and Zürich 1990

[9] European Recommendations for Aluminium Alloy Structures in Fatigue Design. Final Draft 9.0, December 1990,

24 INVESTIGATIONS INTO THE FATIGUE BEHAVIOUR OF CHS-JOINTS SUBJECTED TO SPECTRUM LOADING

F. MANG and Ö. BUCAK
Testing Centre for Steel, Wood and Stone, University of Karlsruhe, Germany

GENERAL

A large number of constant amplitude tests on the fatigue behavior of hollow section structures, especially of their joints, is now available. These investigations resulted in the development of a draft for Eurocode 3 [1] as well as in the publication of a compilation of Wöhler curves (S-N-lines) [2].

A few tests have been carried out with hollow section joints under service fatigue load (block or random) in the scope of the European Offshore Programs and some national research activities. In these tests, either a modified load spectrum from the region of the North Sea or fictitious load spectra and those from the field of crane construction were applied to one type of joint and one or two variations of joint dimensions.

First systematic investigations of hollow section joints under load spectrum have been carried out in Karlsruhe. The report, dealing with this topic, is given in this paper.

In order to make a service life analysis of the structural members and constructions under random loads, their cumulative damage has to be determined. By this, the intensity and the type of the summation of partial damages are meant. The hypothesis of cumulative damage according to Palmgren-Miner is the only hypothesis that is commonly used, because of its simple application. It is based on the assumption that the service life of a structural member decreases linear with the number of applied load cycles. As per definition, the rupture occurs when the cumulative damage ratio reaches the value 1.0. In practice, the values of damage ratio, according to Miner, can considerably deviate from 1.0, ranging approximately from 0.2 to 20.0. In some cases, this can possibly lead to risky optimistic prognostics regarding the service life. In addition to other insufficiencies, this hypothesis does not even consider that those load variations, which are below the initial fatigue strength, take part in damaging the structure.

The investigations presented in the following were carried out as block program tests in order to be able to compare with them tests on flat test specimens known from literature. Differences between the service life by block program tests and by those under random generator can have a factor between 2 and 10.

The mentioned factor decreases if several partial sequences are run up to the fracture. This was the case with the present experimental

Testing of Metals for Structures. Edited by F.M. Mazzolani. © 1992 RILEM.
Published by Taylor & Francis, 2 Park Square, Milton Park, Abingdon, Oxon, OX14 4RN.
ISBN 0 419 15810 3

investigations, so that it can be started out from a smaller up to no difference of the results compared to the real case.

With the investigations in Karlsruhe at least three partial sequences but normally 5 and at most 18 partial sequences have been run through up to the fracture, so that this can be described as a good up to very good mixing of the individual load limits. For this reason, not too big differences of the results are to be expected compared to a random test.

It is desirable, however, to verify this prediction later by means of experimental investigations.

INVESTIGATIONS OF TESTING PROCEDURES ON HOLLOW SECTION JOINTS UNDER SERVICE LOAD (LOAD COLLECTIVES)

Systematic tests on hollow section joints are being carried out at present with X- and K-type test specimens under axial load (see fig. 1).

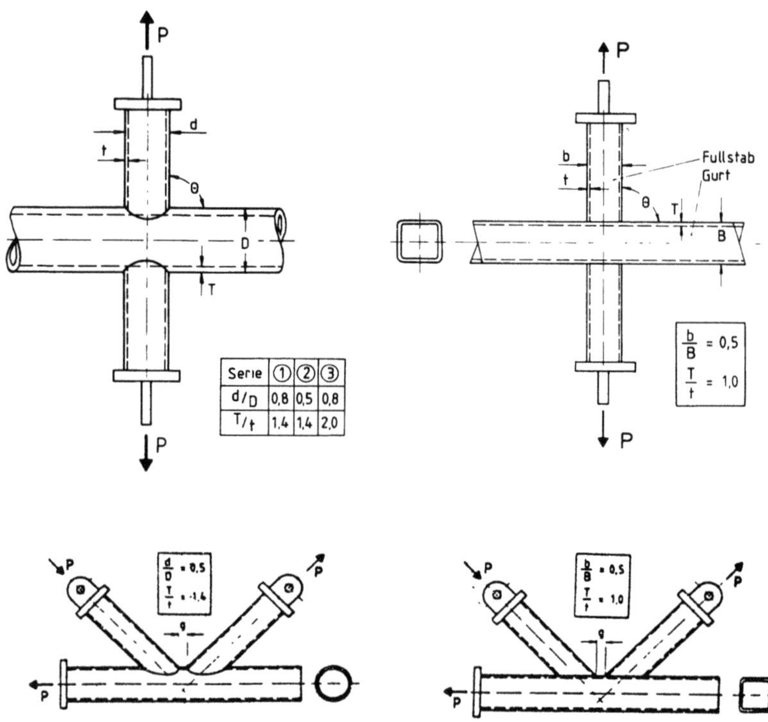

Fig. 1 X- and K-joints of circular and rectangular hollow sections for fatigue tests under spectrum loading.

The application of load spectra known from literature, such as p = 1/3 and 2/3, as well as of collectives from the offshore applications is realized in the scope of a special research project (SFB 210, "Strömungsmechanische Bemessungsgrundlagen für Bauwerke"/Basic flow load design data for structures) in Karlsruhe. Additional collectives are determined by measuring structural data under wind and water loads. Figure 2 shows a compilation of load spectra applied. Test results under load spectra have been compared with the values given by the Wöhler curves, and the applicability of Miner's rule was checked.

The investigations of CHS X-joints were carried out with test pieces of three different dimensions (2 diameter ratios - d/D = 0.5 and 0.8; and two wall thickness rations - T/t = 1.4 and 2.0). Using these specimens, the dependence of such hollow section joints on the existing geometric parameter ratios could be considered. In addition, the given parameter ratios lead to three different types of failure:

a) Joint failure in chord shear (see fig. 3)
b) Crack starting from the weld toe of the web member (see fig. 4)
c) Crack starting from the inside surface of the tube in the non-welded area of the chord (see fig. 5)

The investigated types of joints result in rupture types dependent on the existing geometry of the joint. For the test pieces with a web member diameter of 51 mm (d/D = 0.5; T/t = 1/4), the rupture occurred by ripping out of the chord member. While the test pieces with a diameter ratio of d/D = 0.8 and a wall thickness ratio of T/t = 2.0 failed by a crack of the web member starting from the weld toe, the fatigue failure of the series d/D = 0.8 and T/t = 1.4 occurred either by ripping out of the chord member or by cracks in the weld toe of the web member. Thus, the transition range of both possible types of failures (failure of the web and chord member respectively) could be obtained with the dimensions of this series.

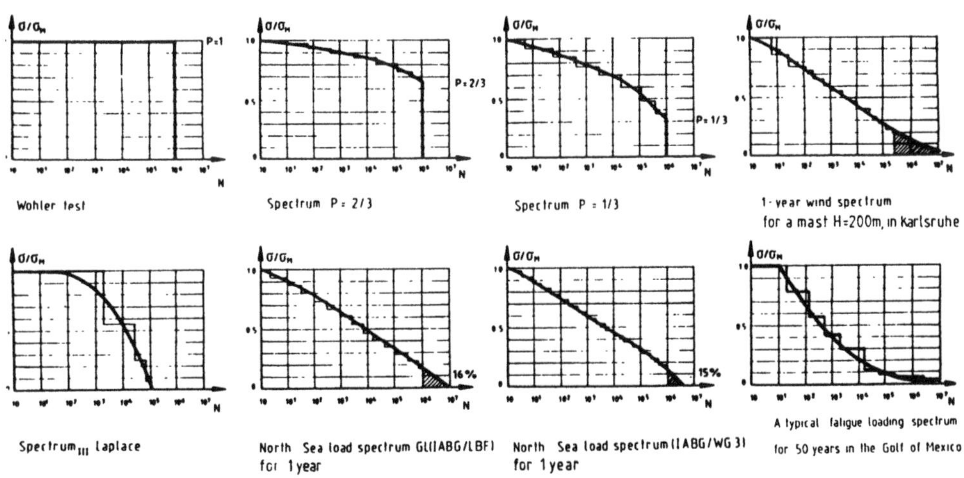

Fig. 2 Compilation of the applied load spectrum

Fig. 3 Joint failure in chord shear

Fig. 4 Crack starting from the weld toe of the web member

Fig. 5 Crack starting from the inside surface of the tube

From figures 6a, 6b, and 6c the results of the investigations on X-type CHS joints, and from figure 6d the results of RHS-joints can be seen. The tests have been carried out with the same testing machines, under the same test conditions without corrosion impact.

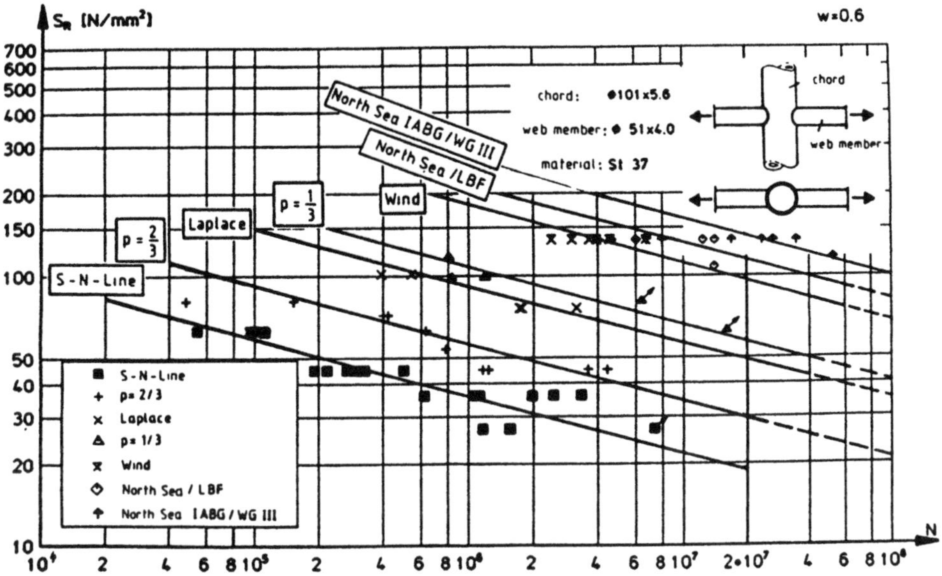

Fig. 6a Results of the fatigue investigations on CHS X-type joints under load spectrum (d/D = 0,5 and T/t = 1,4)

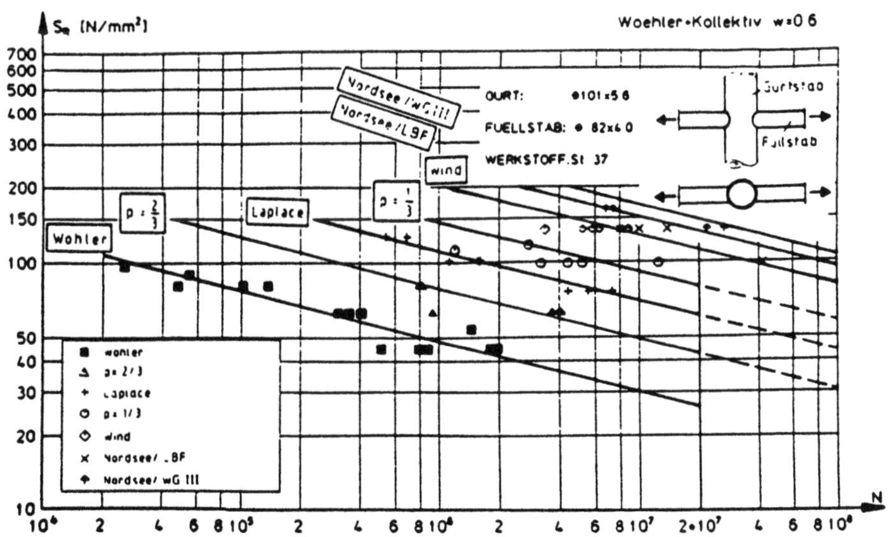

Fig. 6b Results of the fatigue investigations on CHS X-type joints under load spectrum (d/D = 0,8 and T/t = 1,4)

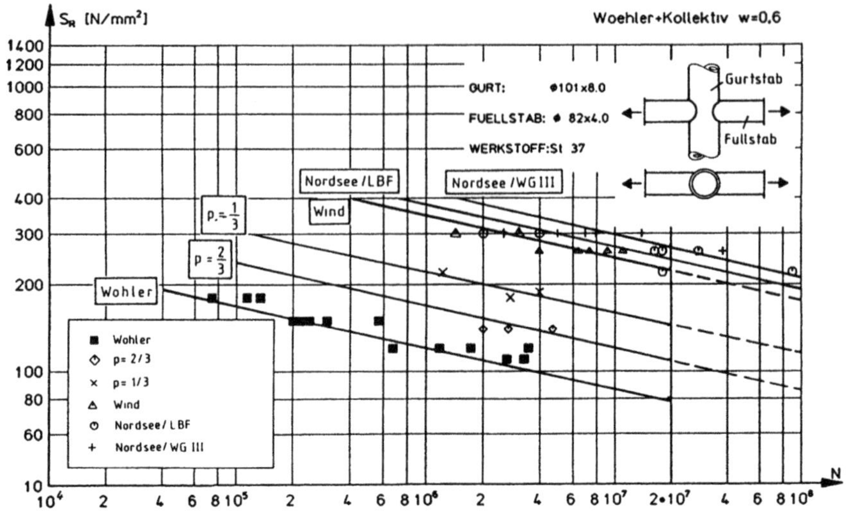

Fig. 6c Results of the fatigue investigations on CHS X-type joints under load spectrum (d/D = 0,8 and T/t = 2,0)

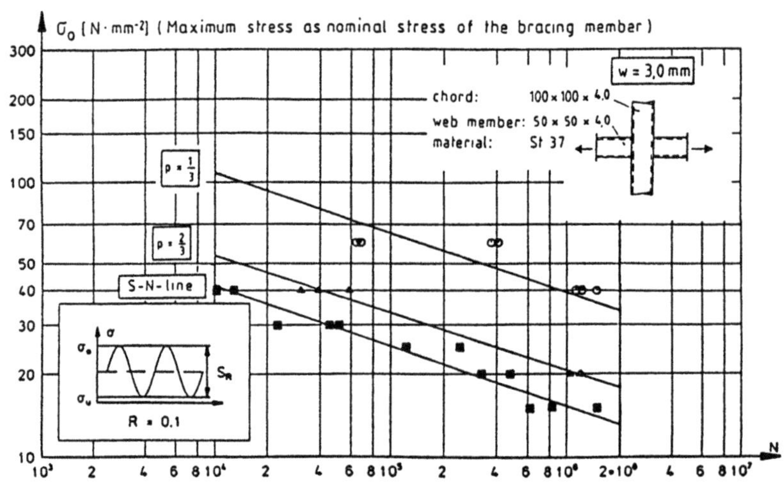

Fig. 6d Results of the fatigue investigation of RHS X-type joints under load spectrum (b/B = 0,5 and T/t = 1,0)

In order to clarify the effect on the service fatigue strengths, besides X-type joints, K-type joints were also tested under identical loading collectives up to now. A testing device, which has been used for previous investigations of K-joints, was applied in order to gain truss forces at the load of the testing machine. The results of these investigations are given in figure 7.

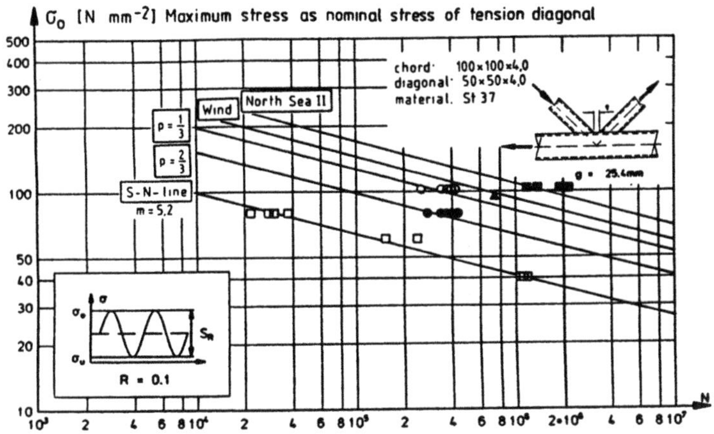

Fig. 7 Results of the fatigue investigations on RHS K-type joints under load spectrum (b/B = 0,5 and T/t = 1,0)

Presently, tests on K-joints made of circular hollow sections are being carried out.

An important question to be answered concerns the size of the crack, when a joint type loses its serviceability. Since the definition of the bearing capacity of the applied testing load proved to be only a bulk value, measurements of the crack growth were conducted. As a criterion for disconnecting the testing machine, an elongation of the test pieces from 0.1 up to 3 mm, divided into various stages, was fixed. The limitation of the maximum elongation of the test pieces to 3 mm proved to be sufficient, since the crack reached over half of the width of the web member when using this value. Nevertheless, even in this failure condition, the test piece could still bear the maximum testing force. Results from the crack growth measurements on a CHS series (series X1WS1a) are shown in figure 8.

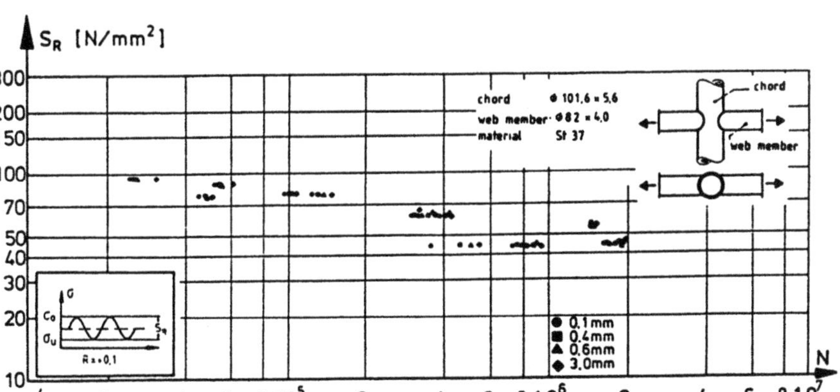

Fig. 8 Measurements of crack growths on circular hollow section joints

By means of the measurements of crack growths, the service lives of structural members and the servicing intervals are to be determined later according to the case of application. A structure under flow load from wind can have a more or less high remaining service life with a crack size corresponding to 0.6 mm and an elongation of 3 mm, whereas a structure under flow load from water with a crack size of 0.3 to 0.4 mm (with this scale of test piece elongation, the crack has already gone through the wall thickness of the chord member with test pieces of a chord member dimension of Θ 101.6 x 5.6, whereas an elongation of approximately 0.6 mm is required for test pieces having a chord member dimension of Θ 101.6 x 8.0, so that the crack goes through the wall of the web member) can result in the collapse of the total structure due to water, flowing into the construction.

APPLICATION OF THE PALMGREN-MINER-RULE

The application of the Palmgren-Miner-Rule (see fig. 9) did not result in a good coincidence in the field investigated.

This statement is applicable to all six possibilities of the procedures (see fig. 10) these are

- constant slope of the S-N-line in the total area
- bisected slope of the S-N-line at load cycles of $2 \cdot 10^6$
- modified slope of the S-N-line beginning at the load cycles of $5 \cdot 10^6$ $(m = 5)$
- cut-off of the S-N-line at load cycles of $2 \cdot 10^6$
- cut-off of the S-N-line at load cycles of 10^7
- cut-off of the S-N-line at load cycles of 10^8

The scatter of the Miner-sums in the investigated parameter range and the collective shapes are between 0,27 and 8,62. The results of the Miner-sums determined by the investigations have been plotted in the diagram of Schutz und Zenner [3] (fig. 11) for checking the evidence. The fundamentals of the diagram of Schutz und Zenner are the results of 348 calculations of Miner, which were conducted on steel, aluminium and titanium samples with block program tests in ambient temperature and without any impact of corrosion. The large scatter shown in figure 11 is decisive for the reliability of this calculation method. The fact that the value, which most frequently occurs, is in the range of 1.0, cannot be seen as a proof for reliability.

From figure 12, the scatter of the Miner sums is evident. The mean values corresponding to the individual experimental series are also plotted (marked with the letter M).

A further way of outlining is illustrated in figure 13. According to this, the dispersions are not dependent (due to the plastification and thus, on the reduction of residual stresses) on the stress levels.

It can be concluded from available tests that the cumulative damage results in values below 1.0 for load collectives with a higher fullness, as for example 2/3, and in values between 3.0 and 4.0 for a smaller fullness. For this reason, it is recommended not to indicate a constant value for the cumulative damage according to Miner's rule. It should be dependent on the loading collective.

Palmgren - Miner - Rule
(cumulative linear damage hypothesis)

$$S = \sum_{i=1}^{n} \frac{n_i}{N_i} \leq 1.0$$

S.......total damage

n_i... is the number of cycles of stress range $\Delta\sigma_i$ which occur during the required design life

N_i ... is the number of cycles of stress rang to failure, which depends on the detail category or S - N - Line

Relative - Miner - Rule

$$N_A = N_S \frac{\Sigma(\frac{n}{N})_S}{\Sigma(\frac{n}{N})_A}$$

N_A......calculated life time under modified load spectrum A

N_S.....calculated life time under load spectrum S

$\Sigma(\frac{n}{N})_S$.....cumulative damage under the load spectrum S

$\Sigma(\frac{n}{N})_A$.....cumulative damage under the modified load spectrum A

Fig. 9 Palmgren-Miner-Rule and relative Miner-Rule

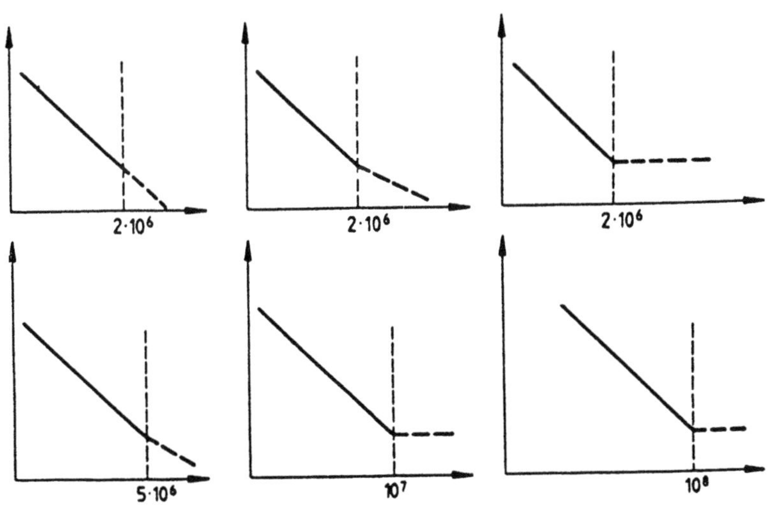

Fig. 10 S-N-lines used when applying the Miner-Rule

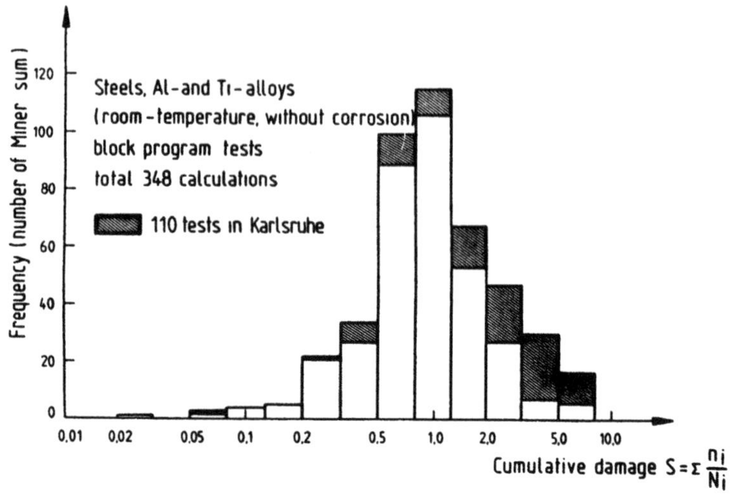

Fig. 11 Frequency of failures

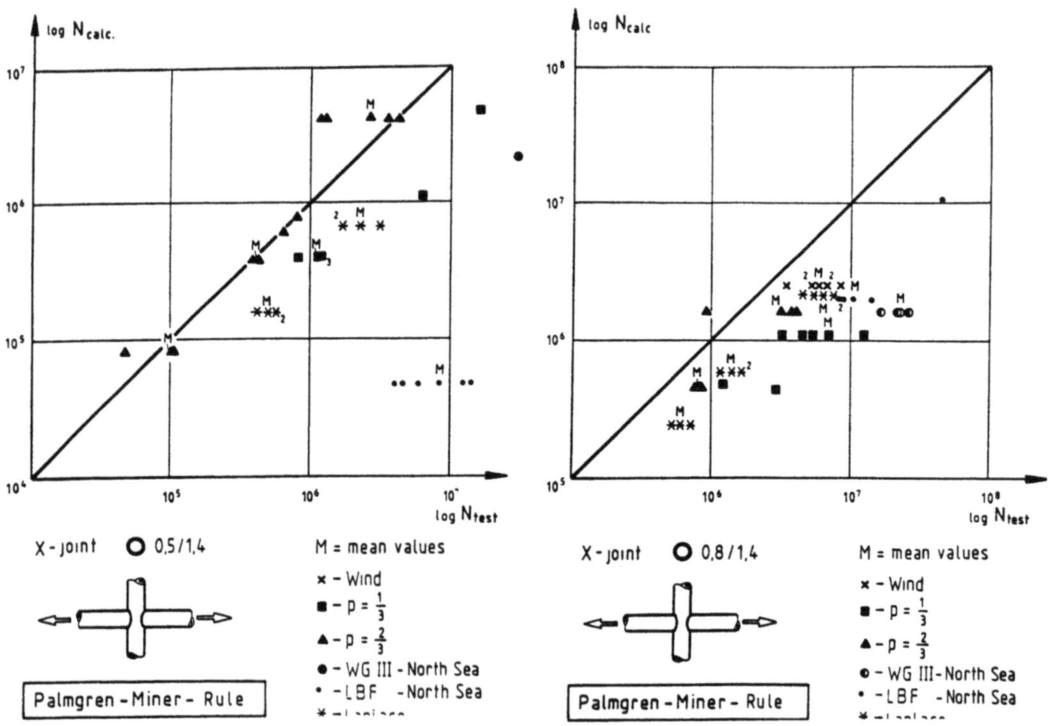

Fig. 12 Comparison of the load cycles (service life) calculated according to the Miner-Rule and determined in the test

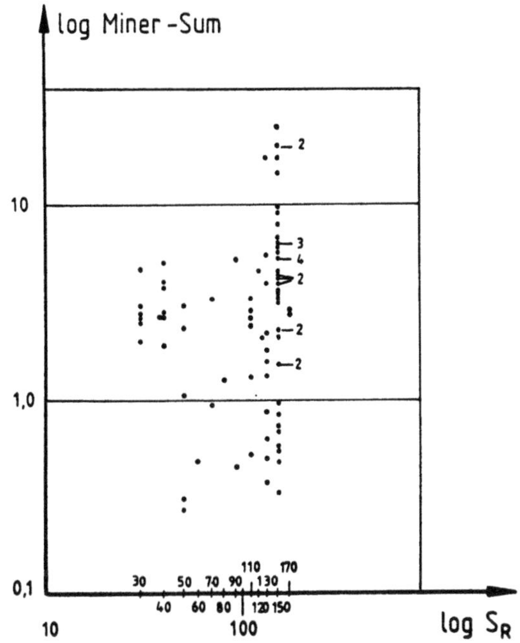

Fig. 13 Dispersion of the Miner-sum in dependence on the height of the stress range (upper load of the spectra)

The application of the relative Miner-Rule did not result in a good coincidence of the individual collectives with test results available. Basing on the present status of the tests, it can be said that the relative Miner-Rule is only valid, when the collectives to be compared have the same or a very similar fullness. This statement was confirmed by the tests conducted with the North Sea spectra and with the wind collective.

Figure 14 shows the results of some calculations according to the relative Miner-Rule in place of all types of specimens and ranges of calculation investigated.

REFERENCES

[1] N.N.: Eurocode 3, Design of Steel Structures, August 88

[2] Mang, F., Bucak, Ö. and Klingler, I.: Wöhlerlinien-Katalog für Hohlprofilverbindungen
Studiengesellschaft für Anwendungstechnik von Eisen und Stahl e.V., Düsseldorf, Juni 1987, Januar 1988

[3] Schütz and Zenner: Schadensakkumulationshypothesen zur Lebensdauervorhersage bei schwingender Beanspruchung
Zeitschrift für Werkstofftechnik 1973
Part 1, No. 1 p. 25-33
Part 2, No. 2 p. 97-102

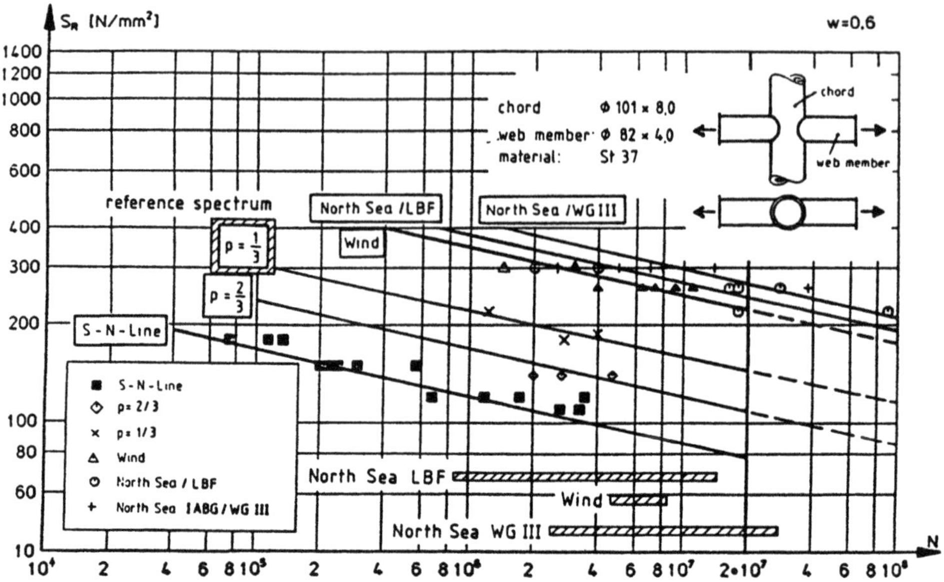

Fig. 14 Application of the relative Miner-Rule on the tests presented (graphical illustration of the results)

25 BEHAVIOUR OF STRUCTURES OF HIGH-STRENGTH STEELS

F. MANG and Ö. BUCAK
Testing Centre for Steel, Wood and Stone, University of Karlsruhe, Germany

1 Introduction

The German DASt-Ri 011 (February 1979) "Hochfeste, schweißgeeignete Feinkornbaustähle StE 460 und StE 690, Anwendung für Stahlbauten" considers the increased use of high-strength fine-grained structural steels suitable for welding. Meanwhile, their field of application covers the construction of tanks, pipelines, vehicles and cranes as well as steel buildings. This development is closely associated with the progresses in the steel manufacturing and welding techniques. Due to their low carbon content and their high strength against brittle fracture, owing to their fine-grained structure, these steels can be welded manually and automatically in all usual welding procedures. The economy of the high-strength steels is due to high allowable stresses compared to the conventional structural steels St 37 and St 52, which is especially advantageous for structural members under tensile stresses and structures under working stresses.

Another advantage of high-strength steels is that sheets and rolled sections of appropriate dimensions can be prefabricated in the rolling mill. The first series of rolled sections of wrought iron were fabricated between 1800 and 1820 in rail mills (Fig. 1).

The first steel structural sections were angle sections produced in 1831. In 1845 welded steel hollow sections were used for the first time and in 1849 the first H-sections (double-T-sections) were rolled.

2 Rolled sections of StE 460 under compressive stress

Large rolled sections of high-strength steel have proven good when used as compression members. For their classification into the buckling stress curves of Eurocode 3 experimental investigations on sections IPE 400 and HE 400B were carried out in Karlsruhe.

So far, rolled sections of high-strength steel with $h/b > 1.2$ had to be classified into the same buckling stress curves as for conventional steel (Fig. 2). For $h/b \leq 1.2$ the classification was better compared to conventional steel.

As a result of the tests carried out in Karlsruhe on large sections and in Lüttich on small sections, the classification of rolled sections of high-strength steel with $h/b > 1.2$ was improved.

Testing of Metals for Structures. Edited by F.M. Mazzolani. © 1992 RILEM.
Published by Taylor & Francis, 2 Park Square, Milton Park, Abingdon, Oxon, OX14 4RN.
ISBN 0 419 15810 3

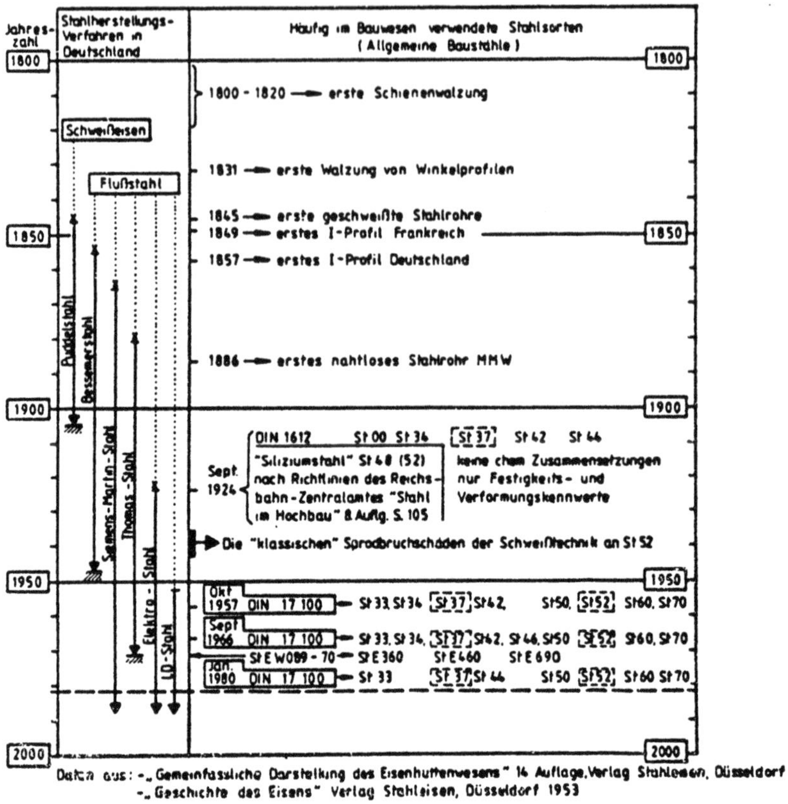

Fig. 1. Historical development of steel manufacture |1|

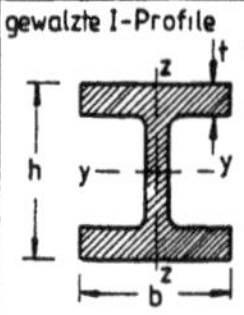

Querschnitt		Knicken rechtwinklig zur Achse	Knickspannungslinie
gewalzte I-Profile	$\frac{h}{b} > 1{,}2$, $t \leq 40$ mm	y - y	a
		z - z	b
	$\frac{h}{b} \leq 1{,}2$, $t \leq 40$ mm	y - y	b (a)
		z - z	c (b)
	$t > 40$ mm	y - y	d
		z - z	d

( ): Für die Stahlsorten St E 460 und St E 690 dürfen die in Klammern angegebenen, günstigeren Knickspannungslinien angenommen werden.

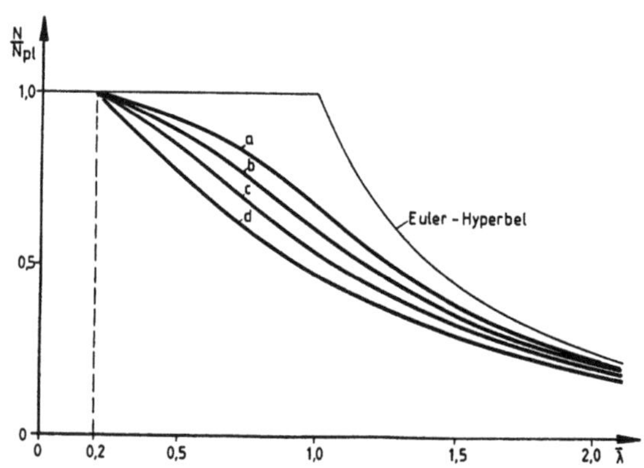

Fig. 2. Classification of rolled sections into the European buckling stress curves before the experimental investigations of Karlsruhe

Fig. 3 shows the valid curves. For high-strength steel the curves in brackets apply.

Fig. 4 shows a section HE 400B (l = 8.0 m) of StE 460 in the 50 MN power press of the "Versuchsanstalt für Stahl, Holz und Steine".

3 Hollow section structures of high-strength steel

3.1 General

The German standard DIN 18 808 "Stahlbauten, Tragwerke aus Hohlprofilen unter vorwiegend ruhender Beanspruchung", published in 1984, does not cover the steels StE 460 and StE 690. The standard includes design regulations for truss joints and knee joints, which consider the specific loadbearing behaviour of the hollow sections. The regulations are based on experimental and theoretical investigations carried out in various countries.

Querschnitt		Knicken rechtwinklig zur Achse	Knickspannungslinien
gewalzte I-Profile 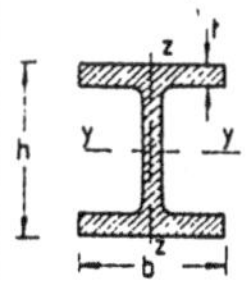	$\frac{h}{b} > 1{,}2$, $t \le 40$ mm	y - y z - z	a (a_0) b (a)
	$\frac{h}{b} > 1{,}2$, 40 mm < t ≤ 100 mm	y - y z - z	b (a) c (b)
	$\frac{h}{b} \le 1{,}2$, $t \le 100$ mm	y - y z - z	b (a) c (b)
	$\frac{h}{b} \le 1{,}2$, $t > 100$ mm	y - y z - z	d d

( ): Für die Stahlsorten St E 460 und St E 690 dürfen die in Klammern angegebenen, günstigeren Knickspannungslinien angenommen werden, jedoch mit t ≤ 40 mm.

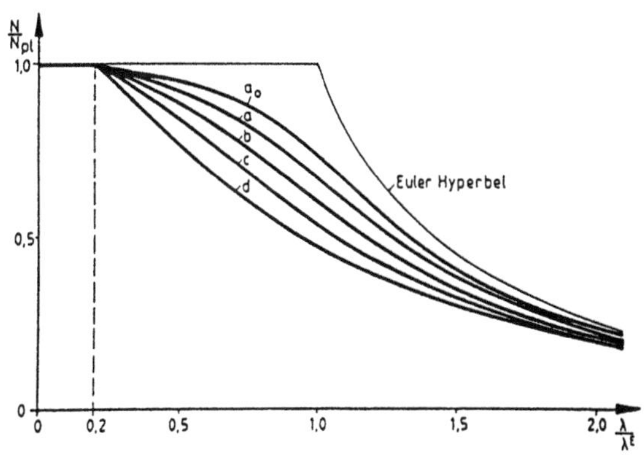

Fig. 3. Classification of rolled sections according to Eurocode 3, ed. Nov. 1979

Fig. 4. Rolled section HE 400B
of StE 460 after buckling

The latest edition of Eurocode 3 "Steel constructions" also does not include regulations for high-strength steel structures. This paper therefore reports on the results of experimental investigations on high-strength steels and compares them to the results of conventional steels.

Series of tests on sections of StE 460 and StE 690 carried out at the "Versuchsanstalt für Stahl, Holz und Steine" and in Japan in some cases resulted in a parallel displacement of the S-N curves, as shown in Fig. 5 for a hollow section knee joint with 100 % overlap.

As the number of test series for StE 460 and StE 690 was small and the results were inconsistent with regard to the course of the S-N curve in the high cycle range, compared to the results of small specimens, these steels have been excluded from the Eurocode 3 work so far.

Based upon the work for draft Eurocode 3 the stress range ($s_R = \Delta\sigma$) is taken as the basis of future diagrams. Fig. 6 shows that the idea of using $\Delta\sigma$ for welded constructions is justified.

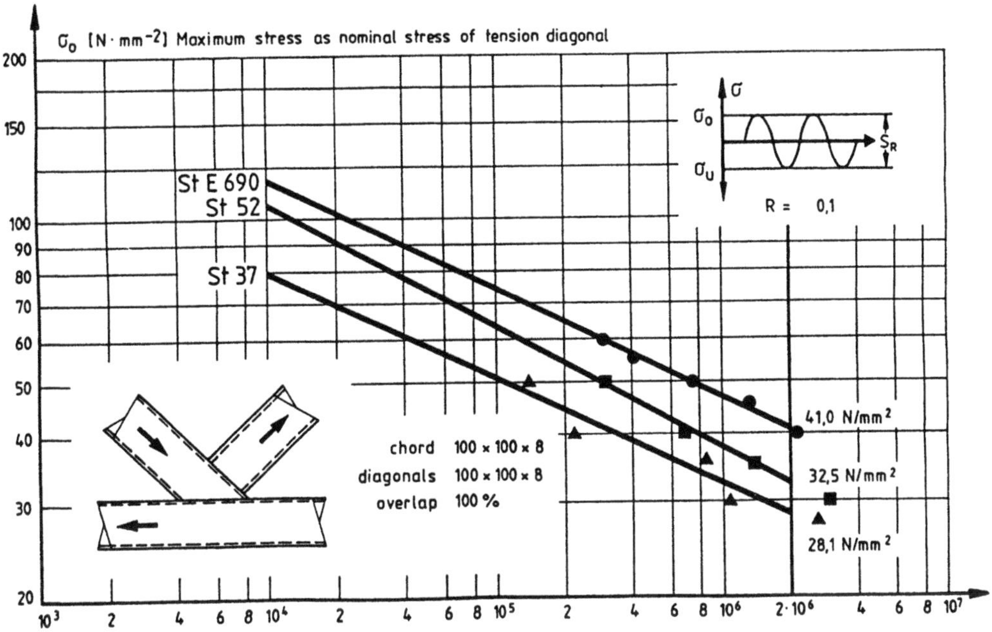

Fig. 5. Tests on joints with equal dimensions and varied steel grades

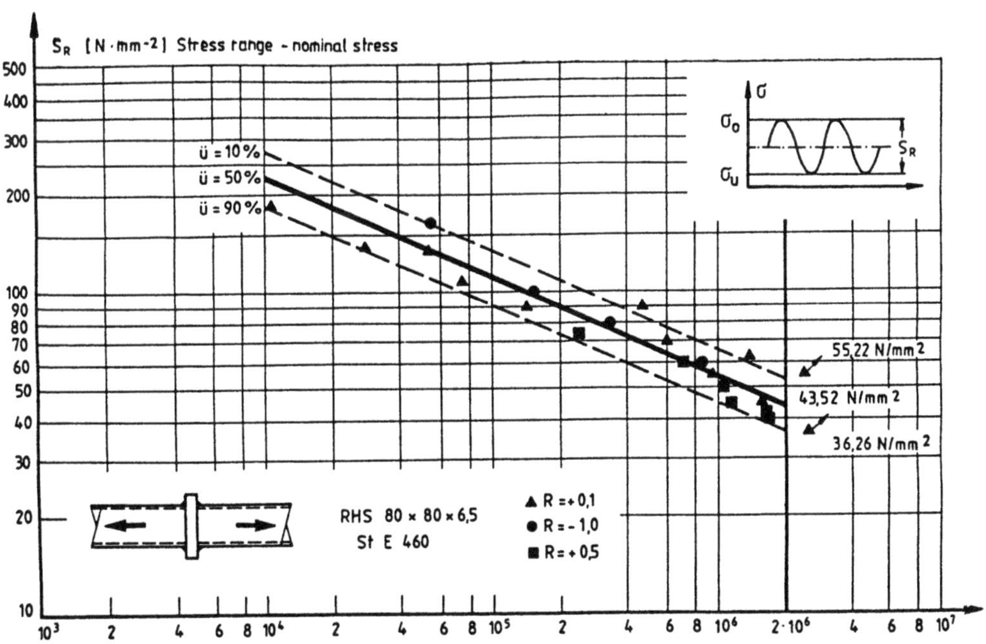

Fig. 6. The $\Delta\sigma$ concept applied to test specimens with fillet welds

3.2 Knee joints

Knee joints of rectangular hollow sections can be designed in various ways. Two of the most usual types are shown in Fig. 7. Fig. 7a shows a rectangular unstiffened knee joint with mitre cut, which is used most frequently for architectonic reasons. In the case of thin-walled sections (large ratio B/T) the loadbearing capacity can be considerably increased by means of stiffening plates (see Fig. 7b).

The following applies analogously to knee joints with $\Theta > 90°$ since in that case the geometric ratios are more advantageous.

The hollow sections of the knee joints were cut on mitre at an angle of 45°; the weld preparation was made manually by means of a grinding machine to obtain perfect fit. Thus, the right angle (Θ = 90°) was ensured in all cases.

The edges of the specimens without stiffening plate were prepared as shown in Fig. 8. All welds are butt welds, except the weld at the lower edge of the hollow section (compression flange). For reasons of manufacturing technique this part was fillet-welded. The specimens with stiffening plate were connected by means of circumferential fillet welds. The dimension a of the fillet welds corresponded to the wall thickness of the connected sections (see Fig. 8).

Metallographic sections for hardness measurements (HV 10) were cut out of different weld areas of the specimens. The results are shown in Fig. 9 for specimens without stiffening plate and in Fig. 10 for specimens with stiffening plate. The peak values ranged up to 330 HV 10 proving realistic production conditions of the test pieces.

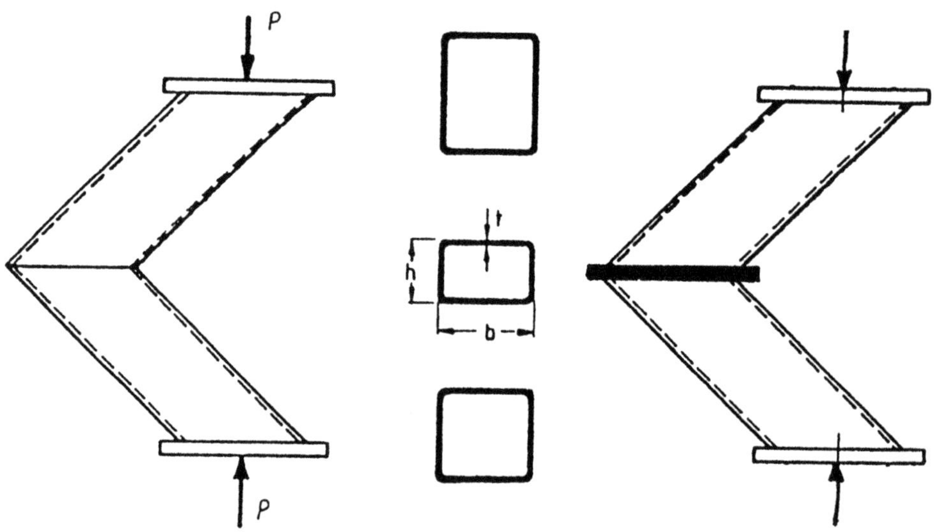

Fig. 7

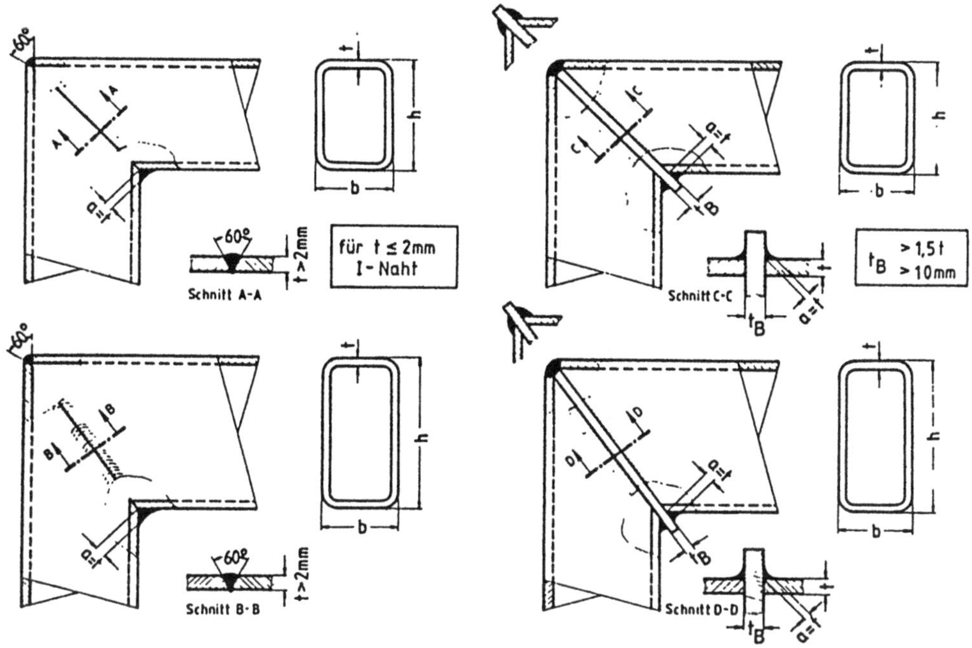

Fig. 8

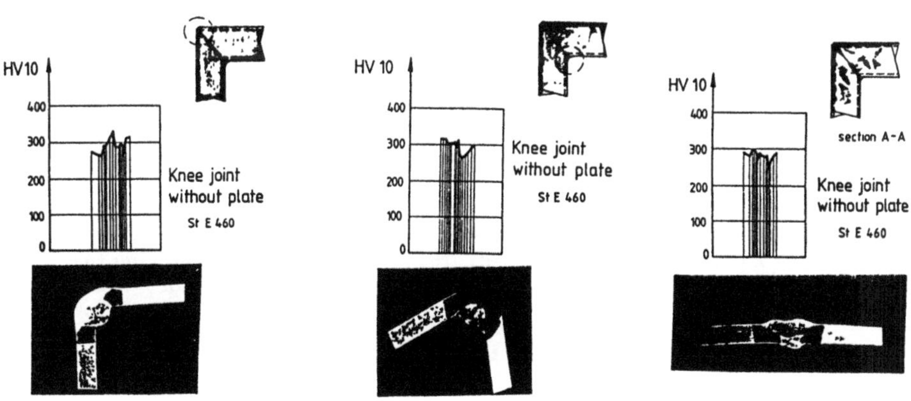

Fig. 9. Metallographic section and hardness distribution on parts of specimens without stiffening plate

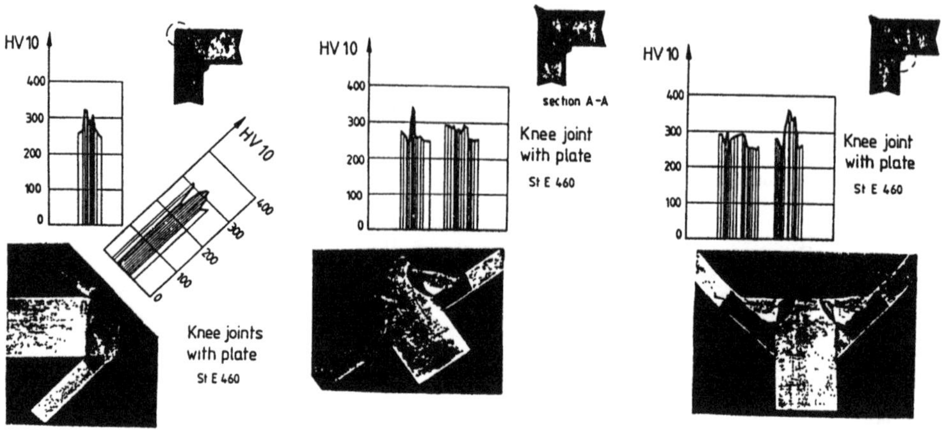

Fig. 10. Metallographic section and hardness distribution on parts of specimens with stiffening plate

Design proposal

The test results and comparative theoretical studies show that the design proposal made in DIN 18 808 for the conventional steels St 37 and St 52 also applies to the high-strength steels. Thus the loadbearing capacity of the types shown in Fig. 7 shall be verified as follows

$$\sigma = \frac{M}{W_{el}} + \frac{N}{A} \leq \alpha \cdot \sigma_{zul}$$

where

- M = bending moment
- N = normal force
- W_{el} = elastic section modulus
- A = cross-sectional area
- σ_{zul} = allowable stress
- α = shape factor ≤ 1 for the unstiffened knee joint

as given in the diagrams of Figs. 11 and 12.

With the mentioned equations and the shape factor α the stresses and the stiffness of the knee joint are verified. This, however, does not consider the loadbearing capacity of the weld. For this purpose, the factor α of unstiffened knee joints is limited to a maximum value based on the ratio of

$$\frac{\text{allowable weld stress}}{\text{allowable member stress}}$$

(according to DASt-Ri 011, ed. Feb. 1979) as shown in table 1.

For knee joints with stiffening plate the welds have to be verified according to DIN 18 800 or DASt-Ri 011, ed. 1979.

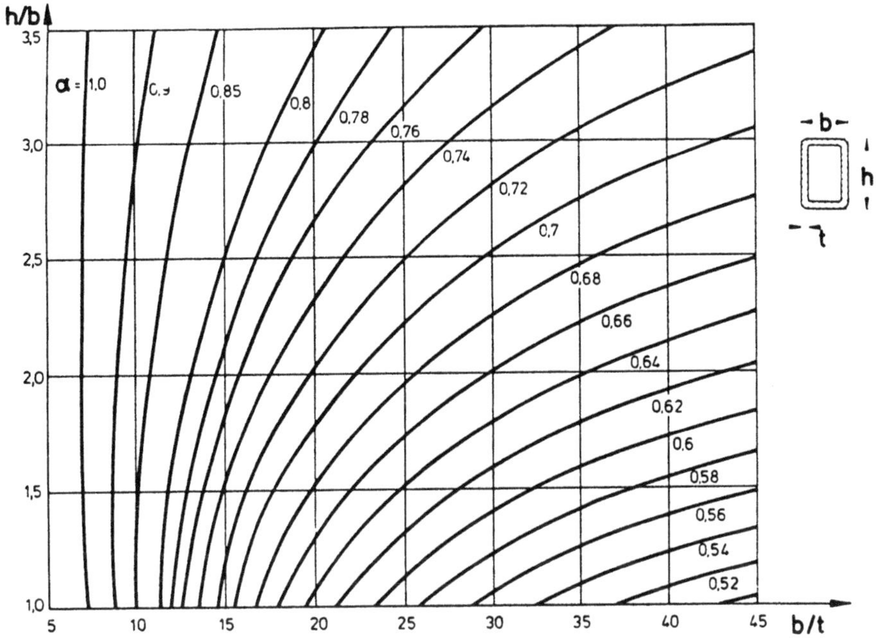

Fig. 11 Formfactors α for unreinforced Knee joints

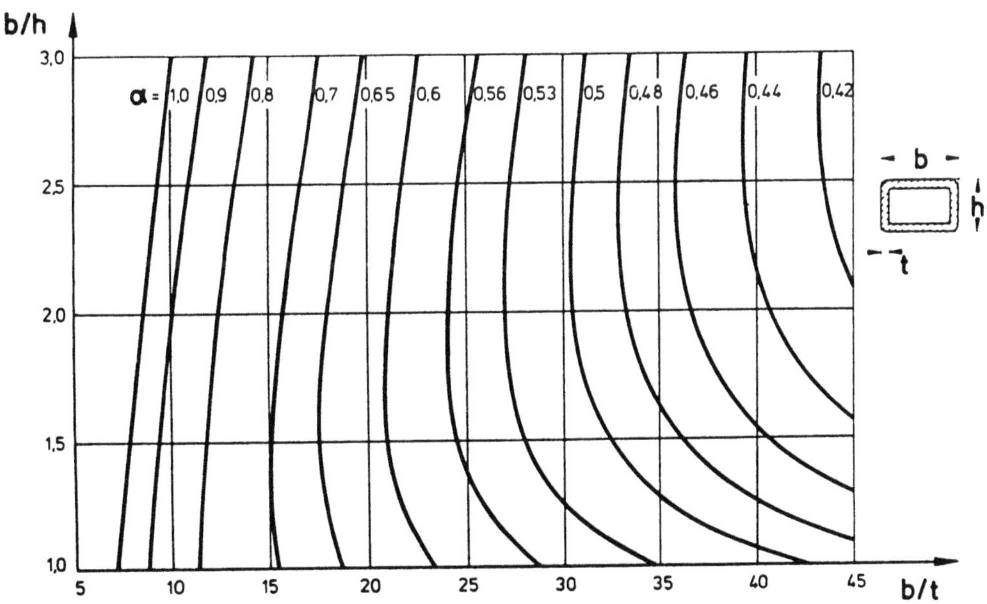

Fig. 12 Formfactors α for unreinforced Knee joints

Table 1: Maximum values of α, if the weld is not verified seperately

material	StE 460	StE 690
shape factor α	0.71	0.59

The plate wall thickness shall be

$$t_B \geq \begin{cases} 1.5\ t \\ 10\ \text{mm} \end{cases}$$

Fig. 13 shows the good agreement between the reduction factors determined on the basis of the test results for specimens of StE 460 and StE 690 and those given in DIN 18 808.

3.3 T-joints of rectangular hollow sections

The main problems which occur with T-joints are the irregular stress distribution (see Fig. 14), which causes reduced ultimate loads, and the reduced joint stiffness, which results from the elastic deformability of the chord flange.

|6| gives design diagrams of the utilization factors μ of the bracing members of T-joints.

In the case of moment load the factor is

$$\mu_M = \frac{\gamma \cdot \text{actual } M}{M_{pl}}$$

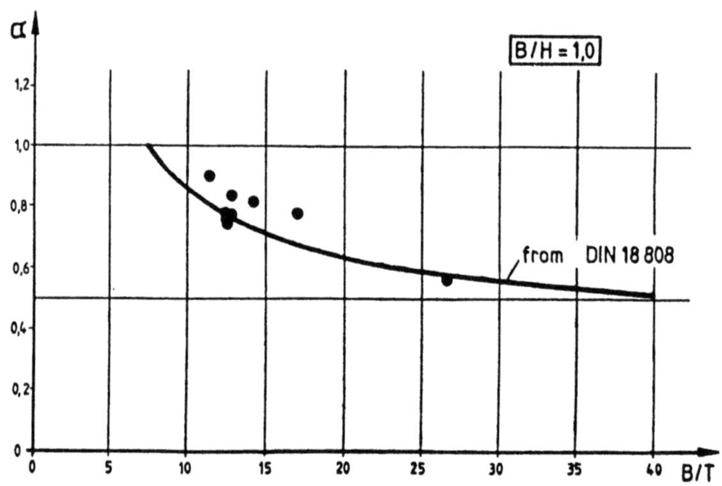

Fig. 13 Shape factors α of unstiffened RHS Knee Joints

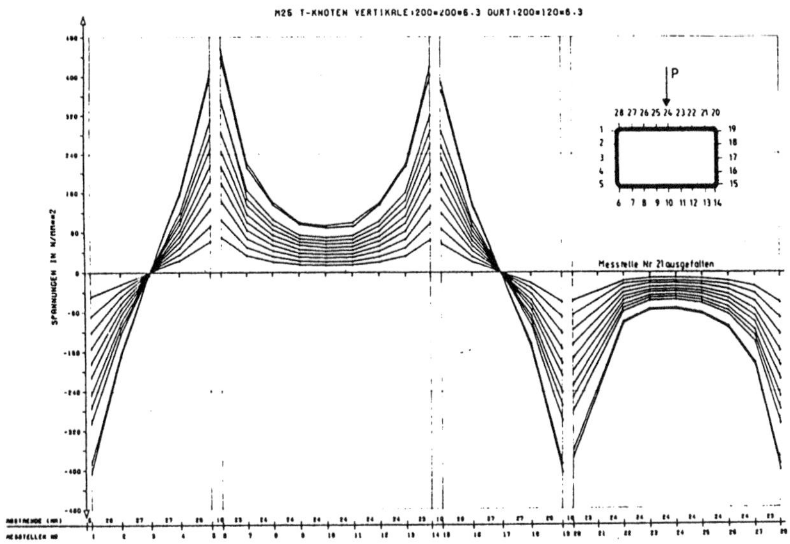

Fig. 14. Stress distribution on a T-joint

and in the case of normal force

$$\mu_N = \frac{\gamma \cdot \text{actual } N}{N_{pl}}$$

Fig. 15 shows the utilization factors for T-joints of StE 460 and StE 690 determined from the test data. This shows that the diagrams also apply to high-strength steels.

Fig. 16 shows the ratio of plastic and elastic joint torsion of T-joints of high-strength steel. For comparison, the diagram also includes the curve determined for conventional steel.

Fig. 17 shows the joint coefficient C* against B/T for T-joints with width ratio b/B = 1.0. The diagram also includes the test results of joints of high-strength steel StE 460 and StE 690.

Metallographic sections for hardness measurements were cut out of different weld areas of the specimens. The results are given in Fig. 18. The peak hardnesses ranged up to 310 HV 10.

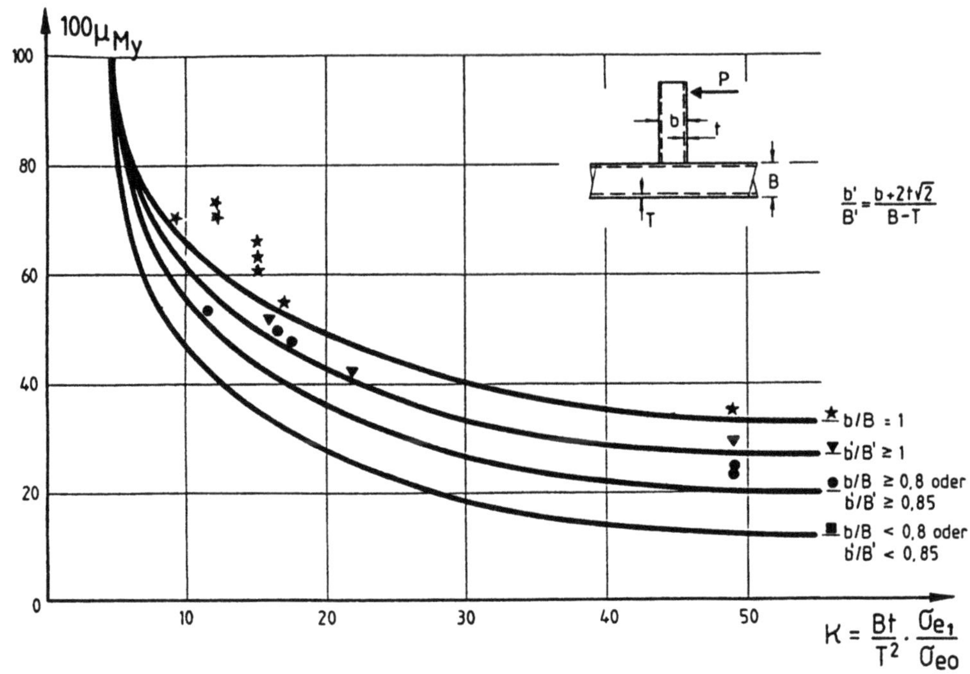

Fig. 15. Design diagram for T-joints under moment load with test data for T-joints of high-strength steel StE 460 and StE 690

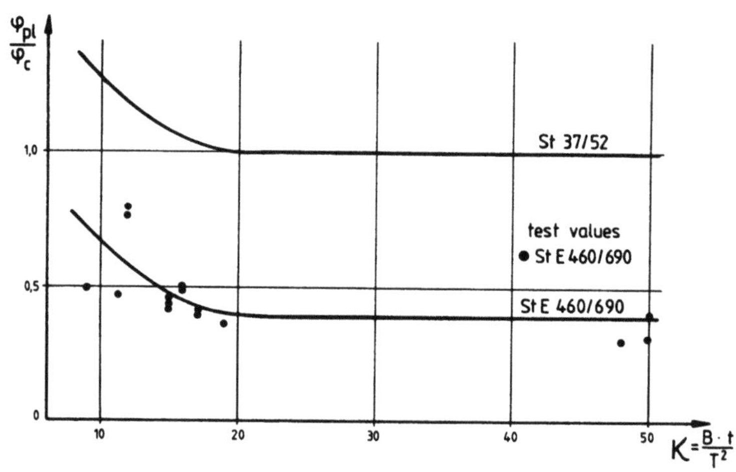

Fig. 16 Ratio of plastic and elastic joint torsion

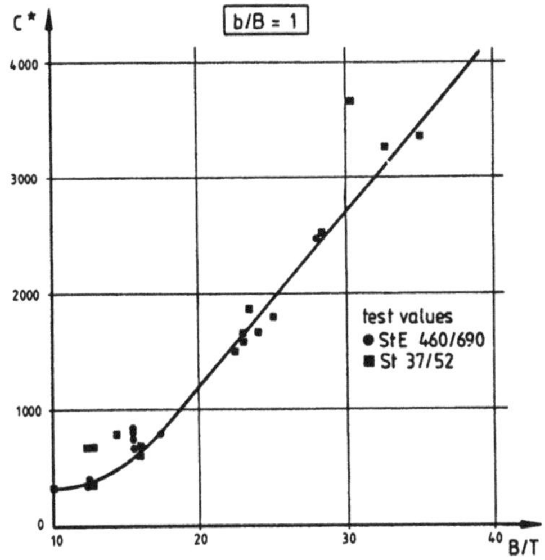

Fig. 17 Joint coefficient C* against B/T for b/B=1

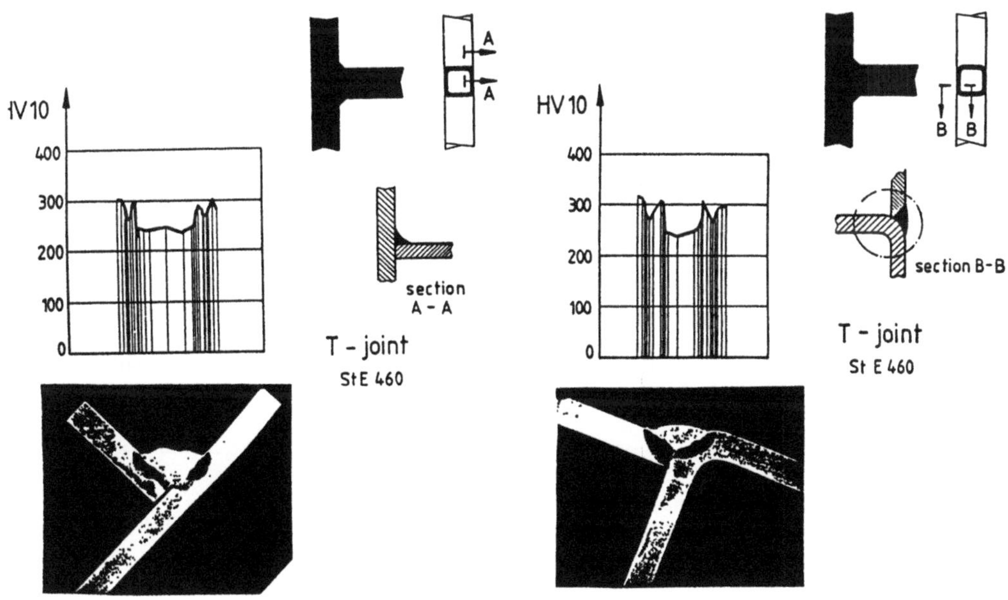

Fig. 18. Metallographic sections and hardness distribution on parts of the specimens (T-joints)

4 Rolled sections with transverse fillet welds (stiffeners)

As fillet weldings on structures under fatigue loading can induce serious notches, the design of stiffeners and connections is decisive of the utilization of the loadbearing capacity. The stiffeners are particularly important with slender sections of high-strength steel. The notch cases can be described by means of design variants, as shown in Fig. 19.

We differentiate between:

- stiffeners ending outside the area of tensile stresses (type I)
- stiffeners with cutouts at the neck (type II)
- stiffeners completely welded to the beam (type III)

With type IIIa the stiffening plates are adjusted to the outline of the rolled section. With type IIIb the stiffening plates are bevelled and filled up with weld metal.

The research program financed by DVS/AIF (German Government) included experimental investigations and finite element calculations. The first results plotted in S-N curves (Figs. 20 and 21) show that

- there is no serious difference between the results of the various designs, so that the "serious" notches (type II) do not have to be degraded,
- the classification of Eurocode 3 is far on the safe side,
- with the investigated wall thicknesses, stress-relieving annealing may bring further advantages.

Parallel tests were carried out on sections of high-strength steel StE 460. The results are shown in Fig. 22. It is shown that the fatigue behaviour is a little better in the range of high stress cycles.

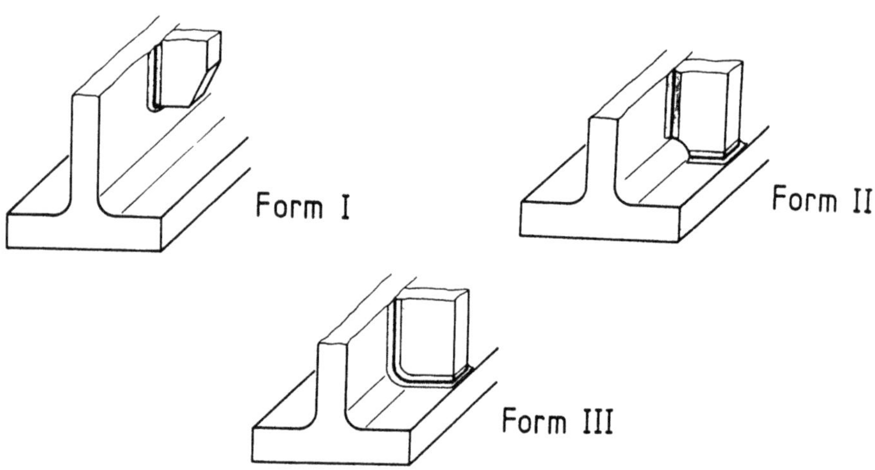

Fig. 19. Design variant of stiffeners of rolled beams

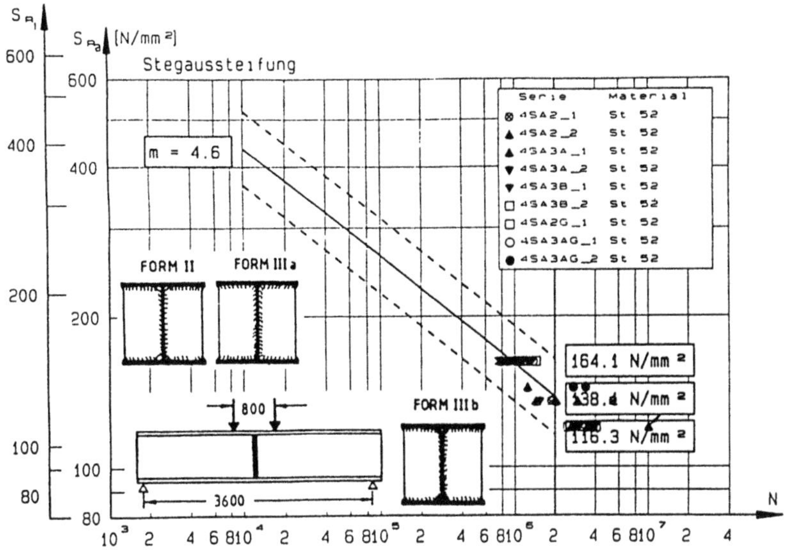

Fig. 20. Test results of 4-point-bending tests on rolled specimens with stiffener

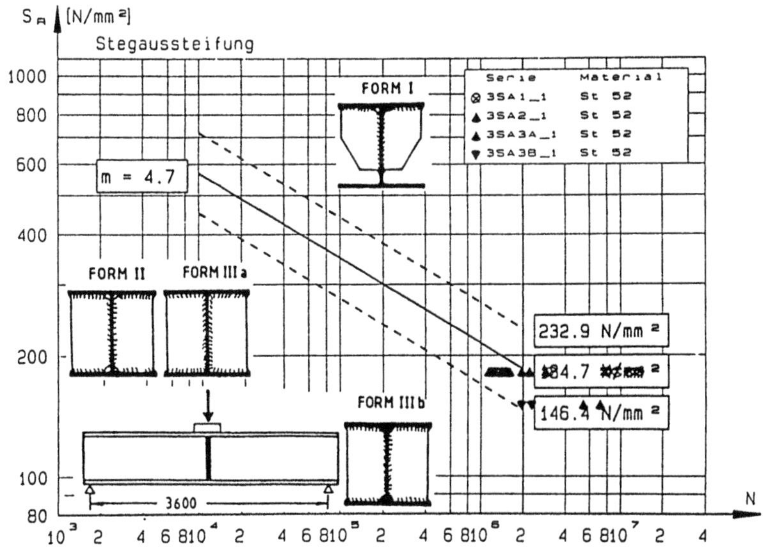

Fig. 21. Test results of 3-point-bending tests on rolled specimens with stiffener

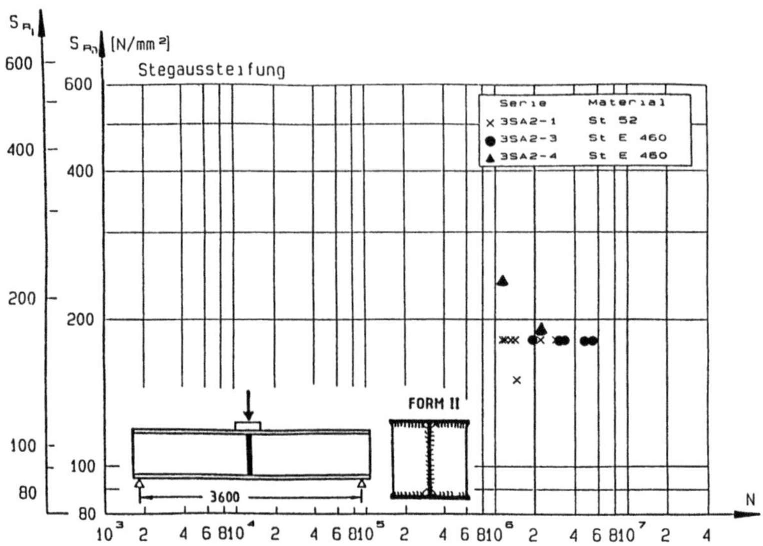

Fig. 22

Literature

|1| Eurocode 3 — "Design of Steel Structures" Part 1 - 9
Nov. 1989
Feb. 1990
Sept. 1989 Eurocode 3, Annex

|2| DASt-Ri 011: Hochfeste, schweißgeeignete Feinkornbaustähle StE 460 und StE 690, Anwendung für Stahlbauten
Stahlbau Verlag Köln, Feb. 1979

|3| DIN 18 808: Stahlbauten, Tragwerke aus Hohlprofilen unter vorwiegend ruhender Beanspruchung (Steel structures of hollow sections under predominantly static load).
Beuth Verlag, Berlin, Okt. 1984.

|4| DIN 18 800, T1: Stahlbauten; Bemessung und Konstruktion (Steel structures; design and construction), März 1988

|5| DIN 18 800, T2: Stahlbauten; Stabilitätsfälle, Knicken von Stäben und Stabwerken (Steel structures; stability, buckling of bars and skeletal structures), März 1988

|6| Mang, F., Bucak, Ö.: Hohlprofilkonstruktionen. Stahlbau-Handbuch, Bd. 1, Kap. 12. Stahlbau-Verlags GmbH, Köln, 1982.

26 REMAINING FATIGUE LIFE OF OLD STEEL BRIDGES

F. MANG and Ö. BUCAK
Testing Centre for Steel, Wood and Stone, University of Karlsruhe, Germany

SUMMARY

This paper describes an investigation of a complete bridge structure of the "museum railway" in the community of Blumberg as well as two bridge structures of the "Deutsche Bundesbahn". Results of component tests are compared with data taken from the literature as well as other results of similar investigations.

1 PRELIMINARY REMARKS

All over the world the industrialization starting early in the 20th century paved the way for numerous constructions, particularly for steel railway bridges. Still today, we find bridges dating from the end of the last century, i.e. about 1850. These bridges are mainly riveted constructions designed for low external loads. The standard service lives of these constructions were short compared to current designs. In many cases, however, they have been exceeded by far. All over Europe we find bridges which have been continuously in service for more than 100 years. Most of these buildings are classified as historical monuments or shall be preserved for economical reasons.

In most cases it is impossible to make a definitive judgement as to whether a construction under fatigue loading can be used further after its standard service life has been reached or exceeded. Further problems occur when the safety of an old bridge has to be recalculated due to increased design loads (e.g. new load spectra for trains) or when the safety of a building damaged in an accident has to be estimated. This can be explained by the poor knowledge of the static strength and the fatigue strength of steels (wrought iron, puddled iron) and constructional details (X lattice girders, rivet joints) used in the 19th century.

The oldest railway bridge, the Koblenz/Waldshut between the Federal Republic of Germany and Switzerland, which is in the charge of the Versuchsanstalt für Stahl, Holz und Steine, dates from 1858 and is still in operation today.
This bridge is shown in Figs 1 and 2.

Extensive measurements under traffic load and investigations on the material allow the operation of this bridge to continue–at least for the next 20 years. Presently the gap in single elements due to rust formation (Fig. 3) seems to

Testing of Metals for Structures. Edited by F.M. Mazzolani.
Published by Taylor & Francis, 2 Park Square, Milton Park, Abingdon, Oxon, OX14 4RN.
ISBN 0 419 15810 3

Figure 1 General view of Koblenz/Waldshut railway bridge.

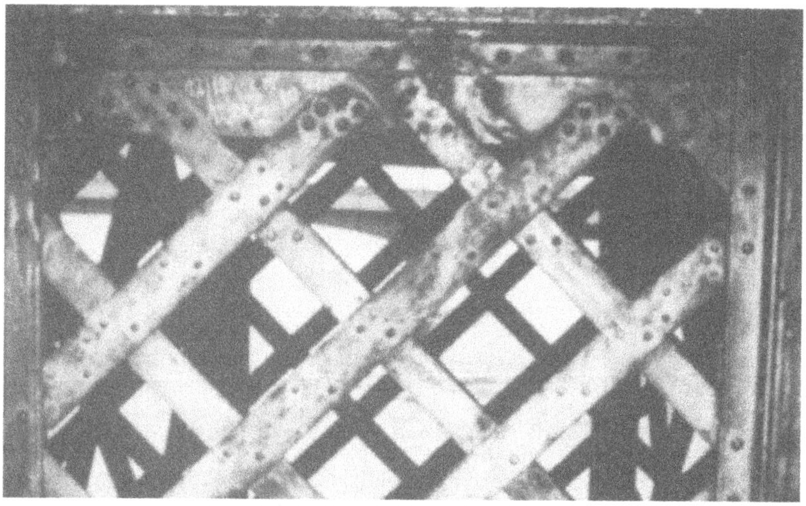

Figure 2 Detail of the Koblenz Waldshut railway bridge.

be a problem. After complete removal of the rust these areas are filled with a flexible lute (Fig. 4). The original values of the forces caused by the rust formation are being investigated and will be reported on later.

The judgement of old bridges is based on the results of material tests on specimens taken from areas under low loading conditions, such as bracings and support areas. Up to the present, these data have been reduced by high safety factors to obtain a safe estimation. The structural design of the connections has a considerable influence on the load-carrying capacity of the

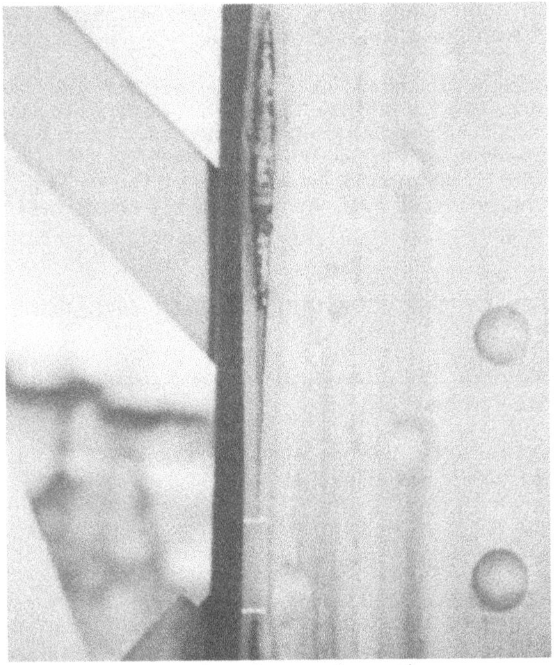

Figure 3 Local rust formation

Figure 4 Flexible lute

whole construction. Till today, tests on whole constructions and their connections have been carried out only sporadically.

"Damage" or "fatigue", incurred during the service life, are determined by means of hypotheses the reliability of which is not quite indisputed. Data on the load history of bridges built during the two world wars and in early post-war times are not available. The judgement is therefore based on mean values and estimated load histories. In order to be on the safe side, we try today to obtain the required data by tests on all structural components under high loading conditions.

2 DETERMINATION OF THE REMAINING SERVICE LIFE OF A FATIGUE-LOADED CONSTRUCTION

Up to a short time ago, the procedure to determine the remaining service life of a bridge was as follows:

- visual inspection of the bridge, outward appearance (corrosion), with some bridges this was the dominating point of judgement

- specification of repairs or reinforcement measures (e.g. due to increased loads from engine and cars)

- determination of points under maximum stress (by means of the design calculation or by new calculation e.g. according to BE 804); determination of points governing the safety of the construction (e.g. tensile members) for highway bridges see DIN 1072 and DIN 18809 taking account of impact loads and crash barrier spacings

- taking of specimens from uncritical, but representative areas of the construction

- tests, e.g. chemical analysis, tension test and fatigue tests, on the original material

- additional calculations based on actual material properties

- determination of the previous load history and future loads

- determination of the remaining service life using Miner's linear cumulative damage rule and fixing of safety factor for the respective bridge construction

- fixing of inspection intervals

- fracture-mechanical tests, e.g. COD tests, determination of the critical crack size, previous damages.

Defective structural components (damaged e.g. by corrosion, by accident (see Fig. 5) or by bullets) may cause a considerable reduction of the life of a construction depending on the situation of the defect and on the stress level.

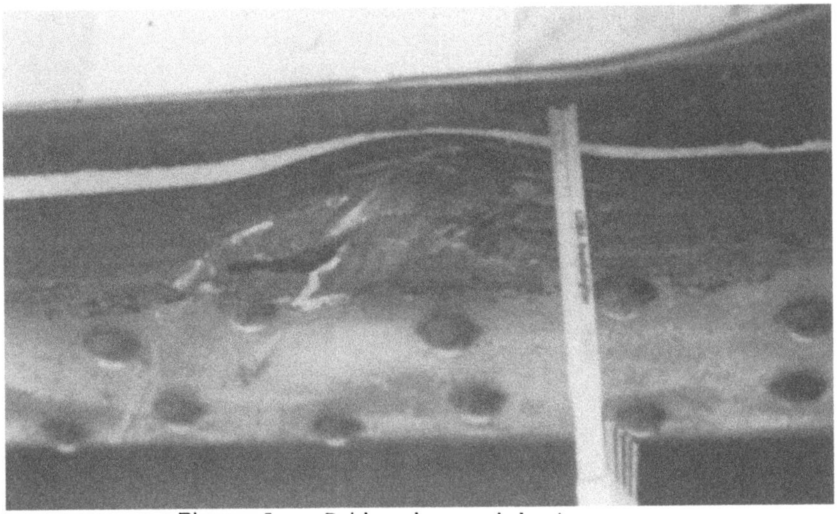

Figure 5 Bridge damaged by impact

We therefore recommend to attach special importance to these points when visiting a construction

Certificates on various constructions under high-cycle fatigue loading and tests carried out for a research program [1] supplied sufficient data on the material properties thus allowing a preliminary estimation. Therefore, only few tests in the final stage are necessary to confirm these presumptions. The load-carrying behaviour of such constructions as a whole and the load-carrying capacity of structural components in the original state are not yet sufficiently known. The points in question are: rivet slip, rivet preloading, displacement under shear load, distribution of the load to other elements of a structural component. Large structural components should therefore be investigated as a whole.

At the "Versuchsanstalt für Stahl, Holz und Steine" tests were carried out on a whole bridge built in 1887 which had been chosen from different bridge systems of the "Museumsbahn" (Fig. 6.)

3 FATIGUE TESTS AND RESULTS

3.1 Tests on the whole bridge

The whole bridge was built in the 50 MN testing machine of the institute. Figure 7 shows the bridge built in according to load configuration LF 1.

The test was interrupted at 108 070 stress cycles as a correct load application was no longer possible due to a large crack in the connection between transverse girder and main girder. Figure 8 shows the point of failure.

Figure 6 Bridge of the "Museumsbahn"

Figure 7 Bridge of the "Museumsbahn" in the 50 MN testing machine

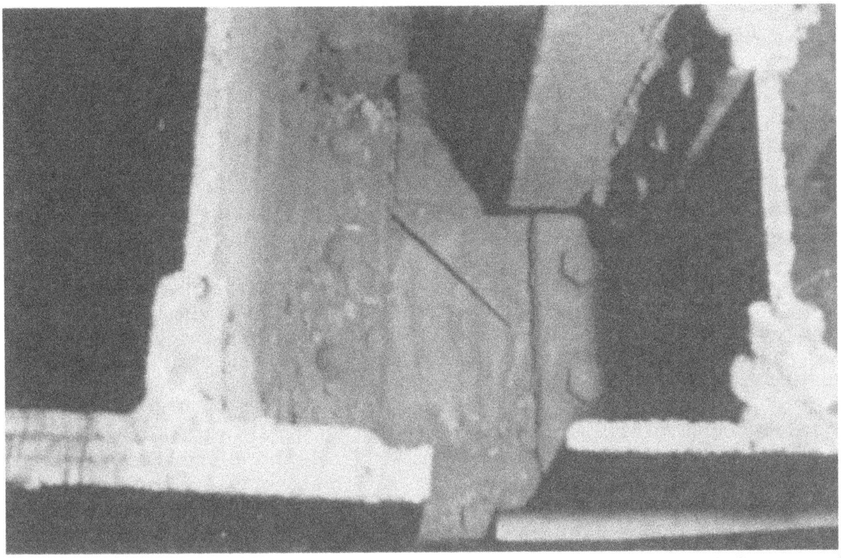

Figure 8 Connection between transverse girder and main girder, crack in web plate

Afterwards, the bridge was sawed up (see Fig. 9) to carry out separate tests on components (main girder, longitudinal girder) and connections (transverse girder, main girder).

Analogously, the Stahringen bridge dating from 1895 was sawed up to investigate the two main girders and the connections between main girder and transverse girder. Figure 10 shows the components of one half of the bridge.

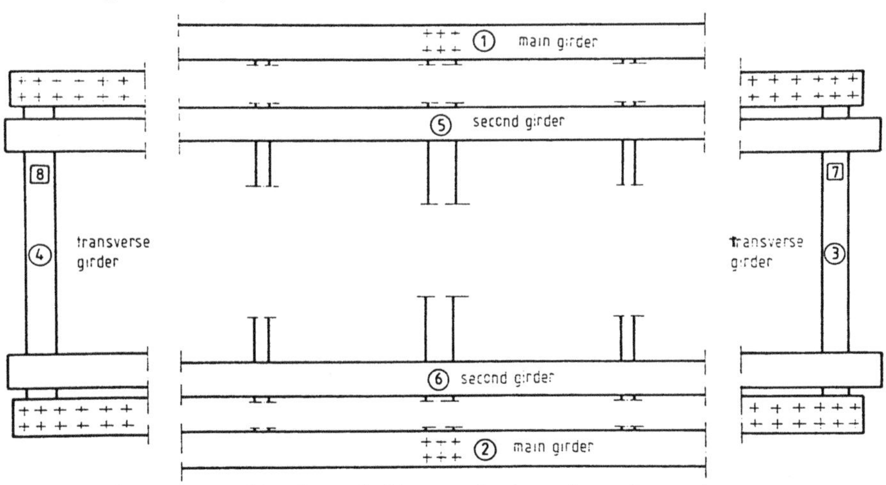

Figure 9 Blumberg bridge - designation of specimens

Punched specimens, riveted specimens as well as specimens of the plain web plate material were taken from the remaining bridge parts.

3.2 Tests on the main girder

The main girders of the two bridges were subjected to four-point-bending tests with a span of 3300 mm. The spacing between the two points of load application was 800 mm. The tests were carried out on two stress levels (135 N/mm^2 and 165 N/mm^2) with an ultimate stress ratio of R = + 0.2.

In these cases, the fatigue crack usually started from a rivet hole in the tension chord due to a stress concentration in the hole rim. The punching and riveting procedures as well as corrosion may also cause damages in the hole rim. An exception to this rule are those cases where components have been seriously damaged e.g. by impact or by bomb splinters.

Under a maximum stress of 165 N/mm^2 and at 1 534 000 stress cycles, the crack occurred in the web of main girder I of the Blumberg bridge. Despite the crack, the main girder carried the full load and the test was continued until the two angles and the cover plate of the main girder tensile chord failed at a total number of stress cycles of 1 572 600. This means that the minimum number of stress cycles between crack appearance and total failure of the main girder was 38 600 (see Fig. 11).

Fracture first occurred in the web, then in the angle section and finally in the cover plate. The cracks were not situated on one level.

Figure 10 Stahringen bridge – structural components

Figure 11 Points of fracture of main girder I

Subsequently, the main girder II of the Museumsbahn bridge and the two main girders of the Stahringen bridge was tested under the same conditions. The first crack occurred in the main girder web. The numbers of stress cycles between crack appearance and total failure of the main girder ranged up to 100 000.

Three additional specimens were taken from the uncracked main girder parts of the Blumberg bridge. These specimens were subjected to three-point-bending tests with a span of 1500 mm. The maximum loads were 450 kN and 600 kN, the ultimate stress ratios were R = + 0.35 and R = + 0.1. These test results will help in judging the influence of the mean stress.

Figure 12 shows the test results in comparison with earlier data obtained by tests on punched specimens of puddled iron. The diagram shows that the results of the main girder and transverse girder tests are within the range of scattering of results for small-scale specimens.

The results of the specimens taken from the remaining parts of the main girder are slightly beyond the P_u = 97.5 % line which is due to previous fatigue loads endured during service and during the girder tests (1.6 to $2.0 \cdot 10^6$ stress cycles).

3.3 Tests on transverse girder connections

In the tests on the whole bridge, failure occurred at the transverse girder connections. These connections were therefore a point of major consideration in the investigation of the service life of bridges. Full details of these investigations are given in [10].

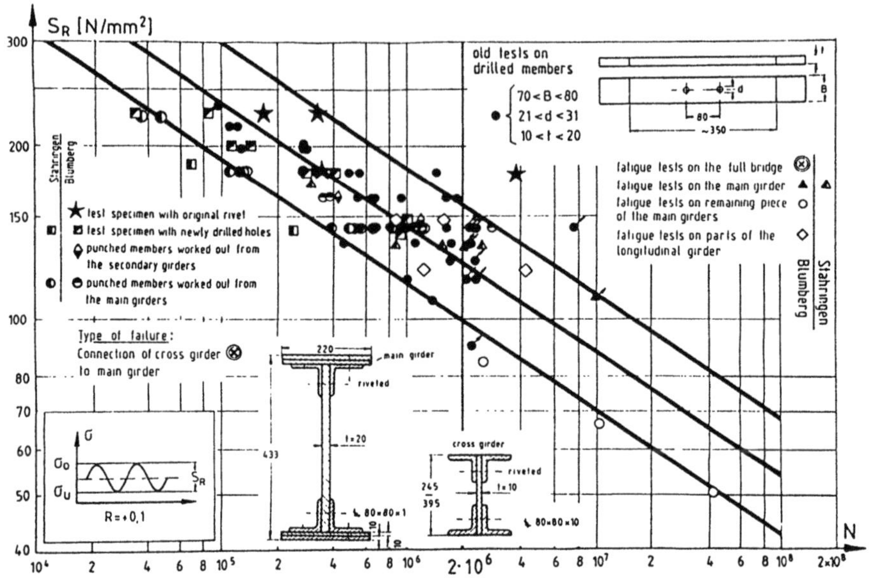

Figure 12 Results of tests on components of the Blumberg bridge compared to earlier test results of punched specimens

3.4 Tests on part of the longitudinal girders

Four specimens were taken from the longitudinal girders of the Museumsbahn bridge and subjected to three-point-bending tests with a span of 1500 mm. For two specimens the maximum stress was 165 N/mm², for the other ones 135 N/mm². The ultimate stress ratio was R = + 0.1. The test results are shown in figure 12 in comparison with earlier data for flat plate specimens of similar age. The cracks usually started at a distance of about 600 mm from the point of support and proceeded diagonally through the web. In some specimens further cracks occurred in the angle sections. Figure 13 gives an example of the failure mode.

3.5 Tests on riveted and punched specimens

The central parts of wind bracings of the bridge were fatigue tested in the original state (with rivets) in a pulsator.

The remaining material for the bracings was used for punched specimens, i.e. specimens with new holes ⌀ 20 mm, which were fatigue tested at different stress levels.

Further specimens with original holes were taken from remaining parts of main girders and longitudinal girders. The results of these tests are plotted in figure 14.

Figure 13 Failure mode for the longitudinal girder of the Blumberg bridge

It is evident that the test results are in good agreement with earlier data, although the data of the Stahringen bridge are comparatively low. The scattering of the results is comparatively large, as it is the rule with old constructions.

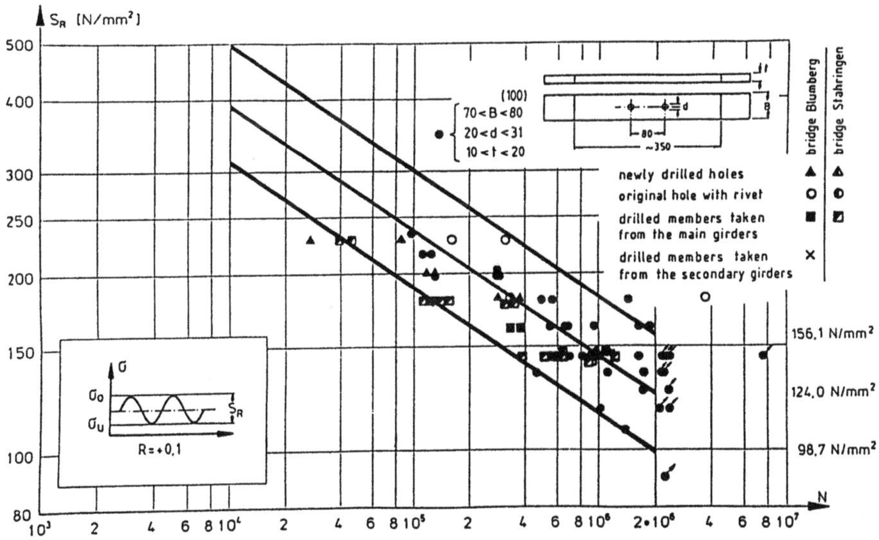

Figure 14 Results of tests on punched specimens and specimens with original rivets

Two specimens taken from the web area are also in good agreement with earlier test data. As expected, the fatigue cracks started from the rivet holes. Figure 15 shows the specimens taken from the main girder of the Stahringen bridge after failure. After the first crack had occurred, the specimens were refixed until a second or third crack started from another rivet hole.

Figure 15 Specimen after failure (punched specimen from Stahringen bridge)

4 COMPARATIVE TEST DATA FROM THE LITERATURE

As mentioned before, information on the fatigue behaviour of old steels, particularly from the 19th century, is very scarce. Hirt and Brühwiler [3] were the first researchers to carry out a methodical evaluation for two old bridge components. They compared their results to those of three other reports. For comparison, the test results of the Blumberg bridge and the Stahringen bridge were entered into the diagrams of figs. 16 and 17.

In figs. 18 and 19, our test results are compared to data for plain specimens dating from 1880 [3]. These diagrams are completed by data for punched and

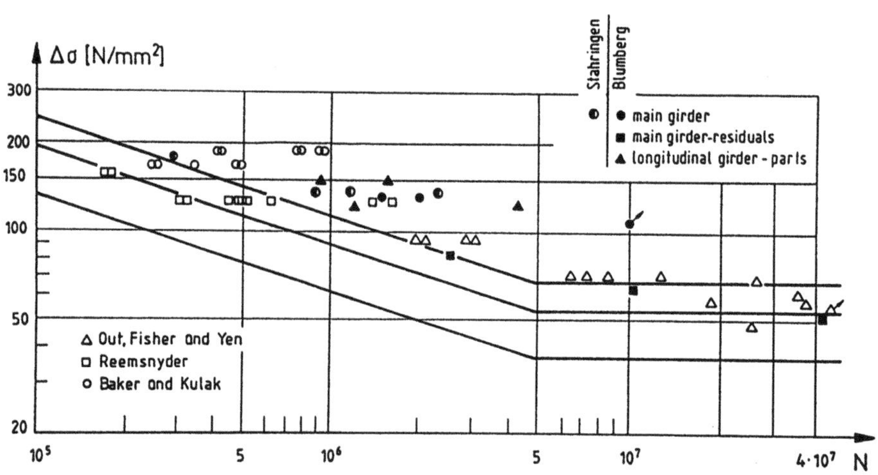

Figure 16 Test results and data from [3]

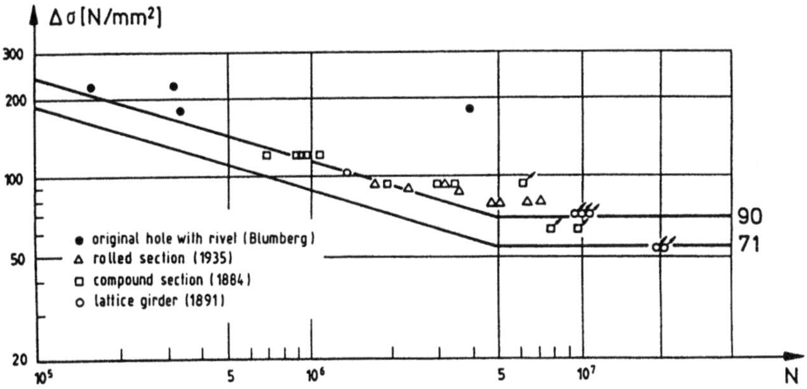

Figure 17 Test results and data from [3]

riveted specimens and for original structural components. Figure 20 includes results of tests on structural components of the Blumberg bridge.

The annual of SFB 315 (special research program) [11] includes further comparisons particularly with the data of the draft standard DS 805 "Bewertung der Tragfähigkeit bestehender Eisenbahnbrücken" from July 1987.

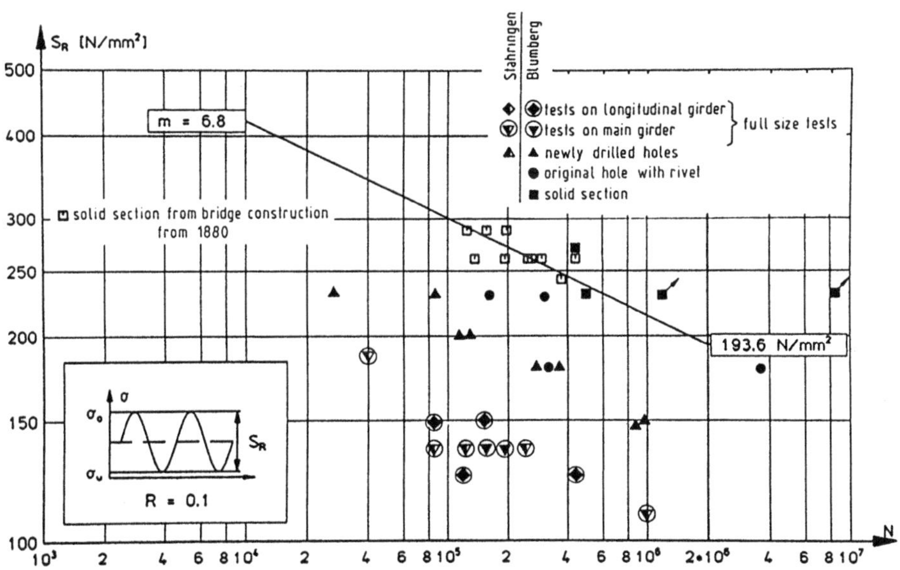

Figure 18 Test results and data from [5], bridges dating from 1880, plain material

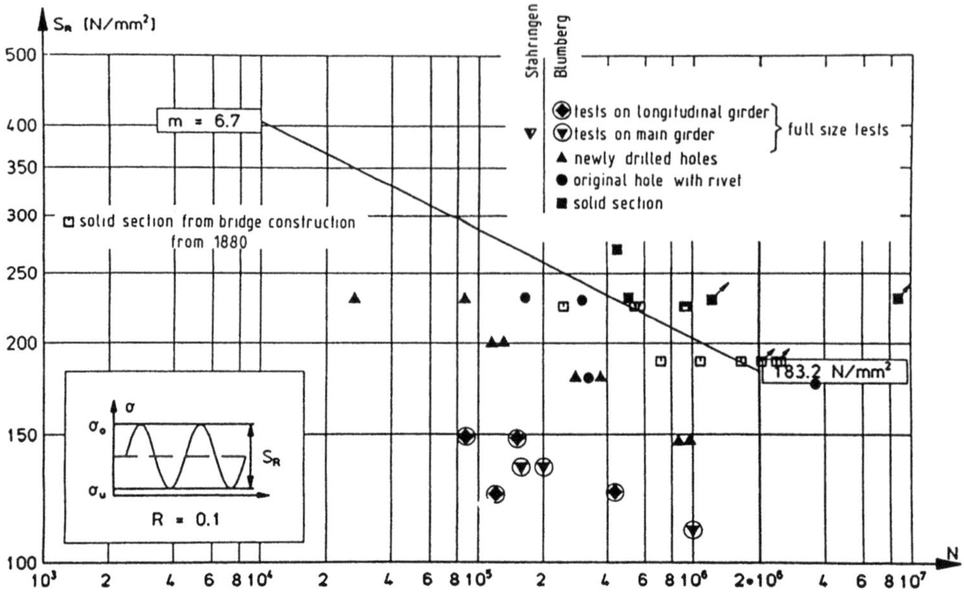

Figure 19 Test results and data from [5], bridges dating from 1880, plain material prestressed by 290 N/mm^2 and 10^5 stress cycles

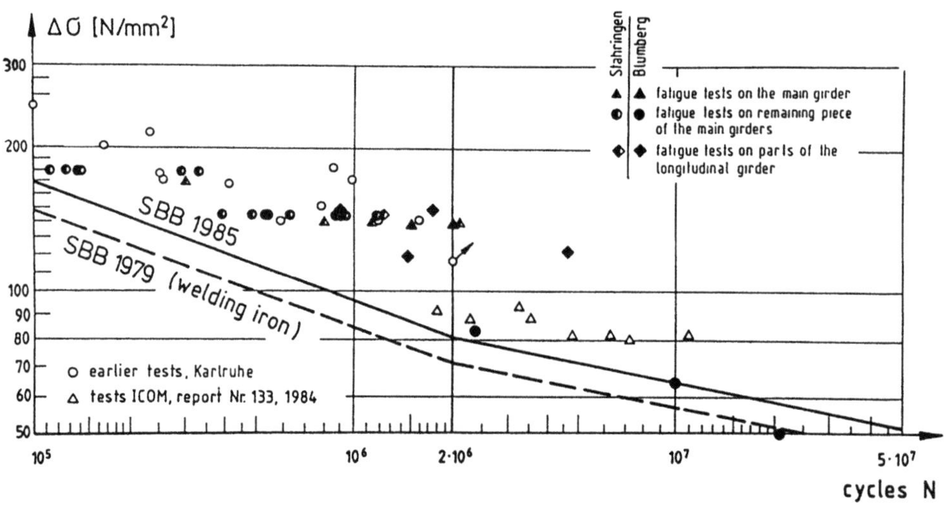

Figure 20 Test results and data from [4]

REFERENCES

1. MANG, F., Stahl im Altbau und Wohnungsbau. Abschlußbericht zur Forschungsstudie des Landes Nordrhein-Westfalen (VBI-72.02-92/77)

2. MANG, F.; STEIDL, G.; BUCAK, Ö., Altstahl im Bauwesen. Schweißen und Schneiden 1985, No. 1, p. 1-5

3. BRÜHWILER, E.; HIRT, M.A., Das Ermüdungsverhalten genieteter Brükkenbauteile. Der Stahlbau 1987, No. 1, p. 1-8

4. HERZOG, M., Erwiderung zur Zuschrift von Tschumi, M. auf Herzog, M., Abschätzung der Restlebensdauer älterer genieteter Eisenbahnbrücken. Der Stahlbau 1986, No. 5, p. 159-160

5. BAEHRE, R.; KOSTEAS, D., Einfluß der Vorbelastung auf die Restnutzungsdauer schweißeiserner Brücken. Bericht Nr. 7496 der Versuchsanstalt für Stahl, Holz und Steine der Universität Karlsruhe, Januar 1979 (not published)

6. STEINHARDT, O., Festigkeitsverhalten von Schweißeisen aus Brückenbauwerken des 19. Jahrhunderts. ETR 26 (1977), No. 6, p. 383-387

7. N.N., Unveröffentlichte Untersuchungen der Versuchsanstalt für Stahl, Holz und Steine der Universität Karlsruhe

8. STIER, W.; KOSTEAS D.; GRAF, U., Ermüdungsverhalten von Brücken aus Schweißeisen. Der Stahlbau 1983, No. 5, p. 136-142

9. BUCAK, Ö.; MANG, F., Untersuchungen an bestehenden Brückenbauwerken im Sinne von E-DS 805 sowie spezielle Institutsversuche an einem Originaltragwerk (Anno 1877) und an dessen Bauteilen. Festschrift Rolf Baehre, 1988. Herausgeber: O. Steinhardt und K. Möhler, Versuchsanstalt für Stahl, Holz und Steine, Universität Karlsruhe

10. MANG, F.; BUCAK, Ö., Theoretical and experimental investigations on railway bridges dating from 1856 to 1895. Int. Conf. on "Bridge Management", University of Surrey, March 28-30, 1990

11. BAEHRE, R.; WITTEMANN, K., Untersuchung der Rißausbreitung in historischen Stählen mit bruchmechanischen Methoden. Sonderforschungsbereich 315 "Erhalten historisch bedeutsamer Bauwerke", Universität Karlsruhe (to be published)

12. BUCAK, Ö.; KÄPPLEIN, R.; MANG, F., Ermüdungsuntersuchungen an einem Brückenbauwerk aus dem Jahre 1878. Veröffentlicht im Jahrbuch 1987 des SFB 315 "Erhalten historisch bedeutsamer Bauwerke". Verlag Ernst und Sohn, Berlin 1987

13. MANG, F.; BUCAK, Ö., Experimental and Theoretical Investigations of Existing Railway Bridges. IABSE-Workshop "Remaining Fatigue Life of Old Steel Bridges", Lausanne 1990

PART FOUR

NON-DESTRUCTIVE METHODS OF INSPECTION

27 AUTOMATIC VICKERS INDENTATION MEASUREMENT

G. BARBATO and S. DESOGUS
CNR, Istituto di Metrologia 'G. Colonnetti', Turin, Italy
P. GRATTONI
CNR, Centro Studi per la Televisione, Turin, Italy

1 INTRODUCTION

The empirical non-metrological origin of hardness measurements often prevents us from obtaining results unaffected by the uncertainties inherent in the different measurement methods adopted; uncertainties which may be high or even unacceptable, when compared with the usual industrial requirements. Moreover, the systematic character of such uncertainties often mislead even the experts in industrial laboratories that daily apply hardness measurement methods. They do not fail, in fact, to show their surprise when learning, for instance, that differences in the Rockwell C scale higher than one whole point may be found between some national hardness standard machines, whereas their experience in the use of workshop hardness testers will indicate an uncertainty in hardness measurement results of only a few tenths of a point (due, in practice, to non-repeatability alone).

The lack of correct information on the magnitude of the differences caused by systematic effects (observable, however, in reference hardness blocks of different manufacturers), often leads to disagreements, misunderstanding, and even legal actions.

The Vickers hardness measurement method, which is admittedly more metrologically formulated than other more frequently used methods, is also affected by systematic measurement uncertainties; these are evidenced not only by the differences in standard machines brought to light by international comparisons, but also and more simply in the results obtained from measurements by different operators /1/. In this case, however, it is not certain that the discrepancy evidenced is a subjective error of the operator or an intrinsic limitation of the measurement method.

Other limitations of the Vickers measurements are the laboriousness and long duration of the measurement process itself. The tiring effects of prolonged observations with hardness testers equipped with microscopes (usually, single-

Testing of Metals for Structures. Edited by F.M. Mazzolani. © 1992 RILEM.
Published by Taylor & Francis, 2 Park Square, Milton Park, Abingdon, Oxon, OX14 4RN.
ISBN 0 419 15810 3

eyepiece microscopes) act adversely on the operator's capacity of concentration and increase, in this case accidentally, the uncertainty of the measurement value.

To obviate these intrinsic limitations, both practical and metrological, of the Vickers method, we have developed and patented a system for indentation image analysis, with which both image area and diagonals can be determined.

2 MEASUREMENT OF VICKERS INDENTATION DIAGONALS

Problems of the uncertainty in Vickers indentation measurements and the different results that are often obtained by different operators have been the object of a number of proposals. The first and most obvious solution, though of dubious validity, which appears in Standards documents /2/ establishes to take the mean value of the measurements by two operators. Another method utilize a number of indentations on reference blocks; the indentation length is declared, so that in checking workshop hardness testers also the operator can determine his own "calibration". Obviously, this latter solution has undeniable practical advantages and gives a clear indication of the accuracy level within which calibration of hardness testers can be meaningful. This method, however, solves only the problem of result uniformity in a calibration system with reference to a national standard machine; it does not settle the point of the accuracy of standard machines, when a hierarchical order is not established.

The magnitude of the discrepancies usually to be found between different operators is of the same order as that of optics resolution; consequently, one might be led to using instruments having higher resolution. An investigation in this connection, however restricted, showed, instead, an intrinsic limitation of the measurand: the image in Fig. 1, which represents a Vickers indentation obtained by a scanning electron microscope, shows that the edges are far from being sharp. The indentation profile, as determined by means of a roughness testers (Fig. 2), shows that contours are not clearly cut, but rather rounded out: in other words, no sharply defined diagonal vertex is detectable.

The mentioned subjective error by the operator might therefore be due to a different way of extrapolating indentation sides to define a vertex that actually does not exist. An analysis in this connection will show that automatic extrapolation (i.e. the definition of the vertex position by a mathematical algorithm), more than an increase in instrument resolution, is advisable when uncertainty has to be reduced.

Fig.1 Indentation of the Vickers type as obtained by a scanning electron microscope.

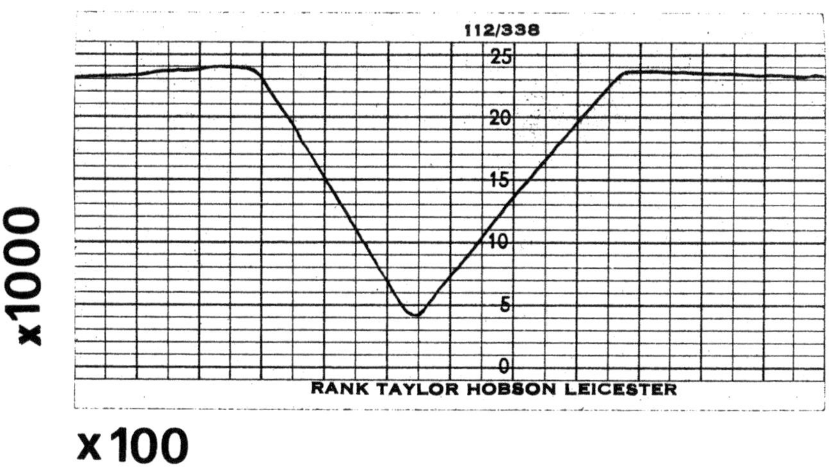

Fig.2 Indentation profile as determined by means of a roughness tester.

Furthermore, the alternative of automatic measurements is obviously technically mature; there exist numerous instruments for other applications with which the area of the surface of a certain intensity of gray can be determined even in very complex images. Nevertheless, the use of such instruments in hardness measurements is not immediate for two kinds of reasons. The area of the image portion darker than a given level of gray can be directly obtained with these instruments. Consequently, this measurement should be subjected in the first place to a sort of calibration, in order to check the effects of light source intensity, relative aperture of the objective used, and the effects of other factors liable to affect the signal level at which the threshold is established. Secondly, even if it is taken for granted that the scanning range can be reduced to a window containing only the indentation and that, consequently, the measured surface corresponds only to the area of the indentation surface and is not affected by imperfections or dust on the block surface, the obtained information could hardly be correlated with the values normally obtained from diagonals measurements: diagonals are, in fact practically independent of the elastic recovery of a material, whereas the shape of the indentation is noticeably affected by ridging, sinking and elastic recovery /3/.

In fact, particularly in the sector of microhardness measurements where this problem is more felt, it is no rare occurrence that the effect of materials behaviour utterly falsifies the indentation shape, by deforming it as is shown in Figs. 3 and 4 to such an extent that a simplified analysis of the image could not yield meaningful results.

Fig. 3 Soft-material indentation shape.

Fig.4 Hard-material indentation shape.

Resort had therefore to be made to a sophisticated procedure for image analysis, which is based on successive phases: contour detection, vertex recognition, and dimension measurement.

3 IMAGE SEGMENTATION

The choice of the most suitable method for segmenting the image into its two main components - indentation and background - was made on the basis of the following considerations:

a) The generic image of a Vickers indentation, as obtained by reflection of a light beam (function of diaphragm aperture) projected on the surface z(x, y) under examination, appears as a dark structure of a "pseudo-square" shape against a lighter, irregular background, where spots may be present. The image irradiance E(x, y) which is converted into an electric signal V(t) by means of a TV-camera, can be described, under certain assumptions, by the equation:

$$E(x, y) = \rho(x, y) \; R(\frac{z}{x}, \frac{z}{y}) \qquad (1)$$

where ρ (x, y) is the reflection coefficient, which is a function of the surface material, and R(z/ x, z/ y) is the reflectance map /4, 5, 6, 7, 8/, which is a function of the gradient space /9, 10/.

By assuming the reflection coefficient ρ(x, y) constant at all points of the surface, E(x, y) and consequently V(t), become a function of surface gradient. Therefore, the established conditions being satisfied,it is thus possible to determine the characteristics of the surface starting from the TV-signal. Fig. 5 is a tridimensional representation of the luminance as referred to the indentation of Fig. 4.

b) The techniques for image segmentation are essentially traceable to two basic methods: the region growing, and the edge detection methods /11/.

The choice between the two methods was conditioned by the particular measurement problems of minimal sensitivity to variations in luminance L, focusing, etc. Accordingly, region growing techniques can immediately be discarded as they are based on laws of pixel aggregation, which depend necessarily on their luminance; such techniques do not supply by themselves general rules for choosing precise contour points. As an example, the threshold method, one of the most elementary methods, can be described as a sectioning of the luminance surface (given in assonometric form in Fig. 5), at an arbitrary threshold level.

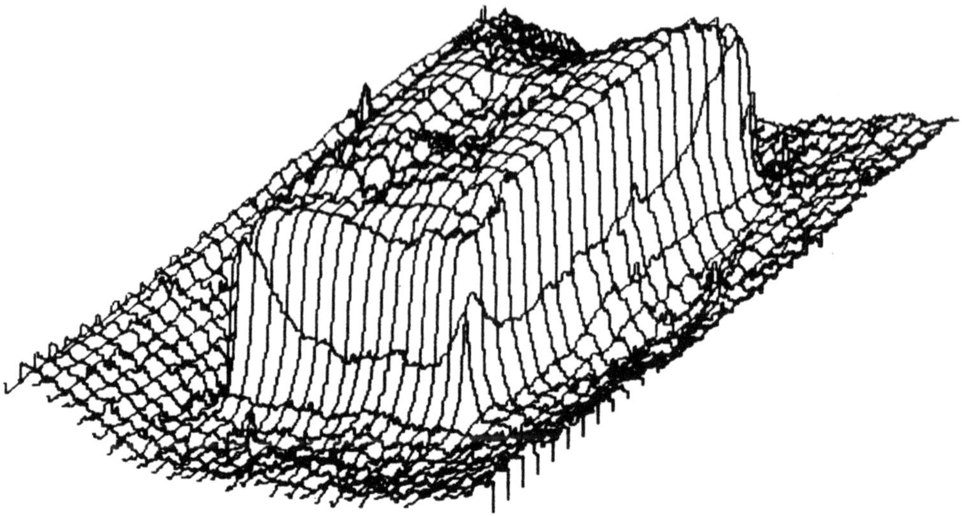

Fig. 5 Assonometric representation of the luminance of the indentation in Fig. 4

Obviously, the dimensions of the resulting structure depend on the threshold level used, and consequently this is apt to introduce a subjective element in the measurement of Vickers indentation. For the foregoing reasons, described by the example supplied above, the most suitable way must therefore be chosen from edge-detection methods.

4 EDGE DETECTION

The literature in the sector of Image Processing describes a great number of techniques for edge detection; these techniques can however, be referred to two basic approaches /12/: enhancement/thresholding and edge fitting.

The former enhances some typical contour characteristics by means of spatial operators. The points where such characteristics are greater than a pre-established threshold, are considered edge points. With the latter method the degree to which the image fits an ideal edge model is determined; the points where fitting exceeds an established value are considered edge points /13/.

We chose the first approach, because the computational simplicity of algorithms of this kind (no imaginary numbers required) makes them faster and easier to implement on a small computer. Edge points are enhanced by a differential operator, and those above a pre-established threshold are individually coded with "gradient intensity" and "gradient direction".

Moreover, the use of a differential operator meets the requirement of making contour identification independent of absolute luminance values, and it can be shown that contour location, corresponding to gradient maxima, results, in a first approximation, to be insensitive to small blurring effects (see appendix A).

We chose the non-linear (differential) Sobel operator (Fig. 6), as its characteristics of isotropy, sensitivity to contour position, signal-to-noise ratio, etc. are better than those of other differential operators /12/.

	j-1	j	j+1
i-1	1	0	-1
i	2	0	-2
i+1	1	0	-1

S_O Horizontal mask

	j-1	j	j+1
i-1	1	2	1
i	0	0	0
i+1	-1	-2	-1

S_V Vertical mask

Fig. 6 Sobel operator masks

The Sobel operator evaluates gradient magnitude G and gradient direction θ at each point of the discrete image array as follow:

$$G(i,j) = \sqrt{G_V^2(i,j) + G_O^2(i,j)} \quad (2)$$

$$\Theta(i,j) = \text{Arctan}\,[G_V(i,j)/G_O(j, j)] \quad (3)$$

with:

$$G_V(i,j) = L(i,j) * S_V(i,j) \quad \text{and}$$
$$G_O(i,j) = L(i,j) * S_O(i,j)$$

where L(i,j) with (i = 1,2 ... M, j = 1,2 ... N) represents the discrete luminance array of MxN elements; "*" denotes two-dimensional spatial convolution and $S_V(i,j)$ and $S_O(i,j)$ are the horizontal and vertical Sobel operator masks shown in Fig. 6.

In the present application, the Sobel operator output greater than the established threshold is coded by associating the direction information (3) to the intensity information (2). The remaining points are set equal to zero.

5 CONTOUR DETECTION AND RECOGNITION

Contour detection and recognition are performed by working in a discrete polar coordinate reference system (r_i, φ_i) with the pole coincident with the indentation centroid B. The contour location r_i for each angular sector φ_i is assumed to be in correspondence of the absolute maximum of gradient G (see Eq. 6). This operation results in a φ_i ordered list of contour points, which is useful for the successive recognition process of the indentation corners.

The "a priori" knowledge of the structure to be identified (an however oriented "pseudo-square" dark object on a lighter background) has been exploited in order to reduce the possible mistakes in contour point detection due to extraneous structures (spot, line, etc.). For this purpose, the search field has been reduced to a band R around a first rough approximation of the indentation contour and constraints on the gradient direction have been set.

The contour detection and recognition procedure is therefore performed in two steps: the first computing the gradient G of the image and the search field R, the second locating the indentation contour points.

<u>Step 1.</u> The image is scanned sequentially row by row and absolute value G(i,j) and phase Θ(i,j) of the gradient are computed. At the same time the one dimensional arrays:

$$W(\Theta) = \sum_{i=1}^{M} \sum_{j=1}^{N} W\,[\Theta(i,j)] + G(i,j), \tag{4}$$

$$I(\Theta) = \sum_{i=1}^{M} \sum_{j=1}^{N} I\,[\Theta(i,j)] + i \cdot G(i,j) \tag{5}$$

$$J(\Theta) = \sum_{i=1}^{M} \sum_{j=1}^{N} J\,[\Theta(i,j)] + j \cdot G(i,j) \tag{6}$$

are computed, where: W(Θ) is the weighted histogram of the gradient directions; I(Θ) and J(Θ) are the non-normalized coordinate of the centroid of all those array elements having Θ(i,j) as gradient direction (MxN being the dimension of the image array). If the indentation contour shape is "pseudo-square", as hypothesized before, then W(Θ) will exhibit 4 peaks, nearly 90° apart, representing the directions (Θ_1, Θ_2, Θ_3, Θ_4) of the four sides (Fig. 7a).

An estimate of each side direction is computed first by globally detecting the orientation of the indentation (this is obtained by determining the phase shift Θ_M at which the convolution between W(Θ) and a suitable comb filter C(Θ) reaches its absolute maximum, (Fig.7a) and then, from this first approximation ($\Theta'_1=\Theta_M$; $\Theta'_2=\Theta_M+90°$; $\Theta'_3=\Theta_M+180°$, $\Theta'_4=\Theta_M+270°$), by locally refining the localization of each single maximum on $W(\Theta)$.

At this point, the estimate of a representative contour point for each side (B_1, B_2, B_3, B_4) and of the indentation centroid B (Fig.7b) are easily computed as:

$$B_i \Rightarrow \left(\frac{I(\Theta_i)}{W(\Theta_i)}, \frac{J(\Theta_i)}{W(\Theta_i)}\right), \text{ where : } i = 1, \ldots 4 \tag{7a}$$

$$B \Rightarrow \frac{B_1 + B_2 + B_3 + B_4}{4} \tag{7b}$$

and the band-width d of the search field R is chosen as

$$d = 0.2 \; \frac{|B_1 - B_3| + |B_4 - B_2|}{2} \tag{7c}$$

Step 2. The contour point locations are determined by re-examining only the points inside search field R.

Only those points whose gradient direction Θ_i is compatible with their angular coordinate φ_j (in the sense that we are only looking for a dark to bright transition moving away radially from the centroid B) are considered. The contour points are assumed to be in correspondence of the absolute maximum of G, for each φ_i.

At the end of these operations all the ordered contour points are available (Fig. 8). The geometric characteristics of the indentation can, therefore, be determined: the area by integration, the vertices either as a local extrapolation of contour points or by interpolation of the whole contour with an interpolating quadrangle.

6 VERTEX POSITION DETERMINATION

The estimate of the ideal position of the indentation vertices is, at present, performed by human operators as we said in the introduction.

Since the vertices are not very sharp, as Figs.1 and 2 shown, to locate them the operator must extrapolate in some way the information conveyed by the concurring sides.

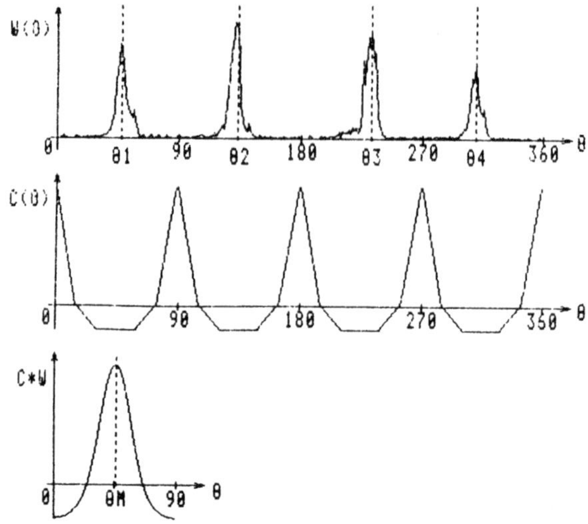

a) W(Θ) is the weighted histogram of gradient directions;
C(Θ) is a comb filter function, used to detect the global orientation of indentation;
C*W is the convolution function between W(Θ) and C(Θ).

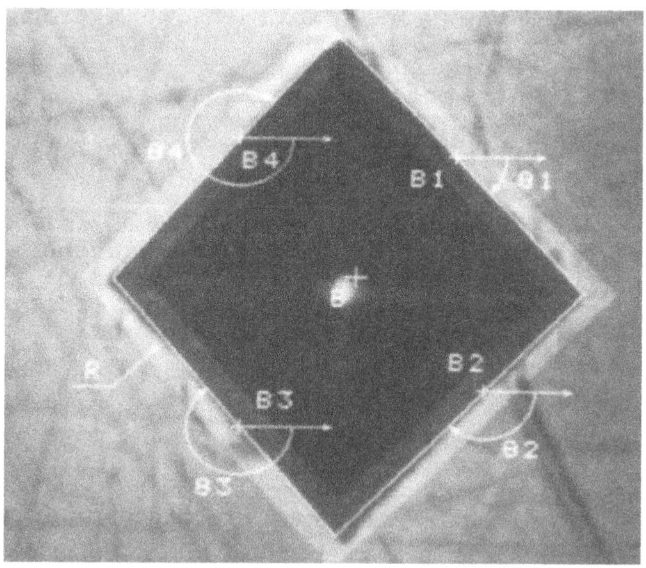

b) Contour search field R superimposed on the original indentation image.

Fig.7 Determination of contour search field R

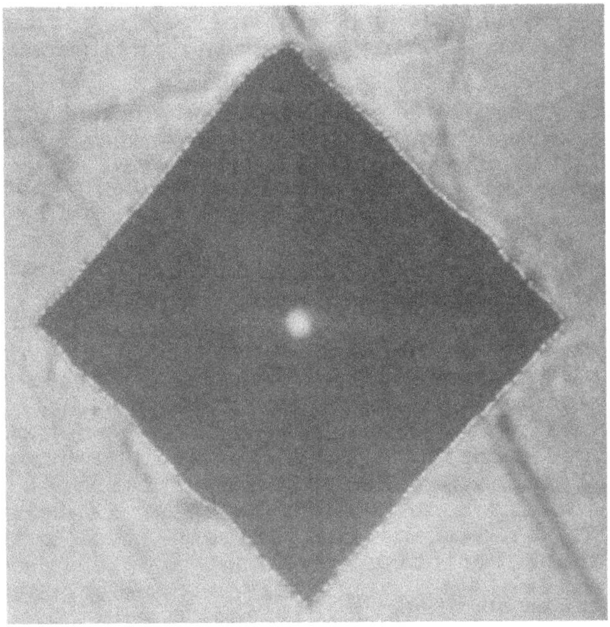

Fig.8 Detected contour points superimposed on indentation image.

In the present work we considered two automatic methods to perform this task. The first one works in a similar way as a human operator is supposed doing, that is, only a little part of the two sides near each vertex is interpolated with a straight line; the vertices are assumed to be at the crossing point of each couple of lines.

The second assumes the vertices of the indentation coincident with those of the quadrangle, composed by 4 second order parabolic sides, that best fits all the contour points detected in the previous process.

While the first method is expected to give results in good agreement with those obtained by a human operator, the second seems to get a better repeatability of results, even though they should depend on the shape of sides.

A solution of the best-fit-problem present in the second method, has been obtained by minimizing the mean square radial error between the detected contour points and the quadrangle, by calculating the 12 parameters unknown (8 for the coordinates of the vertices and 4 for the curvature rise of the parables),

$$\sum_{i=1}^{N} (d_i)^2 \Rightarrow \min \tag{8}$$

Radial error d is defined as the distance between the detected contour point P_i and the correspondent intersection point produced by the radius connecting the centroid B to P_i and a parabolic side of the quadrangle.

Minimization connected with eq.(8) can be dealt with as a non-linear programming problem, for whose solution the algorithm of Fletcher and Powell /14/ proved to be particularly effective.

7 THE ADOPTED MEASUREMENT SYSTEMS

For the reason of obtaining results in principle equal to that of the measuring systems generally utilized, only the first of the two methods described have been utilized for an experimental analysis. It is applicable at two different accuracy levels, namely, at a so called "normal level", developed by Officine Galileo to such a practical way of use, as to equip hardness and micro-hardness testers of normal production and at a so called "primary level", in which the metrological characteristics have been specially considered.

At the former level completely automatic operations are possible by using the digitized image of the whole indentation. This is the practice for hardness block calibrations at IMGC and at the laboratory for hardness block calibration of the Italian Calibration Service (SIT) of Officine Galileo.

At the latter and higher metrological level, though operations are still semi-automatic, the measurement procedure is far slower and is therefore applied only in national and international comparisons. One of the distinctive features of this method is that it is based on a direct reference to primary measurements, so that the measurement values obtained are, as far as possible, stable in time. With the procedure in question, in fact, it is possible to eliminate effects caused by the instability of the TV-camera, by magnification uncertainty and aberration errors of the optics used. This method was implemented also with the purpose of determining the dimensions of reference indentations, by means of which ordinary measurement systems can be kept under control. The uncertainties introduced by traditional ordinary-type system components are therefore very small and it is thus possible to evidence the characteristics of repeatability, reproducibility, and long-term stability of the image-analysis method adopted.

In practice, this measurement system, like the ordinary industrial system, uses a microscope, a telecamera, and a system for image processing, but it is additionally equipped with a translation slide having a motion resolution of 0.1 μm and with a laser system for translation measurements. Moreover the measurement of all the indentations is made with a 50x lens of NA 0.5, which is the maximum usable for indentation observation. This is possible, because it is not necessary to observe the whole indentation, but only a vertex at a time. The procedure for image analysis was modified, in order to determine the position of the vertex under observation alone.

The indentation is aligned to a diagonal along the direction of axis x of the system for image analysis, which is itself aligned along the direction of the translation slide. The slide is moved in such a way as to position the left indentation vertex approximately on the optical measurement axis. With the image-analysis method, the x_s coordinate is then determined. The reading of the laser measurement system is zero reset and the slide is moved to bring the right vertex near the optical axis; its x_d coordinate is measured in the way already described. If L is the value of the translation measured by the laser system, the diagonal is equal to $L+x_d-x_s$.

Subsequently, the indentation is rotated by 90° and the other diagonal is measured.

With this procedure, the optical system serves only as a reference line. Since in the positioning of the indentation vertex the deviation to the optical axis is always less than 1 μm, the largest portion to the diagonal measurement value is the value of the translation from one vertex to the opposite, as measured by means of the laser interferometer with an uncertainty of the order of 0.01 μm, which is therefore very small. The largest contribution to the total uncertainty derives then from the uncertainty with which the vertices are defined by the image-analysis system.

8 EXPERIMENTAL RESULTS

Several tests were carried out to check the repeatability and reproducibility of the method, some of which were repeated after some months, to evaluate stability.

To evaluate repeatability, the measurements of the coordinate of a vertex were repeated without making any change between one measurement and the next. It was observed that result dispersion was of the order of a half pixel; however seldom, it happened that isolated results showed differences amounting to as much as two pixels with respect to the mean value. Such differences are likely to be due to random electrical or mechanical disturbances. Although results of this type ought to be evaluated and, possibly, eliminated according to Chauvenet's principle, since the observed deviations from the mean were limited it was preferred to repeat measurements ten times and average results, in order to avoid making analysis too laborious. Given the rapidity of automatic measurements, the additional time involved is acceptable; the uncertainty of the results thus obtained can be said to be lower than half a pixel.

Measurement reproducibility was evaluated by repeating the sighting of both vertices and the measurement of the diagonal. The difference between two diagonal measurements, repeated in immediate succession, at a 95% confidence level, was of the order of 0.1 µm.

The Table I gives the differences [µm] obtained for indentations made with different loads on blocks of different hardness and different materials. These differences have been grouped so that the rows have equal level of hardness, and the columns have nearly equal diagonal dimensions.

For each row and each column the standard deviation is given as an indicator of the dispersion of the results obtained. For the rows, the deviation depends on the difference in the indentation shapes and materials (steel, brass, and copper) of the hardness blocks; for the columns on the diagonal dimension range.

It can be noticed that the difference of the standard deviation as a function of the diagonal length is not significant. This result can be explained because the

Table I Differences [µm] between two successive measurements

Hardness	Diagonal [µm]					
	9÷10	19÷21	38÷41	60÷70	90÷100	σ
900 HV0.5	0,00 -0,07	0,01 -0,03	-0,02 0,09	0,04 -0,08	-0,05 0,00	0,05
550 HV0.5		0,04 -0,04	0 0,01	-0,03 -0,01	-0,05 -0,07	0,03
250 HV0.5	-0,02 -0,08	-0,03 0	-0,04 -0,04	0,02 -0,05	-0,01 -0,02	0,03
100 HV0.2	0 0,07	-0,02 0,08	0,03 0,1	0,05 0,02	0,06 0,02	0,04
40 HV0.2	-0,05 -0,06	0,05 0	-0,09 -0,09	-0,06 -0,06	-0,1 -0,12	0,05
σ	0,04	0,04	0,06	0,04	0,05	

measurement method adopted considers only the vertex position and therefore is practically unaffected by the length of the diagonal.

More important is the absence of significant effects depending on hardness, i.e., on the different materials used for the construction of reference blocks, and on the indentation shape, which remarkably differs from a 900 HV 0.5 steel block to a copper block of only 40 HV 0.2. This means, in fact that the method is equally reproducible also in unfavourable working conditions.

In some cases, measurements could be repeated after some months, and stability in time could thus be evaluated. The results of the tests were a mean difference value of the order of 0.01 μm and a standard deviation of the order of 0.1 μm, instead of the theoretically predictable 0.08 μm, that is, a little-significant difference. Stability consequently proved to be quite satisfactory.

9 CONCLUSIONS

The characteristics mentioned here indicate that the method in question is suitable to realize a primary Vickers-hardness reference. Nevertheless, since there does not exist a natural hardness standard for traceability, one cannot exclude systematic errors, which can be eliminated only by an independent realization of primary measurement systems. One of the characteristics of the image-analysis system adopted is the attempt at reproducing the physico-psychologic behaviour of human vision; consequently, in this method the characteristics of repeatability, reproducibility, and stability - distinctive of automatic systems - are directly connected with a "natural" definition of the measurement of Vickers hardness.

The comparison tests carried out for the accreditation of the laboratory of Officine Galileo as a SIT centre showed that an agreement, respecting the limits considered by ISO Standards for the systems for the calibration of standard hardness blocks, can be maintained between two measurement systems based on different technology and different calculation programs, if the working principles are the same.

REFERENCE

/1/ Barbato, G., Desogus, S. Problems in the measurement of Vickers and Brinell Indentations. Measurement, to be published

/2/ ISO R640

/3/ O'Neil H. Hardness Measurement of Metals and Alloys. Chapman and Hall Ltd, London, 1967, pp. 24, 29

/4/ Horn B.K.P. Shape from shading: A method for obtaining the shape of a smooth opaque object from one view, MIT Project MAC Technical Report MAC TR-79, 1970.

/5/ Horn B.K.P. Obtaining shape from shading information, in The Psychology of Computer Vision, (P.H. Winston, Ed.), pp. 115-155, McGraw-Hill, New York, 1975.

/6/ Horn B.K.P. Hill shading and the reflectance map, Proc. IEEE 69, 1981, 14-47.

/7/ Horn B.K.P. and Bachman B.L. Using synthetic images to register real images with surface models, MIT Artificial Intelligence Laboratory, Memo 437, 1977.

/8/ Horn B.K.P. and Sjoberg R.W. Calculating the reflectance map, Appl. Opt. 18, 1979, 1770-1779.

/9/ Huffman D.A. Impossible objects as nonsense sentences, in Machine Intelligence (B. Meltzer and D. Michie, Eds.), vol. 6, pp. 295-393, Edinburgh Univ. Press, 1971.

/10/ Mackworth A.K. Interpreting pictures of polyhedral scenes, Artif. Intell. 4, 1973, 121-137.

/11/ Rosenfeld A. and Kak A.C. Digital Picture Processing Academic Press New York, 1982, vol. 2.

/12/ Abdou I.E. and Pratt W.K. Quantitative Design and Evaluation of Enhancement thresholding Edge Detector. Proceeding of the IEEE, vol. 67, n. 5, may 1979, pp. 753-763.

/13/ Hueckel M.H. A local visual operator which recognizes edges and lines. Journal of the Association for Computing Machinery, vol. 20, n. 4, October 1973, pp. 634-647.

/14/ Fletcher R., Powell M.J.D. A rapidly convergent descent method for minimization. Computer Journal 1963, vol. 6, pp. 163-168.

APPENDIX A

Let $l(x)$ be the luminance function along axis x (Fig. 10a) and $(d^2l(x)/dx^2) = 0$ for the contour point $x = x_o$ (maximum of $dl(x)/dx$ (Fig.10b)). Let $h(x)$ be the point spread function representing the blurring effect; then

$$f(x) = \int_{-\varepsilon}^{+\varepsilon} l(x - \alpha)\, h(\alpha)\, d\alpha \qquad \text{(A1)}$$

is the blurred version of $l(x)$ (Fig.10d).

If $h(\alpha)$ is an even function vanishing outside interval $-\varepsilon \vdash\!\dashv +\varepsilon$, as can be assume in many real cases, then, for ε small enough, $l(x)$ can be expanded in the Taylor series around x_o limited to the third-order term, and the equation (A1) can be rewritten as follows:

$$f(x)=\int_{-\varepsilon}^{+\varepsilon}\left\{l(x_o)+\frac{dl(x_o)}{dx}(x-x_o-\alpha)+\frac{1}{6}\,\frac{d^3l(x_o)}{dx^3}(x-x_o-\alpha)^3\right\} h(\alpha)d\alpha \qquad \text{(A2)}$$

(note that $d^2l(x_o)/dx^2 = 0$ by hypothesis).

Then, the actual contour point location is computed by searching for the zero of the second-order derivative of $f(x)$:

$$\frac{d^2f(x)}{dx^2} = \int_{-\varepsilon}^{+\varepsilon} \frac{d^3l(x_o)}{dx^3}\,(x - x_o - \alpha)\; h\,(\alpha)\; d\alpha \qquad \text{(A3)}$$

For $x = x_o$ then:

$$\frac{d^2f(x_o)}{dx^2} = \frac{d^3l(x_o)}{dx^3}\int_{-\varepsilon}^{+\varepsilon} -\alpha\; h(\alpha)\; d\alpha \;=\; 0 \qquad \text{(A4)}$$

because it is the integration of an odd function between symmetric limits. Blur (see Fig.10e) has not changed the location of the contour point, q.e.d.

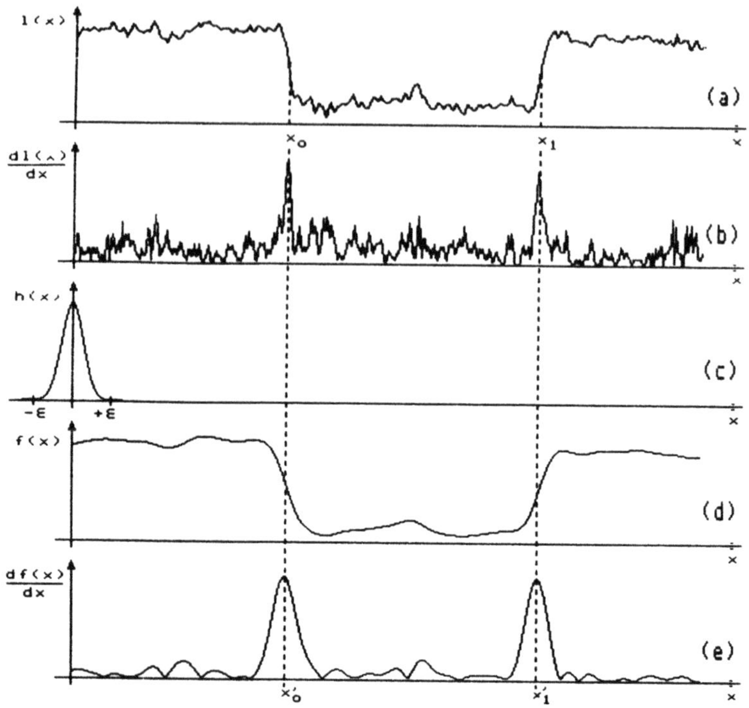

Fig.10 Sensitivity of the differential-based method for contour location to blur

a) cross-section luminance l(x) of an indentation image;
b) absolute value of l(x) first derivative;
c) point spread function h(x);
d) luminance function f(x) after blurring;
e) absolute value of f(x) first derivative.

Location of maxima of dl(x)/dx and df(x)/dx are coincident.

28 OPTICAL METHODS FOR NON-DESTRUCTIVE INSPECTION OF STRUCTURAL COMPONENTS

G. DI CHIRICO
Department of Mechanical Engineering, University of Cagliari, Italy

SUMMARY

Some basic information is provided on the most promising optical methods for nondestructive tests. In particular, this paper deals with hologram, moiré and speckle interferometry for inspection of materials and structures.

INTRODUCTION

Optical methods offer significant contributions to the investigation of structural ccmponents, when complex design geometries are involved and boundary conditions are poorly defined.

The diffusion of techniques based on laser interferometry during the past decades has given new emphasis to solutions of difficult problems involving the control of structural integrity in many sectors of manufacturing (defects in bonding, welding, assembling; fiber breakage and delamination in composite materials; debonding of ceramic coatings; internal discontinuity or dishomogeneity; surface or internal cracks; etc.).

Moreover, it is worth pointing out that, by utilizing standard nondestructive tests (such as X-rays, ultrasonics, etc.), only the location and shape of a defect can be detected; but the knowledge of geometric characteristics is generally not sufficient to define the associated level of risk. Therefore, an attempt must be made to "visualize" the influence of the defect (or imperfection, or discontinuity...) on the mechanical response of the element in question under actual operating conditions. A solution to this problem is often possible by means of suitable optical techniques.

The purpose of this paper is to provide some basic information about the main optical methods for inspection of structural components, such as hologram, moiré and speckle interferometry.

Testing of Metals for Structures. Edited by F.M. Mazzolani.
Published by Taylor & Francis, 2 Park Square, Milton Park, Abingdon, Oxon, OX14 4RN.
ISBN 0 419 15810 3

HOLOGRAM INTERFEROMETRY

Introduced by Gabor in 1948, holography is a special photographic technique in which a beam of light that strikes or passes across an object is made to interfere on the plane of a photosensitive plate, together with a reference beam that is coherent with the first one. The hologram produced in this way preserves all of the information regarding both the amplitude and the phase of the recorded light wavefront. Once the hologram has been developed and illuminated again with the reference beam, it becomes possible to observe a complete, three-dimensional reconstruction of the original image (1) (2).

Holograms consist of generic diffraction grids, and should vibrations or shifting move the grid a distance equal to the size of its pitch, then it will no longer be possible to obtain a faithful reconstruction of the object, meaning that the quality of the hologram must be considered poor.

Normally, during the reconstruction of a hologram of poor quality, dark areas appear that represent interference fringes produced by the different reconstructions of the object. In such cases, the hologram behaves in the same way as an interferometer. This phenomenon drew the attention of a number of different researchers, and the results of attempts to utilize it in a controlled way, as a tool for performing interferometric measurements, constitutes, in the broadest sense of the term, holographic interferometry.

This branch of modern optics, therefore, regards the formation and interpretation of the interference patterns that arise when a wavefront registered in a hologram is reconstructed and made to interfere with a wavefront being used as a reference. The full importance of these phenomena did not become clear until 1965, when they led to a veritable flood of studies and research aimed at developing effective and reliable techniques of investigation.

The fringes can be observed in "real time" or "in double exposure".

In the case of the first option, the virtual image reconstructed from the hologram is placed so that it exactly overlaps the surface of the original object; eventual displacements or deformations in this surface alter the optical paths of the light rays reflected or scattered from the surface, leading to the recording of an image slightly different from the one recorded previously. The two images interfere, as can be observed directly from the fringes, which represent the contours of the areas deformed in the same quantity; in other words, fringes represent the loci of the points where the displacement vector has a constant value. In this way, displacements can be detected of orders of magnitude corresponding to the wavelength of the employed radiation. Any change in the state of the object's surface (increased stress; thermal variations; etc.) causes a simultaneous alteration of the fringe system, explaining why the technique is called "real time" holographic interferometry. This technique requires, however, a high degree of precision in the positioning of the holographic plate, in addition to which the working surface and its components must remain perfectly stable. For this reason, it is not always possible to use this approach.

With the second option, meaning the "double exposure", many of these problems are eliminated. A single plate, appropriately positioned, is used to make two succêssive exposures of the object, before the plate itself is developed. In this way, the interference fringes are "frozen" in the hologram, where they become visible during the subsequent reconstruction process, regardless of the position of the object.

The holographic patterns in fig. 1 and 2 show, as an example, the influence of internal defects on the mechanical behaviour of pressure tubes for a nuclear reactor (3). Using real time hologram interferometry, position and shape of such defects are visualized through the observation of fringe distortion on the outer surface of the tubes, at two different pressure levels. Fig. 1 makes evident the presence of two thin linear marks on the inner wall of the tube; in the second case, one sees the effect of an extended circumferential mark in the central region of the tube specimen.

MOIRE TECHNIQUES

The moiré effect is an optical phenomenon obtained when two grids consisting of parallel lines with a sufficiently small pitch are overlapped and observed both in transparency or by reflection (4).

The phenomenon - which was observed for the first time by Rayleigh - was widely illustrated in 1887 by A. Righi, who also applied the technique to the measurement of relative displacements. Nonetheless, apart from a few applications for miscellaneous purposes, it was only at the beginning of the nineteen-seventies that the technique began to spread as a means of analyzing strains and stresses.

The formation of moiré fringes was used to evaluate relative displacements on the surface of an object, which then led to the strain components. The practical procedure involves placing a grid on the prototype or model to be examined, and then placing a second reference

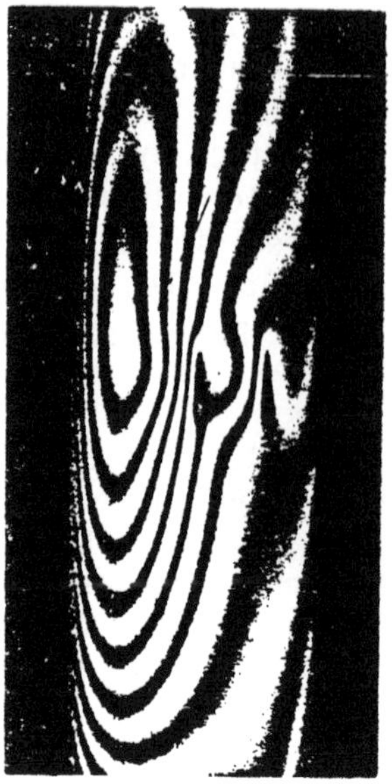

Figure 1
Holointerferometric pattern of the radial displacements of a tube specimen with two thin linear marks on the inner wall.

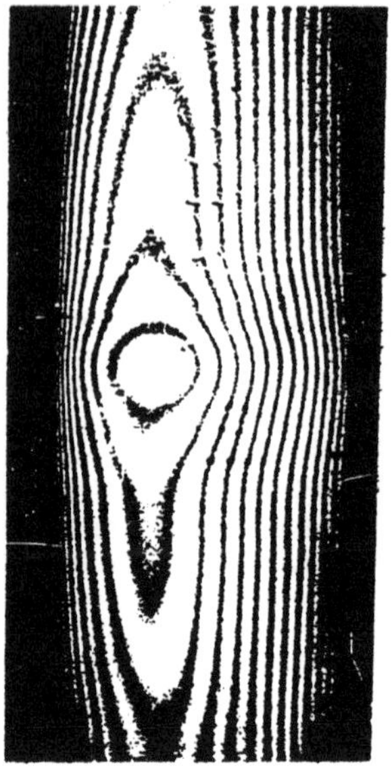

Figure 2
Radial displacements of a tube specimen with a circumferential mark on the inner wall.

grid in front of the first one in such a way that the two match. When the object undergoes a deformation, the grid attached to the surface is also deformed, following the surface displacements, while, by means of the rigid reference grid already in place, the pattern of moiré fringes is outlined. These fringes represent loci of points that have undergone identical relative displacements in the normal direction as regards the lines of the reference grid. They are easily interpreted, given that, unlike the holographic fringes, they always refer to pre-established directions.

Readings can also be made of displacements occurring outside the surface plane (shadow moiré; projected moiré) and, by using transparent models, it is also possible to analyze internal sections.

Fig. 3 illustrates an example of shadow moiré fringes revealing buckling of a large shear panel (5).

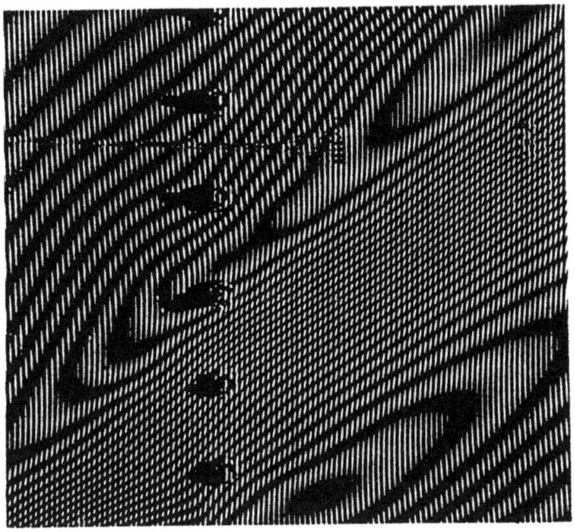

Figure 3
Shadow moiré fringes revealing buckling of a large shear panel.

SPECKLE PHOTOGRAPHY

A laser beam illuminating a rough surface gives rise to a scattered intensity distribution which takes the form of alternately bright and dark spots. The pattern produced is an interference figure, commonly known as a field of speckles (fig. 4). Therefore, if a picture of a coherently illuminated object is taken with a camera, it will not be homogeneous, but filled with speckles, whose size will depend on the aperture of the camera's diaphragm (6).

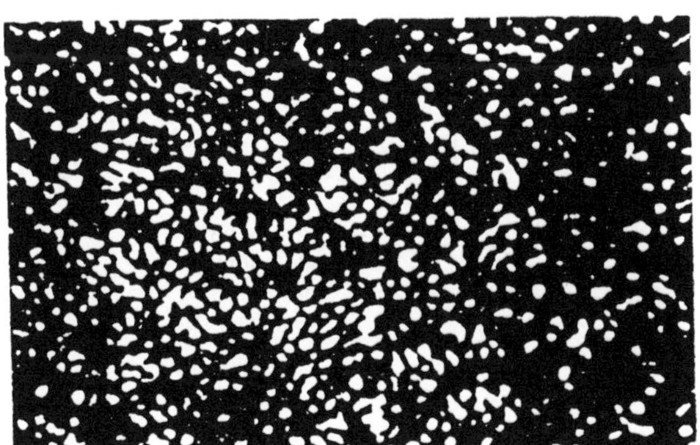

Figure 4
A pattern of speckles

The measurement of the displacements that an object undergoes between two recorded states is performed in two steps: a) recording on a photographic support of the two states of the object; b) processing of the results by an optical filtering technique.

The speckle ruling acts as a grid naturally printed on the object surface. When the subject is deformed, a point on the object will be displaced and the speckle ruling associated with this point will be displaced accordingly. By double-recording the deformed and undeformed speckle rulings on the same photographic film, a moiré fringe pattern is outlined which depicts the in-plane displacement component orthogonal to the ruling lines.

In a Fourier filtering arrangement (fig. 5), the spectra from the areas of constant frequency will contribute to produce couples of symmetrical diffraction orders. By placing an aperture in the plane of Fourier filtering, which allows only one diffracted order to pass, the areas having displacements of an odd multiple of half the pitch will appear dark in the image plane. Therefore, a fringe pattern is obtained that represents the loci of the component of in-plane displacement normal to the ruling lines. The process described is based on a point by point filtering technique. It is also possible to filter the whole field in order to have a global overview.

All speckle methods have the advantage with respect to moiré techniques that they do not require any grid printing on the object surface; on the other hand, some problems arise with the fringe quality and the limited tolerance to rotation and total displacement (7).

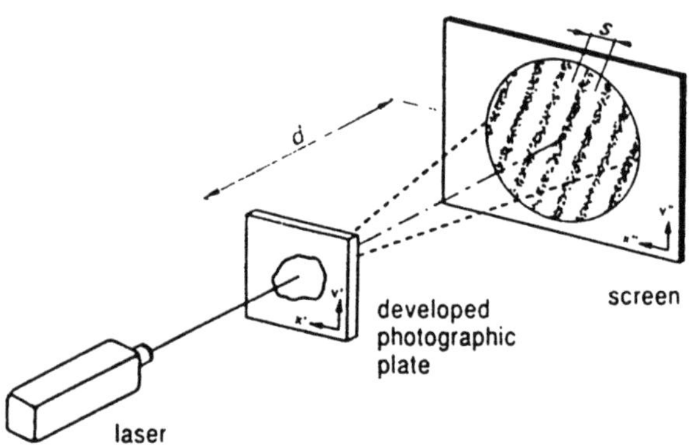

Figure 5
The point by point filtering set-up

CONCLUSIONS

These brief notes are nothing more than an attempt to give some information on the actual possibilities of optical methods for non-destructive investigation on structural components. The area of utilization of laser interferometry is being continuously developed in order to provide substantial contributions to practical solutions for difficult industrial problems. These solutions not only have a direct impact on the improvement of production technologies, but often provide designers with new criteria for evaluating the reliability of the prototypes under true operating conditions.

REFERENCES

1. C. Vest: Holographic Interferometry, J. Wiley, N.Y. 1979.
2. R.K. Erf: Holographic Nondestructive Testing, Academic Press, N.Y. 1974.
3. G. Di Chirico, L. Pirodda: A Real Time Holographic Investigation on the Mechanical Behaviour of Pressure Tubes with Internal Defects, Proceed. Southeastern Conf. on Theor. and Appl. Mechanics, Columbia (USA), April 1986.
4. A.J. Durelli, V.J. Parks: Moiré Analysis of Strain, Prentice Hall, N.J. 1970.
5. A.S. Redner: Shadow Moiré Method, Bull. SM-2 Strainoptic Tech. Inc., July 1985.
6. M. Monti: Speckle Photography applied to the Study of the Mechanical Behaviour of Granular Media, Report of Swiss Federal Inst. of Technology, Lausanne 1989.
7. Y.Y. Hung: Displacement and Strain Measurement, in "Speckle Metrology" (Ed. R. Erf), Academic Press, N.Y. 1978.

29 ELECTROMAGNETIC ACOUSTIC TECHNIQUE AND ITS APPLICATIONS TO METAL INSPECTION

F. TONOLINI
CISE SpA, Segrate, Milan, Italy

SUMMARY

In this paper the potentiality of ElectroMagnetic Acoustic Technique (EMAT) in the inspection of metal materials and structures is outlined. The working principles of EMAT and its capability to generate different wave types in metal without contact with the material's surface are described. The block diagram of special developed instrumentation is also given.

The advantages and limits of EMAT are discussed and some applications are envisaged. Applications as defect detection in thin plates, tubes and rods, thikness determination, phisycal parameter evaluation, residual and applied stresses, are mentioned.

1 INTRODUCTION

The ultrasonic inspection is the most used method in non-desctructive evaluation of materials and structures.

The probe commonly employed in the generation of ultrasonic beams is based on piezoceramic elements. An electric pulse driven by a pulser produces compressions and dilatations of the element generating ultrasonic waves wich are transmitted trough a couplant to material. Because of irreversible character of piezoelectricity, oscillating displacements of the material's surface produce a voltage pulse across piezoceramic element wich may be collected through very thin wires welded to surface ends of the element. The probe may therefore works as transmitter as well as receiver.

In ultrasonic examination specific and severe requirements have to be satisfied. Probe high sensitivity, excellent acoustic contact with materials's surface, negligeable frequency dispersion, low attenuation in material are necessary. Depending on type of material and of defect to be detected and characterized, special wave types, rightly angled beams and proper frequency have to be used.

For piezoceramic probes the acoustic contact with the piece surface has to be assured by means of special pastes or liquid couplants. For

Testing of Metals for Structures. Edited by F.M. Mazzolani. © 1992 RILEM.
Published by Taylor & Francis, 2 Park Square, Milton Park, Abingdon, Oxon, OX14 4RN.
ISBN 0 419 15810 3

this reason is difficult to mantain a good contact for a long time and to obtain its good reproduceability. In the case of rough surface, moving parts, hot bodies, the application of piezo-probes is completely unreliable or quite impossible.

Many of these limits may be overcome using a recently developed techmique (EMAT) based on completely different transducer. In the following, the characteristics, advantages, limits and applications of EMAT will be reported.

2 WORKING PRINCIPLES OF EMAT

An EMAT probe is basically composed of a magnet which induces into the material a static magnetic field, and a coil ran by a suitable RF current pulse in order to produce in the surface region of the material an eddy current field. The interaction between eddy current and static magnetic field generates a Lorentz force field which excites an ultrasonic wave. The phenomenon is reversible and therefore the same probe or type of device can be used as receiver.

3 WAVE TYPE GENERATION

Different types of wave may be generated by EMAT: volume, guided, surface and Lamb waves.

Normal incidence shear polarized volume waves can be obtained using single element transducer. Normal incidence longitudinal volume waves are produced when yoke configuration of the magnet is used. By means of multielement array probe it is possible, changing the frequency of coil current, to generate ultrasonic beams at different angles.

Using plate transducers, 90 angle beams may be produced and both horizontally and vertically guided polarized waves may be established in thin plates, rods or tubes.

Surface waves (Rayleigh and creeping waves) may also obtained taking advantage of limit angle of direct longitudinal waves or of conversion modes of shear waves.

Finally, Lamb waves (symmetrical and antisymmetrical modes) under certain conditions may be generated. Their dispersive character (dependence of propagation velocity of waves on the frequency) however may complicate their application.

4 INSTRUMENTATION

Because of the low sensitivity of EMAT transducer both as transmitter and receiver, a special instrumentation system for probe driving and signal conditioning and analysis must to be used. A special filtering and real time averaging of several signals have to be performed in order to obtain an acceptable signal to noise ratio. The electromagnetic interference have to be also avoided. A compact and limited weight instrumentation system for in field applications has been manufactured by CISE.

5 ADVANTAGES AND LIMITS OF EMAT

Various advatages may be attained using EMAT.

Because physical contact with metal surface is not required, the investigation may be applied to inspection without using any couplant and to examination of bodies with rough, rusted, painted, scaled surfaces. Also hot items and moving elements can be investigated avoiding easy failures and considerable wearing often occuring in the case of piezolelectric probes.

In addition, the generation of different wave types allows to examine defects generally difficult to detect owing their location in the material of the component. Using surface or guided or shear polarized or variable angle waves, it is possible to investigate important defects for the structural integrity as surface and subsurface flaws, cracks positioned at interface between cladding and base material. A unique and useful application is the examination of defects located at very large distances in thin plates, tubes and rods by means of guided waves for which geometrical attenuation is negligeable: the inspection in this case may be performed without moving the probe.

On the contrary considerable and severe limits must be considered. The sensitivity of EMAT is very low in comparison with piezoceramic probes (up to a factor 100) and in any case EMAT method is applicable only to metal materials. Finally the manufacturing of small probes is quite difficult.

6 APPLICATIONS

Owing to above mentioned special characteristics, EMAT may be convenientely used in many applications as Non-Destructive Evaluation method.

Detection, sizing and characterization of defects especially in thin plates, tubes and rods can be obtained. By using guided waves it is possible to detect flaws positioned at greater and greater distances as thinner and thinner are the thickness of the component to be examined. Items as heat exchanger tubes (1.2), coatings, rolled thin sections, reinforcing rods, tie bars, bolts, cladded bodies, etc. may efficientely investigated.

The thikness of plates and tubes can be determined with good precision through time of flight measurement. The overlapping technique of echos is applied to reach good precision in time of flight evaluation and cross correlation algorithm can be applied to improve signal to noise ratio.

Physical parameters determination may be obtained. Elastic modules or other physical bulk parameters related to material condition may be deduced measuring ultrasonic velocities in the medium.

Residual and applied stresses may be measured in the volume of the material. The birefringence method may be applied (3). Polarized shear waves are used. Indipendent measurements at normal and parallel direction to load are performed. Stressed zones induce velocity variations of ultrasonic beams which must to be measured with a precision not less than 10(exp-4). The great advantage here in using EMAT is that contact and couplant to the metal surface are not necessary.

Own to contactless application of EMAT and to unnecessity of a couplant, hot bodies may be examined. Cooling, of course, can be required depending on temperature of the item to be investigated and the lift off distance between probe and metal surface. Pressure vessels, pipes, valves, sheets, during plant operation, may be examined. Inspection of items during production may also performed.

7 CONCLUSIONS

EMAT method and related intrumentation have attained an advanced stage of development.

Systematic measurements, however, have to be performed depending on specific application in order to reach full validation and required industrial reliability.

Other applications to aerospace and civil structures field may be envisaged and convenientely investigated.

8 REFERENCES

1) M. Certo, S. Ghia, V. Regis, F. Tonolini. NDT and Monitoring Innovative Technologies to Improve Power Plant Availability. Proceedings of the 12th World Conference on Non Destructive Testing. Amsterdam, April 23-28 , 1989 Vol.1 pg 610-611.

2)M. Certo, E. D'Alessio, L. Guerini, S. Ghia, A. Grassi. Preheater Inspection by EMAT Transducer. Proceedings of International Conference on Monitoring, Surveillance and Predictive Maintenance of Plants and Structures. Taormina, October 16-18, 1989 Vol.1 pg 125-130.

3)M. Certo, S. Ghia. Residual Stresses Evaluation by Ultrasonic Measurement. Proceedings of the 12th World Conference on Non-Destructive Testing. Amsterdam, April 23.28, 1989. Vol.2, pg. 1481-1483.

PART FIVE

GEOMETRICAL AND MECHANICAL IMPERFECTIONS

30 NEEDS FOR EXPERIMENTAL MEASURES OF GEOMETRICAL FRAME IMPERFECTIONS

A. DE LUCA
Istituto di Scienza e Tecnica delle Costruzioni, Università della Basilicata, Potenza, Italy

SUMMARY

In this paper the problem of evaluating frame geometrical imperfections is analyzed with reference to the following points: first to define, on the basis of available experimental results which are the allowable tolerances for erection purposes second to define the values which can be neglected in order to carry out the analysis on the ideal perfect structure.

Particular emphasis is placed to the Eurocode approach which requires to introduce in any case frame geometrical imperfections in the analysis. Finally, in order to evaluate, by means of numerical simulations, the effects of these imperfections and therefore the magnitude which could be neglected and the interaction between frame and bar imperfections, some results already achieved in this field are presented.

1 INTRODUCTION

The possibilities offered by the numerical models and by the personal computer capacities lead to the attempt to take into account all possible imperfections of the structure, of the element and of the cross section in order to more closely simulate real situations. It has to be said though that still it is needed, in the design office, to adopt simplified models which allow to synthetically design the structure without loosing the physical understanding of the behaviour while performing cumbersome nonlinear analysis of highly imperfect structures.

It is believed that it is should be a task of research to investigate on all possible effects of imperfections by means of experimental testing and by numerical simulation. Research has then the task of formulating equivalent simple models to be used in design. As an example we can refer to the column buckling check. Even though it is well known by any designer which are the geometrical and mechanical imperfections of the compressed bar and in which manner it could be possible to include all these imperfections in a nonlinear analysis, it is very seldom that this type of more accurate analysis could be

Testing of Metals for Structures. Edited by F.M. Mazzolani. © 1992 RILEM.
Published by Taylor & Francis, 2 Park Square, Milton Park, Abingdon, Oxon, OX14 4RN.
ISBN 0 419 15810 3

more effective than a simple check carried out by any code buckling formula.

It has to be said though that, contrarily to what happens in the case of compressed bars, where all possible type of imperfections have been measured and analyzed with respect to their effect on the column load bearing capacity, in the case of frames not enough research has been performed in order first to measure all the frame imperfections and then evaluate their effects on the frame load bearing capacity. Several studies have been performed on specific problems not following a consistent program for evaluating all frame imperfections effects analogously to what was done for columns.

Due to the difficulties which arise when trying to investigate in a systematic manner on the effects of all types of frame imperfections, only recently the concept of *"Imperfect Frame"* [1] has been introduced.

In this paper the need of measuring frame geometrical imperfections which are already present in more recent codes [2], despite of the limited number of available experimental and numerical results, is evidenced. The concept of *"Industrial Frame"* [3] is used and the results of research carried out at the Istituto di Tecnica delle Costruzioni in Napoli [4,5,6] are presented.

2 THE CONCEPT OF IMPERFECT FRAME

The concept of "Imperfect Frame" [1,3], can be derived by an analogy with the "Imperfect Column" affected by geometrical and mechanical imperfections as it is shown in figure 1 [3]. In particular, as in the case of columns an initial camber equal to L/1000 is unavoidable, the industrial frame cannot be unaffected by erection tolerances which result in out of plumb of each story and of the entire structure (fig. 1a). In both cases this geometrical imperfection results in bending stresses deriving from axial loads. It is worthed pointing out that, while for columns the effect of out of straightness is taken into account in design by means of the stability check which makes use of the appropriate buckling curve, in the case of frames there is usually no code allowance for such misalignment effect.

The most effective bar mechanical imperfections are known to be residual stresses which are nonuniformly distributed along the cross section and can be evidenced by a stub column test where the real cross-section constitutive relationship is not coincident with the ideal one. In the same manner also the semirigid connection can be considered as a mechanical imperfection which makes the real moment-rotation relationship different from the ideal one (fig. 1b). If the bar is discretized into a finite number of longitudinal elements, each one of these elements will be affected by residual stresses and therefore

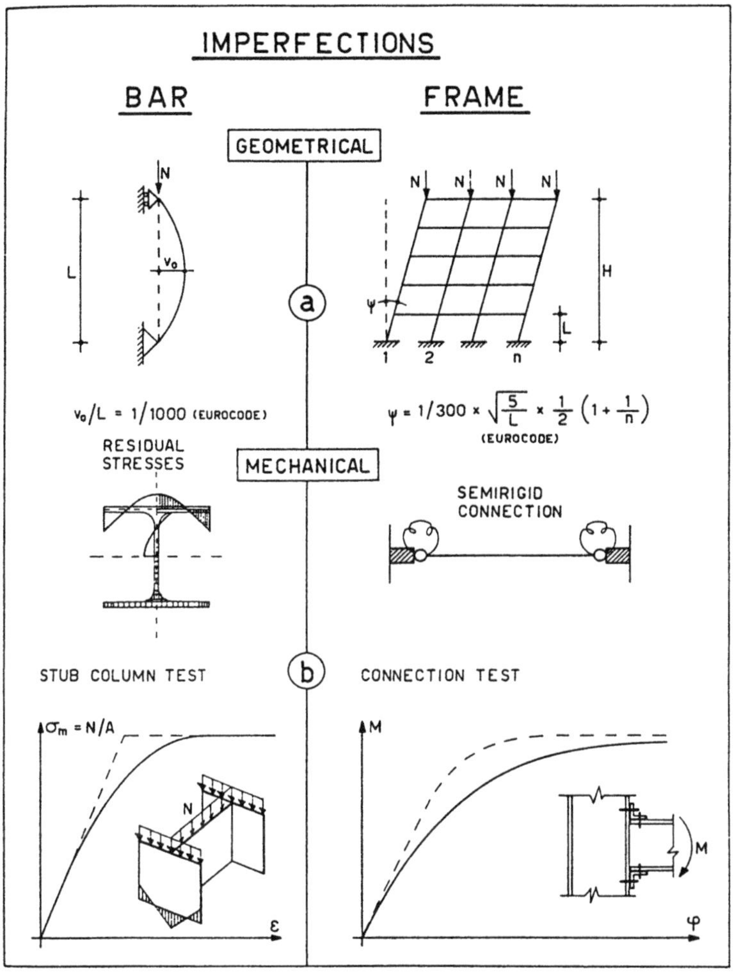

Fig. 1 *"Industrial bar" and "Industrial frame" imperfections [3].*

the moment -curvature relationship of the bar element itself will be modified. Analogously, since a framed structure is formed by several bars, they can be assumed as representative elements of a discretization in which, at each joint, a new imperfection, represented by the nonlinear moment-rotation relationship, has to be introduced in order to simulate the actual beam-to-column connection.

The effects of imperfections, which lead to anticipation of plastic phenomena in the structure, are evidenced in fig.2 where it is shown how geometrical imperfections lead to anticipated yielding of cross-section fibers in the column and

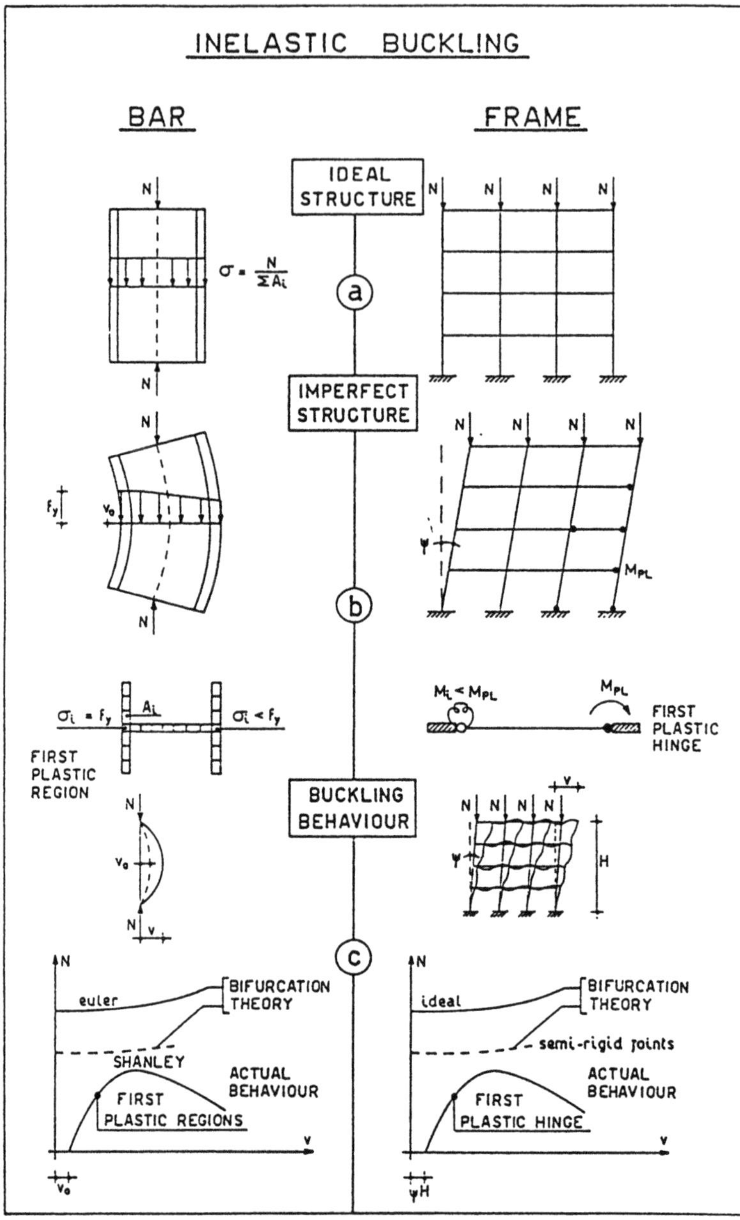

Fig. 2 *Effects of imperfections on plastic behaviour of structures [3].*

to early plastic hinges formation in the frame. In the same figure are also provided the real load deflection curves as compared to the ideal ones.

3 IMPERFECTION TOLERANCES AND EQUIVALENT IMPERFECTIONS

The definition of frame geometrical imperfections (story ang global out-of-alignment) has two implications which affect the different aspects of fabrication and design.

From the fabrication point of view it is important to define the maximum allowable tolerances which can be tolerated for the erection of the structure. This value represents a measure of quality and has no direct implication on the analysis of the structure.

From the design point of view it is necessary to define the maximum value of geometrical imperfection which allows to assume the real imperfect structure equivalent, in *engineering sense*, to the ideal one.

It is obvious that the two problems interact each other since they are dealing with the same physical measure of the deviation of the real structure from the ideal perfectly straight one.

In line to what has been anticipated in the introduction, i.e. to provide simple design methods sufficiently accurate to represent the real imperfect structures without performing cumbersome computations, the following approach is suggested. A first step should be represented by the experimental measures, on different typologies of buildings, of the erection mis-alignment of columns. The second step should be to carry out numerical simulations which allow to evaluate the magnitude of this misalignemnt which can be considered negligible from the design point of view.

This approach is not followed by the present codes. In fact in most codes the problem of frame imperfection is practically neglected. In Eurocode [2] instead, both the erection tolerances and the equivalent geometrical imperfection to be assumed in the analysis are provided. For what concerns allowable tolerances the expression suggested by the Eurocode is the following:

$$\frac{e}{\sum h} = \psi = \frac{0.0035}{\sqrt{NP}} \tag{1}$$

where:

e represents the deviation of a column in a multistorey building at any floor level, from a vertical line through the intended location of the column base

$\sum h$ is the total height from the base to the floor level concerned,

and:

NP is the number of storey from the base to the floor level concerned.

For what concerns the frame imperfections to be introduced

in the analysis the Eurocode, within section 5.2.4 on allowance for imperfections, provides at section 5.2.4.2 the method of application at clauses (1) and (2):

Imperfections shall be allowed for in the analysis by including appropriate additional quantities, comprising frame imperfections, bracing systems imperfections and member imperfectios.

The effects of frame imperfections (see 5.2.4.3) shall be included in the global analysis of the structure. The resulting forces and moments shall be used for member design.

It is clear then that frame imperfections cannot be neglected when adopting Eurocode. For what concerns the values to be used the Eurocode, at section 5.2.4.3 Frame imperfections, provides the following indications:

The effects of imperfections shall be allowed for in frame analysis by means of an equivalent geometric imperfection in the form of an inital sway imperfection determined from:

$$\psi = K_c K_s \psi_0 \tag{2}$$

with: $\psi_0 = 1/200$

$$K_c = \sqrt{0.5 + 1/NC} \quad but \quad K_c < 1.0$$

$$K_s = \sqrt{0.2 + 1/NP} \quad but \quad K_s < 1.0$$

where: NC is the number of columns per plane

and: NP is the number of storeys.

In figure 3, together with the suggested values of equation (2), are also given the imperfections suggested by an older version of Eurocode 3 [7] and the tolerances suggested by the ECCS Recommendations [8].

From figure 3 it can be observed that the present version of Eurocode 3 has practically doubled the values of the equivalent imperfections to be used in the analysis with respect to the 1984 version which was closer to the recommended ECCS tolerances. It is also worthed pointing out that the ECCS values for tolerances were based on studies devoted to experimental measures and subsequent statistical evaluations of frame imperfections [9, 10, 11, 12].It seems though that these studies were not so extensive to allow an extrapolation to all possible typologies. Some of these studies were in fact concentrated with core-braced buildings, as represented in fig.4, with small coefficient of correlation (fig.5).

Another point to be underlined is that the Eurocode requires to include both the frame out-of-plumb and the cross section and member imperfections as it was reported previously. This concept, evidenced in figure 6, should suggest that the Eurocode imperfection of the frame is not an equivalent one but should

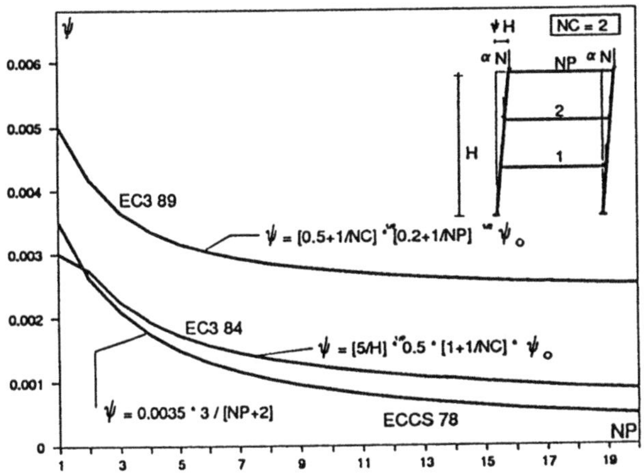

Fig.3 *Geometrical imperfections suggested by different codes.*

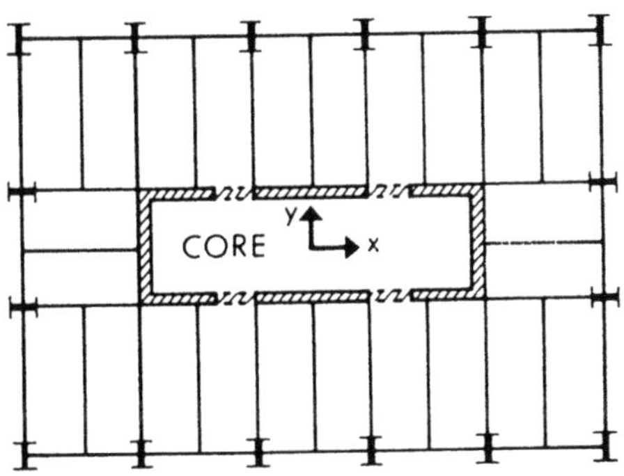

Fig.4 *Plan view of building with measured out-of-plumbs [9].*

be closer to the actual measured one and therefore to the values suggested in the same code as erection tolerances.

It is well known in fact that the concept of equivalent geometrical imperfection, first introduced by Perry-Robertson for evaluating the buckling load of columns, allows to take into account of geometrical as well as mechanical imperfections by means of an appropriately increased geometrical imperfection.

4. NUMERICAL MODELS AND RESULTS

Several studies [1,4,5,6] have been performed for evaluating the behaviour of frames affected by cross-section, bar and frame imperfections. In these studies the frame slenderness:

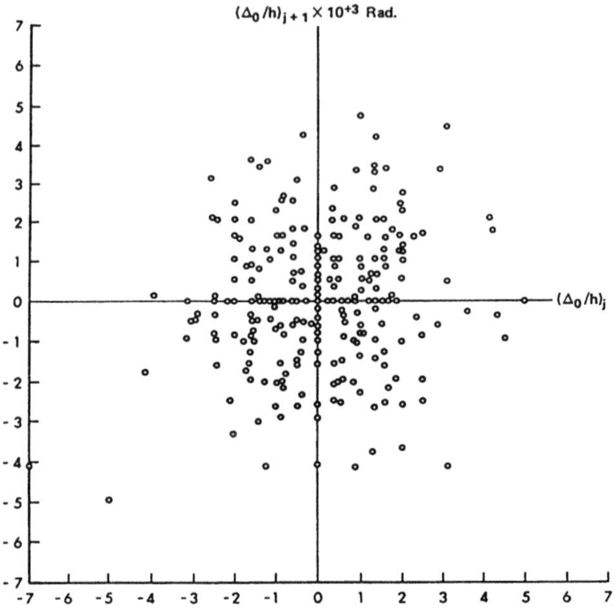

Fig. 5 *Correlation between two out-of-plumb columns adjacent in a storey [10].*

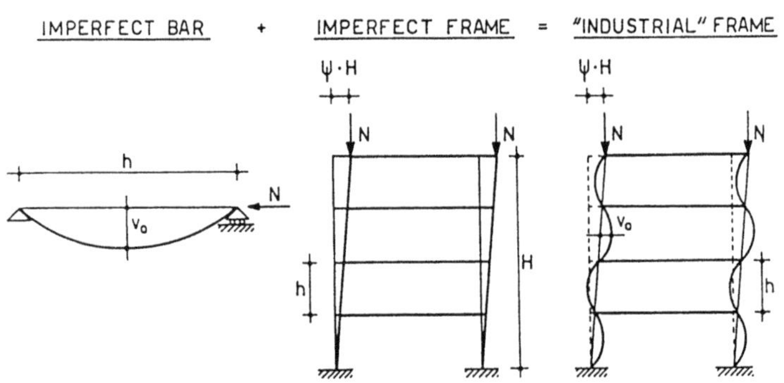

Fig.6 *Combination of bar and frame imperfections [5].*

$$\lambda_f = \sqrt{\frac{\alpha_y}{\alpha_c}} \tag{3}$$

has been considered as a main governing parameter in analogy to the nondimensional column slenderness:

$$\bar{\lambda} = \frac{1}{\pi}\ \frac{h}{r}\sqrt{\frac{f_y}{E}} = \sqrt{\frac{N_y}{N_c}} \tag{4}$$

and parametric analysis have been carried out in order to evaluate on the following effects:

1 the ratio between bending stiffness of the beam (modified for taking into account the semirigid connection) and bending stiffness of the column;

2 the law which defines variation of stiffness along the height;

3 the topology of the frame.

4 the ratios between columns and beams plastic moments, having assumed full strength connections;

5 the nonlinear law characterizing the semirigid connection;

6 the conventional slenderness of the frame which can be evaluated by means of the ratio between the elastic critical multiplier and the multiplier of vertical loads which leads to yielding of the highly stressed columns;

7 the amount and the distribution of local imperfections (residual stresses and out-of-camber) and global imperfections (out-of-plumb, misalignment).

A typical load-displacement curve of a frame affected by cross-section, bar and frame imperfections is the one represented in figure 7.

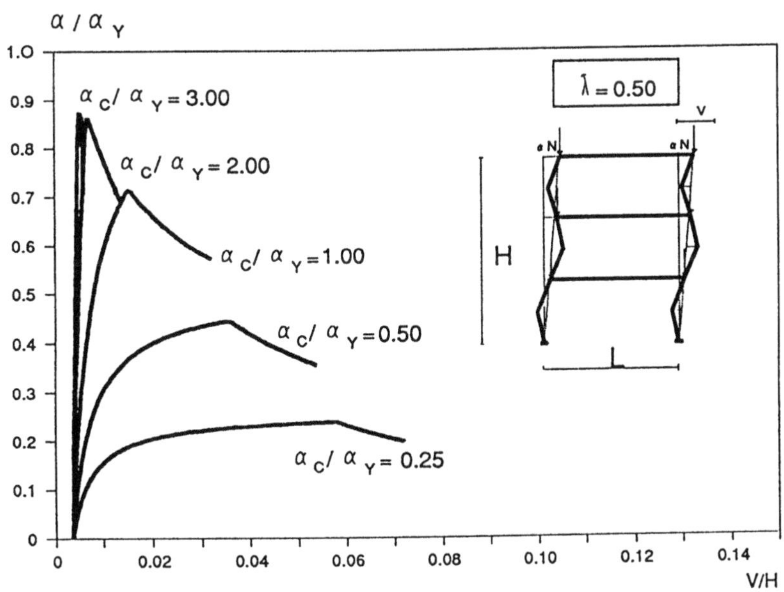

Fig.7 *Nonlinear load-displacement curve of imperfect frame [6].*

The extension of the results to the case of semirigid frames has also been analyzed in [5] with reference to portal frames. In figure 8 are provided the ultimate values of multiplier of vertical loads, nondimensionalized with respect to the ones obtained in the case of rigid frames, versus connection stiffnesses.

For what concerns the effect of interaction between bar and frame imperfections, some results can be found in [6] where it has been shown that different patterns of bar imperfections affect the ultimate multiplier of the frame by an amount smaller than 10%. In figure 9 are synthetically represented these results. The four frames examined for this purpose are provided

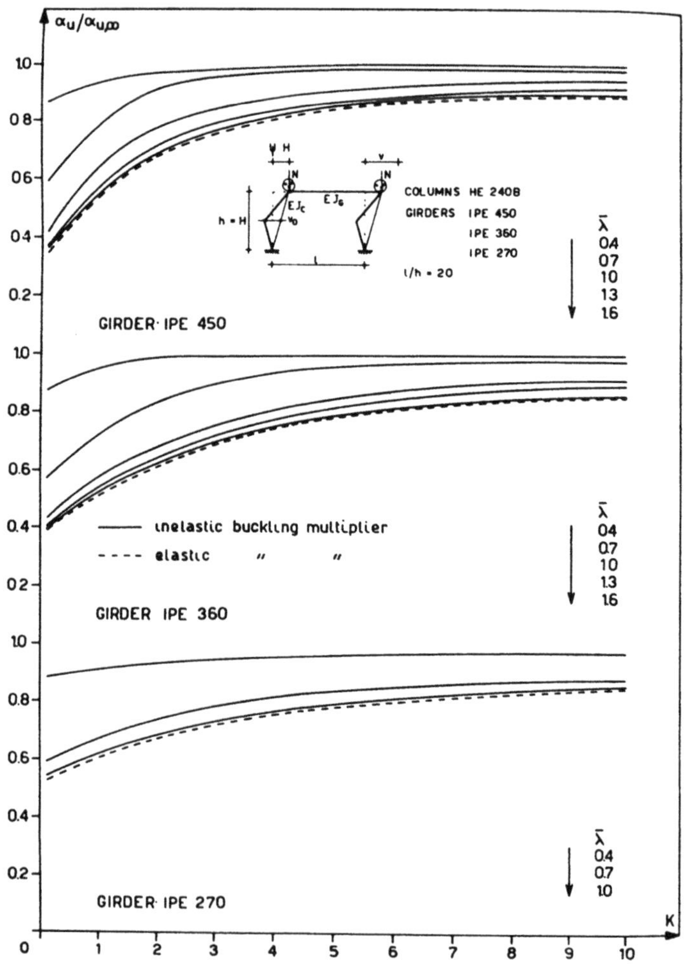

Fig. 8 *Inelastic buckling multiplier normalized with respect to 'rigid framing' case [5].*

in the upper part of the figure and they respectively refer to the continuous, dashed, dotted and dashed-dotted lines of the bottom part of the figure. Different values of the global slenderness of the frame, previously defied by equation (3), have been examined. In the figure is diagrammed the value of the ultimate multiplier normalized with respect to the squash load of the highly stressed column for different values of nondimensional column slendernesses.

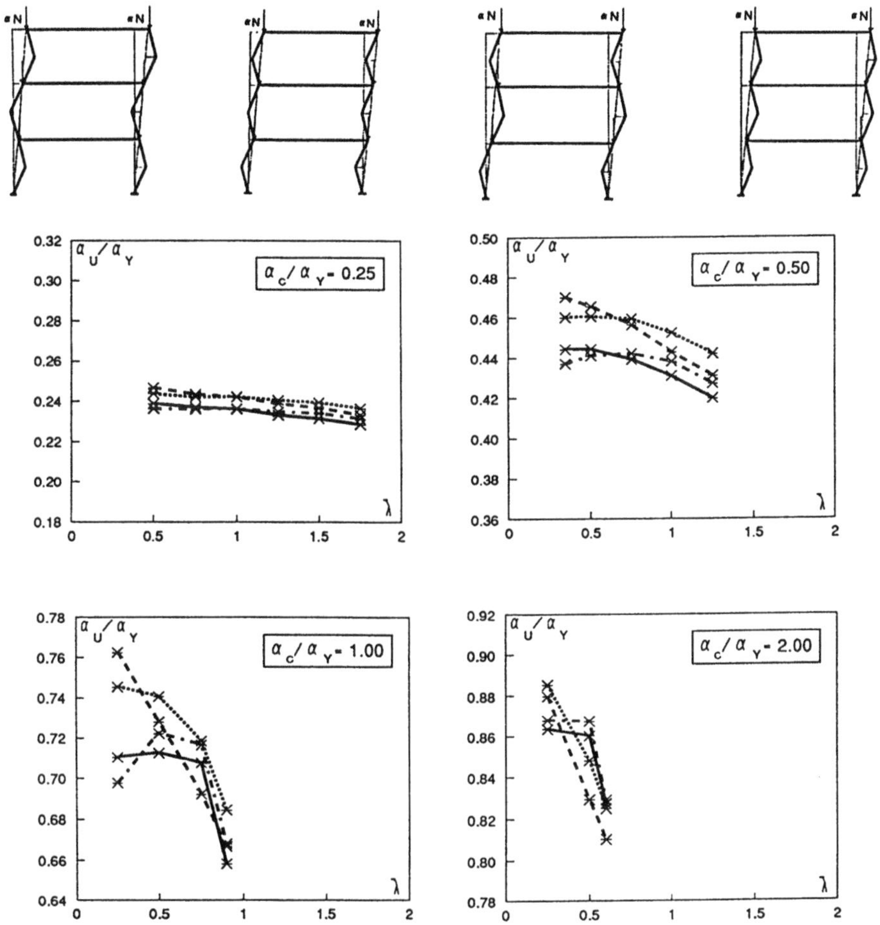

Fig.9 *Effects of bar imperfections on inelastic buckling multiplier of imperfect frames.[6]*

5 CONCLUSIONS

It seems that extensive research is needed in the field of geometrical imperfections of frames with particular reference to the following points:

1. In field measures of out-of-plumbs of frames pertaining to different typologies i.e. core-braced frames, unbraced frames, industrial frames and subsequent statistical evaluation of experimental data.
2. Numerical simulation of frame affected by cross-section, bar and global imperfections for different typologies and load cases.
3. Definition of equivalent frame geometrical imperfections to be used in the analysis.
4. Formulation of new simplified design methods which do not separate column check from global frame verification.

6 REFERENCES

[1] De Luca, A., Effetti delle imperfezioni sulla capacita' portante di telai industriali metallici - *Tesi di Dottorato*, Universita' di Napoli, Aprile 1987.

[2] Commission of the European Communities, Eurocode No.3: Design of steel structures Part 1 - General rules and rules for buildings, Edited draft Issue 4 July 1990.

[3] Cosenza, E., De Luca, A., Faella, C., Mazzolani, F.M., Imperfection sensitivity of "industrial" steel frames - *Steel structures : Advances, design and constructions*, Elsevier Applied Science, Cardiff, July 1987, 326-336.

[4] Cosenza, E., De Luca, A., Faella, C., The concept of "frame slenderness" for the definition of "frame stability curves", SSRC 1988 *Annual Technical Session-Minneapolis*, SSRC, Bethlehem, 1988, 59-71.

[5] Cosenza, E., De Luca, A., Faella, C., Inelastic buckling of semirigid sway frames, Chapter 9 in: *Structural connections: stability and strength*, London, Elsevier Applied Science, 1989, 297-333.

[6] De Luca, A., De Stefano, M., Faella, C.,L'effetto delle imperfezioni dell'asta sul carico di stabilita' inelastico di strutture intelaiate, *Proceedings XII Congresso CTA*, Capri, 1989, 509-520.

[7] Commission of the European Communities, Rapport EUR 8849, Eurocode No.3: Common unified rules of steel structures, Bruxelles, (1984) 95.

[8] European Convention for Constructional Steelwork, European recommendations for steel construction, Bruxelles, ECCS, (1978) 234.

[9] Beaulieu, D., Adams, P.F. The results of a survey on structural out-of-plumbs *Canadian Journal of Civil Engineering*, Vol.5, N. 4, (1978), 462-470.

[10] Beaulieu, D., Adams, P.F. Probability considerations in assessing the effects of structural out-of-plumbs *Canadian Journal of Civil Engineering*, Vol.6, N. 4, (1979), 610-616.

[11] Lindner, J., Ungenwolte Sciefstellungen von Stahlstutzen, *IABSE 12th Congress*, Vancouver, (1984).

[12] Baehre, R. Carlsson, G. Stalstommars mattnoggrannhet i hallbyggnader, *Byggforskningen*, Stockholm, Rapport R39:1972.

31 MECHANICAL IMPERFECTIONS: TESTS AND MODELS

F.M. MAZZOLANI
Istituto di Tecnica delle Costruzioni, Faculty of Engineering,
University of Naples, Italy

SUMMARY

The last 25 years were devoted to the experimental investigation on mechanical imperfections in metal elements. The interpretation of these results by means of mathematical models allowed the numerical simulation of the buckling phenomena of steel and aluminium columns.
The main features of this activity are presented here with particular reference to hot-rolled steel profiles, extruded and welded aluminium alloy profiles, welded tubular steel profiles.

1 HISTORICAL BACKGROUND

The beginning of sixties led to the unquestionable abandonment of the concept of a perfect straight "theoretical bar" made of isotropic and homogeneous material and to the adoption of the more realistic concept of an "industrial bar" affected by inevitable imperfections of accidental nature. These imperfections were subdivided - as it is well known - into two main categories: geometrical and mechanical imperfections.

Geometrical imperfections are usually understood to be the differences between the nominal and the actual geometry of the structural element, either for the longitudinal dimensions or for the transversal dimensions of the cross-section.
Mechanical imperfections can be grouped into two different types:
- residual stress distribution;
- inhomogeneous distribution of mechanical properties.

Since that time, the evaluation of the load bearing capacity of steel members started to take into consideration such imperfections - As a consequence in the analysis of buckling problems, the bifurcation methods were gradually substituted by simulation methods, allowing a greater versatility in introducing all kind of imperfections in the mathematical models.

It was, therefore, necessary to plan the development of wide experimental research in order to find out the most common distributions of imperfections, and to assess mechanical models based on testing results to be introduced in the simulation methods.

As far as residual stresses are concerned, the universally used experimental technique is also now the sectioning test method, first used in the USA by the Column Research Council

Testing of Metals for Structures. Edited by F.M. Mazzolani. © 1992 RILEM.
Published by Taylor & Francis, 2 Park Square, Milton Park, Abingdon, Oxon, OX14 4RN.
ISBN 0 419 15810 3

(CRC), now Structural Stability Research Council (SSRC), and afterwards adopted in Europe - This method - as it is well known - consists of slicing different parts of the profile into longitudinal, parallel strips and measuring deformation through strain gauges (fig. 1). The resulting strips can be after used for tensile tests in order to obtain the geographical distribution of mechanical properties of materials.

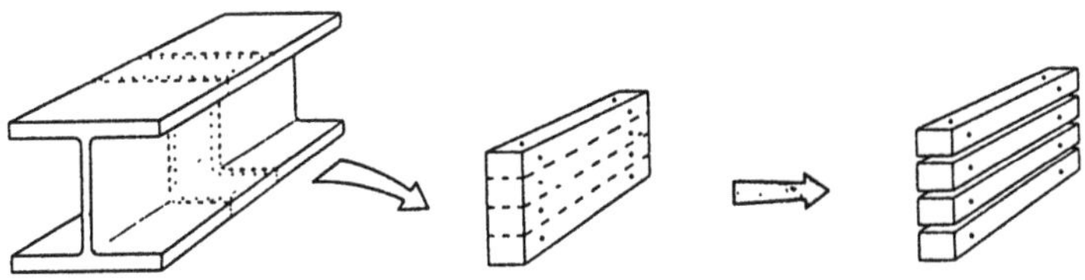

fig.1

An indirect way to have a global answer of the presence of both residual stresses and variable distribution of yield stress over the cross section is given by the "stub column test" (fig. 2). This approach was widely applied to structural profiles made of steel and aluminium alloys in order to evaluate their load bearing capacity and to codify the corresponding results [1,2].

The personal activity of the Author was devoted to these problems in the last 25 years. In particular, the main chapters can be characterized as follows:

- period 1965 - 1975: hot rolled steel profiles (references from 3 to 10)
- period 1970 - 1985: extruded and welded aluminium alloy profiles (references from 11 to 23)
- period 1983 - 1990: welded tubular steel profiles (references from 24 to 29).

The mechanical imperfection tests were made at the laboratory of the "Istituto di Tecnica delle Costruzioni" of the Engineering Faculty of the University of Naples. The exploitation of testing results was done by means of appropriate simulation methods which allow the evaluation of the influence of such imperfections on the buckling performance of metal bars.

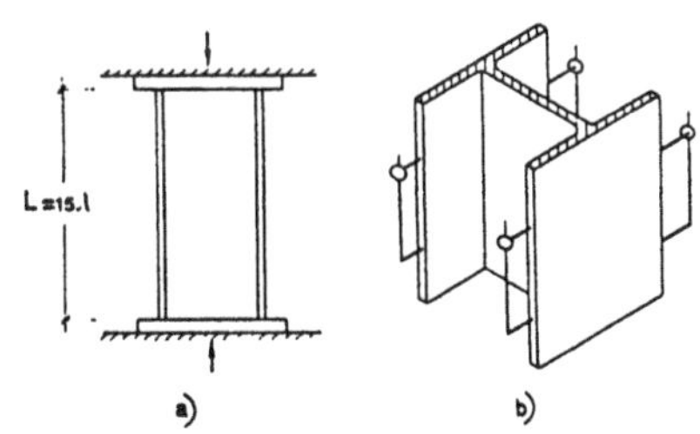

fig.2

2 HOT-ROLLED STEEL PROFILES

In these sections residual stresses are of a thermal nature, due to the cooling process which follows the hot rolling (fig. 3). Their distribution can be afterwards modified because of the mechanical process of straightening (fig. 4).

In case of I sections, the experience has so far shown that longitudinal residual stresses due to cooling in the flanges and in the web are closely related to the geometrical ratios which define the cross-sectional shape, due to the physical fact that the compactness of each cross-section influences the distribution of temperature during cooling.

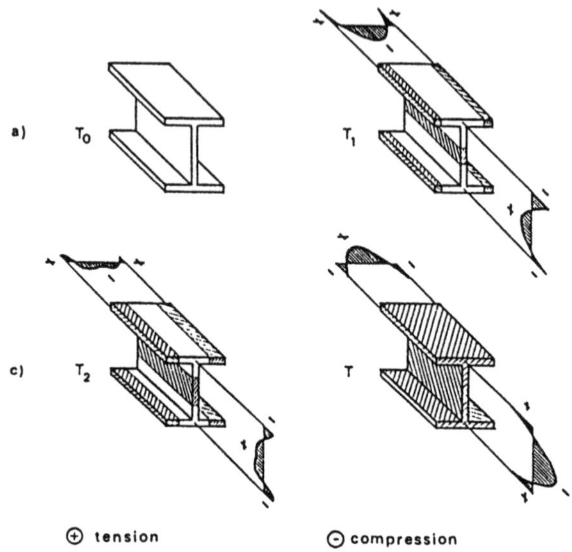

fig.3

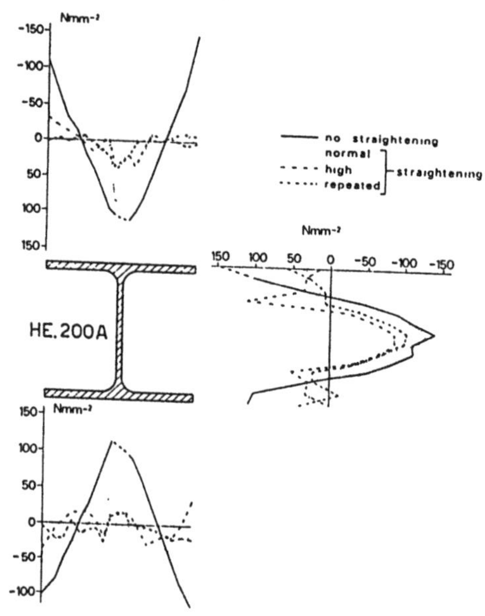

fig.4

Typical distributions of residual stresses for various type of I sections are shown in fig. 5: they synthesize the results of many US and European tests. The straightening procedure most commonly used in the steel industry consists of introducing the cold bars into a train of rolls which submits them to alternate bending. This process is useful in relieving thermal residual stresses (fig. 4).

The mechanical imperfection measurements made in Italy at the University of Naples were devoted to the analysis of the following shapes:
HEA 100, HEB 100, HEM 100, HEA 200, HEB 200, HEM 200, HEA 300, HEB 300, HEM 300, IPE 200.

The model of distribution of residual stresses for H - shapes was based on testing data and it was characterized by a linear distribution at the ends and a parabolic one in the middle both of flanges and web (fig. 6). This model was assumed to be internal to the area of testing results.

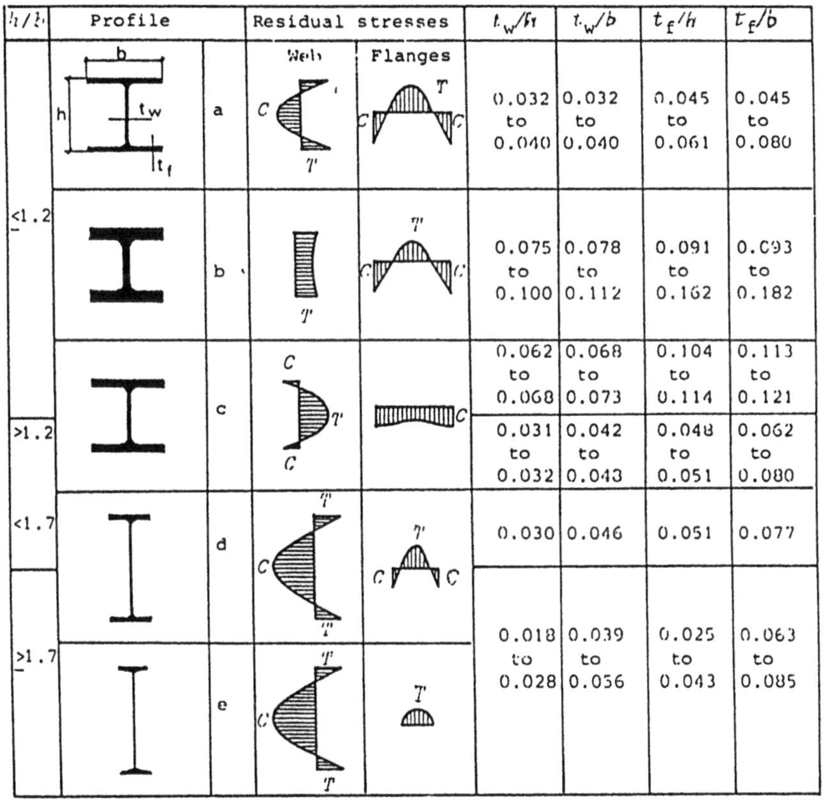

h/b	Profile	Residual stresses	t_w/h	t_w/b	t_f/h	t_f/b
≤ 1.2	a	Web / Flanges	0.032 to 0.040	0.032 to 0.040	0.045 to 0.061	0.045 to 0.080
≤ 1.2	b		0.075 to 0.100	0.078 to 0.112	0.091 to 0.162	0.093 to 0.182
≤ 1.2	c		0.062 to 0.068	0.068 to 0.073	0.104 to 0.114	0.113 to 0.121
>1.2	c		0.031 to 0.032	0.042 to 0.043	0.048 to 0.051	0.062 to 0.080
<1.7	d		0.030	0.046	0.051	0.077
≥ 1.7	d–e		0.018 to 0.028	0.039 to 0.056	0.025 to 0.043	0.063 to 0.085

fig.5

Different distributions of residual stresses coming from this investigation are given in fig. 7; these models were introduced into the simulation program in order to emphasize the different influences played by residual stresses on the shape of buckling curves (fig. 8):
- the same profile with different distributions of residual stresses (fig. 8a);
- the same depth in different profiles (fig. 8b);
- the same series with different depths (fig. 8c).

On the same profiles, the geographical distribution of mechanical properties was measured. The results in term of yield stress values and scatters are given in fig. 9, by confirming that the average values in the flanges are always less than in the web from 7 to 26 %.

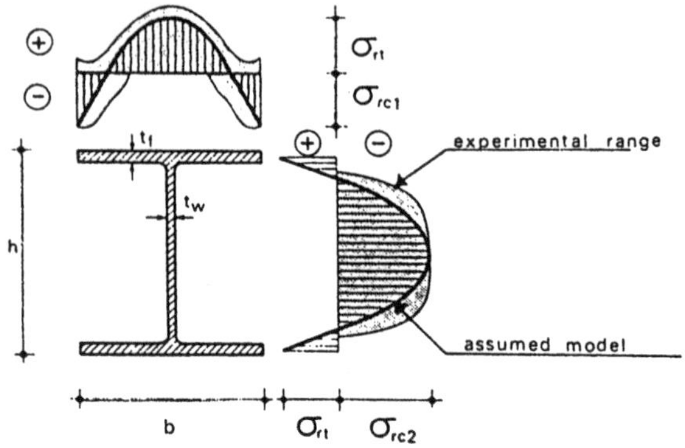

fig.6

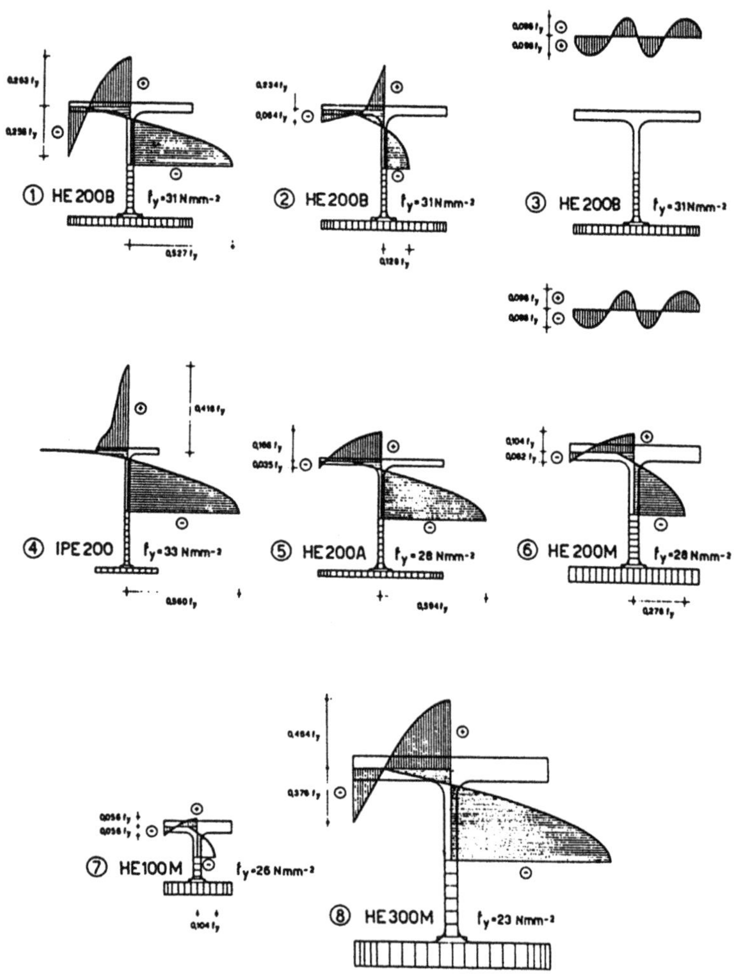

fig.7

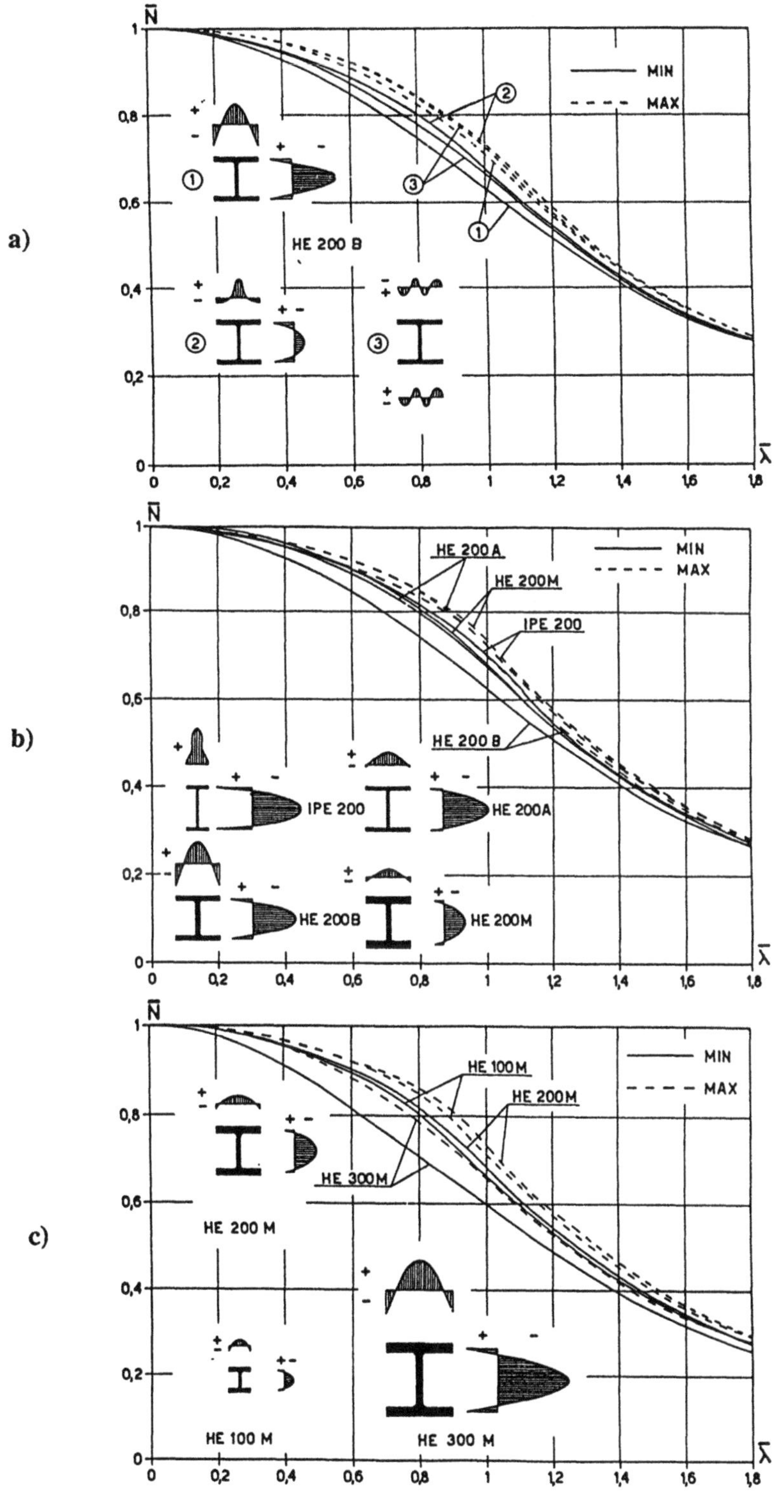

a)
MIN
MAX
HE 200 B
b)
HE 200 A
HE 200M
IPE 200
HE 200 B
IPE 200
HE 200A
HE 200B
HE 200M
c)
HE 100 M
HE 200M
HE 300M
HE 200 M
HE 100 M
HE 300 M

fig.8

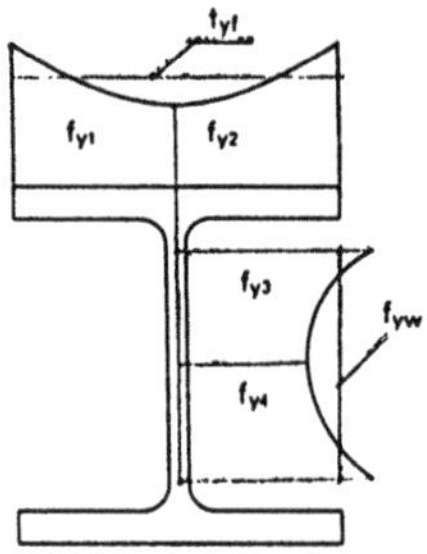

a)

f	HEA 100	HEB 100	HEM 100	HEA 200	HEB 200	HEM 200	HEA 300	HEB 300	HEM 300	IPE 200
1) f_{y1}/f_{yf}	1,06	1,07	1,07	1,04	1,09	1,06	1,07	1,15	1,80	1,02
2) f_{y2}/f_{yf}	0,90	0,95	0,95	0,99	0,95	0,94	1,10	0,97	0,99	0,97
3) f_{y3}/f_{yw}	0,99	1,02	1,02	1,23	1,31	1,36	1,08	1,08	1,00	0,95
4) f_{y4}/f_{yw}	0,96	0,99	0,99	0,97	0,85	0,82	0,98	1,00	1,05	1,03
5) f_{yf}	30,02	31,84	25,35	26,78	28,36	24,49	27,06	24,69	23;08	30,20
6) f_{yw}	33,94	35,38	28,33	30,42	34,30	32,96	29,45	26,56	24,69	36,80
7) f_{yf}/f_{yw}	0,88	0,90	0,89	0,88	0,82	0,74	0,92	0,93	0,93	0,82
8) f_{ym}	31,98	33,61	26,84	28,60	31,33	28,72	28,25	25,62	23,88	33,50
9) f_{yc}	30,98	32,65	25,95	27,49	29,73	26,30	27,66	25,13	23,40	32,87

b)

fig.9

3 ALUMINIUM ALLOY STRUCTURAL ELEMENTS

During the seventies, the main activity of the ECCS Committee "Aluminium Alloy Structures" was devoted to characterize the aluminium alloy members as "industrial bars", carrying out extensive studies and research in accordance with the most recent trends. The research programs were developed at the Universities of Liege and Naples, together with the Experimental Institute of Ligth Metals (ISML) of Novara.

The computational methods of the European Recommendations for Aluminium Alloy Structures (first edition 1978) have been based on this activity. In particular the buckling curves have been defined on the bases of the results of experimental and numerical analysis on the influence of mechanical imperfections, which were completely unknown up to few years before.

3.1 EXTRUDED PROFILES

There are several physical reasons for predicting that thermal residual stresses caused by cooling in extruded aluminium alloy profiles are smaller than those in similar hot-rolled steel profiles.

The confirmation of such intuition was obtained by some oriented tests. The influence of the kind of alloy was examined on the double T 63 x 63 x 4 extruded shape made of four different alloys (fig. 10).

The conclusions were that:

- the mechanical properties of the material as well as its chemical composition do not affect the intensity and distribution of residual stresses;
- the distribution is very irregular and does not follow any defined law, but the maximum values of residual stresses are always small.

The second investigation was directed to identify the influence of the usual manufacturing phases on the formation of residual stresses, by considering the profile in the following conditions:

- extruded and cooled by air;
- extruded and straightened (about 1 %);

- extruded, straightened and artificially tempered (100° C for 4 hours and 140° C for 24 hours).

The corresponding results are given in fig. 11. Also in this case residual stresses presented irregular distributions with very low values.

After appropriate simulation analysis based on these data, it was decided to neglect residual stresses in the evaluation of load bearing capacity of extruded members for practical purposes.

3.2 WELDED PLATES

In contrast to the case of extruded profiles, residual stresses represent an important mechanical imperfection in welded profiles.

A first investigation was done on simple groove butt - welded joints made of two alloys: one work - hardened (5086) and one heat - treated (7020). The results of measurements are given in fig. 12, where both residual stress and elastic limit ($f_{0.2}$) distribution are shown.

The maximum tensile residual stresses at the weld are higher for the work - hardened alloy (0.69 $f_{0.2}$) than for the heat treated ones (0.31 $f_{0.2}$). On the contrary, the reduction in strength in the so-called heat-affected zone around the weld is greater for the heat-treated alloy (0.56 $f_{0.2}$) than for the work-hardened ones (0.92 f $f_{0.2}$), as physically espected.

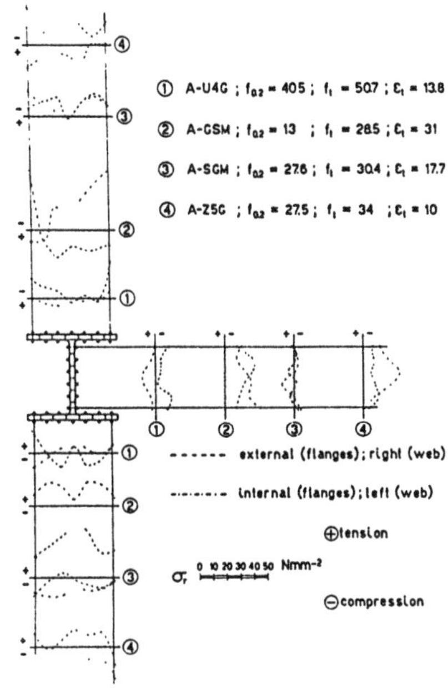

fig.10

3.3 WELDED PROFILES

Three welded profiles made of 6082 alloy were examined:

- T profile, obtained by building - up three plates, as in case of steel;
- P profile, obtained by building - up two extruded Ts as flanges and a plate as web;
- C profile, obtained by building - up two extruded channels and two web plates.

The results of both residual stress and elastic limit distributions are given in fig. 13. We can observe that:

- residual stress distributions are very regular;
- the highest values of residual stresses are in tension at the welds, where they reach the elastic limit of the weld material (about 140 Nmm-2)
- the reduction of elastic limit in the heat-affected zones confirms the results obtained on welded plates.

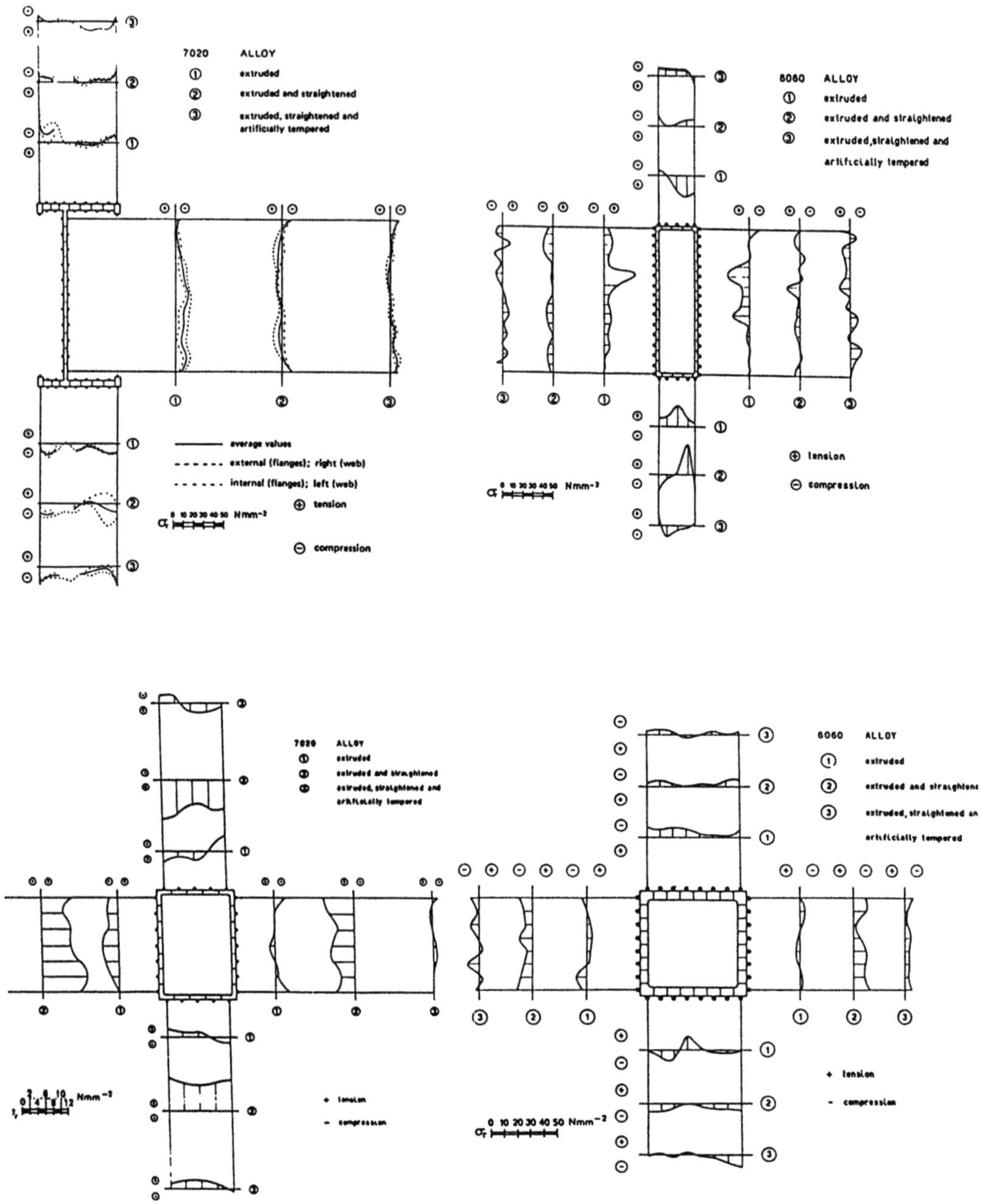
7020 ALLOY
① extruded
② extruded and straightened
③ extruded, straightened and artificially tempered
average values
external (flanges); right (web)
internal (flanges); left (web)
⊕ tension
⊖ compression
Nmm⁻²
6060 ALLOY
① extruded
② extruded and straightened
③ extruded, straightened and artificially tempered
⊕ tension
⊖ compression
0 10 20 30 40 50 Nmm⁻²
+ tension
- compression

fig.11

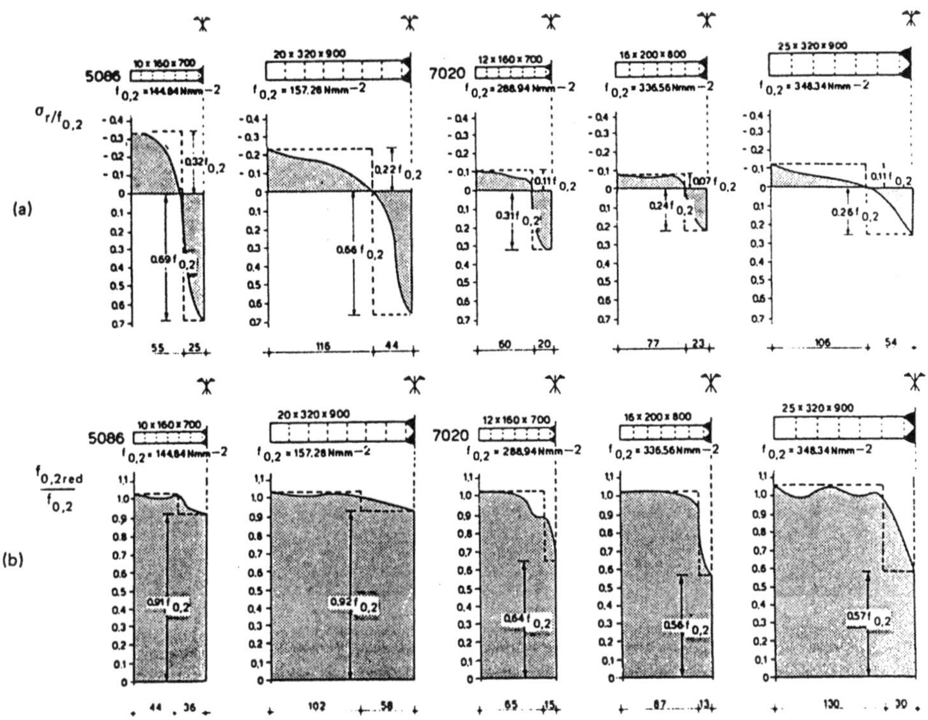

fig.12

The corresponding distribution models were assessed on the bases of testing results (see dotted diagrams of fig. 13). They were used in the numerical simulation which led to the definition of the calculation method for welded columns codified in the ECCS Recommendations (fig. 14).

4 TUBULAR STEEL PROFILES

In 1983, Cometube in cooperation with the Committee of European Communities and with the financial aid of CECA, sponsored a common research program with the participation of several European laboratories (TNO, Delft; University of Karlsruhe; University of Naples; Polytechnic of Kingston; CETIM of Senlis, France; University of Liege).

The main purpose of this research was to compare the different methods for measuring residual stresses in square hollow sections made of steel with different forming processes. The discussion on the final results led to a complete judgment on the reliability of each method used in the different laboratories.

The square cross-section 200 x 200x 8 were delivered in three different types:

- type A, cold-formed after welding with releasing;
- type B, cold-formed after welding without releasing;
- type C, hot-formed.

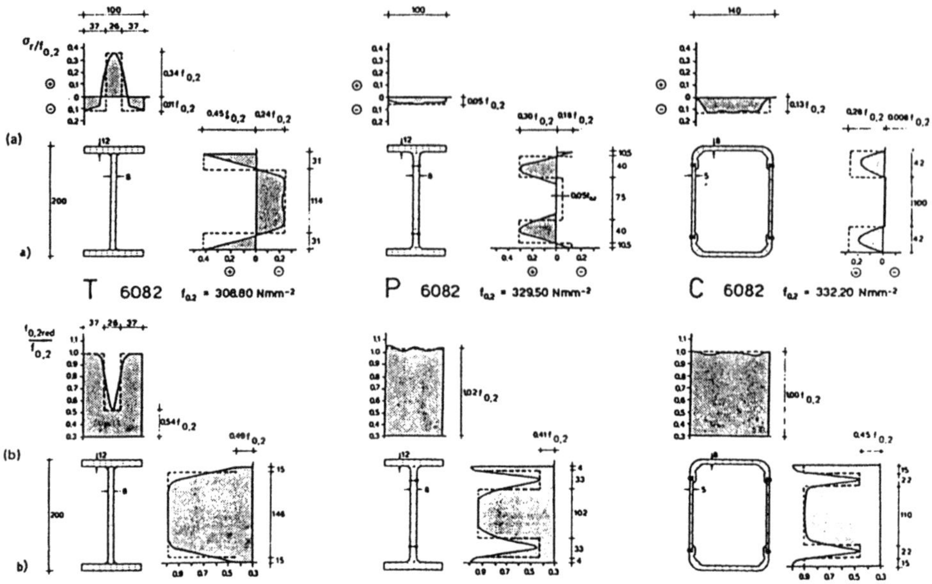
T 6082 $f_{0,2}$ = 308.80 Nmm-2
P 6082 $f_{0,2}$ = 329.50 Nmm-2
C 6082 $f_{0,2}$ = 332.20 Nmm-2

fig.13

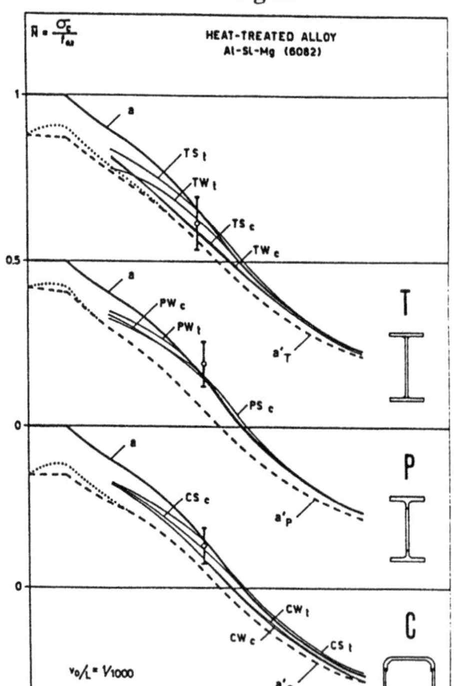
HEAT-TREATED ALLOY
Al-Si-Mg (6082)
T
P
C
v0/L = 1/1000

fig.14

The results of the sectioning tests made at the University of Naples (fig. 15) have shown that:

- the internal stress state is mainly flexural with a large variation through the thickness;
- the highest longitudinal residual stresses at the edges of the thickness were of about 100 Nmm-2 for A sections and of about 400 Nmm-2 for B sections, showing the strong effect of releasing;
- the mean values are always very low (about 5 % of the edge values);
- the transversal residual stresses are around 10 % of the longitudinal ones.

From the above results, the stresses has been assumed as shown in fig. 16, with a linear variation along the thickness from distribution model of residual external compressive to internal tensile values which ratios σ_r/fy are : 0.26 for type A and 0.77 for type B.

For type C the distributions were quite irregular and difficult to modelize, but the maximum values were always less than in the other types.

The above model was used in simulation calculations, which results were compared with the ECCS curves (fig. 16), showing that the use of curves b and c for welded and cold-formed box-sections with and without releasing respectively is properly on the safe side.

5 CONCLUSIONS

In the last 30 years the field of measurement of mechanical imperfections on mono-dimensional structural element made of steel and of aluminium alloy has grown rich of several applications, which were performed separately and sometime independently in different research activities by scientists and technical organisms.

The common goal was to generalize testing results into physical models which had to be introduced into mathematical models in order to predict the performance of structural members under different loading conditions.

Now, the acquired experience is enough to give sufficient bases to assess unified rules. They must be directed:

- to define the experimental procedures both for direct and indirect measurement of mechanical imperfections;
- to interpret the testing results by means of appropriate physical models;
- to codify the use of simulation results according to different purposes.

REFERENCES

[1] **Ballio G., Mazzolani F.M.**: "Theory and Design of Steel Structures". Chapman and Hall - London, 1983;

[2] **Mazzolani F.M.**: "Aluminium Alloy Structures". Pitman - London, 1985;

[3] **Como M., Mazzolani F.M.**: "Ricerca teorico-sperimentale sullo svergolamento nel piano e fuori del piano dei profilati in presenza di tensioni residue". Costruzioni Metalliche, N°3-1969;

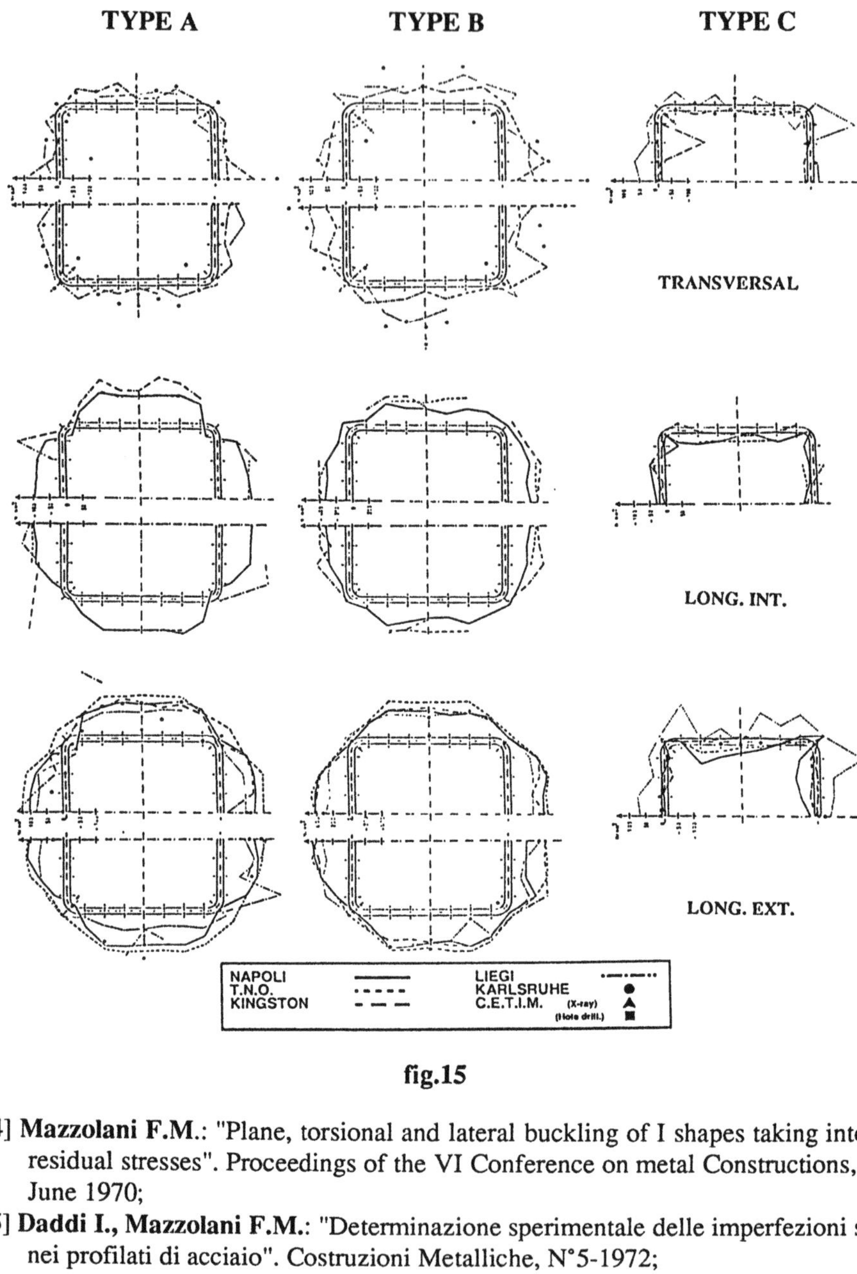

fig.15

[4] **Mazzolani F.M.**: "Plane, torsional and lateral buckling of I shapes taking into account residual stresses". Proceedings of the VI Conference on metal Constructions, Warsaw, June 1970;

[5] **Daddi I., Mazzolani F.M.**: "Determinazione sperimentale delle imperfezioni strutturali nei profilati di acciaio". Costruzioni Metalliche, N°5-1972;

[6] **Mazzolani F.M.**: "Analisi sperimentale delle tensioni residue nei profilati metallici". Atti del 1° convegno nazionale A.I.A.S., Palermo,1972;

[7] **Mazzolani F.M.**: "Buckling curves of hot rolled steel shapes with structural imperfections". Proceeding of International Colloquium on Column Strength, Paris, 1972;

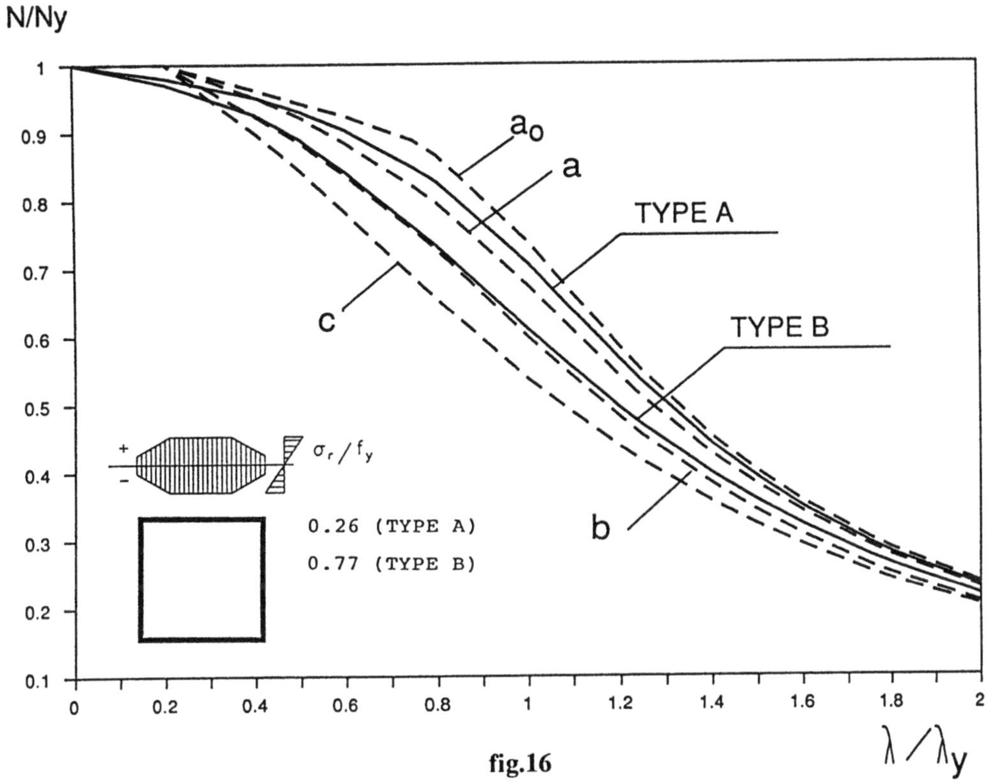

fig.16

[8] **Daddi I., Mazzolani F.M.**: "Il livellamento delle tensioni residue sotto carichi oscillanti". Costruzioni Metalliche, N°2-1973;

[9] **Mazzolani F.M.**: "Influenza delle imperfezioni strutturali sulla instabilità delle colonne in acciaio". Costruzioni Metalliche, N°6-1973;

[10] **Mazzolani F.M., Corsaro I.**: "Indagine sperimentale sulle autotensioni di tubi a gomito di caldaie di recupero". Bollettino Tecnico Finsider, N°330/331-1974;

[11] **Mazzolani F.M.**: "Le imperfezioni strutturali nei giunti di alluminio saldati". Costruzioni Metalliche, N°6-1974;

[12] **Mazzolani F.M.**: "Les imperfections structurales dans les assemblages soudés en aluminium". Revue de l'Aluminium, n. 431, 1974;

[13] **Mazzolani F.M.**: "Residual stress tests on french profiles 63 x 63 x 4 in aluminium alloys". ECCS Committee 16, Doc. 16-74-3, 1974;

[14] **Mazzolani F.M.**: "Residual stress tests on austrian profiles". ECCS Committée 16, Doc 16-75-1,1975;

[15] **Mazzolani F.M.**: "Inelastic behaviour of metal bars". Stavebnicky Casopis, September 1975;

[16] **Mazzolani F.M., Frey F.**: "Buckling behaviour of aluminium alloy extruded members". Proceedings of the Second International Colloquium on Stability of Steel Structures, Liège, Aprile 1977;

[17] **Gatto F., Mazzolani F.M., Morri D.**: "Experimental analysis of residual stresses and of mechanical characteristics in welded profiles of AL-SI-MG alloy (Type 6082)". Italian Machinery and Equipment, N°50-1979

[18] **Faella C., Mazzolani F.M.**: "European buckling curves for aluminium alloy extruded members". Alluminio, N°10-1980;

[19] **Faella C., Mazzolani F.M.**: "European buckling curves for aluminium alloy welded members". Alluminio N°11-1980;

[20] **Mazzolani F.M.**: "Welded construction in Aluminium European Recommendations: welded members". I.I.W. Colloquium on Aluminium and its Alloys in Welded Construction, Porto,1981;

[21] **Mazzolani F.M.**: "The influence of mechanical imperfections on the structural behaviour of welded aluminium alloy members". 2nd International Conference on Aluminium Weldments, Munich-1982;

[22] **De Luca A., Mazzolani F.M.**: "Models of distribution of mechanical imperfections in aluminium alloy welded members". 35 th Meeting of the Committee ECCS-T2 "Aluminium Alloy Structures", Cambridge,1983;

[23] **Mazzolani F.M., Frey F.**: "ECCS stability code for aluminium alloy members: present state and work in progress". Third International Colloquium on Stability of Metal Structures, Paris, November 1983

[24] **Mazzolani F.M.**: "Comparative study of measuring methods of residual stresses in hollow sections in steel". ECSC 7210-SA-308 Research report, 1985;

[25] **Faella C., Mazzolani F.M., Talarico G.**:"La variazione delle tensioni residue nello spessore dei tubi in acciaio: determinazione sperimentale e conseguenze sulla capacità portante". Atti del X Convegno C.T.A., Montecatini, 1985;

[26] **Landolfo R., Mazzolani F.M., Talarico G.**: "Misura delle tensioni residue nelle sezioni a cassone in acciaio". Atti del XI Convegno C.T.A., Capri, 1989;

[27] **Landolfo R., Mazzolani F.M.**: "The influence of the variation through the thickness of residual stresses in tubular columns". International Colloquium on Stability Steel Structures, Budapest, 1990.

32 STUB COLUMN TEST PROCEDURE

D.A. NETHERCOT
Department of Civil Engineering, University of Nottingham, UK

Introduction

Compression tests on short lengths of steel members are an established subsidiary test conducted in support of the large-scale testing of components e.g. columns, and complete structural assemblies e.g. frames. The objective is to provide basic information on the behaviour of the cross-section under compressive loading. This may then be used to supplement material property data obtained from the conventional tensile test (1) when interpreting the results from the large scale test. The information will also be of use when undertaking numerical simulations of the test behaviour.

This brief paper discusses certain aspects of stub column testing within the contact of a universally accepted procedure for conducting a test, measuring and recording the key information produced, interpreting the findings and reporting on the outcome. It suggests that a worldwide representative body such as RILEM might be the appropriate mechanism for the production of such a procedure. The paper presents no new data as its purpose was to stimulate discussion at the May 1990 Naples workshop.

History

A procedure for stub column tests was first provided in Lehigh University, Fritz Lab. Report No. 220A.36, authored by Tall and issued in February 1961. This was revised by an IIW group, who approved it in 1962. In 1975 Task Group 6 of the American Structural Stability Research Council again revised the document. It is reproduced as Technical Memorandum No. 3 : Stub-Column Test Procedure in the Fourth Edition of the Structural Stability Research Council's "Guide to Stability Design Criteria for Metal Structures" (2).

Althrough no actual survey has been conducted, it is known from personal contacts that this document forms the basis for steel column testing in many laboratories around the world. However, testing equipment, instrumentation and data processing, as well as the reasons for conducting such tests have changed significantly since its production.

Testing of Metals for Structures. Edited by F.M. Mazzolani.
Published by Taylor & Francis, 2 Park Square, Milton Park, Abingdon, Oxon, OX14 4RN.
ISBN 0 419 15810 3

The current position

The view given in ref. 2 is that stub column testing is largely about determining the cross-section's tangent modulus as affected by the presence of residual stresses for subsequent use in a tangent modulus type of column design formula. Althrough local buckling is mentioned (within the context of cold-formed sections only), the possibility of obtaining the effective squash load and/or of investigating cross-section classification (3) is not covered.

A second aspect of the advisability of updating the document is the changes in testing equipment, instrumentation and data processing facilities that ought to be reflected in the procedure. The present document is written in terms of relatively unsophisticated testing machines, dial gauges, manual recording, etc.

Viewed as a complete procedure it is also short of material on the need to measure and record such fundamental items as the specimen dimensions. With the greatly increased use of test data on a worldwide basis for calibrating design rules, the need for a full set of measurements and for complete documentation of the test should be highlighted rather more. Indeed an ideal document would contain a fully detailed sample set of results as well as a checklist for each stage : specimen selection, testing, instrumentation and reporting.

Proposal

Given the position of stub column testing as a fundamental material property type of test, together with the various changes that have combined to make the present "standard" (2) increasingly out of date, RILEM consider the production of a new comprehensive test procedure. This could be accomplished as a task for the existing TC 83, providing it received a more wide-ranging brief.

Conclusions

The status of the existing SSRC Technical Memorandum on steel column testing has been reviewed against a background of both the use to which the results of such tests are likely to be put and the greater sophistication of present day test equipment. A case for a new document, prepared by RILEM, has been proposed.

References

1. Kato, B et al, "Tension Testing of Metallic Structural Materials for Determining Stress-Strain Relations under Monotonic and Uniaxial Tensile Loading", RILEM Journal, 1990.

2. Galambos, T.V., ed. "Guide to Stability Design Criteria for Metal Structures", 4th edition, Wiley, New York, 1986.

3. Eurocode 3, "Design of Steel Structures", Predraft, April 1990, CEC, Brussels.

33 RESIDUAL STRESSES: MEASUREMENTS, THEORETICAL PREDICTIONS AND THEIR USE IN THE DESIGN OF STEEL STRUCTURES

J. RONDAL
Department M.S.M., University of Liege, Belgium

SUMMARY

Residual stresses play an important role in the design of steel structures. They can have a significant effect on brittle fracture, fatigue, stress corrosion and buckling of structural steel members.

The paper is aimed to discuss three points which deal with residual stresses :
- the methods of measurement of residual stresses ;
- the theoretical prediction of these stresses ;
- the role of the residual stresses in the design of steel structures.

Some points which are not yet properly solved are highlighted in the conclusions.

INTRODUCTION

Residual stresses are stresses that exist in materials independently of any external forces. They are produced in metals by most forming and fabrication processes of structural components: forging, rolling, welding, heat treating, grinding, machining,... These processes cause residual stresses by inducing plastic deformation of the metal through severe mechanical forces or temperature gradients. Localized permanent expansion or contraction of the metallic lattice in processes such as nitriding, carburizing or heat treatment which induce phase transformations can also generate residual stresses [1].

Residual stresses can be classified in three categories :
1. In the first category, macroscopic stresses are distributed over large areas and are due to plastic deformation or heat treatments.
2. In contrast, the residual stresses of the second category affect small areas such grains; they are due to the yield strength anisotropy in the crystals of materials undergoing plastic deformations.
3. Third category concerns residual microstresses ; these are due to heat treatments and result from heterogeneous microstructures, like perlite and ferrite, causing different elongation coefficients for various grains.

Amongst these three categories, the first one is the most important for structural purposes. Depending on their distribution and magnitude, residual stresses can have a significant negative effect on brittle

Testing of Metals for Structures. Edited by F.M. Mazzolani. © 1992 RILEM.
Published by Taylor & Francis, 2 Park Square, Milton Park, Abingdon, Oxon, OX14 4RN.
ISBN 0 419 15810 3

fracture, fatigue, stress corrosion and buckling of structural steel members.

The distribution and the magnitude of the residual stresses depend very much on the fabrication process of the profile, e.g. (fig. 1) :
- in hot-rolled profiles, the residual stresses arise from non-uniform temperature distributions during the cooling stage of forming a hot-rolled shape. The part of the section which cools first is in residual compression, whilst the region which cools last is in residual tension. The residual stresses are mainly oriented in the direction of the rolling and have magnitudes rather constant through the thickness of the profiles (at least for profiles of moderate to medium thicknesses);
- in contrast to ho-rolled members, cold formed profiles are affected by deformational residual stresses in both longitudinal on transversal directions. These residual stresses can vary very much through the thickness of the profiles.

METHODS OF RESIDUAL STRESS MEASUREMENT

There are many scientific papers which deal with the methods of measuring residual stresses. The most frequently used method for determining the residual stress pattern in steel members is slicing a portion of the member into longitudinal strips. Slicing relieve the restraint causing the stresses and corresponding changes in strains can be measured and are then used to calculate the original residual stress in each slice [2]. Mechanical or electric strain gauges can be used for the strain measurement before and after slicing.

The slicing method can not be used directly for heavy sections where the variation of the residual stresses through the thickness has to be taken into account. However, an extension of the method with slices through the thickness has been proposed by ASCHENDORFF et al [3]. Difficulties arise also when the method is apply to tubular profiles because strains are often impossible to measure on the inner surface which is generally inaccessible for initial readings before slicing. Procedures have however been proposed for an extension of the slicing method to hollow sections [2, 4].

Non-destructive methods of residual stress measurement have been developed these last years. Though some of these methods are not yet applicable for a complete determination of the pattern of residual stresses, in steel members (ultrasonic, electromagnetic, positron annihilation, nuclear hyperfine and other methods), some others have been used successfully (e.g.: X-Ray diffraction, indentation, chemical etching and hole drilling method) [1].

The question of the reliability and the reproducy of the results given by the different methods is, in practice, of great importance. A comparative study, with measurements carried out by different methods on a same steel profile, has been performed, some years ago, by a group composed of six european universities and research centres [4] :
- University of Napoli (Italy) : slicing method ;
- Kingston Polytechnic (United Kingdom) : slicing method ;
- University of Liège (Belgium) : Strain gauges and slicing ;
- University of Karlsruhe (West Germany) : X-Ray method ;
- CETIM (France) : X-Ray and hole drilling methods ;
- TNO (The Netherlands) : hole drilling method.

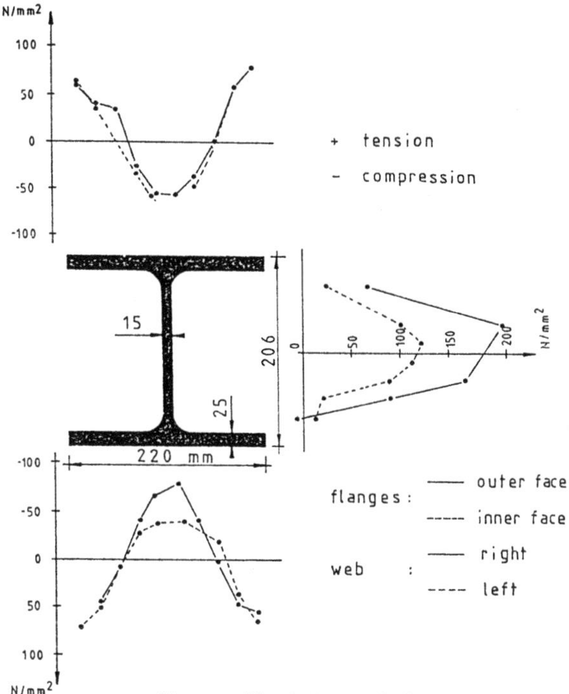

a. Hot-rolled H profile

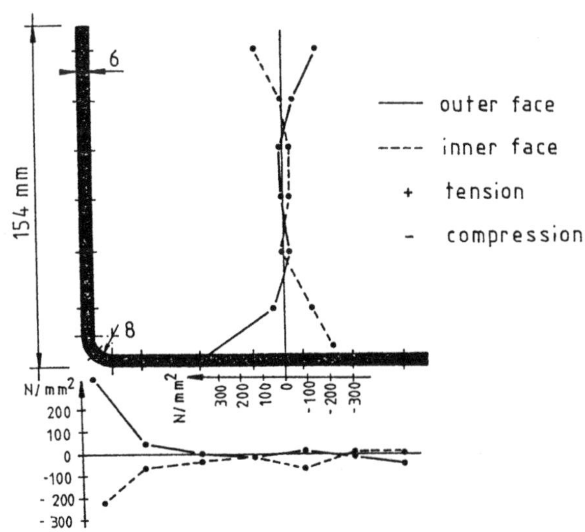

b. Cold-rolled angle

Fig. 1 - Examples of longitudinal residual stresses in steel members

A square hollow section of 200 x 200 x 8 mm, with three different fabrication processes, has been used for the tests :

- specimen A : welded hollow section, cold-formed and stress relieved ;
- specimen B : welded hollow section, cold-formed ;
- specimen C : unwelded hot-formed hollow section.

This comparative study has shown that the different methods lead to measurements in good agreement provided the magnitude of the residual stresses be large enough (fig. 2.a). On the contrary, the reliability of all the measurements is questionable for low levels of residual stresses (fig. 2.b).

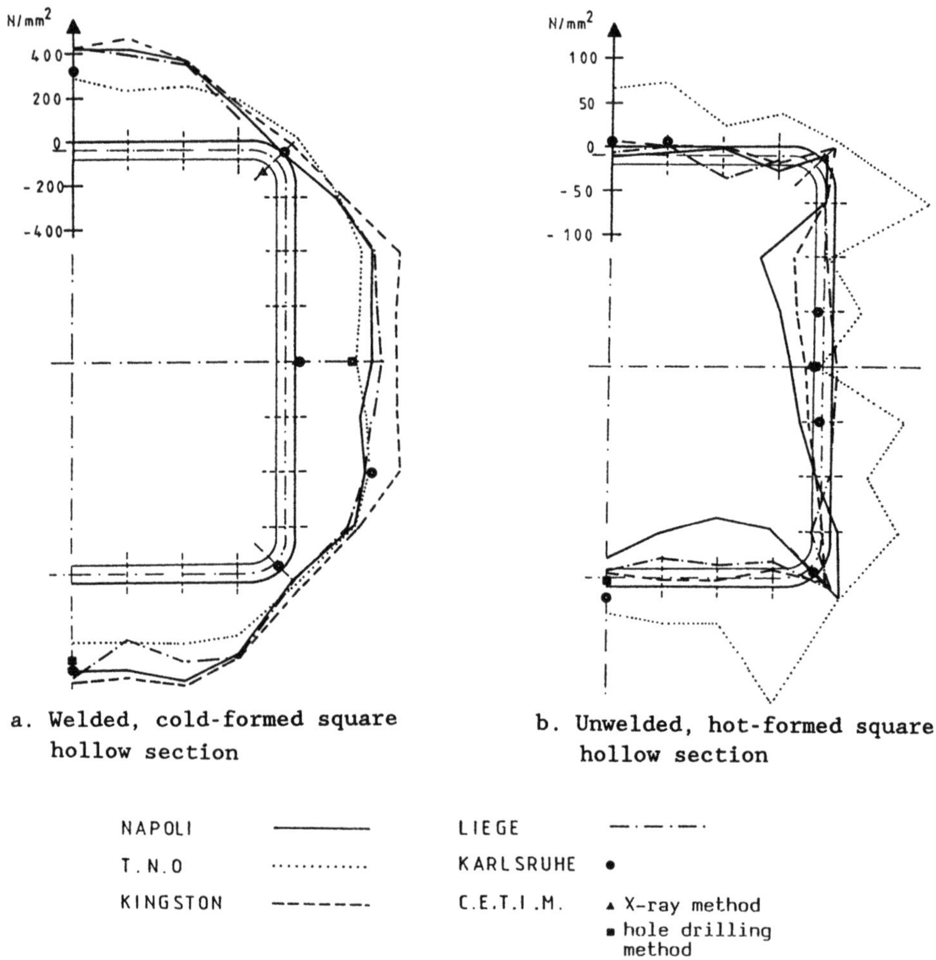

Fig. 2 - Longitudinal residual stresses (outer face)

If it is accepted to restrict the knowledge to the sole longitudinal stresses - they play the main role in stability problems -, the slicing method, which is simple, economic and easy to use, can be recommended. However, if the complete field of residual stresses has to be known, the hole-drilling method and the method of slicing using strain gauges must be preferred. Figure 3 shows, the goos agreement obtained between these

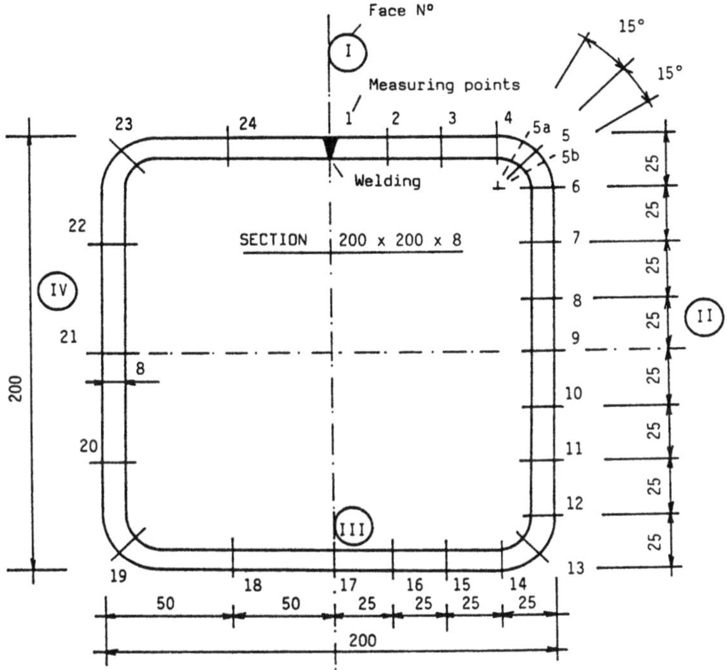

a. points of measurement

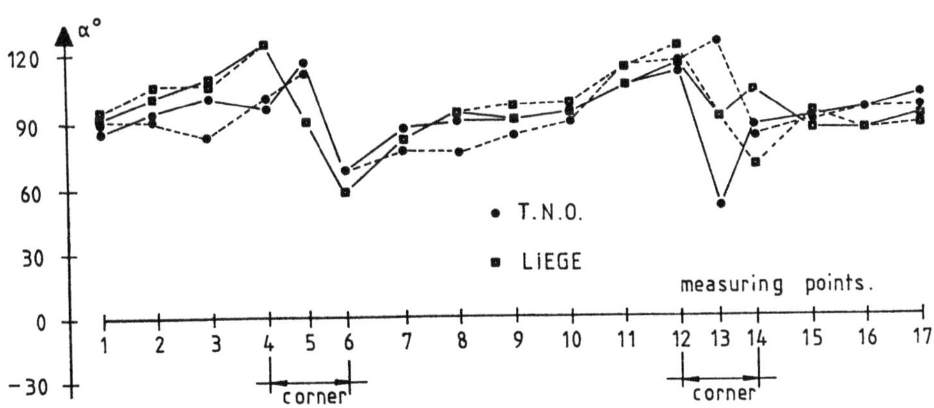

b. Angle of orientation of the first principal residual stress

Fig. 3 - Orientation of the principal residual stresses (outer face)

two methods in determining the orientation of the principal residual stresses. On the outer face, for a welded cold-formed square hollow section, the first principal residual stress is noticeably oriented to 90° of the longitudinal axis in the flat faces and to 45° in the corners. For the inner face, the orientations are, respectively, of 0° in the faces and of 45° in the fillets [4].

THEORETICAL PREDICTIONS OF THE RESIDUAL STRESSES

Any attempt to predict residual stresses theoretically is, of course, welcome. The theoretical methods aimed to this purpose shall not depend, in contrast to experimental methods, on the size of the specimens; they will allow for parametric studies, with the result of a better understanding of the influence of the respective variables on the one hand and of the history of the onset of residual stresses on the other hand. However such theoretical methods are hard to derive and, therefore, generally based on rough simplifications. Accordingly, the validity of any theoretical method must be checked by comparing theoretical to experimental results. Prediction methods of residual stresses have been suggested by several authors for various fabrication procedures, e.g. : PLUMIER and BAUS [5], for welded sections, KATO and AOKI [6], for welded hollow sections, INGVARSSON [7] and WENG and WHITE [8], for cold-bending and RONDAL [9], for cold-rolling. Figure 4 taken from reference [9], shows the theoretical longitudinal stresses obtained for three types of cold-rolled sections.

Modern numerical methods can also be used for the prediction of residual stresses. However, at least when the residual stresses are due to plastic deformations, it is necessary to use non linear finite element computer programs so as to account for large deformations of the material. That is much time computing, even with fast computers. An example of a numerical simulation of the cold-bending of a plate [10] is presented in figure 5 ; it must be considered as a first attempt in the field but the development of more elaborated finite elements and numerical procedures is still required before to envisage a practical use of these sophisticated tools.

RESIDUAL STRESSES IN THE DESIGN OF STEEL STRUCTURES

All the recent design specifications dealing with steel structures take account of the effect of the residual stresses on fatigue and buckling behaviour of the structure. For buckling purposes principally, typical residual stress patterns have been published which can be used for computer simulation of the buckling behaviour. For example, figure 6 gives the residual stress patterns which are recommended by ECCS for the ultimate limit state calculation of sway frames with rigid joints [11].

Unfortunately, sufficiently accurate patterns of residual stresses can not be provided for all types of steel profiles. This is, for example, the case for cold-formed profiles. Recent papers have shown that the residual stress patterns exhibited by cold-formed profiles lead to buckling behaviours which differ in some ways from these shown by hot-rolled sections [12, 13, 14].

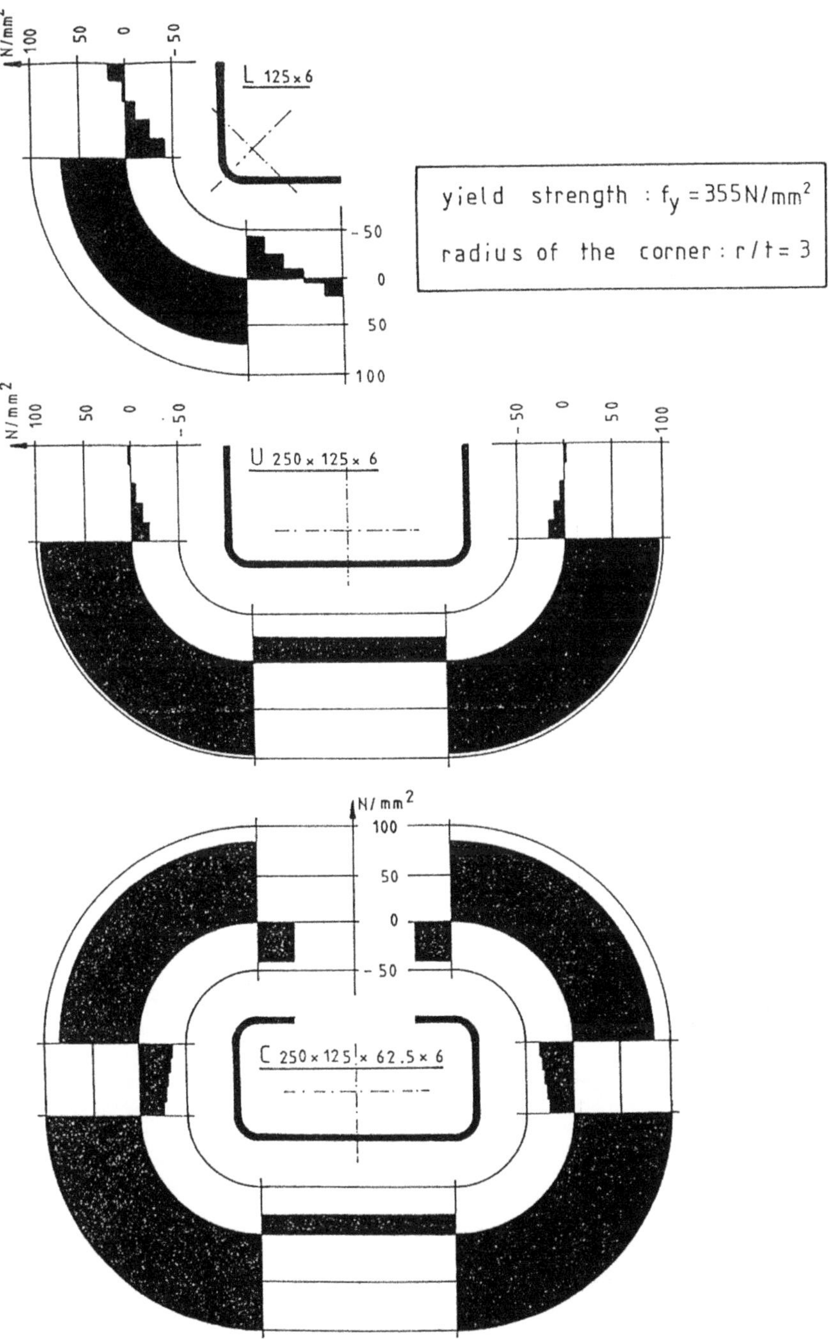

Fig. 4 - Longitudinal residual stresses for three types of cold-rolled sections (outer face).

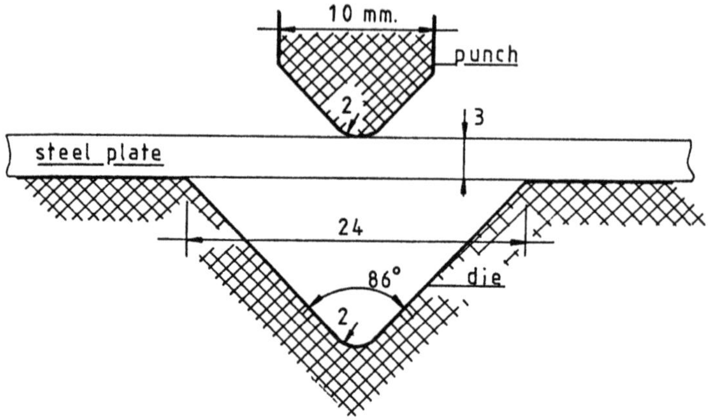

a. Definition of the problem

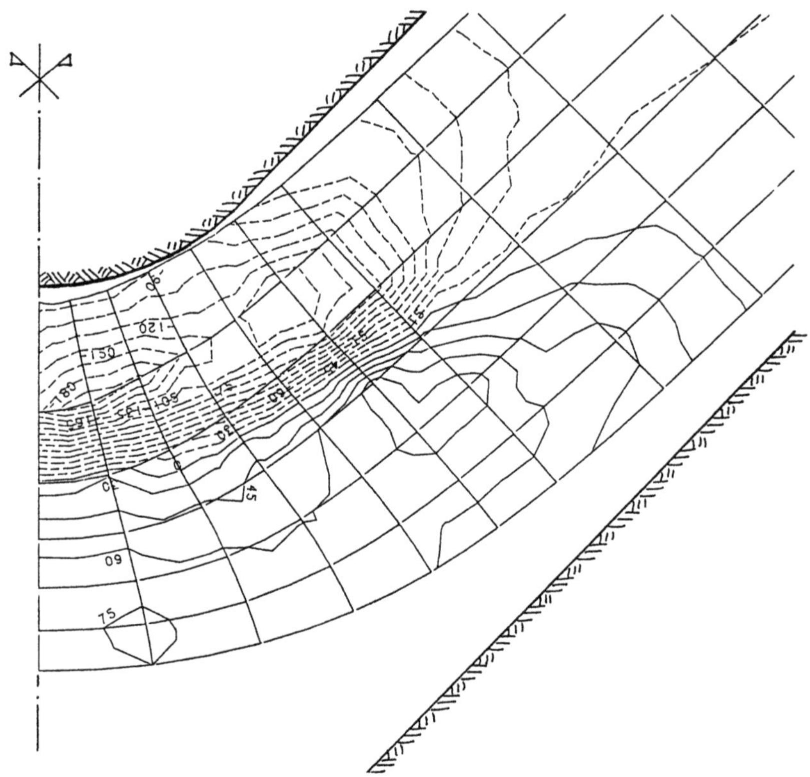

b. Longitudinal residual stresses after cold-bending (N/mm^2)

Fig. 5 - Numerical simulation of the cold-bending of a steel plate.

b1) TUBES

(rolled tubes and welded tubes, hot finished)

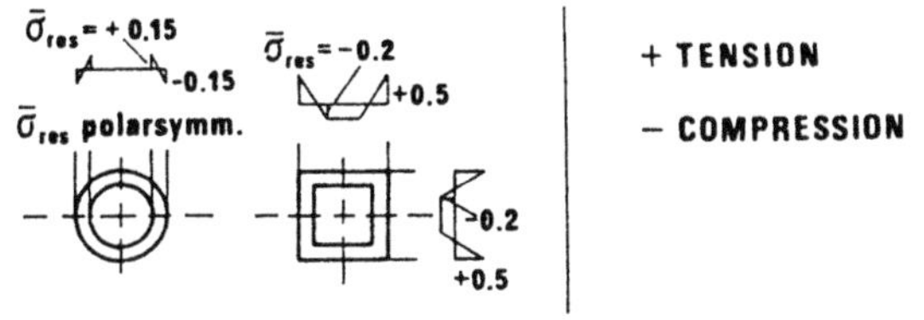

b2) ROLLED I-SECTIONS

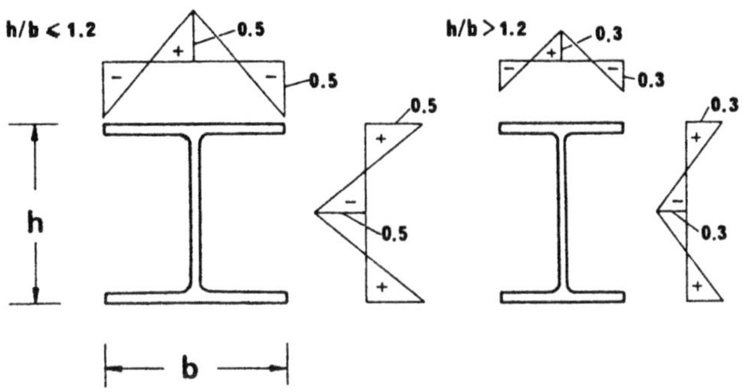

b3) WELDED I SECTIONS

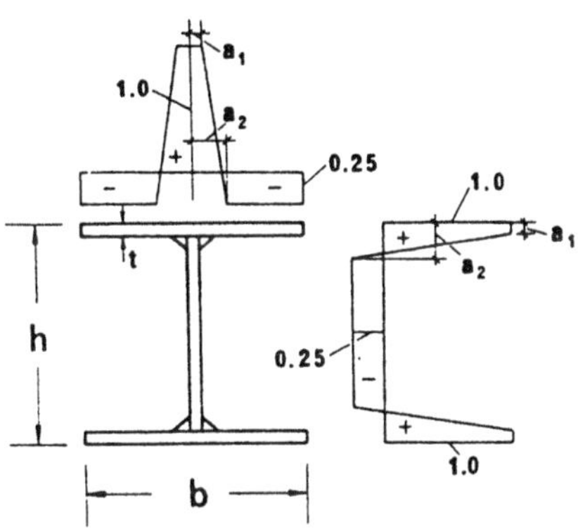

	a_1	a_2
FLANGE	0.075 b	0.125 b
WEB	0.075(h−2t)	0.125(h−2t)

b4) WELDED BOX SECTIONS

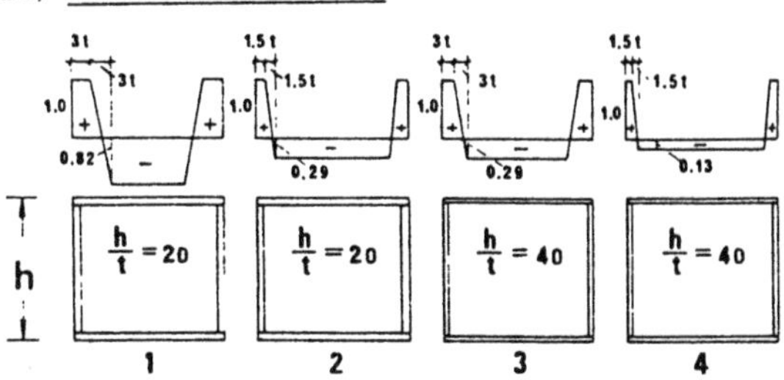

1,3 HEAVY WELD　　　2,4 LIGHT WELD

Fig. 6 - Residual stress patterns (after ECCS, $\bar{\sigma}_{res} = \sigma_{res}/235\ N/mm^2$).

CONCLUSIONS

Residual stresses play an important role in the modern design of steel structures.

Significative progress can be noticed, in the last years, regarding the experimental methods of measurement and the use of these stresses in the design methods - either by means of formulae given in specifications or by numerical techniques. However, there is an urgent need of codification of the experimental procedures to be recommended in practice for the measurement of residual stresses in all types of steel profiles, in the first hand, and of the residual stress patterns which must be used for numerical simulation of the behaviour of steel structures, in the second hand, to ensure the safety of these computations.

REFERENCES

1. ASTM, Residual Stress Effects in Fatigue, Proceedings of the ASTM Symposium, Committee E-9 on Fatigue, Phoenix, May 11, 1981.

2. SHERMAN, D.R., Residual Stress Measurement in Tubular Members, ASCE, Journal of the Structural Division, Vol. 95, N° 4, April 1969, pp. 635 - 647.

3. ASCHENDORFF, K.K., BERNARD, A. and PLUMIER, A., Le flambement de profilés européens de forte épaisseur, Construction Métallique, N° 3, 1983, pp. 21 - 29.

4. GRIMAULT, J.P. and RONDAL, J., Etude comparative de différentes méthodes de mesure des contraintes résiduelles dans les profils en acier, Commission of the European Communities, Research 7210.SA.308, Final Report, 1986.

5. PLUMIER, A. and BAUS, R., Etude paramétrique de la résistance au flambement de colonnes en H d'imperfection structurale élevée, Construction Métallique, N° 1, 1981, pp. 3 - 20.

6. KATO, B. and AOKI, H., Residual Stresses in Cold-Formed Tubes, Journal of Strain Analysis, Vol. 13, N° 4, 1978, pp. 193 - 204.

7. INGVARSSON, L., Cold-Forming Residual Stresses and Box Columns Built-up by Two Cold-Formed Channel Sections Welded Together, Royal Institute of Technology, Dept. of Building Statics and Structural Engineering, Stockholm, Bulletin N° 121, 1977.

8. WENG, C.C. and WHITE, R.N., Residual Stresses in Cold-Bent Thick Steel Plates, ASCE, Journal of Structural Engineering, Vol. 116, N° 1, January 1990, pp. 24 - 29.

9. RONDAL, J., Residual Stresses in Cold-Rolled Profiles, Construction and Building Materials, Vol. 1, N° 3, September 1987, pp. 150 - 164.

10. GRISARD, L., Modélisation du pliage d'une tôle, University of Liège, Dpt. of Mechanics of Materials, Stability of Constructions, Mechanics of Structures, Liège, Report N° 162, 1987.

11. ECCS, Ultimate Limit State Calculation of Sway Frames with Rigid Joints, European Convention for Constructional Steelwork, Technical Committee 8 - Structural Stability, Publication N° 33, 1984.

12. COSTA FERREIRA, C.M., and RONDAL, J., Influence of Flexural Residual Stresses on the Stability of Compressed Angles, Journal of Constructional Steel Research, Vol. 9, N° 3, 1988, pp. 169 - 177.

13. CHAN, S.L., Strength of Cold-Formed Box Columns with Coupled Local and Global Buckling, The Structural Engineer, Vol. 68, N° 7, April 1990, pp. 125 - 132.

14. LANDOLFO, R. and MAZZOLANI, F., The Influence of the Variation Through the Thickness of Residual Stresses in Tubular Columns, International Colloquium on Stability of Steel Structures, Budapest, Preliminary Report, Vol. 3, April 1990, pp. 109 - 114.

PART SIX

BEHAVIOUR OF COMPOSITE STRUCTURES

34 SLIP AND UPLIFT MEASUREMENTS ALONG THE STEEL AND CONCRETE INTERFACE OF VARIOUS TYPES OF COMPOSITE BEAMS

J.M. ARIBERT
Institut National des Sciences Appliquées, Rennes, France

SUMMARY

The aim of this paper is to show that measures issued from specimen tests (as push-out and pull-out tests) may give dubious results to characterize slip and uplift distributions along the steel-concrete interface of real composite beams. For that, different comparisons are presented between measures from specimens, direct measures on tested beams and calculation results with the help of a numerical model. Some recommendations are suggested to improve testing procedures in the concerned topic.

1 INTRODUCTION

Experimental testing of beams and other structures in laboratory may appear nowadays as an expensive and laboured procedure to develop knowledge in steel and concrete composite construction, by comparison with the capabilities of modern computation means. Particularly, if the investigated topic depends on several mechanical parameters and its final aim consists in proposing simplified design methods, numerical approach is often the most efficient procedure. But obviously using of numerical models implies to have at one's disposal some correct experimental data, so that in fact numerical calculation and testing in laboratory must advance together. The present paper is devoted to measurement problems in relation to the connection behaviour in composite beams ; the case of partial shear connection is only considered here because it offers more experimental and theoretical interest than full connection : connectors are subjected to greater deformations when slip and uplift occur along steel-concrete interface. For information, partial connection is allowed for buildings (but not for bridges) and may save some cost because often design is governed by serviceability requirements or other criteria (architectural aspects, ease of construction) ; moreover, using of this kind of connection is almost inevitable when precast floor slabs are joined (leading to wider connector spacing) or when the slab is cast on profiled steel sheeting (because of weaker resistance of connectors located inside corrugations perpendicular to the steel beam flange).

The main questions about connection in such context are the following ones : is it sufficient founding upon ordinary tests on specimens (like push-out and pull-out tests) to obtain correct data for a numerical model ? Can specimen tests give reliable information in comparison with the real behaviour of connectors working inside a composite beam ? Are there new testing

Testing of Metals for Structures. Edited by F.M. Mazzolani.
Published by Taylor & Francis, 2 Park Square, Milton Park, Abingdon, Oxon, OX14 4RN.
ISBN 0 419 15810 3

procedures which may be suggested ? It is not easy to answer accurately the above questions for the very reason that connection behaviour may depend on the type of composite beam, which compels us to distinguish in this presentation between beams with solid slab and beams with slab cast on profiled steel sheeting.

Our approach is going to consist in showing various experimental results issued from beam tests in laboratory and in comparing them with theoretical results supplied by numerical simulation ; to make shorter the presentation, only the case of simply supported beams will be considered. The data useful to calculation will be deduced whether from push-out and pull-out tests or from direct measures taken on beams. First of all, the numerical model concerned here is briefly presented in its main lines.

2 ELEMENTS ON THE NUMERICAL MODEL

A detailed presentation of the numerical model is given in ref. [1]. We introduce the following symbols :

- x : co-ordinate of any cross-section of the composite beam,
- $v_a(x)$, $v_{(b)}(x)$: bending deflections respectively of the steel member and the slab,
- ε_a, ε_b : longitudinal strains at each centroidal axis respectively of the steel member and the concrete slab,
- $\gamma(x)$, $\Delta(x)$: slip and uplift distributions along the steel-concrete interface,
- $F(x)$: longitudinal interaction force (equal to the slab compressive force).

Then the basic equations and relationships to satisfy are :

(i) for any cross-section between connectors : two equilibrium equations in bending moment and vertical shear force, and behaviour relationships in generalized displacements (to be noted that axial and flexural rigidities of steel and concrete parts are depending on the interaction force $F(x)$ and the beam curvatures $\frac{d^2v_a}{dx^2}$ and $\frac{d^2v_b}{dx^2}$;

(ii) along the steel-concrete interface : two compatibility equations, the first between slip and strains :

$$\frac{d\gamma}{dx} = \varepsilon_b - \varepsilon_a + d_a \frac{d^2v_a}{dx^2} + d_b \frac{d^2v_b}{dx^2} \quad \textbf{(1)}$$

(where d_a and d_b are distances between interface and respectively each centroidal axis of the steel profile and the slab), and the second equation between uplift and deflections :

$$\Delta(x) = v_b(x) - v_a(x) \geqslant 0 \ .$$

(iii) At crossing of any connector : three discontinuous equilibrium equations, according to **fig. 1** :

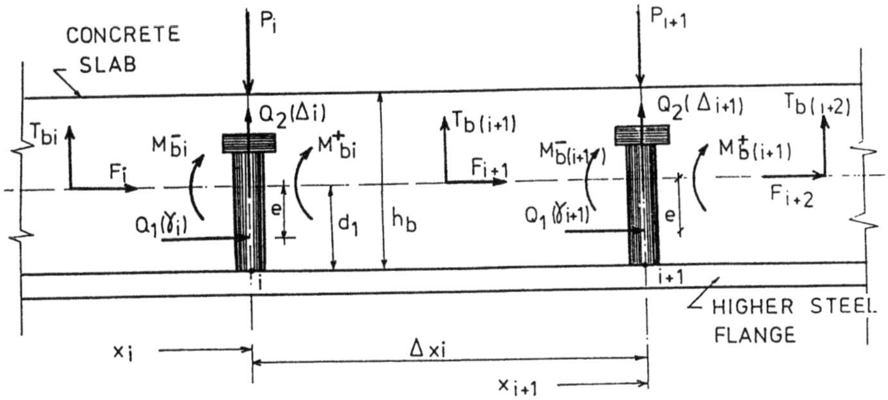

***Fig. 1** - Discontinuity relationships at connector crossing*

- equilibrium of longitudinal forces :

$$F^+ = F^- + Q_1(\gamma) \tag{3}$$

where Q_1 is the shear force on the connector related to slip by some non linear relationship ;

- equilibrium in transversal forces :

$$T_b^+ = T_b^- + Q_2(\Delta) - P \tag{4}$$

where T_b is the vertical shear force on concrete part in cross-section and Q_2 the tensile force on the connector related to uplift by another non linear relationship.

- Equilibrium in moment :

$$M_b^+ = M_b^- - Q_1(\gamma)\, e \tag{5}$$

where M_b is the part of bending moment only applied to concrete.

To solve the above equations, an iterative procedure must be used because of the non linearities due to the behaviour of connectors, steel and concrete. Taking into account the discontinuous character of equations **(3)**, **(4)** and **(5)**, it is logical to transform the other equations into discrete relationships, which can be done more easily if the connector spacing is small ; for that, it is always possible to introduce virtual connectors with zero strengths. Then, the equations can be integrated analytically or numerically inside the small spacings so obtained, which allows to formulate the problem in terms of a classical finite element method. The author has shown [1] that a formulation by transfer matrixes is interesting to use because the boundary conditions of a simply supported beam ($v_a = v_b = F = 0$) are located at the beam ends ; also, such formulation is well adapted to continuous composite beams [2]. To be noted that equation **(4)** is valid only if $\Delta \geq 0$, which is unknown at any step of loading ; to tide over that difficulty, a relaxation iterative procedure (index j) may be used : a contact force $C^{(j)}$ is added to the right member of equation **(4)**, according to :

$$C^{(j)} = \begin{cases} C^{(j-1)} + Q_2\,(\Delta^{(j-1)}) & \text{if } \Delta^{(j-1)} < 0 \\ 0 & \text{if } \Delta^{(j-1)} \geqslant 0 \end{cases} \tag{6}$$

Recently, the numerical model has been generalized to compute beams with a slab cast on profiled steel sheeting [3] ; moreover, relatively sophisticated laws for material behaviour and non monotonic relationship $Q_1(\gamma)$ can be introduced in it. Generally, the profile steel has been characterized by a symmetrical diagram in tension and compression **(fig. 2-a)** with a yield plateau and a strain-hardening region (assuming a constant modulus $\alpha_a\ E_a$) ; a similar diagram has been used for reinforcement. Formulation for concrete in compression **(fig. 2-b)** has been deduced from experimental data published in the literature [4, 5] and is valid for compressive strain ε_b up to 1 % (fairly often, ε_b is bounded at 0.4 %) ; sometimes, slight concrete strength f_t in tension can be considered also.

3 CASE OF BEAM WITH SOLID SLAB

Now, we are going to present a first investigation dealing with a very partially connected beam which has been tested in Rennes ; its geometry is shown in **fig. 3** and the exact material properties can be found in ref. [6]. The whole connection is realized by means of 16 welded studs with head (diameter : 19 mm, f_y = 350 MPa) ; these studs are regrouped per block of four connectors so that spacing between adjacent blocks is very wide (not less than 1.5 m). Concerning load-deflection curves, **fig 4** shows

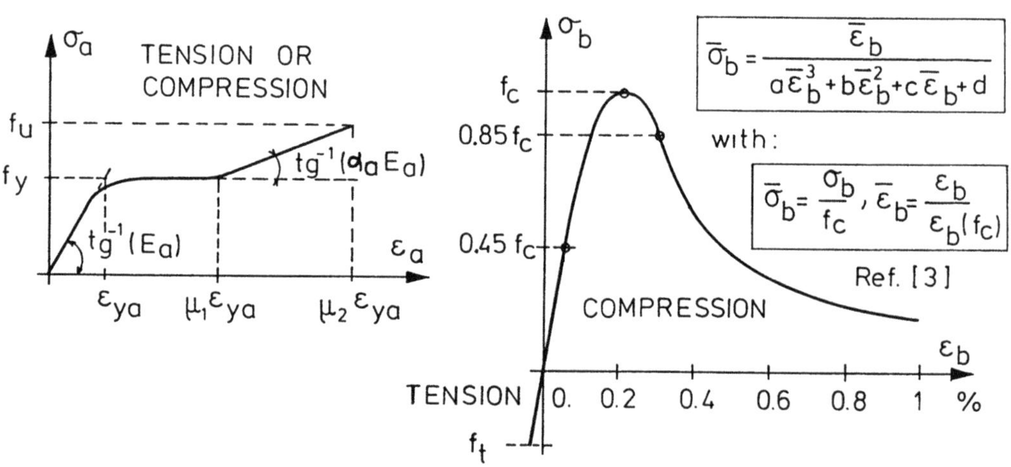

Fig. 2

a/ Steel behaviour *b/ Concrete behaviour*

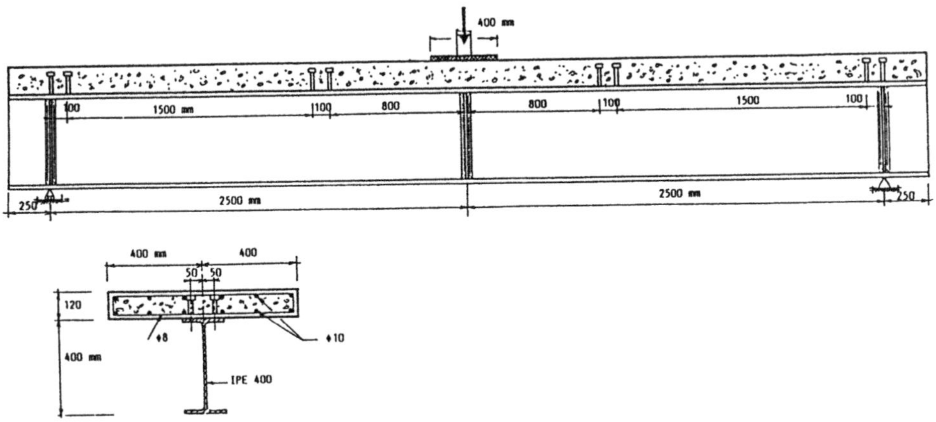

Fig. 3 - Tested composite beam with partial and spaced shear connection

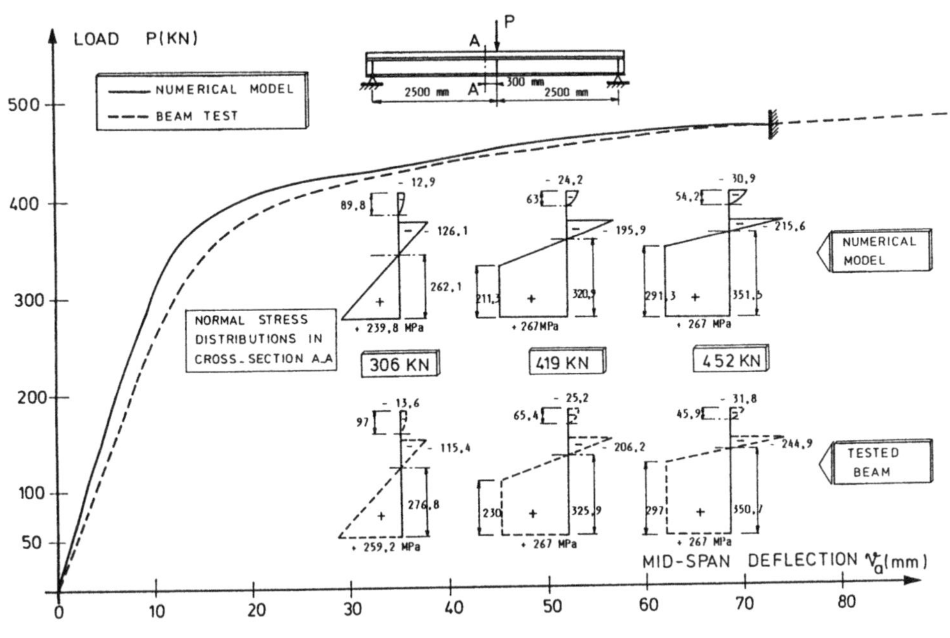

Fig. 4 - Load-deflection curves and normal stress distributions in cross-section

good agreement between the numerical curve (represented by a continuous line) and the experimental one (discontinuous line) ; besides this agreement is confirmed by comparing the evolution of the normal stress distributions in the cross-section located at 300 mm from the median one (measurement was impossible in median cross-section because of obstacle due to stiffeners). In order to understand why the above comparisons are satisfying, it is necessary to study thoroughly some local aspects along the steel-concrete interface.

3-1 Slip behaviour of the connection

Preliminary push-out tests carried out in accordance with the standard requirements of Eurocode 4 ([7]n-chapter 12) have allowed to determine the following relationship per connector (with small scattering) :

$$Q_1(\gamma) = Q_\infty(1 - e^{-0.7\gamma})^n \quad \text{with : } \gamma\ (\text{mm}) < 3.5\ \text{mm} . \tag{7}$$

The asymptotic value Q_∞ is equal to 130 kN for a concrete strength f_c near to 30 MPa. The exponent n appears depending on the contact conditions at interface and on the connector spacing ; but under the conditions of the tested beam (with bond prevented by greasing the steel flange), n is practically equal to 0.8. By introducing relationship **(7)** in the numerical model, evolution of the slip distribution has been computed for the concerned beam ; **fig. 5** shows again good agreement between numerical and experimental distributions at least for different loads up to 440 kN (for information, every experimental curve corresponds to mean measures on each side of the point load and is corrected by numerical smoothing). Beyond the loading value 440 kN, the measured slip on the beam is slightly greater than the computed one, probably because of some difference between the real connector behaviour inside the beam and that deduced from specimen tests which are known on principle as more severe.

In order to verify the validity of law **(7)**, a special procedure to investigate the stud behaviour has been tried by means of direct measures on the loaded

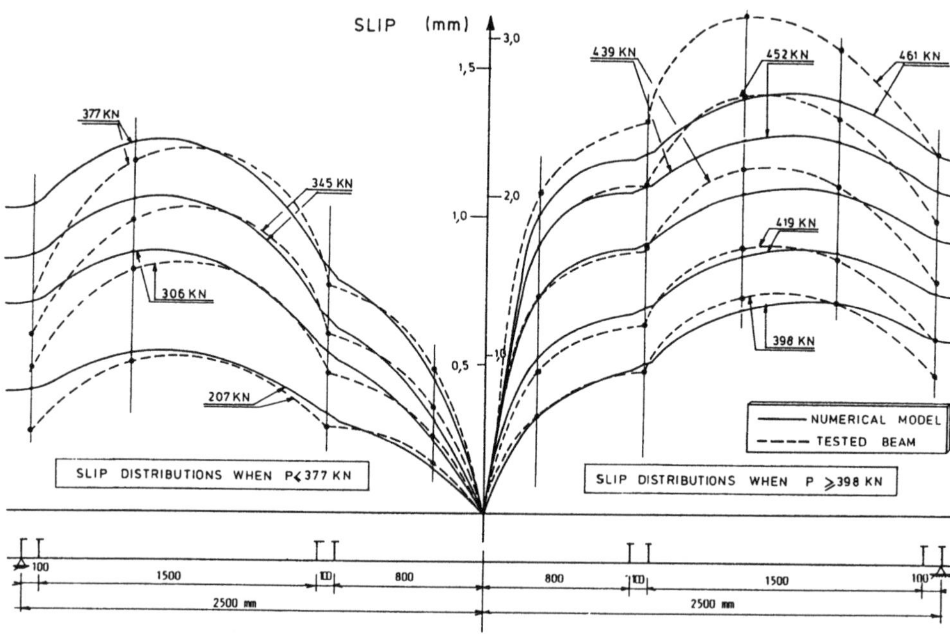

Fig. 5 - Slip distributions along any mid-span

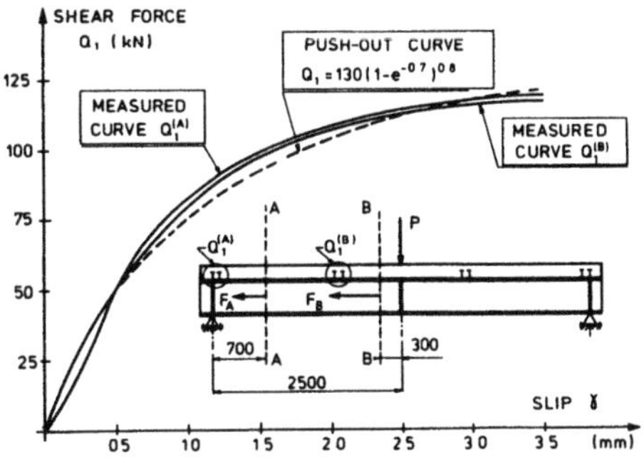

Fig. 6 - Determination of connector behaviour directly on the tested beam

beam. By using strain gages glued on web anf flange of the steel member in two different cross-sections (i.e. cross-sections AA and BB in **fig.6**), we have been able to determine at each loading step the real distribution of the longitudinal force F along this member and then the shear force on a particular stud (see equation 3). Simultaneously, we have measured the slip of the concerned stud by means of potentiometric transducers regularly put along the line between the slab and the steel flange edge. The procedure result is illustrated by **fig. 6** where we must ascertain that the so-obtained curves $Q_1(\gamma)$ set up a background information for equation **(7)** ; to be noted the same accuracy for cross-section AA as cross-section BB. So, it appears that ordinary push-out testing constitutes a sufficient approach to characterize well the shear behaviour of studs working inside a composite beam with solid slab ; such procedure can be used fruitfully to supply data for numerical modelling.

3-2 Uplift behaviour of the connection

Taking into account the complexity of the real internal forces inside the beam slab, two types of pull-out testing have been carried out, as shown in **fig. 7** : on the one hand a symmetrical test where two opposite tension loads are applied on the specimen, and on the other hand an unsymmetrical test where the specimen slab (having the same reinforcement rate as the beam slab) is subjected to some bending because a tension load alone is applied. To be noted that for the first type of testing the rigid handles used to hold the potentiometric transducers required for external measurement of uplift can be set in a parallel direction to web plan of the two profiles in tension loading ; on the contrary for the second type, lack of space has compelled us to set the rigid handles normally to web plan, which may false partially the uplift measurement because of some prying effect of the steel flange in contact with the slab. The curves $Q_2(\Delta)$ obtained by using the two types of testing are drawn in **fig. 8** ; it appears immediately a great difference between the failure modes : the first type of testing leads to brittle failure whereas the second type tends to demonstrate ductile behaviour. In spite of such dubiousness of behaviour, some compromise had to be found for numerical calculation and the following safe formulation was adopted without appeal :

$$Q_2(\Delta) = \begin{cases} 50\,\Delta & \text{if} \quad 0 \leqslant \Delta < 0.5 \text{ mm} \\ 25 & \text{if} \quad \Delta \geqslant 0.5 \text{ mm} \end{cases} \qquad \textbf{(8)}$$

(units : kN, mm).

Now, another problem must be pointed out about uplift. In our laboratory, several fine observations along interfaces of the concerned beam and other composite beams have led to remark to following fact : even if any stud

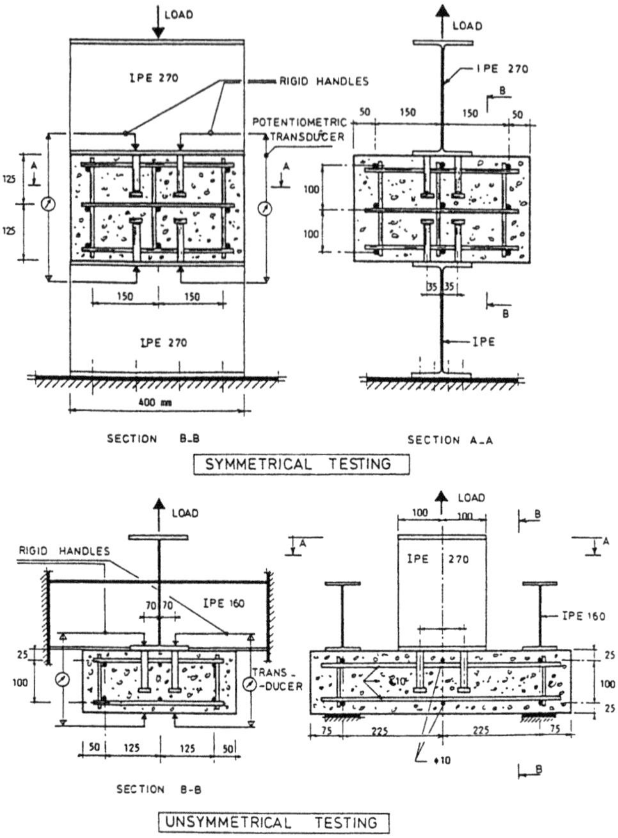

***Fig. 7** - Different types of pull-out tests*

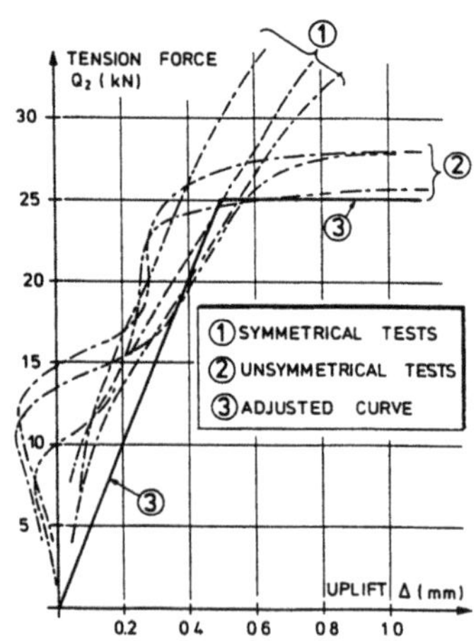

***Fig. 8** - Results of the pull-out tests*

in a solid slab is only subjected to shear force, systematically its slip induces some uplift because of kinematic compatibility of stud deformations. This phenomenon remains to investigate more thoroughly, but for the concerned studs of 19 mm diameter, an approximate value of the uplift kinematic component seems to be :

$$\Delta_k \simeq 0.25\ \gamma \tag{9}$$

In short, this component added to that resulting from the tension force Q_2 give the real value of uplift. Taking into account the above remark and using formulation **(8)**, we have computed uplift distributions with the numerical model ; comparison with distributions directly measured by means of transducers along the interface of the tested beam has been proving in relatively correct agreement at all the loading steps, as illustrated by **fig. 9.**

As partial conclusion, we can say that uplift measurement is a real problem because the connector behaviour in tension appears depending on the slab bending and on the slip of the connector itself. Therefore, testing procedure on specimens remains to be improved very much ; personally we suggest that an attempt might be made to carry out combined push-out and pull-out tests.

3-3 Needs of further investigation

Numerical investigations of the uplift parameter [1] have clearly demonstrated that this phenomenon has only a second order effect upon the general

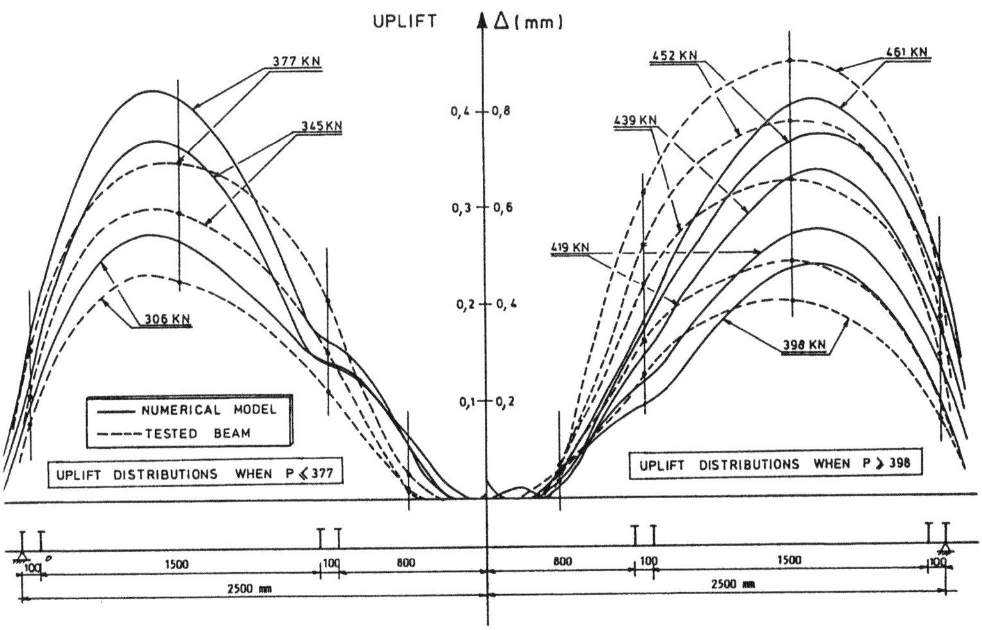

Fig. 9 - Uplift distributions along any mid-span

beam behaviour ; that means the uplift intensity does not modify the load-deflection curve practically, nor the normal stress distributions in cross-sections and the slip distributions. Alternatively, the uplift distribution may change strongly in function of the geometrical and mechanical properties of the composite beam and its connection. But there is much interest in continuing to analyse uplift by means of appropriate testing. For instance, we can mention the case of partially connected beams where a better evaluation of uplift would allow to define better the maximum spacing which prevents some corrosion problems at serviceability limit state. Also , connection design concerning uplift and corrosion problems should be considered very carefully in the case of connectors placed in groups, with a large spacing of groups ; the same recommendation should be applied to connectors having very low strength in tension, for instance studs without head, angles with L section, indentations on beam flange, etc.

4 CASE OF BEAMS WITH SLAB CAST ON PROFILED STEEL SHEETING

In this last part of presentation, deliberately the uplift parameter is cast aside because the main problem is raised here simply by the shear behaviour of connectors. In the contractor practice, it is often recognized that studs welded through profiled steel sheeting do not constitute always the best procedure to make connection, especially in countries where the atmospheric conditions are unfavourable for welding. That explains a present tendency to replace this procedure by developing other means of fixation, even if the new connectors are more flexible than studs. The examples presented hereafter take place in such context and two different systems of fixation are considered.

4-1 First connection system

Here the connector is a self screwing shear stud which is recently dealt by a french manufacturer [8, 9]. The setting procedure is applied in two times (see **fig. 10**). First an hole whose diameter is just smaller than the stud shank diameter is pre-drilled through the steel support. Secondly, the connector is screwed by self tapping through the support, while the torque required to this phasis is directly produced by the drilling machine ; to do that, the drill piece is replaced by a special driving tool which is compatible with the stud head including an hexagonal hollow. For information, steel support thicknesses up to 20 mm can be concerned by such procedure if studs of 12 mm diameter are used, with steel yield limit f_y and strength f_u equal respectively to 590 and 770 MPa.

Several push-out tests have been carried out by associating the concerned connector to a profiled steel sheeting Hi-bond of 1 mm thickness and 55 mm rib depth (the choice was imposed by the stud manufacturer for commercial reasons). Due to the positive action of interlock ensured by indentations and embossments distributed over the rib faces, it was not necessary to provide reinforcement in the specimen slabs, excepted a minimum one in median plan just to limit the effect of concrete shrinkage. Generally design codes do not give accurate specifications to test unusual situations (for instance [7]). However, the arrangement we have chosen for our specimen tests is illustrated to the right of **fig. 11**, with slabs having systematically 4 ribs of 350 mm length. Slip displacements have been measured at two plane levels of studs (number 2 and 4 on figure) by means of 4 potentiometric transducers. In addition to that, two points in each plane for slip measurement have been distinguished : one exactly at the contact zone of the sheet with the profile flange (slip called γ_I) and the other at the concrete fibre tangential to rib tops (slip called γ_{II}). From 6 specimens with ribs connected one time out of two, the experimental investigation has shown that there was no great difference between the slip values γ_I and γ_{II}, as illustrated by the $Q_1(\gamma)$ curves of **fig. 11**. So, a mean relationship can be proposed, that is :

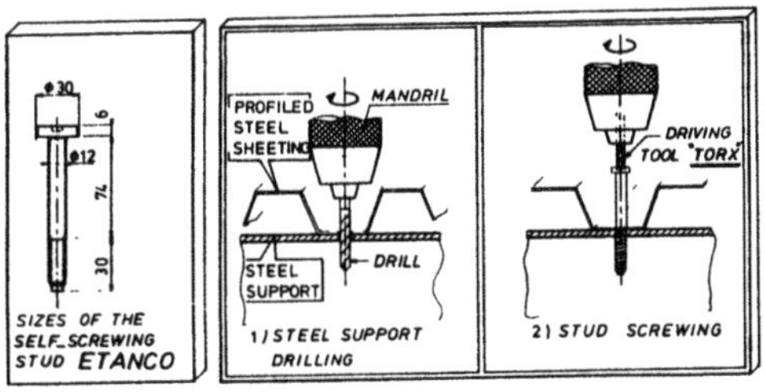

***Fig. 10** - Setting of self-screwing stud connectors*

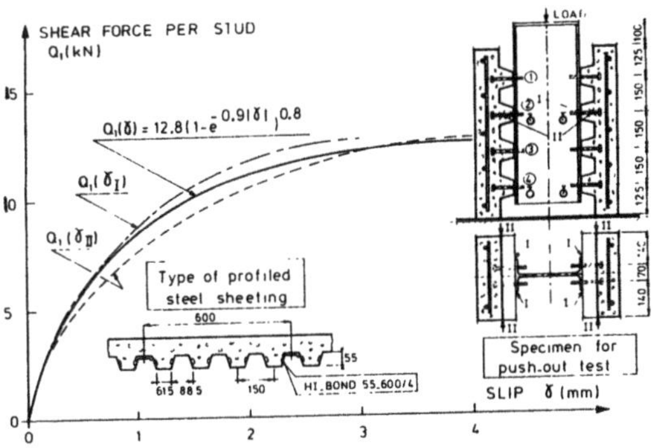

***Fig. 11** - Behaviour of one self-screwing stud associated with profiled steel sheeting*

$$Q_1(\gamma) = 12.8\,(1 - e^{-0.9\gamma})^{0.8} \qquad \textbf{(10)}$$

with $\gamma < 4$ mm .

Another tests on specimens with all connected ribs have led to a relationship very near to (10). For information, collapse has been achieved by concrete shearing in a slightly undulated surface which was defined by the stud heads and rib tops ; simultaneously a partial detachment of concrete from upper faces of the ribs could be observed, even for the unconnected ones. Such brittle collapse is explainable by the too short overall height of studs.

On the other hand, we have tested in Rennes a beam erected by using the above stud-sheet association, the connection being ensured by two self screwing studs spaced all the 300 mm. The beam span and characteristics of its cross-section are mentioned in **fig. 12**. The same procedure as in paragraph 3-1 for <u>direct measurement</u> of the behaviour law $Q_1(\gamma)$ between two cross-sections (AA and BB on the figure) has been used. **Fig. 12** shows immediately that the so-obtained curve is much more favourable than the push-out one, which may be explained by the composite action maintaining the structural integrity (here the rib length is 800 mm) and by a different mode of working for the connectors (it seems that in the beam, the connectors are subjected to lesser shear force and greater bending moment than in push-out tests). To be noted that it has been possible to register some part of the decreasing branch of the $Q_1(\gamma)$ curve because of effect of self-regulation in slip automatically brought by the connection acting as a whole. Finally, the following approximative relationship has been proposed to characterize the real behaviour of the self-screwing stud associated with sheet Hibond 55 :

$$Q_1(\gamma) = \begin{cases} 22.3\ (1 - e^{-2\gamma})^{0.4} & \text{with } \gamma < 1 \text{ mm} \\ 22.0 - \gamma & \text{with } \gamma \geqslant 1 \text{ mm} \end{cases} \qquad \textbf{(11)}$$

Provided that imperatively the new relationship **(11)** is taken into account, it is possible to find correct agreement between numerical calculation and experimental behaviour of the previous beam. So, **fig. 13** shows some comparisons related to the mid-span deflection, the stress state in the median cross-section and the slip distribution along the connection. On the whole, differences remain small, excepting some slips when loading is near to collapse ; at this state, the sheet deformation due to distorsion is too important and cannot be simulated by the model.

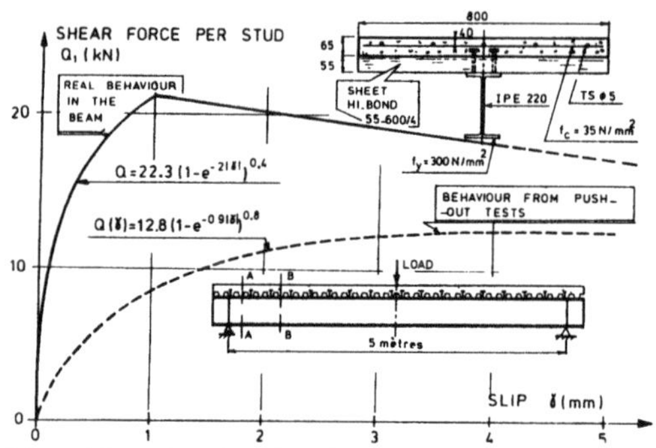

Fig. 12 - Stud behaviour inside a beam of composite floor

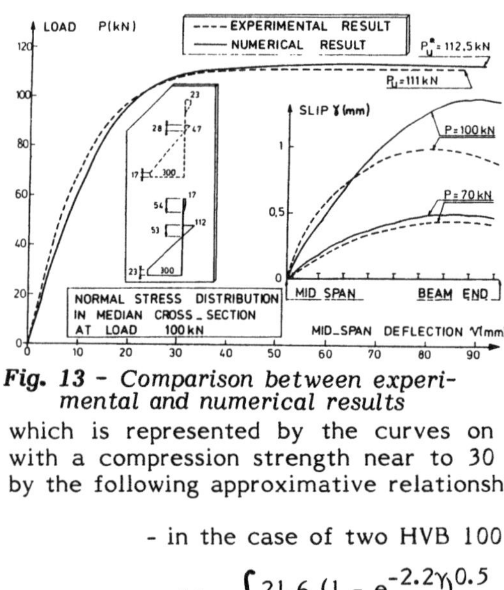

Fig. 13 - Comparison between experimental and numerical results

4-1 Second connection system

Here the connector is an angle connector made by cold-forming from mild steel. It is L-shaped and its foot is fixed to the steel flange by two hardened steel fastening pins which are driven through the connector and into the flange of the steel member using a powder-actuated tool (**fig. 14**). More exactly the connector type that we are going to consider hereafter is Hilti HVB 100 associated to sheeting Hi-bond 55/0.75 ; to be noted the feasibility of placing one or two connectors in any trough if the mid-plan of each connector is parallel to the trough axis. For such arrangement of connectors, we can find in the literature recent results deduced from push-out tests [10, 12]. So, it is possible to propose a mean behaviour of the connectors which is represented by the curves on the right of **fig. 14** ; using concrete with a compression strength near to 30 MPa, this behaviour can be expressed by the following approximative relationships :

- in the case of two HVB 100 per trough :

$$Q_1(\gamma) = \begin{cases} 21.6\ (1 - e^{-2.2\gamma})^{0.5} & \text{with : } \gamma < 4 \text{ mm} \\ 22.77 - 0.29\ \gamma & \text{with : } 4 \leqslant \gamma \leqslant 18 \text{ mm} \end{cases} \qquad \textbf{(12)}$$

- in the case of one HVB 100 per trough :

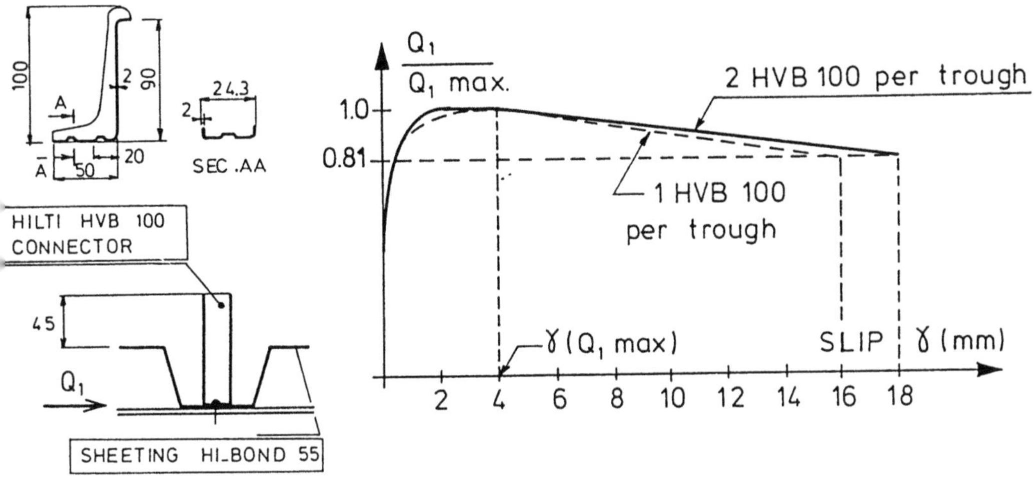

Fig. 14 - Behaviour issued from push-out tests [10]

$$Q_1(\gamma) = \begin{cases} 32.1\ (1^{-2.5\gamma})^{0.5} & \text{with : } \gamma < 0.5\ \text{mm} \\ 5.2\ [1 - e^{-0.7(\gamma - 0.5)}]^{0.8} + 26.4 & \text{with : } 0.5\ \text{mm} \leqslant \gamma \leqslant 4\ \text{mm} \\ 33.27 - 0.52\ \gamma & \text{with : } 4\ \text{mm} \leqslant \gamma \leqslant 16\ \text{mm} \end{cases} \quad \textbf{(13)}$$

To be noted again the large size of the decreasing branch with ultimate slip equal to 16 or 18 mm.

By another way, we can find also in the literature some measures directly taken on three composite beams using exactly the type of the above connection. **Fig. 15** shows the general geometry and loading system of two beams of them ; they differ from one another in the number of connectors per trough.

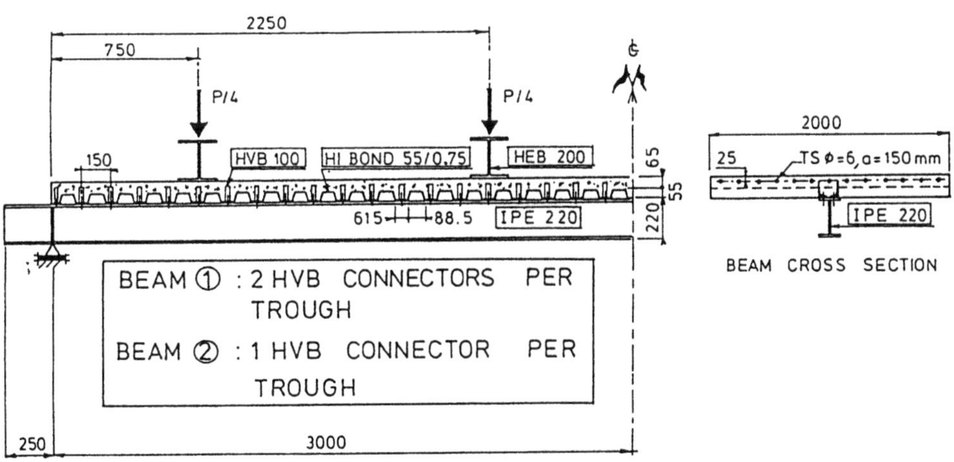

Fig. 15 - Geometry of two tested beams [11]

Beams testing was carried out in Germany [11] and more detailed information on material properties is mentioned in reference [12]. Introducing in our numerical model the relationship **(12)** for beam (1) and relationship **(13)** for beam (2), we have been able to compare the experimental and numerical results. **Fig. 16** and **17** respectively for beams (1) and (2) illustrate this comparison only based on the criterion of load-deflection curve ; through that criterion is generally the easiest to satisfy (stress or slip criterion would be more severe), its appears immediately undeniable differences between calculation (numbered 1 on the figures) and experimental measures. To explain such differences, other attempts of calculation have been carried out by modifying some parameters. For instance, we have considered law $Q_1(\gamma)$ as relative displacement whether between steel flange and rib top of the sheeting or between steel flange and rib bottom ; also, the characteristics of concrete behaviour have been slightly modified with a view to get stronger slab resistance, but in vain. Finally, the alone explanation we have found lies in the connector behaviour adopted for calculation : it is very credible that the behaviour deduced from push-out tests is two unfavourable compared

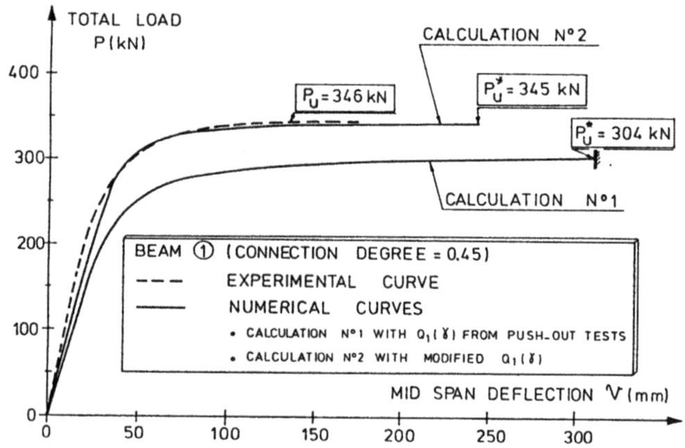

Fig. 16 - Comparison of load-deflection curves for beam (1)

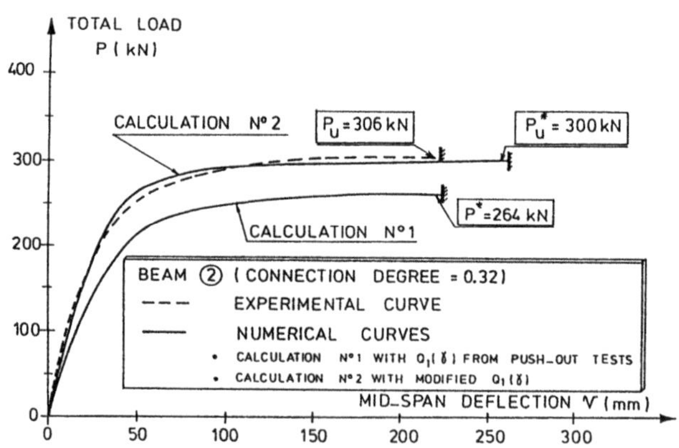

Fig. 17 - Comparison of load-deflection curves for beam (2)

with that of the connectors inside the concerned beams. So, to confirm this assumption, new calculations have been made with modified laws $Q_1(\gamma)$; as final result of such groping numerical investigation, we have established that a good agreement for the load-deflection curves can be obtained if relationship **(12)** and **(13)** are simply multiplied by factor 1.45 (see calculation numbered 2 on **fig. 16 and 17)**.

4-3 Needs of further investigation

The main conclusion we can draw from the experimental observations and numerical adjustments above presented for two systems of connection with steel profiled sheeting is the following : in some cases of composite floor (but may be not all the cases ?), push-out tests are unable to supply correct evaluation of the real connection behaviour in beams ; in particular, connector strength may be much underestimated. Such affirmation may appear categorical in the context of this short presentation. However, we must point out that other examples of beam published in the literature have been computed by means of our model ; in fact, experimental global behaviour of these beams has been simulated correctly as often as the used law $Q_1(\gamma)$ was extracted from direct measurement on beam (for instance, Robinson's beam (1969) and Fisher's beam (1977) in our reference [3]). In consequence, we tend logically to believe that any push-out testing procedure should be criticized thoroughly, even if it seems very sophisticated (this last sentence means that some arrangements are adopted such as to choose large-sized specimens including several ribs, to define the same spacing of connectors in specimens as in the beam, to use displacement control of loading, etc). But any critical investigation needs to have a sure reference at one's disposal ; for the case considered, it seems difficult to do otherwise than to carry out tests on complete beams in parallel way and to take direct measures on them. Obviously, direct measurement must form also the subject of critical examination ; however to do that, the detailed information supplied by numerical calculation may be very useful (so, possible effect of residual stresses in the steel member [3] may be considered with respect to the experimental determination of the longitudinal force in this member).

From the point of view of design, it is important to recall the low connection degree met in most composite beams with profiled steel sheeting. Consequently, there is much interest in knowing a significant shape and the magnitude of curve $Q_1(\gamma)$ with maximum accuracy ; indeed the decreasing branch of this curve can be turned to the best account when the beam span becomes large (but less than 20 m). In particular, we have shown recently [3] how an appropriate definition of connector resistance can be drawn from the whole curve $Q_1(\gamma)$ in function of the beam span. This definition allows to bring down the minimum connection degree near to 0.2 (whereas Eurocode 4 requires connection degrees greater than 0.5), and that without changing the usual design method based on the concept of reduced ultimate bending moment in critical cross-section.

5 CONCLUSION

The present paper is only a modest contribution with more a survey of personal observations and raised questions than accurate proposals. However, it appears clearly that improvement of testing on specimens needs confrontation with parallel tests on complete beams. For beams with solid slab, pull-out tests are mainly concerned and interaction between uplift and slip must be taken into account to solve applied problems as corrosion ones along steel-concrete interface and design of connectors placed in groups with large spacing of groups. On the other hand, for beams with slab cast on profiled steel sheeting, the main question is raised with the standard push-out tests which may supply very conservative results ; it is obvious that a better determination of the shear force - slip relationship could lead to significant economical advantages in composite floor design. About this topic, we inform that a specific experimental program (named Corgec Alphabeta 3) will be developed in France on 1991.

REFERENCES

[1] Aribert J.M. and Abdel Aziz K. - Calculation of composite beams up to ultimate state with the effect of uplift at steel - concrete interface Construction Métallique - n° 4 - 1985, p. 3 - 36.

[2] Aribert J.M. and Abdel Aziz K. - General model for the ultimate state design of statically undeterminate composite beams. Construction Métallique - n° 4 - 1986, p. 3 -41.

[3] Aribert J.M. and Al Bitar A. - Optimization of the design in partial connection of beams in composite floors with thin profiled steel sheeting Construction Métallique - n° 4 - 1989, p. 3 - 33.

[4] Kabaila A. - Equation for the stress strain curve of concrete. ACI Journal - September 1964, p. 1227 - 1229.

[5] Basu A.K. and Somerville W. - Derivation of formulae for the design of rectangular composite columns - Proc. ICE, Supplementary Volume, December 1969.

[6] Aribert J.M. - Sur un modèle général de fonctionnement mécanique de la connexion acier-béton pour divers types de poutres et planchers mixtes, avec comparaison à l'expérience. Proceedings of the First International RILEM Congress - Paris - September 1987 - Chapman and Hall, Vol. 2, p. 871 - 878.

[7] Commission of the European Communities - Eurocode 4 - Design of composite steel and concrete structures - First draft - EUR 9886 FR - 1985.

[8] Aribert J.M. and RIVAL J.C. - Behaviour characteristics of a self screwing shear connector for composite steel-concrete connections - Construction Métallique - n° 3 - 1988 - p. 123 - 135.

[9] Aribert J.M. and Rival J.C. - Characteristics and behaviour modeling of a particular screwed stud for shear connections in composite steel concrete structures - International Colloquium "Bolted and special structural connections" - Moscow - May 1989 - Vol. 4, p. 20 - 25.

[10] Turley, B and Murray, T.M., Research report Hilti HVB 105 shear connectors pushout - tests - FEARS Structural Engineering Laboratory, University of Oklahoma, Res. Rep. FSEL/Hilti 80-01, 1986.

[11] Biegebalkenversuche, Hilti Verbunddübel HVB 100 auf Stahlträger IPE 220, Profilblech Hibond 55/0.8 und Betonplatte aus Normalbeton - Hilti AG, Schaan, Bericht IB 04/88, 1988.

[12] Crisinel M. - Partial-interaction analysis of composite beams with profiled sheeting and non-welded shear connectors - J. Construct. Steel Research 15 - 1990 - p. 65 - 98.

35 METAL–CONCRETE COMPOSITE STRUCTURES: TESTS AND MODELS

E. BRUZZESE
Istituto di Scienza delle Costruzioni, Faculty of Engineering, Naples University, Italy
A. GHERSI
Istituto di Tecnica delle Costruzioni, Faculty of Engineering, Naples University, Italy

SUMMARY

This theoretical and experimental investigation aims at contributing to the knowledge of the ultimate limit state of metal-concrete composite beams. Flexural tests have been carried out on steel-concrete and aluminium-concrete specimens. The test results have been compared with the theoretical values based on several stress-strain relationships for all materials. The wide geometrical and mechanical range of steel-concrete specimens allows to evaluate the influence of some parameters, such as the ratio of slab thickness over section height and the amount of lateral and longitudinal reinforcements. The tests confirm the theoretical assumptions for both steel and aluminium-concrete specimens and give appropriate trends for future investigations on aluminium-concrete beams.

1 STEEL-CONCRETE COMPOSITE BEAMS

1.1 Specimens, test conditions and material features

Tests [1] were carried out on specimens with a 1200 mm span (fig.1), divided into 4 sets, each one with a different slab thickness (70, 90, 110, 130 mm) and the same slab width (250 mm) and steel section (a double T, 80 mm high cross-section). Geometrical characteristics are synthesized by means of the ratios d/h of slab over composite section height and $p_s=A_s/A_c$ of steel over concrete area.

The beams of each set are divided into 4 subsets, each one collecting four homogeneous specimens. The first subset has no lateral and longitudinal reinforcements and the others have 100 mm tie spacing (fig.2) and different longitudinal reinforcements, expressed by means of the ratio $p=A_r/A_c$ of reinforcement over concrete area (tab.1). Table 2 shows the minimum

Table 1

set	d (mm)	d/H	p_s (%)	p (%)
1	70	0.47	4.40	0 - 0.65 - 1.15 - 1.79
2	90	0.53	3.42	0 - 0.70 - 1.14 - 1.79
3	110	0.58	2.80	0 - 0.73 - 1.10 - 1.87
4	130	0.62	2.37	0 - 0.79 - 0.97 - 1.93

Testing of Metals for Structures. Edited by F.M. Mazzolani. © 1992 RILEM.
Published by Taylor & Francis, 2 Park Square, Milton Park, Abingdon, Oxon, OX14 4RN.
ISBN 0 419 15810 3

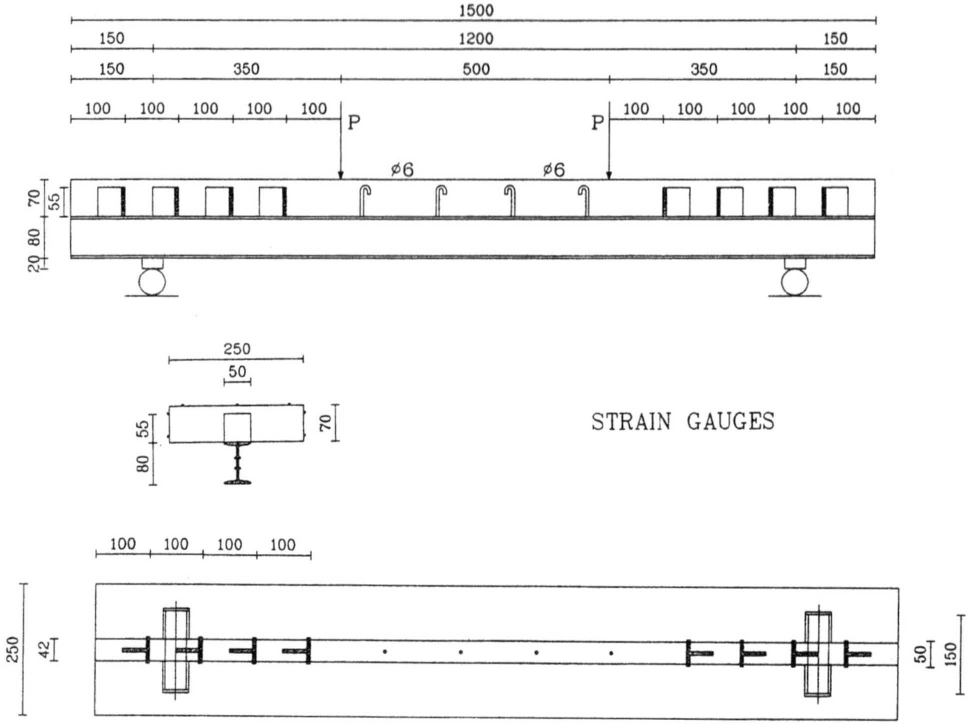

fig.1

and maximum values of the mechanical features of the materials of each set: concrete cylinder strength R_c at the time of testing (40-50 days), modulus of elasticity of concrete E_c, yielding stress of steel f_y for flanges and web.

fig.2

Table 2

set	R_c (Nmm^{-2})	E_c (Nmm^{-2})	f_y (Nmm^{-2}) flange	f_y (Nmm^{-2}) web
1	35.2 41.4	31000 35100	271.0 301.4	270.8 309.0
2	40.0 49.5	36000 37400	290.4 330.0	326.6 346.6
3	30.0 37.4	28200 31300	286.0 296.1	285.0 320.5
4	36.2 55.9	32900 40000	340.0 350.0	340.0 350.0

The connectors consist of T sections welded to the upper flange in the lateral zone, while in the central one the connection is obtained by means of bars 6 mm in diameter. The test conditions are also shown in fig.1: two forces have been symmetrically applied, as commonly used to obtain a purely flexural behaviour in the central zone, where longitudinal strains were measured by means of mechanical strain-gauges along gauge lengths of 100 mm, whose distribution is shown in the same figure.

1.2 Analysis of tests

The ultimate limit state was reached with small consecutive increasing of loads. Strains were measured under stabilized conditions, up to about 95% of ultimate load. Last increasing of load led to great displacements which did not allow further measurements.

During each test, thin cracks appeared on the lower slab edge in an advanced rate of loading (after steel yielding), becoming more wide and extending to the upper edge at the increasing of load (fig.3). Concrete crush on the upper slab edge (fig.4) preceded collapse close to the ultimate load. Such a failure was sometimes preceded by the appearing of longitudinal cracks through the whole height of the slab, due to the localized stresses caused by connectors (fig.5). For few models, having no reinforcements in the slab, connectors caused the full parting of the slab along their alignment (fig.6).

2 ALUMINIUM-CONCRETE COMPOSITE BEAMS

2.1 Specimens, test conditions and material features

Tests were carried out on two specimens having the same geometrical configuration: a 1500 mm span, an 800 mm wide by 80 mm high slab, a double T, 140 mm high, aluminium section (fig.7). The beams differ in slab tie spacing: 65 mm in the first one, 130 mm in the second one. The longitudinal reinforcements in the slab are only for ties support and are not significant to the elastic or ultimate strength (fig.8). Table 3 shows the mechanical features of the materials of each specimen.

The connectors consist of through bolts 16 mm in diameter jointed by means of a lock-nut to the upper flange of the beam, and of bolts 8 mm in the gauge length.

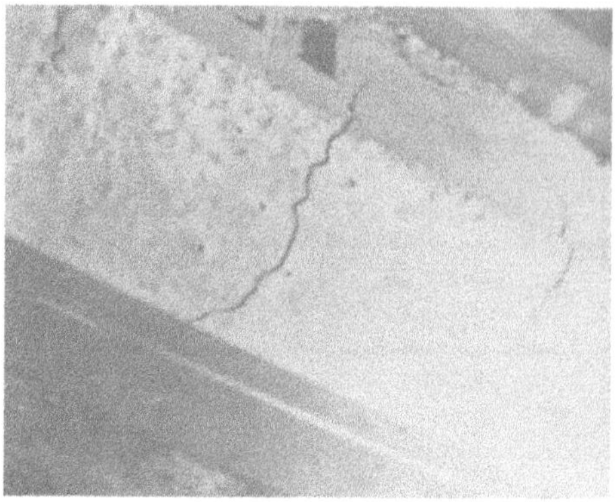

fig.3

fig.4

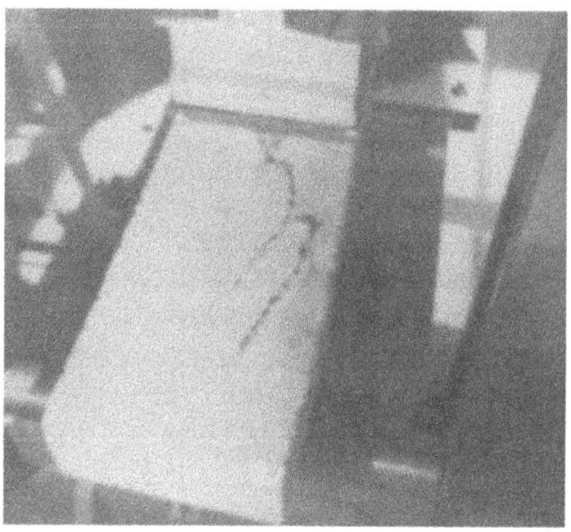

fig.5

fig.6

Table 3

specimen	concrete		aluminium alloy			
	R_c (Nmm-^{2})	E_c (Nmm-^{2})	f_{02} (Nmm-^{2})	f_t (Nmm-^{2})	ε_t (%)	E (Nmm-^{2})
1	43.3	31490	339.9	394.9	1.51	70000
2	55.7	35711	352.3	410.5	1.52	70000

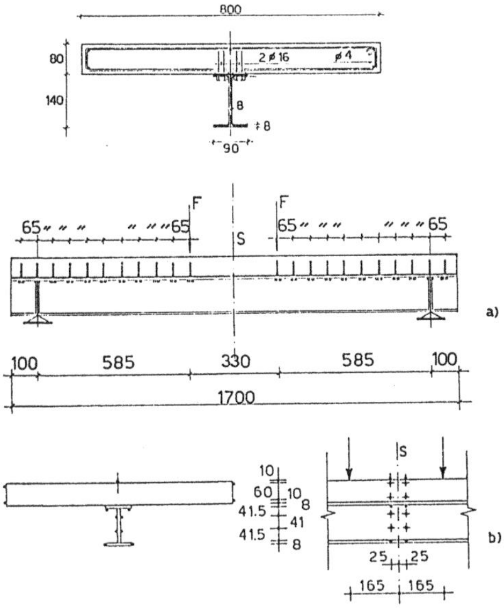

fig.7

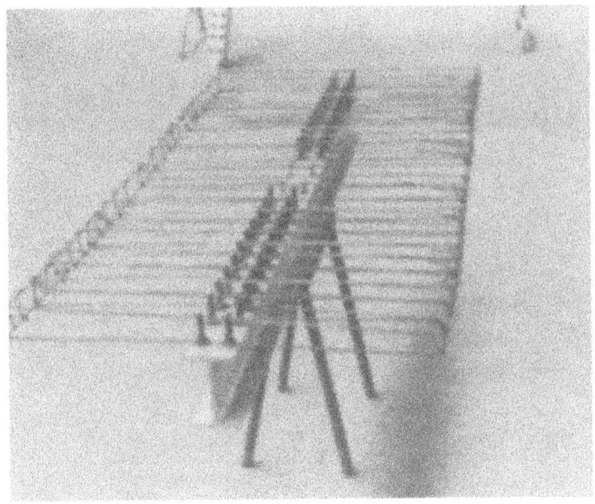

fig.8

The test conditions and the strain measuring, also shown in fig.7, are almost identical to the ones described for steel-concrete specimens.

2.2 Analysis of tests

The ultimate limit state was reached with consecutive loading and complete unloading cycles. Strains were measured under stabilized conditions.

During both tests, thin cracks on the lower slab edge appeared in an early rate of loading (about 0.33 of aluminium elastic range limit), becoming very wide close to the ultimate load (fig.9). The beams collapsed as a result of the concrete crushing (fig.10). In the first model, thin longitudinal cracks appeared at 0.85 of ultimate load, due to the localized stresses caused by connectors. In the second model, having a larger tie spacing, longitudinal cracks occurred early and one of them widened and deepened so as to affect the entire slab thickness (fig.11), leading to the same failure mode noted for some unreinforced steel-concrete beams, with a marked decrease in the load value at the elastic limit and in the ultimate load as compared to the first model.

3 THEORETICAL MODELS FOR MOMENT-CURVATURE RELATIONSHIPS

3.1 Stress-strain relationships for concrete, steel and aluminium

Stress-strain relationship for concrete ought to take into account the effects of concrete grade, the amount of lateral reinforcement, the strain gradient, the size of the core enclosed within ties; at the same time, it ought to be enough simple so as to be used in the practical applications. Among the proposals made by several authors, we have selected four models having different grade of complexity (fig.12). The simplest ones are the bilinear elastic-perfectly plastic and the CEB proposal (a parabola with a maximum at ε=0.2 %

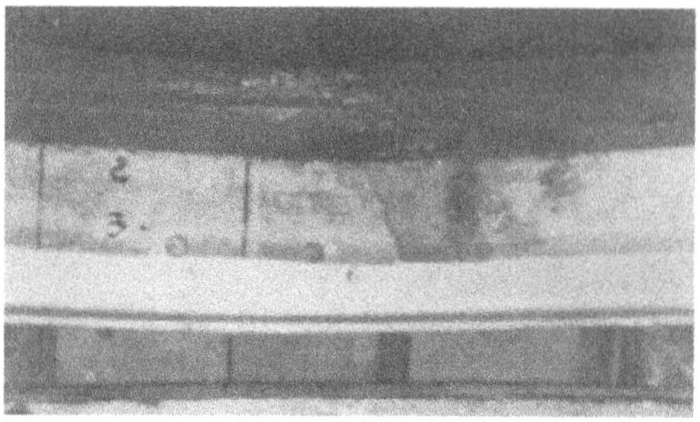

fig.9

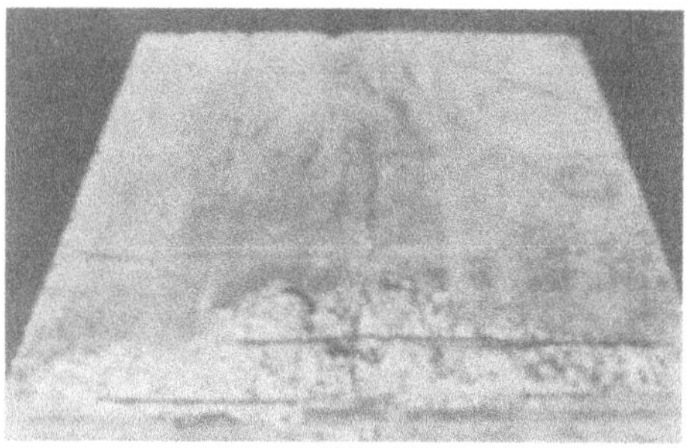

fig.10

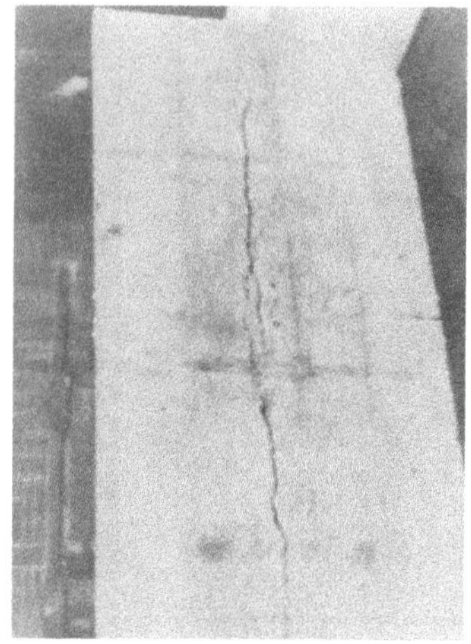

fig.11

followed by a perfectly plastic branch), both with a 0.35 % strain limitation. More consistent are the Sargin's curve and the Kent and Park's law (a parabola followed by a linear falling branch with a slope determined by the before mentioned parameters). Tensile strength of concrete is neglected by all models.

Stress-strain relationship for steel has been mainly interpreted through the elastic-perfectly plastic model, with no hardening. The Ramberg and Osgood's curve with a low hardening factor (fig.13) has been used in order to obtain a better correspondence to some experimental values. The same curve, with a greater hardening factor, has been used to represent the stress-strain relationship for aluminium (fig.14).

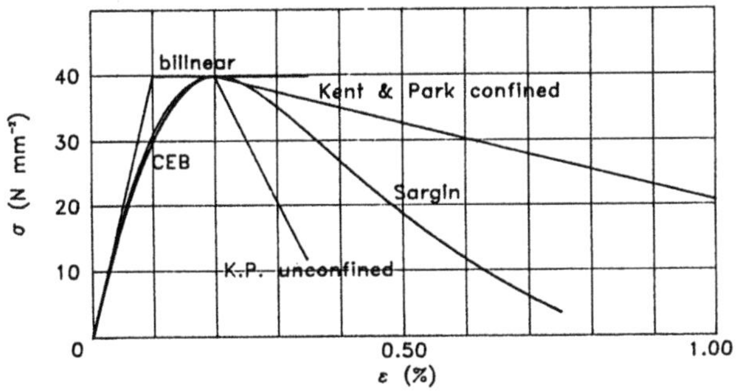

fig.12

CONCRETE

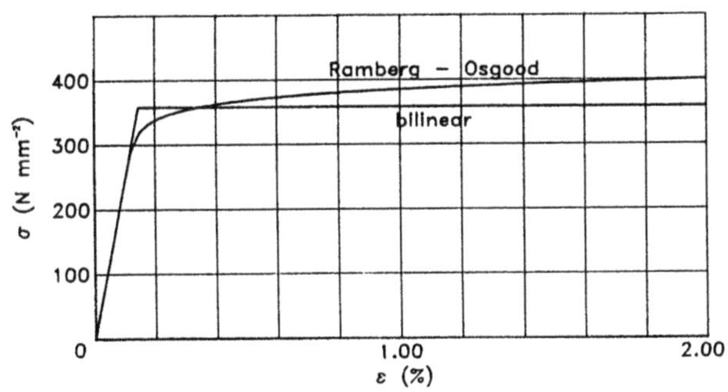

fig.13

STEEL

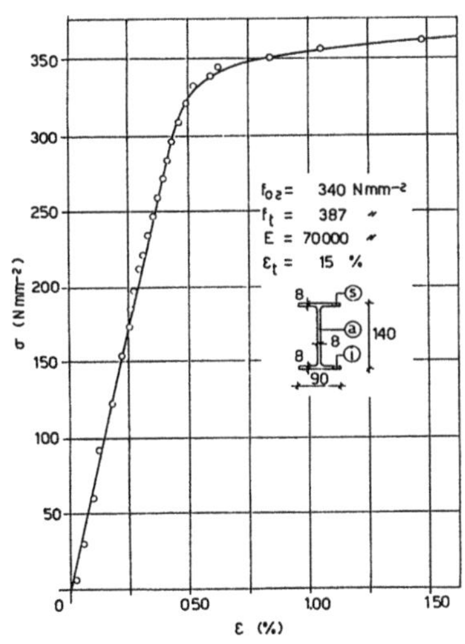

fig.14

ALUMINIUM

3.2 Moment-curvature relationship

The equations expressing force and moment equilibrium in the section at any loading stage are based on the assumption that plane sections remain plane after deformation also in the inelastic range and that the bond at the interface of metal and concrete is complete. Their solution has been found by means of a numerical iterative method which, for a given value of curvature, finds by successive approximation a position of the neutral axis such that the force equilibrium is satisfied within a specified tolerance. In each step the stresses in concrete and metal are evaluated according to the prescribed stress-strain relationships.

Fig.15 shows the theoretical moment-curvature curves obtained for the steel-concrete composite beams (both for slab in plain concrete and with the maximum longitudinal reinforcement) by means of the different stress-strain relationships for concrete. The shape of all curves is nearly identical and the values of the ultimate moment are very similar, while significant differences are attained in the ultimate curvature and in the falling branch using Sargin's and Kent and Park's models. The figures make also evident the changes in slope corresponding to the yielding of lower and upper flange and, when present, of lower slab reinforcements. Fig.16 shows the influence of flange width and of amount of longitudinal reinforcements on such a change in slope.

The curves in fig.17, corresponding to the different stress-strain relationships for steel together with a bilinear model for concrete, show also a change of shape and of ultimate moment.

4 COMPARISON BETWEEN TESTS AND THEORETICAL MODELS - STEEL-CONCRETE BEAMS

Experimental values of curvature at any load rate are obtained as an average of symmetrical measurements on both sides of the specimen. The values related to the four specimens of each subset are reproduced by means of different symbols in fig.18 to 21, so as to display the behaviour of specimens during test, up to about 95 % of ultimate load. The dashed curve, showing the average of experimental values for each subset, is evaluated by numerical regression and is expressed by means of a linear equation for the elastic branch and an exponential equation for the plastic one.

The subsets having a plain concrete slab are marked by a shorter plastic branch. The increasing of longitudinal reinforcements leads to a significant growth of the ultimate moment and a considerable plastic capacity, owed to the yielding of the full steel section before concrete failure. The remarks about plastic branch of experimental curves must be considered mainly as trends of behaviour, to be confirmed with different test conditions (loading steps based on deformation increments instead of direct load increments).

The theoretical curves, referred to bilinear models both for concrete and steel, show a good correspondence with experimental values. A better agreement is obtained by means of Ramberg-Osgood's curve with a low hardening factor, for sets with an higher slab.

The number of strain measurements in steel and concrete made possible a thorough investigation on section deformation (fig.22). The planarity of section after deformation has been confirmed by most tests. In spite of the null value of shear in the central zone, some specimens revealed slips along the steel and concrete interface, owed to the partial failure of connectors in the lateral zones. The subsequent strain diagrams show discontinuities in value and sometimes also in slope. However, such phenomenon seems to have been little influence

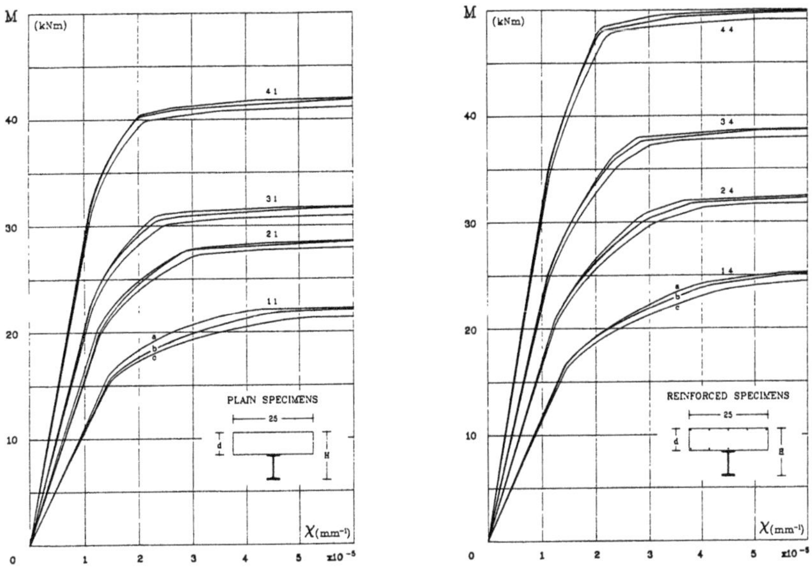

fig.15 - concrete: a = bilinear, b = CEB, Kent-Park, c = Sargin

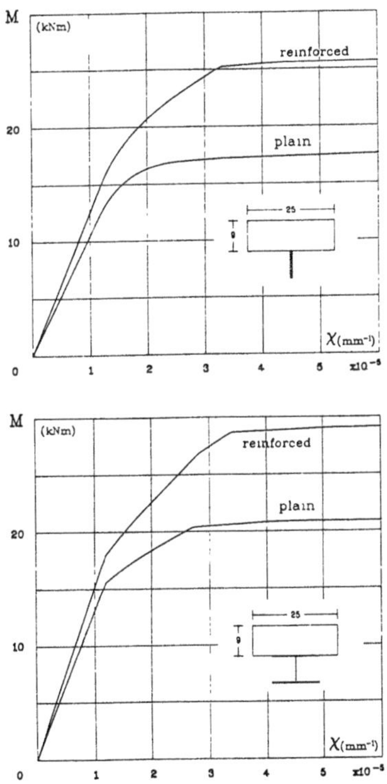

fig.16

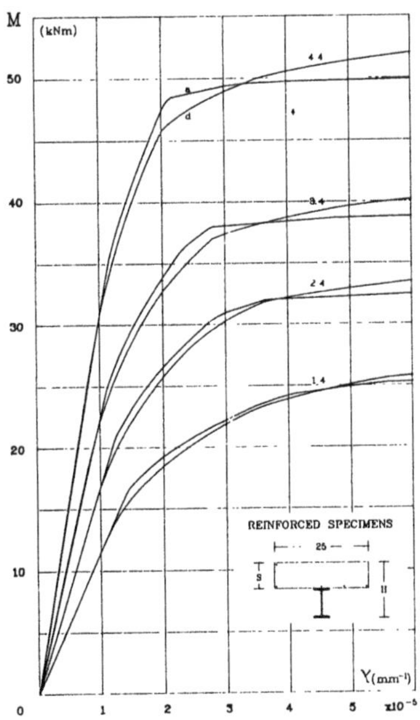

fig.17 - steel: a = bilinear,
b = Ramberg-Osgood

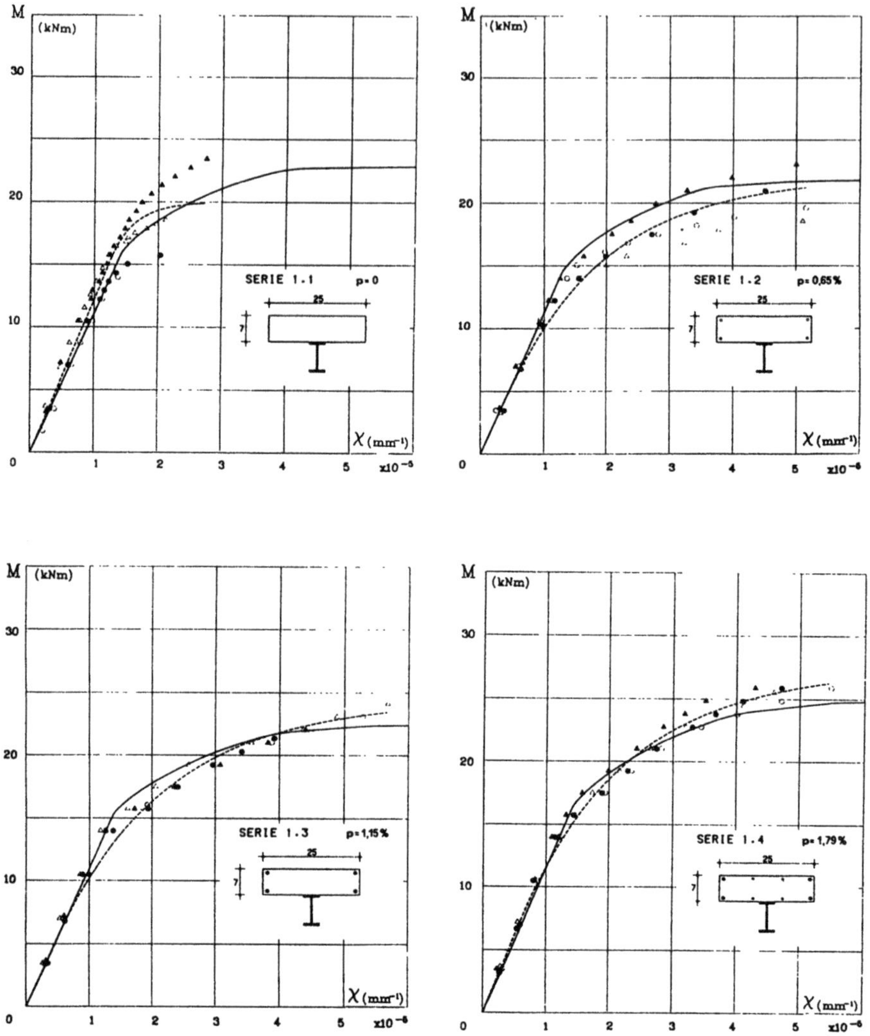
M (kNm)
χ (mm⁻¹)
x10⁻⁵
SERIE 1.1 p=0
SERIE 1.2 p=0,65%
SERIE 1.3 p=1,15%
SERIE 1.4 p=1,79%
25
7

fig.18

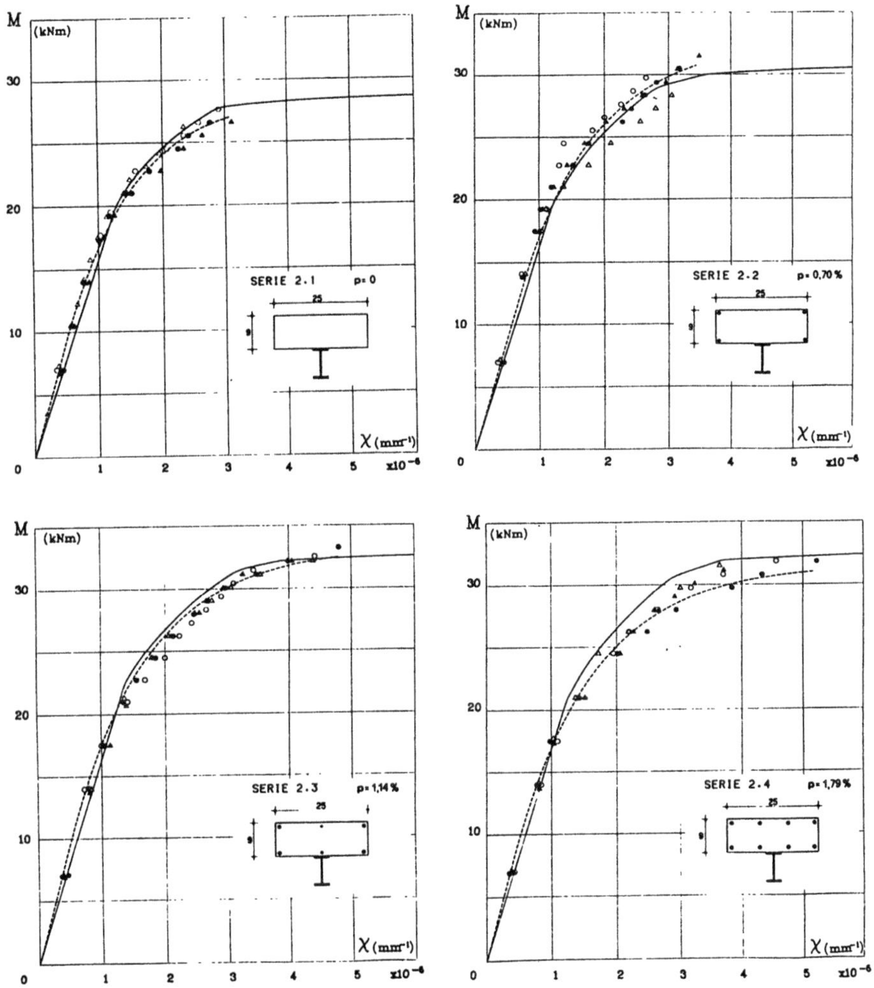
M
(kNm)
SERIE 2.1
p= 0
SERIE 2.2
p= 0,70 %
SERIE 2.3
p= 1,14 %
SERIE 2.4
p= 1,79 %
χ (mm⁻¹)
x10⁻⁶

fig.19

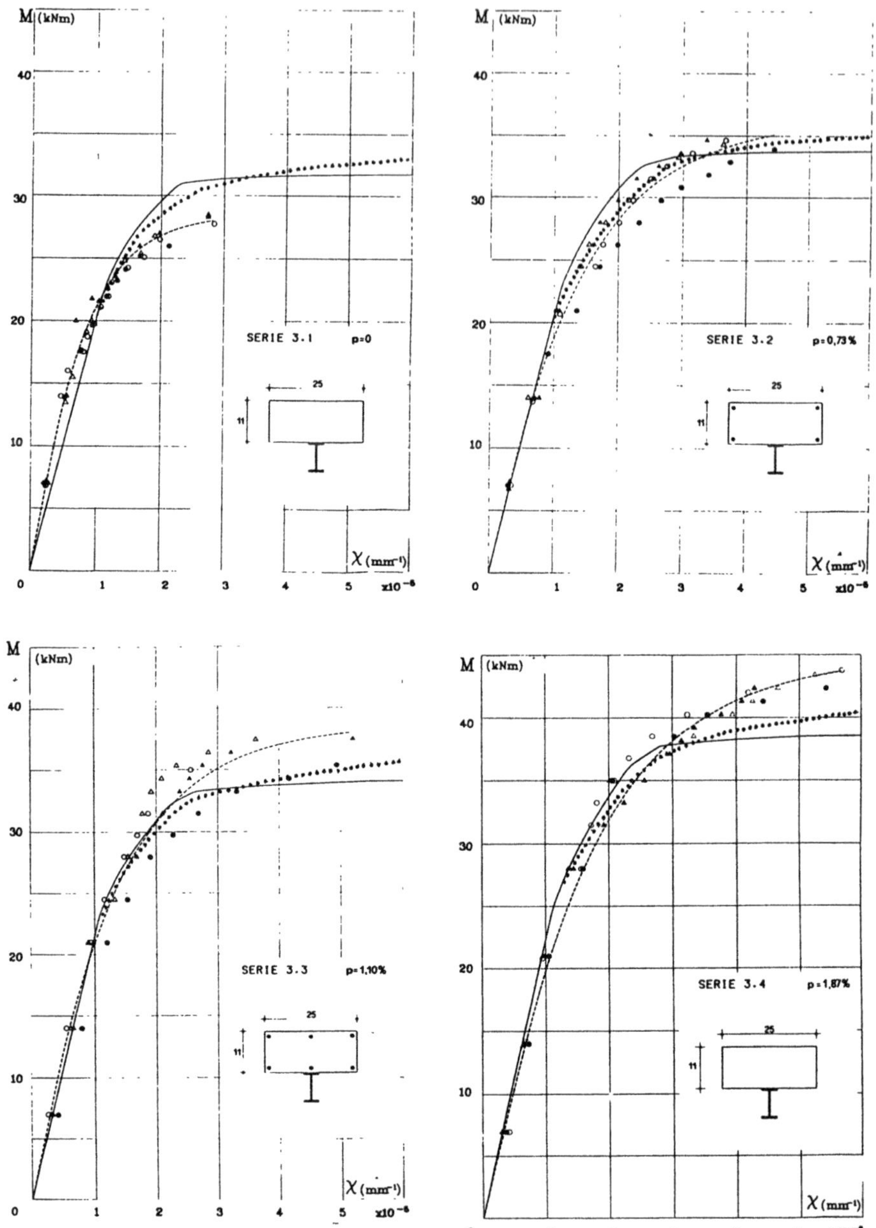
SERIE 3.1
p=0
SERIE 3.2
p=0,73%
SERIE 3.3
p=1,10%
SERIE 3.4
p=1,87%
M (kNm)
χ (mm⁻¹)
x10⁻⁵

fig.20

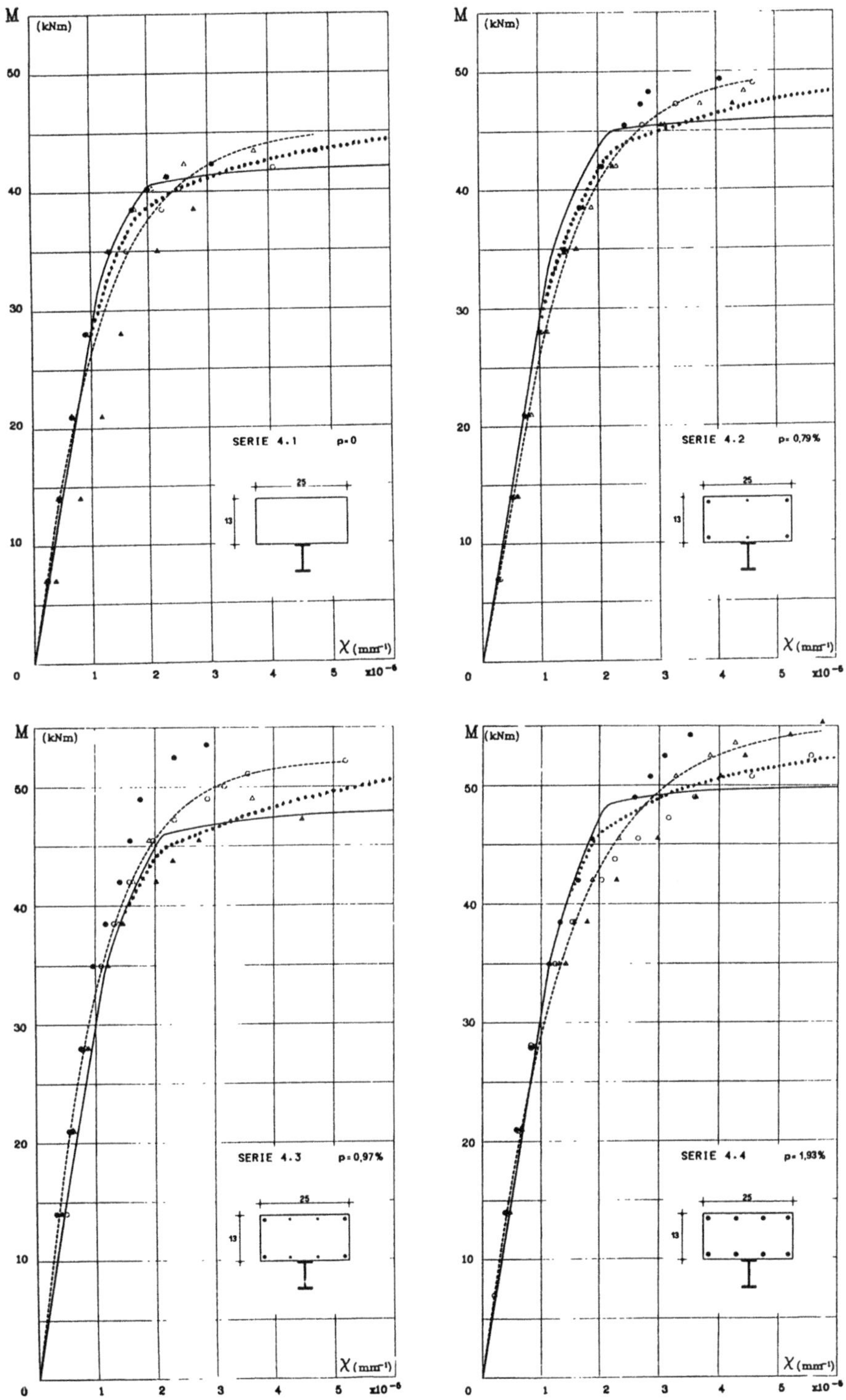

SERIE 4.1 p=0
SERIE 4.2 p=0,79%
SERIE 4.3 p=0,97%
SERIE 4.4 p=1,93%
M (kNm)
χ (mm⁻¹)
x10⁻⁵

fig.21

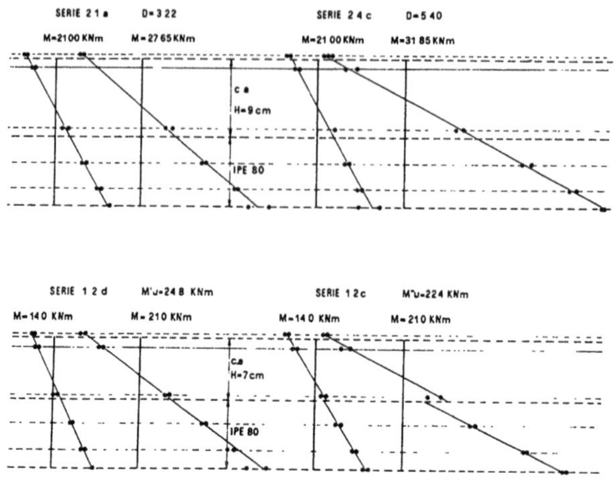

fig.22

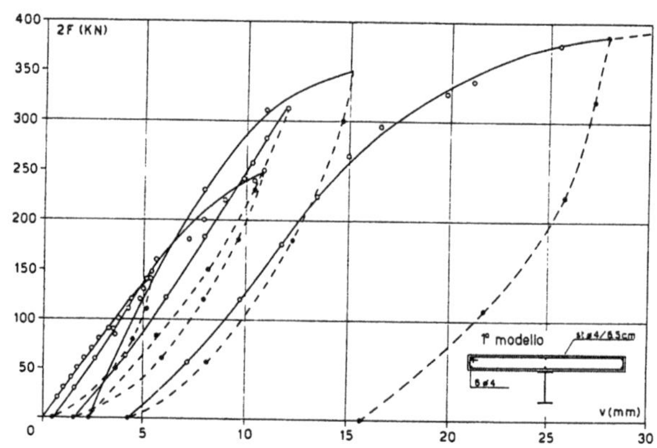

fig.23

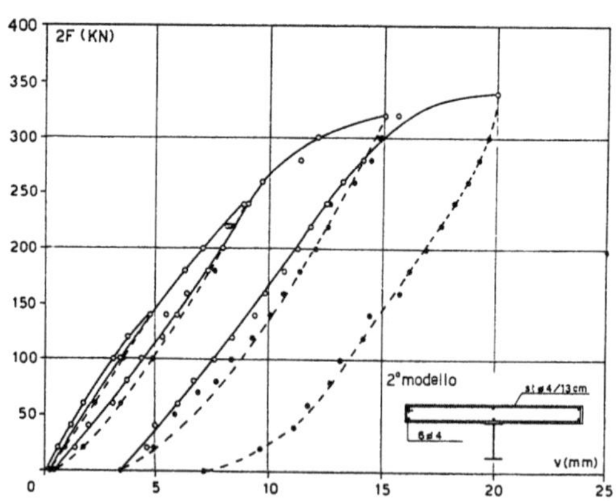

fig.24

on the global behaviour of these specimens: the last diagrams of fig.22 are referred to two identical specimens showing different strain diagrams (with and without slip) and only small differences in the ultimate moment values.

5 COMPARISON BETWEEN TESTS AND THEORETICAL MODELS ALUMINIUM-CONCRETE BEAMS

Fig.23 and 24 show the experimental displacements of the two beams, related to the applied forces. Curves interpolating these points were introduced in order to put in evidence progressive cycles of loading. It is clearly identified an initial elastic behaviour and the subsequent entrance in the inelastic range, with a broad change in slope of the curves. The first specimen presents a marked development into the plastic range. Plastic capability is less accentuated in the second specimen, because of the lower amount of lateral reinforcements and the subsequent reduced ability of the section to withstand transversal strains, which led to the above mentioned longitudinal crack due to the localized stresses caused by connectors.

Fig.25 and 26 compare experimental and theoretical values of moment and curvature for the two models. Elastic and plastic behaviour is still evident; for the first model the theoretical values are in a good agreement with the experimental ones in the elastic range, while showing an earlier entrance in the inelastic range.

Fig.27 displays the experimental strains (small circles), an interpolation line (dashed line) and the theoretical diagram (solid line). Maximum values of strains obtained by testing are next to the theoretical ones for loads up to 340 kN (83 % of ultimate load), corresponding to concrete strains lower than 0.2 %, while are significantly smaller for greater values of load. Last strain measured for concrete before collapse is about 0.33 %, which is far lower than is envisaged in the Kent and Park's stress-strain relationship. Further tests will be carried out to investigate this problem under different test conditions i.e. with loading steps based on deformation increments.

The planarity of section after deformation is not kept for great values of load. A first small slip between aluminium and concrete interface was noted for a load of 180 kN (44 % of ultimate load) and became more significant in the following steps of loading.

6 CONCLUSIONS

The comparison among theoretical models based on several stress-strain relationships shows that, for the examined geometries, the choice of the concrete model has no importance in the evaluation of the ultimate moment and influences only the ultimate curvature values and the shape of the fully plastic (horizontal or falling) branch of moment-curvature curves.

The experimental values for steel-concrete composite beams are in a good agreement with the numerical ones. The behaviour of section is marked by a wide plastic range which is in agreement with the simplified model of plastic hinge. The ultimate curvatures obtained by loading steps based on direct load increments are significantly lower than the theoretical ones. This investigation emphasizes the advisability of tests based on deformation increments to obtain further results to compare with those here referred, in order to have more realistic informations about the ductility of composite sections.

The assumption that sections remain plane after deformation up to the failure has been confirmed by many steel-concrete specimens. The slips noticed at an high rate of loading in some composite beams have no great influence on ultimate load values. The possibility of longitudinal cracks owed to the localized stresses caused by connectors emphasizes the

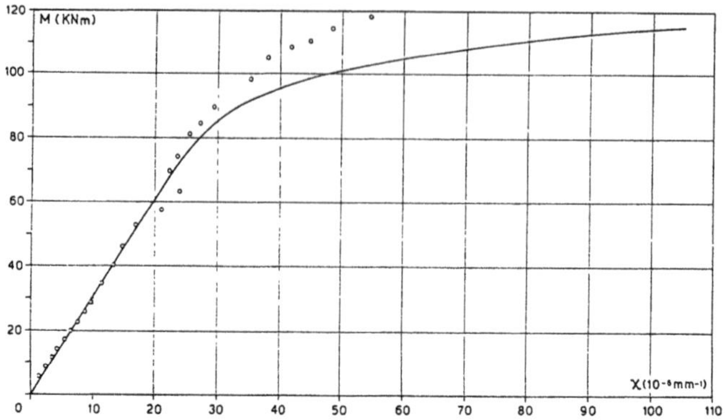

fig.25

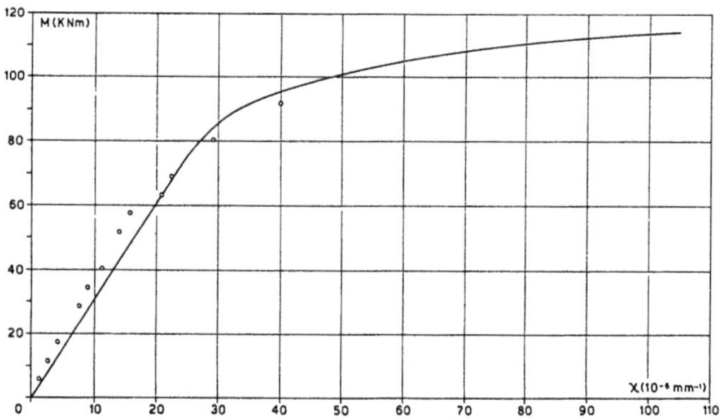

fig.26

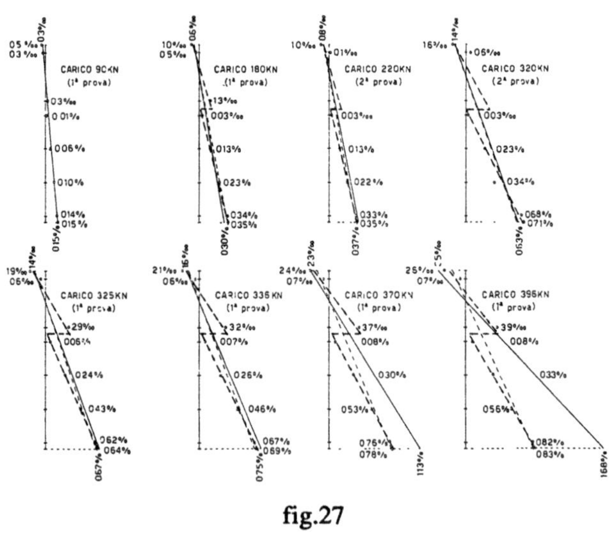

fig.27

opportunity of using an high amount and a spread distribution of lateral reinforcement.

The effect of lateral and longitudinal reinforcements in the slab, in order to increase the section ductility, has been emphasized for steel-concrete beams both theoretically and experimentally but is still to be examined for aluminium-concrete beams. For these ones remains also to be explored the effect of the low modulus ratio, which appears to appreciably affect their behaviour.

References

1. Aiello S., Bruzzese E., Majorano C., Analisi teorico-sperimentale del comportamento allo stato limite ultimo di travi composte di acciaio e calcestruzzo, Pubblicazione n.13, Cattedra di Costruzioni Metalliche, Facoltà di Ingegneria, Napoli, June, 1984.
2. Bruzzese E., Cappelli M., Mazzolani F.M., Experimental investigation on aluminium-concrete beams, Costruzioni Metalliche, n.5, 1989, pp. 265-282.
3. Sargin M., Stress-Strain Relationships for Concrete and the Analysis of Structural Concrete Sections, Study no.4, University of Waterloo Press, 1971.
4. Kent D.C., Park R., Flexural members with confined concrete, Journal of the Structural Division, ASCE, July, 1971, pp. 1969-1990.
5. Ramberg W., Osgood W.R, Description of Stress-Strain Curves by three parameters, NACA Techn. n.902, 1943.
6. Desayi P., Krishnan S., Equation for the Stress-Strain Curve of Concrete, ACI Journal, Proceedings vol.61, n.3, 1964.
7. Barnard P.R., A series of Tests on Simply Supported Composite Beams, Journal of the American Concrete Institute, Proceedings v.62, no.3, Apr., 1965, pp.443-455.
8. Bruzzese E., Esperienze sul comportamento a rottura di travi a sezione composta di acciaio e calcestruzzo, communication to Seminario Gruppo Plasticita', CNR, Pisa, 1965.
9. Johnson R.P., Research on steel-concrete composite beams, Journal of the Structural Division, ASCE, vol.96, no.ST3, March, 1970, pp.445-459.
10. Carputi U., Noto A., Russo Ermolli E., Per la progettazione agli stati limite delle strutture miste acciaio-calcestruzzo, Costruzioni Metalliche, n.6, 1971.
11. Pagnini L., Palazzo B., Sul comportamento allo stato limite ultimo delle sezioni miste acciaio-calcestruzzo, Costruzioni Metalliche, n.2, 1977.
12. Eurocode n. 4, Common unified rules for composite steel and concrete structures, Commission of the European Communities, 1985.
13. Mazzolani F.M., La progettazione delle strutture in leghe di alluminio, Liguori, Napoli, May, 1986.
14. Bruzzese E., De Martino A., Mandara A., Mazzolani F.M., Esame dei parametri comportamentali del sistema misto alluminio-calcestruzzo, Giornate Italiane della Costruzione in Acciaio, Capri, Oct., 1989, pp.223-239.

36 EXPERIMENTAL CONTROLS ON SHEAR CONNECTORS IN COMPOSITE STEEL–CONCRETE STRUCTURES

M. MELE
Department of Structural Engineering and Geotechnics,
University of Rome, Italy

SUMMARY

The importance of testing in order to define the behaviour of shear connections is discussed, and the most significative kinds of tests are briefly presented even for design or for quality controls.

Experimental analysis and controls on shear connectors are fundamental to correct design and to ensuring good performance of the composite structure during its life.

In fact the connectors are very important in order to define structural behaviour, but unfortunately an analytical approach to their study is not easy owing to the difficulties that arise when defining a realistic and reliable model; only tests in such cases provide sound information.

Furthemore some problems could occur in welding, especially when studs are used, and so some testing to define welding checks must be adopted.

Therefore we really need to test connectors both before and after designing composite structures.

1 TESTING FOR DESIGN

The bearing capacity of a connecting system is usually evaluated by push-out tests as in fig. 1:
according to the specifications of several national and international codes.

As an alternative the scheme of figure 2 could be used, where a horizontal testing machine is being used:
the advantage being had of reducing the probability of error owing to load eccentricities.

From these tests the load/slip relationship can be derived, it being graphically described by curves having the typical form of figure 3
these curves showing the collapse load Pr, the strain capacity, γ_u, and the limit load, Pl, defined as the load value

Testing of Metals for Structures. Edited by F.M. Mazzolani. © 1992 RILEM.
Published by Taylor & Francis, 2 Park Square, Milton Park, Abingdon, Oxon, OX14 4RN.
ISBN 0 419 15810 3

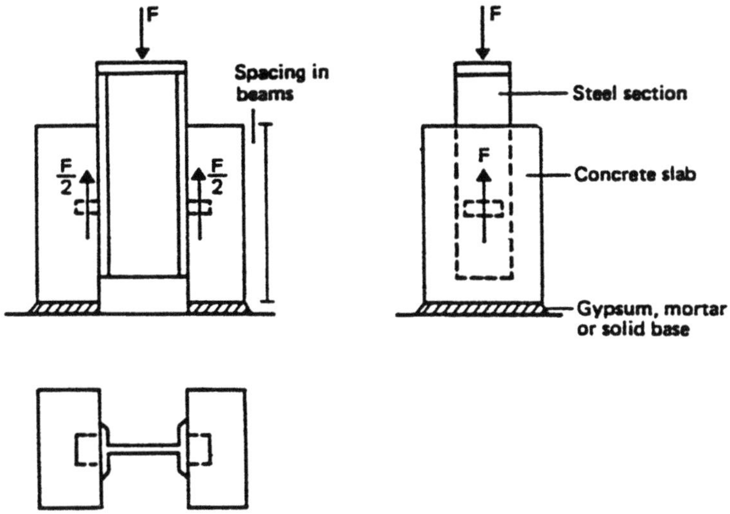
F
Spacing in beams
Steel section
Concrete slab
Gypsum, mortar or solid base

Fig. 1

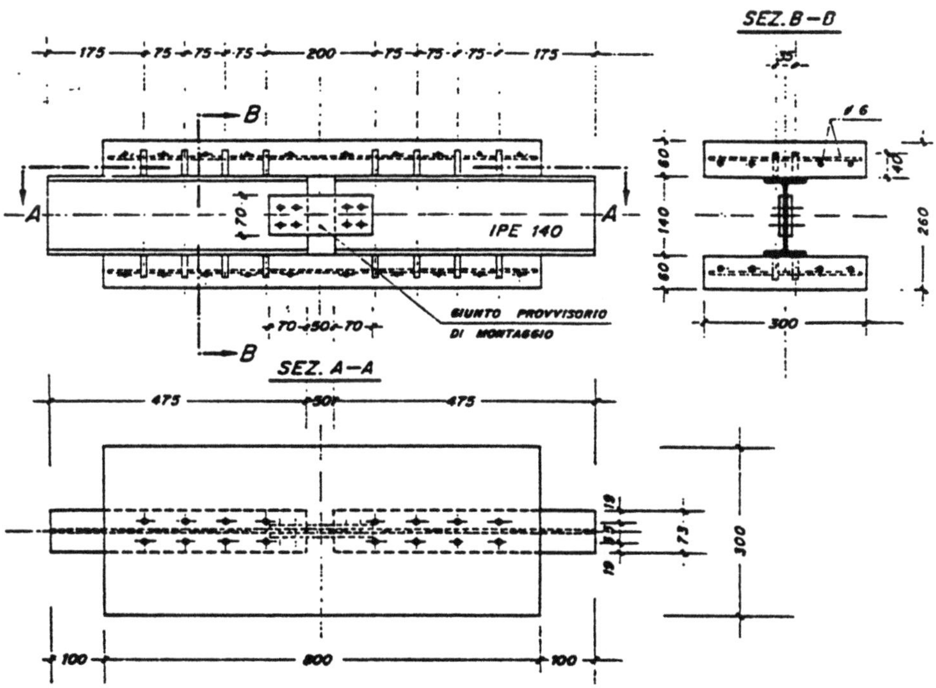
SEZ. B–B
SEZ. A–A
IPE 140
GIUNTO PROVVISORIO DI MONTAGGIO

Fig. 2

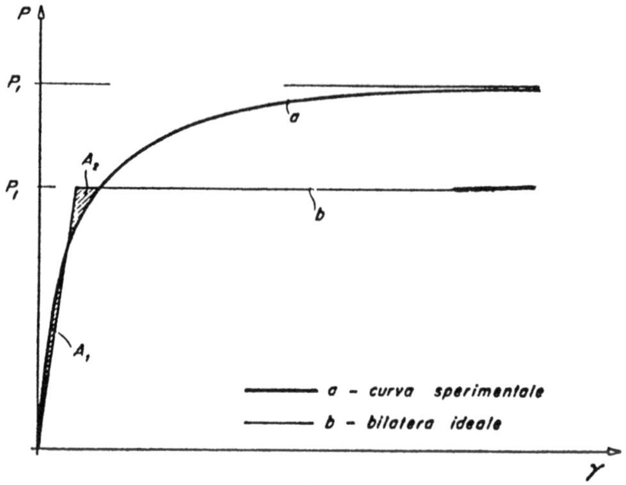

Fig. 3

corresponding to the appearance of large relative displacements between the concrete slab and the steel beam, and the corresponding slip value pl.

Pl is the proper load value to base the design on, in order to achieve a good composite behaviour, which is to say, a quasi-monolithic behaviour; experimental research in this field has shown that an empirical value

$$Pl = 0.65\ Pr$$

can be adopted with very good approximation, the value being about 0.1 mm in any case.

A complete understanding of the behaviour under bending of a connector system needs, however, beam models in addition to the push-out ones.

The first reason for this is the difference between the load/slip curves relative to push/out tests and bending tests, not qualitatively but quantitatively only (fig.4).

In fact, the slips found in bending tests are smaller than for axial-loaded models because:

- the push-out models become mechanisms when the connection is eliminated and the curve tends toward the abscissa, while the beam are in any case adequately con strained, the limit curve in this case being the one corresponding to the beams' not working;
- in the beam test radial forces arise between the concrete slab and steel beam, caused by the curvature induced by bending and consequently frictional forces reduce the slips.

The experimental researches performed in the recent past on this subject have shown, however, that the limit loads in bending tests correspond, to a very good approximation, to the collapse loads risulting from push-out tests; consequently the collapse load derived from the

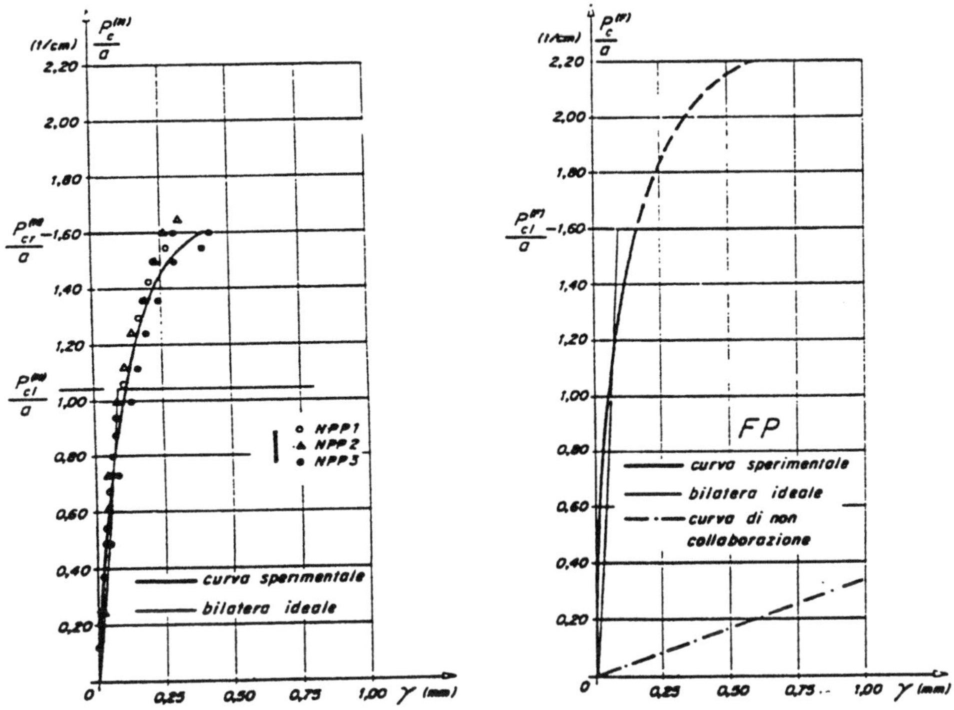

a) b)

Fig. 4

push-out tests can be used to predict the limit loads to base the design of beam elements on.

Furthermore it must be said that only the beam models can give satisfactory and clear information on:

- the influence of slips on diflections,that is on the degree of interaction of a given kind of connectors;
- behaviour up to collaps of structures having a partial degree of shear connection.

Regarding the last-mentioned point, it must be remembered too that the degree of shear connection depends on the load scheme too; only bending tests should be used to determine the minimum value of this degree in order to have good behaviour up to collapse, depending on the ductility properties of the shear connection.

In this field, recent studies have brought out very satisfactory performance even with degrees of connection lower than the minimum demanded by several codes (< 0.5) if the connectors have sufficient ductility (fig.5) of course. These circumstances may act as an invitation to re-think the suitability of imposing limits on partial composite action without any correlation with connection ductility and load conditions. Desirable in order to better clarify

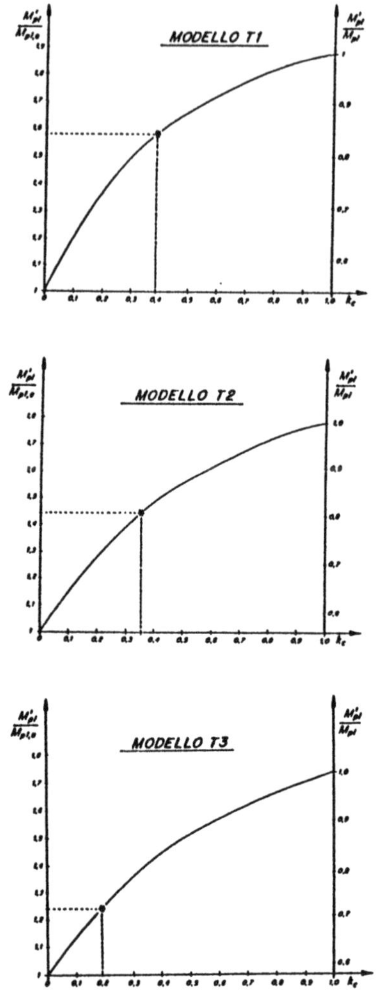

Fig. 5

the relationslip between the degree of shear connection and the demands of ductility.

For the reasons outlined above the use of bending tests is fundamental when particular kinds of connection and/or particular arrangements of shear connectors are investigated (figs. 6,7).

2 TESTING FOR CONTROLS

The welding of studs is a very sensitive operation and demands serious quality control.

There are, unfortunately, very few indications in the codes on this matter at present while more attention should be devoted on it.

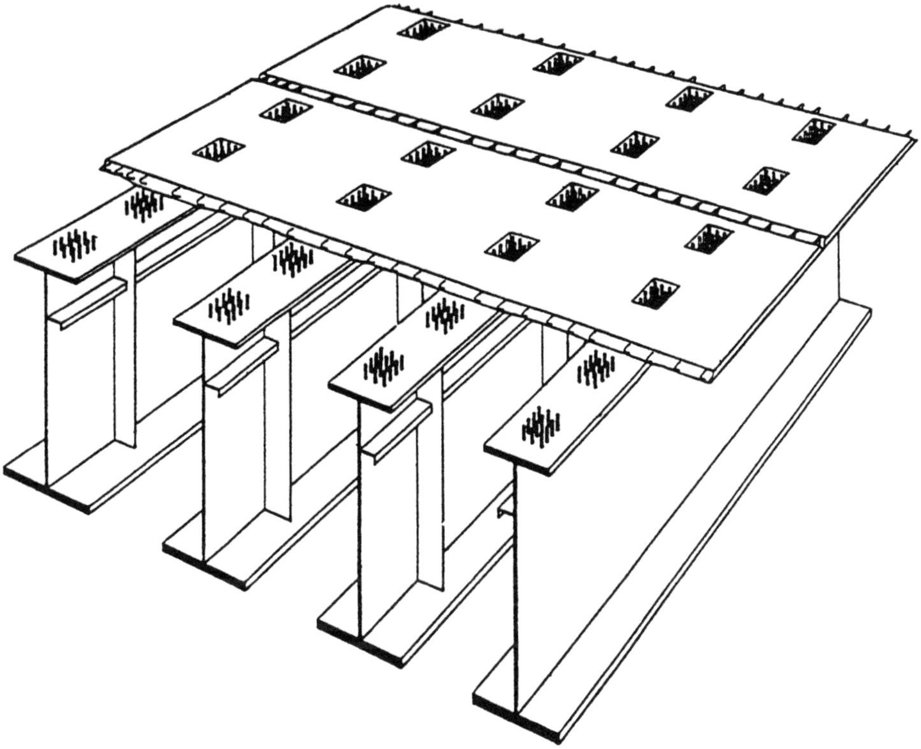

Fig. 6

Given the present situation, full of lacunas as it is, until further research has indicated more appropriate kinds of checks and inspections, the simple following procedure could be adopted:

- visual examination of the welded studs to ascertain the presence and regularity of the melted collars;
- all studs for which the melted collars are not complete are to be bent to a 30 deg. angle by striking them with a hammer. This test should be given to 5% of the studs that passed the visual inspection; for each broken stud found during this last check, two more studs are to be bent. If at the end of the test more than 5% of the studs in the plate are broken, all studs in the plate shall be tested;
- defective studs shall be removed and the wall plate they were joined to must be examined in order to ascertain the presence of surface tear to be repaired.

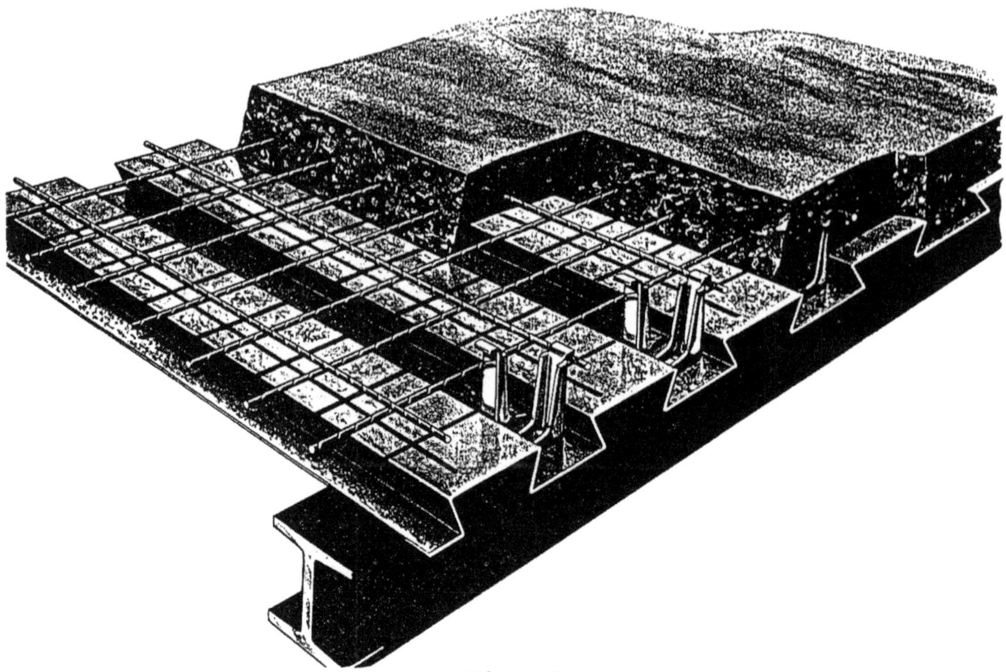

Fig. 7

REFERENCES

1. P. Matildi, M. Mele, R. Puhali: "Analisi sperimentale e criteri di calcolo dei collegamenti trave-soletta nelle strutture composte di acciaio e calcestruzzo. I connettori a taglio diffusi" - Costruzioni Metalliche, 4-5, 1974.
2. P. Matildi, M. Mele, R. Puhali: "Analisi sperimentale e criteri di calcolo dei collegamenti trave-soletta nelle strutture composte di acciaio e calcestruzzo. I dispositivi a taglio concentrati in solette prefabbricate" - Costruzioni Metalliche, 1-2, 1975.
3. G. Frisolini, M. Mele, R. Puhali: "I collegamenti trave-soletta con bulloni ad alta resistenza nelle strutture composte di acciaio e calcestruzzo" - Costruzioni Metalliche , 5-6, 1976.
4. M. Mele, R. Puhali: "Connettori profilati a freddo per il collegamento a taglio delle travi composte di acciaio e calcestruzzo" - Acciaio, 9, 1984.
5. CNR 10016: "Travi composte di acciaio e calcestruzzo. Istruzioni per l'impiego nelle costruzioni" - 1985.
6. BS 5400 - Part 5
7. Composite Structures - ECCS - The Construction Press - London and New York - 1981.
8. Eurocode No.4 - Comuon unified rules for composite steel and concrete structures (draft) - C E C - Bruxelles - 1990.

PART SEVEN

BEHAVIOUR OF THIN-WALLED STRUCTURES

37 TESTING OF PROFILED METAL SHEETING

R. BAEHRE
Testing Centre for Steel, Wood and Stone, University of Karlsruhe, Germany

SUMMARY

Testing of building elements, structural members or complex structures must be based on experiences concerning the functional behaviour of the test specimens, testing procedures and the reliability of test results. For mass products such as, for example, trapezoidal sheeting, standardized testing equipment, test procedures and the evaluation of test results have been established. As the numerous test results are backed up by consistent analytical design models, the number of tests for new products can be limited. The need for testing and the execution of test procedures concerning trapezoidal sheeting are discussed in this paper.

1 General

The aim of testing in general is to investigate the true behaviour of building elements, structural members or structures under conditions which must conform to, or at least be similar to, the respective conditions of the actual structure. Since the actual structure in all its complexity, and the various conditions (e.g. loading types due to different types of sources) cannot be used for testing, both actions on and resistance of structures are simulated, based on acceptable and consistent mechanical and mathematical models. It is thus obvious that testing in general requires qualified experience with regard to the choice of an adequate modelling of all test parameters and accurate analysis of the test aims and data to be recorded during the test.

Tests may be required under various circumstances in order to, for example:

* evaluate the mechanical properties of materials
* take into account changes of these properties due to performance processes
* check or develop analytical design procedures
* perform a design by testing

Testing of Metals for Structures. Edited by F.M. Mazzolani.
Published by Taylor & Francis, 2 Park Square, Milton Park, Abingdon, Oxon, OX14 4RN.
ISBN 0 419 15810 3

* check the performance of existing or planned structures
* confirm the consistency of the production
* study the interaction with other structural components
* predict the life-time of a structure

It is obvious that the wide range of set objectives implies that different test set-ups are necessary, but also that the load-bearing behaviour must be recorded in order to define adequate safety levels. This problem, however, has to be considered with respect to the reliability and scatter of test results, the number of tests and the consequences of a failure.

In addition to the requirements regarding the reliability of test results, it must be ensured that similar types of products will be tested in the same manner - regardless of the location of the testing institute - and that the characteristic values of resistance of the product are based on the same evaluation procedure.

This means that for clearly defined mass products such as, for example, trapezoidal sheeting, which are produced in nearly every country, both uniform testing rules and evaluation procedures must be established in order to avoid differing test results or interpretations of the results.

The testing rules for trapezoidal sheeting emanating from the European Recommendations for the testing of profiled metal sheets [1], worked out in 1978 by the ECCS Committee TC7, have been included in Annex A [2] of EUROCODE 3 [3], in an updated version.

2 The need for testing of trapezoidal sheeting

In the sixties, more and more countries established their own production of trapezoidal sheeting, but exportation of these products also increased. As a consequence of increased competition and trade, product development quickly led to new types of profiles with increased load-bearing capacity, provided by, for example, intermediate stiffeners and higher material strength or geometrical measures (see Fig. 1).

This product development, however, was based more on experiences made regarding the functional behaviour of the products than on analytical methods. This meant that the lack of analytical design methods at the beginning of the development necessitated "design by testing", but the growing understanding of the functional behaviour then created analytical design methods. Theoretical (semi-empirical) design formulae could then be created based on the evaluation of test results. This type of interaction of analytical and experimental results is

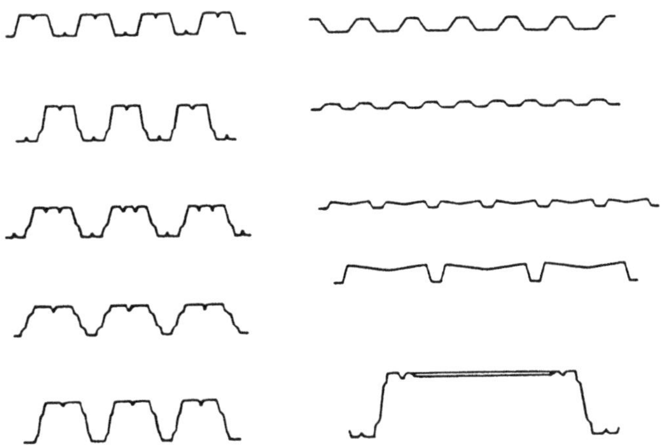

FIG.1. TYPES OF TRAPEZOIDAL SHEETING

typical, whenever special phenomena are responsible for uncertainties in the prediction of design values for strength (ultimate load state) or deformations (seviceability state).

The calibration of design expressions is, as a rule, based on a sufficient number of test results, allowing a statistical analysis to be made. Here, however, a qualified judgement of the different parameters which influence both mechanical phenomena as well as statistical values such as, for example, the variations of basic variables. Thus, and this is a very significant factor, this judgement is consequently an interpreted evaluation, which requires knowledge and experience regarding respective strength and testing problems.

In the past thousands of tests - as described below - have been executed and evaluated, and design formulae have been developed which predict sufficiently the load-bearing capacity of trapezoidal sheeting. Nevertheless it was necessary to calibrate the design formulae on a relatively low level so that the predicted load-bearing capacity is lower as a rule than the actual amount. This is due to the fact that, depending on the geometrical shape, different types of failure can occur and that a wide range of the geometrical properties has to be covered by the design rules.

Another problem is the scatter of the testing results, depending on the sensitivity of various failure modes, which implies that analytical design values have to be "on the safe side " in order to compensate the scatter and to estimate a characteristic strength (e.g. the 5%-fractile value) close to the calculated value.

As an example of the correlation of experimental (r_e) and calculated (r_t) values of the bending moment capacity of trapezoidal sheeting, an appropriate statistical evaluation of test results is shown in Fig. 2 where, based on 274 test results:

$r_e = r_t$ represents the ideal correlation,

$r = \bar{b} \cdot r_t$ is the mean-value-correction line,

$r_k = b \cdot \delta_k \cdot r_t$ and represents the 5%-fractile value (characteristic bending moment)

This Figure shows that, taking into account the scatter of the test results, the design formula indicates the precise behaviour of the sheeting. ($r_k = 1.02 \cdot r_t$)

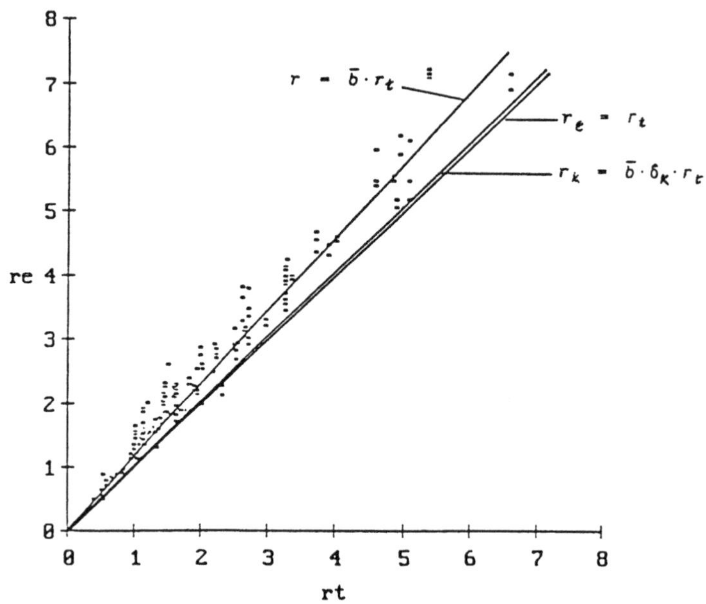

FIG.2. EXAMPLE OF COMPARISONS BETWEEN CALCULATED AND TESTED BENDING MOMENT CAPACITY

Figure 3 shows the result of the interaction between bending the moment at the interior support and the support reaction referred to crippling failure. In this case the characteristic value is lower than the calculated strength ($r_K = 0.91 \cdot r_t$) which has to be taken into account when considering safety levels.

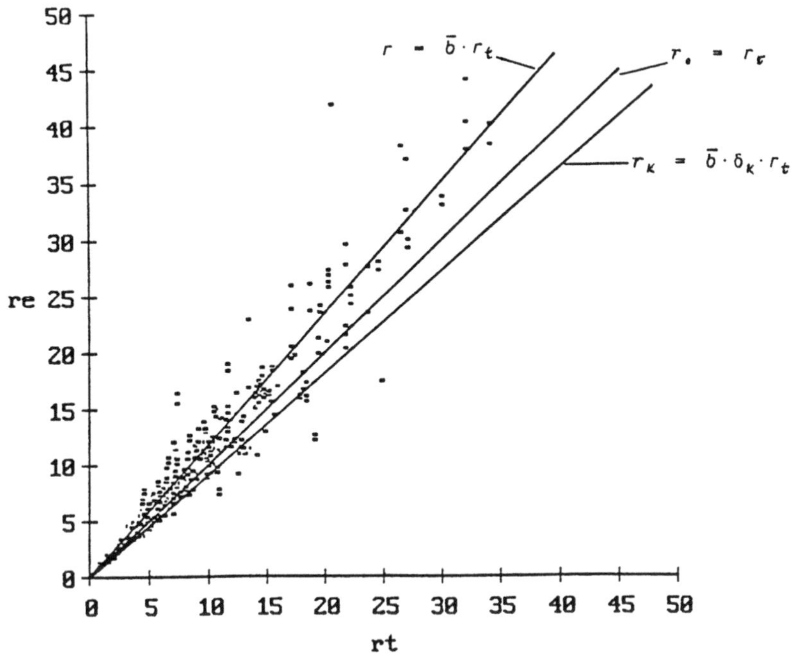

FIG.3. EXAMPLE OF COMPARISONS BETWEEN CALCULATED AND TESTED INTERACTION OF BENDING MOMENT CRIPPLING LOAD CAPACITY

3 Types of testing

Testing methods and the evaluation of test results are described for the determination of moment capacity, bearing capacity and stiffness of sheet metal panels whose primary purpose is to carry uniformly distributed loads.Testing under concentrated loading is also required in order to verify that the performance of the product is satisfactory under local loading caused by, for example, access to a roof for the purpose of erection or maintenance.

Testing of trapezoidal sheeting under in-plane loading as normal forces and/or shear forces is also important, but due to the special testing equipment required for such tests, in-plane loading tests have been limited to a few adequately equipped testing stations.

With regard to the main purpose of the trapezoidal sheeting, knowledge is required concerning the

* bending moment capacity and bending stiffness, estimated by "one span beam tests"
* combined bending moment and shear or crippling resistance, estimated by "intermediate support tests" or "two span beam tests" alternatively
* shear force capacity at the end-support, estimated by "end-support tests"
* concentrated loading resistance both during and after erection, representing "walkability"

Normally the tests have to be performed for both gravity loading (e.g. dead weight and snow load) and uplift loading (wind suction), and, in addition, depending on the application, in two positions, if the sheeting is unsymmetrical. Practical support and loading conditions have to be taken into account, which implies, for example, that for uplift loading the resistance of the fasteners and connections at the end and interior supports must be estimated. More detailed information is presented in ANNEX A to EUROCODE 3.

The tests should simulate the real behaviour of the sheeting under practical conditions. It is important, that the testing equipment and the testing procedures are kept simple in order to achieve reliable and comparable results.

The rate of load application should usually be selected such that the stresses can be treated as quasi-static and be approximately equal in a test series. Furthermore a sufficient number of load ranges should be provided until the load capacity is reached.

The load-bearing tests must be accompanied by the unified tensile tests, where test specimen are taken from the plane parts of the profiles. The above-named test results must then be corrected with respect to the actual values of the core thickness (t_t) and the yield strength (f_{yt}) and the specified values (t, f_y) respectively.

The corrected test result R_n is derived from the actual test result R_t from:

$$R_n = R_t \ (F_y/f_{yt})^{\alpha} \cdot (t/t_t)^{\beta}$$
$$\beta = 1 \text{ for } t \geq t_t$$
$$\beta = 2 \text{ for } t < t_t$$
$$\alpha = 0 \text{ for } f_y \geq f_t$$

$\alpha \quad = 0.5$ for $f_y < f_{yt}$ if failure is caused by local buckling, otherwise $\alpha = 1$

Using this procedure, the test results are transformed into values, with reference to nominal or specified values of sheet thickness and yield strength.

4 Some remarks concerning the statistical evaluation

A statistical evaluation of test results requires a sufficient number of test results and a qualified preknowledge of the behaviour provided by, for example, earlier or similar test series or design models.It is justified, that these conditions are available for trapezoidal sheeting, which implies that the number of tests for new and similar products can be limited. This is also supported by the fact that different tests with similar failure modes can be gathered statistically in the same test population.

If, however, these conditions do not comply with the actual case, then a statistical evaluation should not be made and the evaluation of test results should be based on a rational analysis of the failure modes and the parameters involved. As full-scale tests are often very expensive, it may be advisable to estimate design strength values by mathematical modelling in connection with small-scale or partial tests, before full-scale tests are developed.

5 References

[1] European Recommendations for Steel Construction,"Testing of profiled metal sheeting"ECCS Committee TC 7,1978, Publication No. 20

[2] EUROCODE 3, ANNEX A "Cold formed steel sheeting and members" Draft December 1989

[3] EUROCODE 3, ANNEX Z "Procedures for the determination of design resistance from tests", Draft 1989.

38 BEHAVIOUR OF THIRD GENERATION TRAPEZOIDAL STEEL SHEETINGS

R. LANDOLFO and F.M. MAZZOLANI
Istituto di Tecnica delle Costruzioni, Faculty of Engineering,
University of Naples, Italy

SUMMARY

The third generation profiles are trapezoidal unit with both longitudinal and transversal stiffeners, which provide suitable solution for spans up to 12 m without purlins. In order to investigate the main aspects of bending behaviour of the so-called TRP 200, one of the most representative profiles belonging to the third generation, two series of tests were carried out at University of Naples. In the first serie, 20 specimens of naked sheeting were tested under concentrated and distributed loads. In the second serie, 7 specimens of composite structures were tested under concentrated loads. In this paper,the main test results and their analitical and mechanical interpretation are reported.

1 INTRODUCTION

The development of steel sheetings over the past years has been characterized by three generations of profiles illustrated in fig.1 [1].

The first generation profiles are plain trapezoidal without stiffeners (fig.1a), allowing spans between secondary members of about 3 m. In the second generation profiles the trapezoidal sheets are stiffened in longitudinal direction by appropriate folding (fig.1b) and may span up to 7 m. The third generation profiles are trapezoidal unit with both longitudinal and transversal stiffeners (fig.1c), which provide suitable solution for spans up to 12 m without purlins.

Among the profiles belonging to the third generation, the so-called TRP 200, manufactured by Plannja of Lulea, Sweden, is one of the most representative.

This trapezoidal profile, with depth 199 mm and width 800 mm, is made from hot-dipped galvanized steel sheet with thickness range from 0.75 to 1.50 mm. TRP 200 is also produced in perforated form, with approximately 17% of the surface of the profile corresponds to small holes, for improving acoustic performance.

The mentioned profile may be used for roofing in the field of industrial buildings (fig.2) or in floor structures of multistory buildings by completing its cross section with casted concrete (fig.3). Some applications have been also proposed for the rehabilitation of old roofs and floor structures.

Testing of Metals for Structures. Edited by F.M. Mazzolani.
Published by Taylor & Francis, 2 Park Square, Milton Park, Abingdon, Oxon, OX14 4RN.
ISBN 0 419 15810 3

According to above applications, especially when the building is in seismic areas, an appropriate structural behaviour is required to TRP 200. As a consequence, the analysis of the load bearing capacity of this profile in different working conditions represent the item of primary importance [2/3/4/5/6].

In order to investigate on the main aspects of bending behaviour of TRP 200, two series of tests were carried out at the Laboratory of Istituto di Tecnica delle Costruizioni, University of Naples [4/5]. In the first serie, 20 specimens of naked sheeting were tested under concentrated and distributed loads, for different thicknesses and spans [4].

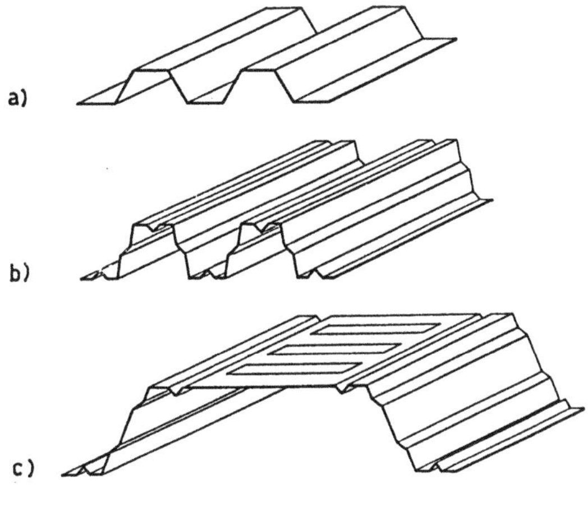

fig.1

In the second serie, 7 specimens of composite structures were tested under concentrated loads [5].

These test results allowed to develop mechanical models for the evaluation of the ultimate bending moment. Further on, a simulation program has been implemented and calibrated on the same testing data both for thin-walled profiles and for composite sections [7].

fig.2

fig.3

2 GEOMETRICAL AND MECHANICAL FEATURES

The main dimensions of the cross-section of the TRP 200 are shown in fig.4 together with some geometrical properties.

The test results have shown that for longitudinal bending the only effective parts of the profile are the bottom flange, the web and the portion of the top flange adjacent to the web. In fact, the transversal corrugations in the top flange serve only to distribute load to the stiffened corners.

Tab.I

t	A	y_g	I_{xx}	I'_{xx}	W_{xx}	i_{xx}
mm	mm^2	mm	cm^4	cm^4/m	cm^3	mm
0.75	545	94.3	314	392	29.99	76
0.90	660	95.2	382	477	36.80	76
1.00	733	95.2	424	530	40.85	76
1.25	916	95.2	531	664	51.16	76
1.50	1100	95.2	637	796	61.37	76

Starting from this hypothesis, the geometrical and mechanical featureses of the cross-section of tab I have been determinated according to effective width approach used in Italian Code for cold-formed steel design [8].

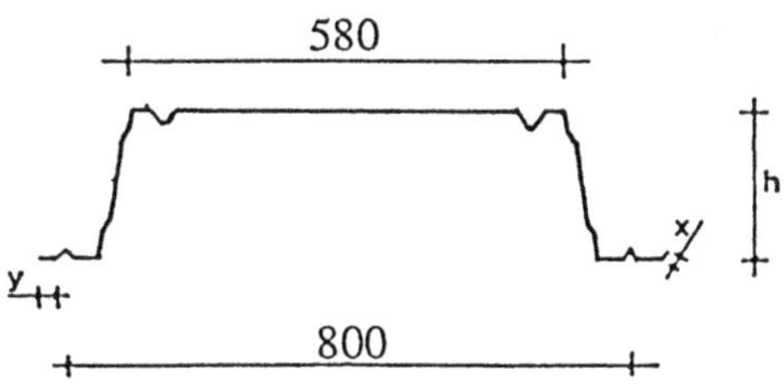

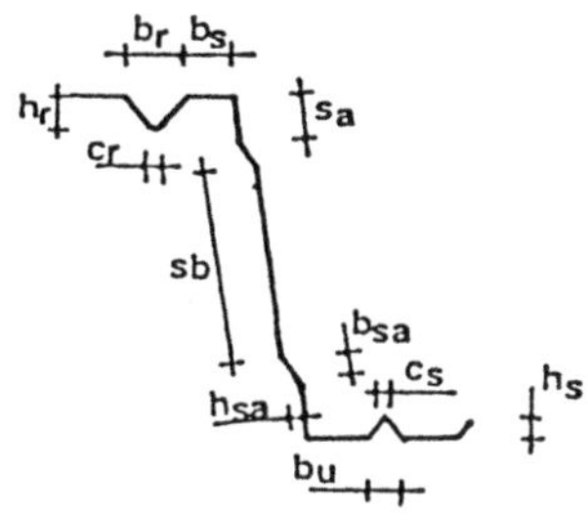

NOMINAL VALUES (mm)

h	b_r	b_s	h_r	c_r	s_a	s_b	b_{sd}	h_{sa}	c_s	b_u	h_s	x	y
199	28	30	14	2	43	111	6	8	9	28	13	12	20

fig.4

3 TESTING PROCEDURES AND RESULTS

3.1 Naked sheeting

The specimens cover the range of thicknesses and spans shown in following table II.

Tab II

TYPE	NORMAL				PERFORATED			
SPAN (m)	2.0	3.0	4.0	6.0	2.0	3.0	4.0	6.0
t (mm)	Number of tests							
0.75		2		2				
0.90	3				2			1
1.00		1		2		1		
1.50			2	2			1	1

Two loading conditions have been investigated in simple support conditions:
a) concentrated load at midspan of specimens with span of 2, 3 and 4 m (fig.5);
b) distributed load on specimen with span of 6 m (fig.6).

Testing results have been successively elaborated according to [9]. The procedure described in these Recommendations may be applied to the determination of the characteristic moment M_k for a given sheet thickness by means following relation:

$$M_k = M_m(1 - c\delta)$$

where: M_m = mean value of the test values of the ultimate moment for the given nominal sheet thickness

c = coefficient which value depends on the number of test observations

δ = population variance (i.e. the standard deviation divided by the mean value)

fig.5

fig.6

When tests have been carried out on a number of different sheet thicknesses, M_m will usually be obtained as a point on empirical curve. The population variance δ for this curve may then be determined from:

$$\delta^2 = \frac{\Sigma\left(\frac{M_i}{M_m}\right)^2 - \frac{1}{n}\left(\Sigma\frac{M_i}{M_m}\right)^2}{n-1}$$

where:
n = total number of observations for all sheet thicknesses
M_i = measured value for test numbered i at a particular sheet thickness
M_m = mean value of test results at the same sheet thickness as test i

The characteristic values of bending moment, calculated according to above procedure, are compared in table III with the theoretical ones given by:

$$M_d = \frac{W_{eff} f_d}{B} \qquad \left(\frac{kNm}{m}\right)$$

being:
W_{eff} : effective strength modulus given in tab.I
f_d : design yield strength (275 Nmm^{-2})
B : 0.8 m

Tab. III

t	$M_{u,k}$	M_d
mm	kNm/m	
0.75	8.45	10.31
0.90	11.52	12.65
1.00	13.63	14.04
1.50	23.42	21.09

From the previous comparison we can observe a satisfactory agreement between experimental and theoretical data but only the last case of t=1.50 mm gives conservative results. The maximum scatter, equal to 20%, may be observed for thinness sheeting.
As far as perforated sheetings are concerned a small reduction of bending capacity, approximately 5%, is observed.

3.2 Composite system

In order to emphasize the influence of main phenomena on the bending behaviour of the TRP 200 used as a composite system, three types of specimens have been prepared (fig.7).

Specimen type A has been obtained by casting concrete inside the sheet up to 4 cm of slab over the shape.

In specimen type B two steel bars φ16 have been arranged on the bottom of the cross-section.

Finally, specimen type C has fasteners placed on the bottom flange of sheeting in order to improve the adherence between steel and concrete. Fasteners type Hilti have been used with a 15 cm spacing.

Seven specimens have been tested under concentrated load at midspan (fig.8). The range of mechanical and geometrical features of the specimens is shown in the following table IV, together the values of the ultimate bending moments given by experiments.

The limitated numbers of tests make unpossible the statistical analysis of the results but, on the other hand, it allows to emphasize some important aspects which influence the structural behaviour of this peculiar system.

Tab.IV

TYPE	SPAN	t	$M_{u,exp}$
	m	mm	kNm/m
A	2.0	0.75	22.74
	3.0	0.90*	32.12
	6.0	0.75	34.43
B	2.0	0.75	61.74
	3.0	0.90*	47.69
	6.0	0.75	77.18
C	6.0	0.75	35.30

* Perforated sheeting

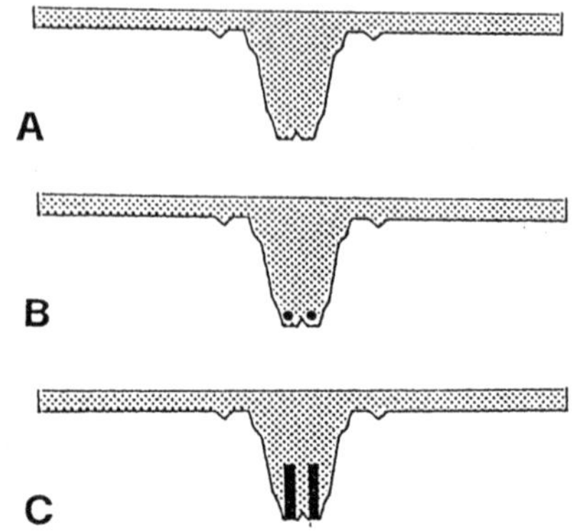

fig.7

Specimens type A have shown a brittle behaviour and collapse occurred because of the formation of a plastic hinge at midspan. Higher strength and a more ductile behaviour have been exhibited by specimens type B due to the reinforcements. The only test performed on the specimen type C has shown almost the same strength of type A but a quite different collapse mechanisms. In fact, failure occurred due to the instantaneous formation of three plastic hinge in three cross-section equally spaced of L/4.

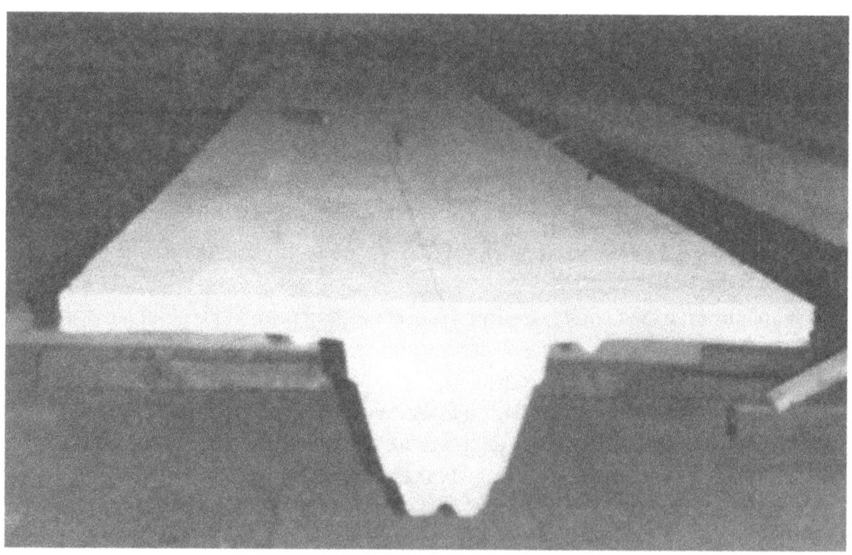
fig.8

The limitated numbers of tests make unpossible the statistical analysis of the results but, on the other hand, it allows to emphasize some important aspects which influence the structural behaviour of this peculiar system.

Specimens type A have shown a brittle behaviour and collapse occurred because of the formation of a plastic hinge at midspan. Higher strength and a more ductile behaviour have been exhibited by specimens type B due to the reinforcements. The only test performed on the specimen type C has shown almost the same strength of type A but a quite different collapse mechanisms. In fact, failure occurred due to the instantaneous formation of three plastic hinge in three cross-section equally spaced of L/4.

4. NEEDS AND CONCLUSIONS

When we interpret testing results on sheeting elements, one of the main difficulties is the necessity to take into account that in the constructions the actual structure is composed by several elements working together in a bi-dimensional assessment and the loading conditions are often far from the experimental one.

During tests, we must avoid the effects of local buckling due to concentrated loads and support reactions. In addition to the case of simply supported edge conditions, also continuous schemes with unsymmetrical loads and different spans have to be tested.

The ECCS Recommendations (1977) partially cover these problems, but the present need is to revise them in order to improve procedures and to introduce appropriate rules for this kind of 3rd generation sheetings.

When these profiles are used in cooperation with the steel skeleton according to the so-called "stressed skin design", they are submitted to diaphragm actions with introduce axial and shear forces. Appropriate testing procedures for analysing axial and shear behaviour must be assessed separately in order to evaluate the corresponding deformabilities to be introduced in the calculations.

In case these profiles are used in composite system, the main problem at the basis of the design methods is the evaluation of the strength of connecting elements. For this kind of mixed steel-concrete structures, appropriate testing procedures must be also defined, starting from the flexural behaviour in close relation with the connection system which is used to assure the suitable cooperations between both materials.

REFERENCES

[1] Bryan E.R. :"Long-span steel decking", Industrial News, 1979

[2] Landolfo R, Mazzolani F.M. :"Il comportamento strutturale delle lamiere grecate di terza generazione", Acciaio n.2, 1988;

[3] Pekoz T., Zhang Y. :"TRP 200 cantilever shear test", Depaetment of Structural Engineering, Cornell University, 1983;

[4] Landolfo R, Mazzolani F.M. :"Indagine sperimentale su lamiere grecate tipo TRP 200", Acciaio n.1, 1989;

[5] Landolfo R,Mazzolani F.M. :"Sistemi misti acciaio calcestruzzo con l'impiego di lamiere grecate di terza generazione: indagine sperimentale", Acciaio n.1, 1990;
[6] Landolfo R, Mazzolani F.M.:"Criteri di calcolo per elementi strutturali in TRP 200", Atti del XII Congresso C.T.A., Capri, 1989;
[7] De Martino A., Ghersi A., Landolfo R, Mazzolani F.M. :" Bending behaviour of long-span steel sheeting: test and simulation", International Conference on Steel e Aluminium Structures, Singapore, Maggio 1991
[8] C.N.R. 10022 :"Profili formati a freddo: istruzioni per l'impiego nelle costruzioni", 1984
[9] ECCS: "European Recommendations for the testing of profiled metal sheets", 1977;

39 TESTING OF THIN-WALLED STRUCTURES

D.A. NETHERCOT
Department of Civil Engineering, University of Nottingham, UK

This presentation was based on a combination of two papers that were published in Vol. 26 No. 1. 1991 of the Journal of Strain Analysis.

The Experimental Assessment of the Force Components Within Thin Walled Structural Steel Members

by

C Gibbons, P A Kirby and D A Nethercot

SUMMARY

A technique is described for manipulating strain gauge data obtained from tests on thin walled members such that individual force components may be identified. The technique recognises the different structural actions present when open sections deform in a full three-dimensional fashion (compression, major and minor axis bending and warping) covering both elastic and elastic-plastic behaviour. Its application in beam-column testing is illustrated, the results exhibiting excellent agreement with the corresponding predictions of a sophisticated non-linear finite element analysis.

A Mechanical Electrical Device for Measuring the Three-Dimensional Deformation of Steel Columns

by

C Gibbons, P A Kirby and D A Nethercot

SUMMARY

As part of an extensive experimental study into the influence of semi-rigid connections on the behaviour of steel columns, a measurement device has been developed which permits the six individual components of three-dimensional deformation to be monitored. The theory used to process the measured deformations and the subsequent testing of a prototype device is described. A comparison between the deformations measured using the device, and those predicted using a column subassemblage finite element model is also presented.

Testing of Metals for Structures. Edited by F.M. Mazzolani.
Published by Taylor & Francis, 2 Park Square, Milton Park, Abingdon, Oxon, OX14 4RN.
ISBN 0 419 15810 3

40 TESTING OF THIN-WALLED STRUCTURES

T. PEKÖZ
Department of Structural Engineering, Cornell University, Ithaca, NY, USA

The behavior of thin-walled steel and aluminum members is quite complex. Thus, testing is an important part of the research and design process for such members. In general, the configuration of the specimen, the means of supporting the specimen and the method of loading influences the test results significantly. The tests therefore require careful planning. Costly large scale tests may in certain cases be avoided by carefully designed smaller scale tests that represent the behavior of the full scale test assemblies.

The philosophy of the use of test results vary from country to country. In some countries such as the United States design has to be carried out analytically according to the specifications. For those parameters and range of parameters where the specifications do not have coverage, design by testing is allowed. Tests may also be carried out to confirm the analytical calculations. Test results may be used directly in design or they may be used to determine some of the parameters that are used in the design equations.

Other cases where tests may be required or desirable are discussed in Reference 1. Due to the possible inaccuracies in the tests in representing the actual behavior, factors of safety or resistance factors to be applied to the test results are greater than those used in the analytical specification provisions. In other countries, tests are mandatory for some types of structural members, systems and behavior.

The evaluation of test results to draw conclusions for design and research purposes also deserves careful scrutiny. The test conditions in general may have variations from the conditions in the actual use of the member or structural system being tested. These have to be considered in the evaluation of the test results.

Quite often due to time and cost constraints there are limited number of tests to draw conclusions from. In such cases, the evaluation of test results includes some intuitive judgement.

In this paper, considerations in testing thin-walled members will be reviewed briefly.

Testing of Metals for Structures. Edited by F.M. Mazzolani. © 1992 RILEM.
Published by Taylor & Francis, 2 Park Square, Milton Park, Abingdon, Oxon, OX14 4RN.
ISBN 0 419 15810 3

TYPES OF TESTS

Stub Column Tests

Usually stub columns are tested to study the effects of local buckling and residual stresses. The information gained from stub column tests are used in the design of columns and beam columns. Perforations and cross-sectional dimensions such as very long lips may influence local buckling behavior. These effects may not estimated accurately through the use of specification provisions.

Originally, stub column test procedures were developed by a task group of the Structural Stability Research Council for heavier hot-rolled columns. These procedures that remained basically unchanged for more than a decade are given in Reference 2. The test and evaluation procedures used in AISI Cold-Formed Steel Manual (3) were developed in Reference 4 specifically for thin walled-members. Specimen requirements are given in these references.

In the design of cold-formed steel columns according to the AISI Specification (5), the effect of local buckling on the overall buckling is accounted for as follows:

- Determine the failure stress ignoring local buckling
- Determine the effective area at the failure stress calculated
- Calculate the failure load multiplying the failure stress by the effective area calculated above.

Thus one of the results expected out of the stub column test is a means of calculating the effective area at any stress. Formulas for such calculations are derived in Reference 4.

A stub column specimen should be long enough so that the local buckling behavior and the effects of residual stresses are properly exhibited. However the specimen should be short enough so that overall column buckling influences the results as little as possible. Reference 4 gives approaches for correcting the results if the column has to be too long to have negligible reduction in the overall column failure load.

Future studies should include the determination of the effective section centroid location for any axial load level.

Column Tests

Column testing may be called for less frequently than stub column testing. In column testing some of the considerations include the following:

- Representation of the end conditions through the use of end fixtures
- Centering of the end loads
- Selection of lengths to be tested

Some information is presented in the recent References 6 and 7. There are no standardized test fixtures and procedures at the moment. It is desirable to gather information on the subject.

Flexural Members

There are several reasons for conducting tests on flexural members. Flexural members include beams and panels. Test results may be used either directly in design or to determine design parameters. For example panels may be tested subjected to vacuum or air-bag loads to determine capacities. The panels are also tested as beams to determine the moment-rotation characteristics over the supports in order to use the results to determine the capacity of continuous panels by analytical approaches.

It should be noted that the method of testing influences the capacity observed greatly. Some of the tests at Cornell University show that even for a simply supported panel, the ratio of the failure moment using uniform pressure to that using line loads can be very high. This is partly due to the fact that the cross-sectional distortion is unfavorable in case of line loads and favorable in case of uniform pressure.

A point of disagreement among the test approaches for panels is the type of supports provided. In the actual structure considerable amount of membrane action is mobilized. Some testing authorities allow the mobilization of membrane action in the tests, others do not.

Several documents on the testing of flexural members and panels exist world-wide. Some examples can be found in References 8, 9 and other documents of the ECCS Technical Committee 7.

The above points deserve further attention.

Connections

In thin-walled structures, mechanical connectors such as screws play an important role. Due to the large number of variables involved in determining the strength, the design specifications take a conservative approach. Again the fixtures and types specimens used do make a significant difference in the out come of the tests. For this reason both the ECCS Recommendations in Reference 10 and the national codes in various countries contain detailed instructions. A critical review of these different approaches is desirable.

Joints

The joints in thin walled structures are in general semirigid due to the distortions that take place around the connectors and in the plate elements being connected. Examples of these are the joints in cold-formed steel industrial racks and joints between purlins and roof panels. Analytical determination of the joint characteristics is not feasible at the present time. References 11 and 12 provide testing and evaluation procedures for semirigid rack joints based on the studies conducted at Cornell University. For purlin to roof deck connections, Reference 3 gives detailed information based on Reference 13. International harmonization of these approaches is desirable.

Full Scale Testing

When the analytical methods are inadequate or when it is desired to verify analytical calculations, full scale tests are carried out. An example is the current full scale building test being carried out at Cornell University on multistory buildings using cold-formed steel studs and prestressed concrete plank floors. This testing program is reported in Reference 14. Shear diaphragm tests on floors, roofs and other planar elements are examples of full scale testing.

EVALUATION OF TEST RESULTS

Evaluation of test results in research and design should be done on the basis of the probability theory and statistics. Such approaches are discussed in Reference 15. It is recognized that there are basically three groups of uncertainties that should be considered in calculating the nominal strength of a member. These are the material properties, geometric parameters and the professional factor. The latter is the accuracy of the predictive equations. In a laboratory test, the material and geometric factors are reasonably well known. The correlation of the test results with the predicted results indicate the professional factor.

The resistance factor to be used in design has to be determined combining the professional factor estimates from the laboratory tests with the statistics of uncertainties of the material and geometric parameters. When the test basis, namely the number of tests, is not sufficient then an intuitive and empirical approach has to be taken. Such an approach is given in Reference 5. An international effort to discuss and harmonize the test evaluation methods is desirable.

COMMITTEES IN CHARGE OF TEST PROCEDURES

There are several national and international committees that discuss test procedures and evaluation methods for thin-walled members and structures. Some of these committees do prepare specifications and codes for tests. For example, in the United States Structural Stability Council has a Task Group that considers test methods. The committees of the American Iron and Steel Institute and the American Society for Testing Materials do consider test procedures and specify provisions for tests.

Internationally European Convention for Constructional Steelwork Technical Committee TC7 makes recommendations for testing. A committee within the framework of RILEM will be very valuable to provide a broader international basis and a different point of view than the ECCS Technical Committee.

REFERENCES

1. Baehre, Rolf, "Testing of Profiled Metal Sheeting," Proceedings of the RILEM Workshop, Naples, Italy, May 1990

2. Galambos, T. V., Editor, "Guide to the Stability Design Criteria for Metal Structures," Fourth Edition, Wiley Interscience, 1988

3. American Iron and Steel Institute, "Cold-Formed Steel Design Manual," Washington, DC, 1986

4. Pekoz, Teoman, "Development of a Unified Approach to the design of Cold-Formed Steel Members," Report CF87-1, American Iron and Steel Institute, March 1987

5. American Iron and Steel Institute, "Specification for the Design of Cold-Formed Steel Structural Members," 1986 Edition

6. Weng, C. C. and Pekoz, T., "Compression Tests of Cold-Formed Steel Columns," Journal of Structural Engineering, ASCE, Vol. 116, No. 5, May 1990, pp. 1230-1246

7. Miller, T. H. and Pekoz, T., "Studies on the Behavior of Cold-Formed Steel Wall Studs," Proceedings of the Tenth International Specialty Conference in St. Louis, Mo in October 1990

8. ECCS Technical Committee 7, "European Recommendations for the Design of light Gage Steel Members," ECCS, 1987

9. ECCS Technical Committee 7, "European Recommendations for the Design of Profiled Sheeting and Sections - Part 1," European Convention for Constructional Steelwork, February 1980.

10. ECCS Technical Committee 7, "European Recommendations for the Testing of Connections in Profiled Sheeting and other Light Gauge Steel Components," ECCS, 1976

11. Rack Manufacturers Institute, "Specification for the Design, Testing and Utilization of Industrial Steel Storage Racks," Rack Manufacturers Institute, June 1990

12. Rack Manufacturers Institute, "Commentary on the Specification for the Design, Testing and Utilization of Industrial Steel Storage Racks," Rack Manufacturers Institute, June 1990

13. Pekoz, T. and Soroushian, P., "Behavior of C- and Z-Purlins under Wind Uplift," Proceedings, Sixth International Conference on Cold-Formed Steel Structures, St. Louis, Mo., November 1982.

14. E. R. diGirolamo, T. Pekoz and T. Bond, "Development and Verification of a Structural System using Cold-Formed Steel Wall Studs," Proceedings of the Tenth International Specialty Conference in St. Louis, Mo in October 1990

15. Pekoz, T. and Hall, W. B., "Probabilistic Evaluation of Test Results," Proceedings, Ninth International Conference on Cold-Formed Steel Structures, St. Louis, MO, November, 8-9, 1988

AUTHOR INDEX

SUBJECT INDEX

This index has been compiled using the index terms assigned to the papers by Materials Information, the information service of the Institute of Metals and ASM International (USA). The assistance of Materials Information is gratefully acknowledged. The numbers are the page numbers of the first page of the paper referred to.

For Product Safety Concerns and Information please contact our EU
representative GPSR@taylorandfrancis.com
Taylor & Francis Verlag GmbH, Kaufingerstraße 24, 80331 München, Germany

www.ingramcontent.com/pod-product-compliance
Ingram Content Group UK Ltd.
Pitfield, Milton Keynes, MK11 3LW, UK
UKHW022006180926
475701UK00003B/219
9780367863654